TRENDS
IN MOLECULAR
ELECTROCHEMISTRY

TRENDS
IN MOLECULAR
ELECTROCHEMISTRY

Edited by
Armando J.L. Pombeiro

Co-Editor
Christian Amatore

F O N T I S
M E D I A

CRC Press
Taylor & Francis Group
Boca Raton London New York

CRC Press is an imprint of the
Taylor & Francis Group, an **informa** business

CRC Press
Taylor & Francis Group
6000 Broken Sound Parkway NW, Suite 300
Boca Raton, FL 33487-2742

First issued in paperback 2020

ISBN 13: 978-0-367-57829-9 (pbk)
ISBN 13: 978-0-8247-5352-8 (pbk)

Visit the Taylor & Francis Web site at
http://www.taylorandfrancis.com

and the CRC Press Web site at
http://www.crcpress.com

Library of Congress Cataloging-in-Publication Data
A catalog record for this book is available from the Library of Congress.

FontisMedia S.A.
FontisMedia S.A.
Avenue Vinet 19
CH-1004 Lausanne, Switzerland

Contents

MOLECULAR ELECTROACTIVATION AND ELECTROCATALYSIS

CHAPTER 3

THE SURFACE ELECTROCHEMISTRY OF METALLO-
PHTHALOCYANINES – DIOXYGEN REDUCTION   99
A. B. P. Lever and Yonglin Ma

CHAPTER 4

FROM ELECTRON TRANSFER TO CHEMISTRY: ELECTROCHEMICAL
ANALYSIS OF ORGANOMETALLIC REACTION CENTERS AND OF
THEIR INTERACTION ACROSS LIGAND BRIDGES   127
Wolfgang Kaim

SUPRAMOLECULAR ELECTROCHEMISTRY

SPECTROELECTROCHEMISTRY

CHAPTER 11
SPECTROELECTROCHEMICAL TECHNIQUES   339
Enzo Alessio, Simon Daff, Marie Elliot, Elisabetta Iengo, Lorna A. Jack,
Kenneth G. Macnamara, John M. Pratt and Lesley J. Yellowlees

UNCONVENTIONAL ELECTROCHEMISTRY

CHAPTER 12
ELECTROCHEMISTRY AT ULTRAMICROELECTRODES:
SMALL AND FAST MAY BE USEFUL   385
Christian Amatore, Stéphane Arbault, Emmanuel Maisonhaute,
Sabine Szunerits, and Laurent Thouin

Introduction

Chemists increasingly apply electrochemical methods to the investigation of their systems, in particular towards a better understanding of molecular properties, the exploration of chemical reactions involving electron-transfer (ET), the initiation of further reactions by ET, the kinetic measurements and the establishment of the reaction mechanisms, as well as the synthesis (electrosynthesis) of desired products.

This book highlights the state-of-the-art in the application of electrochemistry, by taking an interdisciplinary approach to the study of both static and dynamic molecular properties, mainly of coordination compounds, but including inorganic, bioinorganic and organometallic complexes. Supramolecular systems and metalloenzymes are also covered. The principles and approaches are often also valid for organic systems which are illustrated in various contexts.

This book is addressed to those chemists interested on the application of electrochemical techniques to their systems, as well as to electrochemists who have had their attention drawn to chemical problems. This work is therefore of certain value for experienced scientists as well as for relative newcomers. Addressed to senior researchers, it is also of interest to postgraduates and, in some aspects, to advanced undergraduates.

To some extent, it updates the previous book entitled "Molecular Electrochemistry of Inorganic, Bioinorganic and Organometallic Compounds" (NATO ASI Series C385, Kluwer, Dordrecht, 1993), edited by one of us (AJLP), that appeared approximately 12 years ago following a NATO Advanced Research Workshop (ARW) on that field in Sintra, Portugal, in 1992. This pioneering conference, also organized by one of us, was followed, since 1998, by other regular meetings (1998, 2000, 20002), held in Siena, Italy (organized by Prof. P. Zanello), on the same overall field, which also comprised educational purposes.

There is undoubtedly a rapidly increasing interest on the field and on its expansion through the use of modern concepts and unconventional approaches, which in our view justified the preparation of this new volume which provides an updated overview of the field about one decade after the publication of the above NATO ARW book.

The contents of this new book were recently presented at a symposium on "New Trends in Molecular Electrochemistry", organized by the Academy of Sciences of Lisbon within its "Frontiers of Knowledge" series, held at its premises.

The book deals, in an interdisciplinary approach, with the developments achieved in areas such as:

(i)   Redox and other molecular properties (electronic, optical and magnetic properties; electrochemical properties – structure relationships; metal-metal interactions; mixed valence compounds, etc.).

(ii)   Molecular electroactivation and electrocatalysis (reactions induced by ET).

(iii) Bioelectrochemistry (metalloenzymes).

(iv)   Supramolecular electrochemistry (supramolecular architectures, metallodendrimers and molecular devices)

(v)   Spectroelectrochemistry.

(vi)   Unconventional electrochemistry (ultrafast electrochemical techniques, new types of supporting electrolytes, electrochemistry without added electrolyte, ET through the liquid–liquid interface).

Therefore it covers both static and dynamic aspects, the former dealing mainly with structures and their relationships with particular properties and the latter concerning reactivity, kinetics and mechanisms of a wide variety of systems mainly within the broad filed of "inorganic"coordination chemistry (including also organometallic, supramolecular, bioinorganic and metalloenzyme chemistries) and, in a number of cases, also with illustrations in organic chemistry.

We acknowledge the authors of the various chapters for having accepted our invitation and for their so valuable and high quality contributions, as well as Dr. Fred Fenter (FontisMedia) for his continuous encouragement and highly professional assistance in the publication of the book. Thanks are also due to the Academy of Sciences of Lisbon, in particular to its Presidents, Prof. J.M. Toscano Rico and Prof. J.V. de Pina Martins (of the Classes of Sciences and Letters, respectively, who have been swapping the Presidency and Vice-Presidency along the last years) and Treasurers, Prof. F.R. Dias Agudo (also coordinator and representative of the international affairs board) and, recently, Prof. A. Torres Pereira, and to the President of the High Studies Institute (Prof. M. Jacinto Nunes), for approving and kindly supporting the initiative.

Armando J.L. Pombeiro   *Christian Amatore*
The Editor   *The Co-Editor*

Preface

This volume is the first of the resumed "Frontiers of Knowledge" series, initiated to mark the bicentennial anniversary of the foundation (on December 24th, 1779, by Queen Maria I) of the Academy of Sciences of Lisbon. For the presentation and discussion of the contents of the book, still in preparation, a symposium on "New Trends in Molecular Electrochemistry" was held, within that series, at the Academy (September 16-20th, 2003), together with the XII Meeting of the Portuguese Electrochemical Society covering all fields of Electrochemistry.

The "Frontiers of Knowledge" series first appeared in the late 70s within the activity of the High Studies Institute (Instituto de Altos Estudos) of the Academy, at the suggestion of General Luís Maria da Câmara Pina, its Treasurer at that time, and was expanded also under the initiative of other members of the Academy, namely Professors José Pinto Peixoto (President), Fernando Dias Agudo (Teasurer), Manuel Jacinto Nunes (President), José Tiago de Oliveira (Secretary), João Fraústo da Silva, R.J.P. Williams, António Xavier and Armando J.L. Pombeiro. The decision to resume this series was recently taken by the Administrative Council of the Academy: Professors José Manuel Toscano Rico (President), José Vitorino de Pina Martins (Vice-President), Fernando R. Dias Agudo (Treasurer), Armando J.L. Pombeiro (Secretary-General) and João Bigotte Chorão (Vice-Secretary-General). The series includes such representative titles as: "New Trends in Bioionorganic Chemistry" (1979); "New Trends in the Chemistry of Nitrogen Fixation" (1982); "Some Recent Advances in Statistics" (1983); "Advances in Geophysics", Vol. II, "Theory of Climate" (1983); "Estudos de Álgebra, Geometria e Análise" (1978). Usually both national and international editions were published, the latter by Academic Press, and the publications were based on the corresponding symposia held at the Academy to discuss the state-of-the-art in those fields.

Other representative symposia and publications, within the bicentenary celebration programme, include the following subjects: (i) History and Development of Science in Portugal: "A Actividade Pedagógica da Academia das Ciências de Lisboa nos séculos XVIII e XIX" (1981); "The History and Development of Science in Portugal" (up to the 19th century, 2 vols., 1986;

20th century, 3 vols., 1992); "Colóquio sobre Termodinâmica e Reactividade de Sistemas Moleculares" (1994). (ii) Humanism in Portugal: "Erasmo na Academia das Ciências" (1987); "Humanism in Portugal 1500-1600" (1988). (iii) Problems of Modern Society (drug abuse, tobacco addiction, alcoholism, traffic accidents, euthanasia, peace, etc.): "Colóquio sobre a Problemática da Droga em Portugal" (1988); "Colóquio sobre a Problemática do Tabagismo em Portugal (1988); "Colóquio sobre a Problemática do Alcolismo em Portugal" (1989); "Colóquio sobre a Eutanásia" (1993); "A Bioética e o Futuro" (1995); and "Colóquio sobre Portugal e a Paz" (1989).

Other major publications of the Academy include:

(i) Facsimiles of manuscripts and of old books (limited editions) such as the "Livro das Armadas" (The book of the Armadas, from 1497 to 1563) (1979), "Atlas" (the collection of the Viscount of Santarém), the "Atlas de Lázaro Luís" (1563) (1990), the "Crónica dos Reis de Espanha" (Chronicles of the Kings of Spain, 13th and 14th centuries) (in preparation), the "Livro de Horas" (Prayer Book) of the Countess of Bretiandos (XVI ? century) (in preparation), "Colóquios dos Simples e Drogas e Cousas Medicinais da India" by Garcia d'Orta (1563) (1963), "Obras de Pedro Nunes" (16th century, the complete re-edition and translation, with recently found new data, is now under way), "Os Lusíadas" by Luís de Camões (fac-simile of the 1st editions, 1572) (1980), "Dicionário da Língua Portuguesa" of the Academy (vol. I, 1793) (1993) and "A Santa Casa da Misericórdia de Lisboa" by Victor Ribeiro (1902) (1998). Critical editions of old books: "Crónicas dos Reis de Portugal e Sumários de Suas Vidas" by Gaspar Correia (1996), "Gramática da Linguagem Portuguesa" by Fernão de Oliveira (1536) (2000), and " A Arte da Grammatica da Lingua Portugueza" by António José dos Reis Lobato (1770-1869, various editions) (2000).

(ii) The "Memórias", regular proceedings currently in two series (Sciences and Letters, until now with 39 and 32 volumes); some of the volumes concern symposia held at the Academy: "Paleoambientes do Jurássico Superior em Portugal" (vol. 37, 1998), "Últimos Neandertais em Portugal" (vol. 38, 2000) and "Geoquímica e Petrogénese de Rochas Granitóides" (vol. 39, 2001). The "Memórias Económicas" (5 volumes, 1789-1815) and the "Memórias Económicas Inéditas" (1987) concerning the period 1780-1808.

(iii) The "Dicionário da Língua Portuguesa Contemporânea" (Dictionary of the Contemporary Portuguese Language), recently published (2001) within the activities of the Instituto de Lexicologia e Lexicografia da Língua Portuguesa (Institute of Lexicology and Lexicography of the Portuguese Language) of the Academy; other dictionaries are currently in preparation.

(iv) Catalogues: "Catálogo dos Manuscritos" (Série Vermelha, 2 volumes) (1976, 1978); "Livros Quinhentistas Espanhóis da Biblioteca da Academia das Ciências de Lisboa" (1989); "Livros Quinhentistas Portugueses da Biblioteca da Academia das Ciências de Lisboa" (1990); "Livros Quatrocentistas da Biblioteca da Academia das Ciências de Lisboa" (1992); "O Material Didáctico dos séculos XVIII e XIX do Museu Maynense da Academia das Ciências de Lisboa", Rómulo de Carvalho (1993); and "O Material Etnográfico do Museu Maynense da Academia das Ciências de Lisboa", Rómulo de Carvalho (2000). The preparation of a data base with the old books of the Library of the Academy is underway.

(v) Other books with scientific, historical, literary or political interest: "Reconhecimento científico de Angola" (1979); "Portugaliae Monumenta Histórica" (Nova Série, vol. II/1 e vol. II/2 æ Livro de Linhagens do Conde D. Pedro) (1980); "Bibliografia mais relevante sobre Botânica pura e aplicada referente aos países de expressão portuguesa" (1982); "D. João Carlos de Bragança, 2° Duque de Lafões, fundador da Academia das Ciências de Lisboa" (1987); "Estudos sobre a projecção de Camões em Culturas e Literaturas estrangeiras", vol. III (1984); "Colóquio de Estudos Camilianos" (1993); "Descriptive List of the State Papers of Portugal 1661-1780 in the Public Record Office London" (1983).

Apart from the editorial activity and the organization of symposia, the Academia, guided by its original motto "Nisi utile est quod facimus stulta est Gloria" (unless what we do is useful, our fame is foolish), also promotes analyses and discussions of problems of national relevance (Portuguese language, ethics in science, etc.). In the past, it was (i) from the "Instituto Vacínico" (Vaccinic Institute), an initiative of the Academy, that the Council for Public Health emerged, (ii) from the Geological Commission of the Academy that the Geological Services of Portugal were created (only recently they left the Academy premises which however still house the Geological Museum) and (iii) from the Higher Course of Arts (organized by the Academy) that the Faculty of Letters of Lisbon emerged. After this Faculty has left the Academy premises, the freed space was temporarily occupied by the Biblioteca Popular (Popular Library) but is now being renovated for conference, library and museum rooms of the Academy.

The Academy also houses, apart from a museum, one of the four most valuable libraries in the country, with over one million items and which includes the former library of the Convent of Jesus, a magnificent 16th-17th century building which was donated by Queen Maria II, in 1838, to the Academy for its premises. The peripheral legacy libraries (Júlio M. Fogaça and Vasco Magalhães Vilhena) have recently been installed herein.

The museum has several sections, namely Natural History, Physics and Etnography (Indian culture from the Amazonas). It is under reorganization, but the Maynense Gallery, with a valuable collection of scientific instruments, has already been opened.

The Academy awards prizes for outstanding contributions to both sciences and humanities.

Moreover, the Academy also promotes international scientific and cultural relationships and exchanges with other countries. It is the Portuguese representative organization to the International Council of Science (ICSU), former International Council of Scientific Unions, since the foundation of this Council in 1931, and it is one of the Portuguese members of the European Science Foundation, and a member of the Union Académique Internationale (UAI) , of ALLEA (All European Academies) and of the recently created European Academies' Science Advisory Council (EASAC).

The preparation of this book and the organization of a symposium on Molecular Electrochemistry are an expression of this internationally oriented policy of the Academy, applied in this case towards the development of this promising scientific field.

For further information, see: (a) Peixoto, J.P.; Pombeiro, A.J.L. "The Academy of Sciences of Lisbon", in "International Encyclopedia of Learned Societies and Academies", Kiger, J.C. (ed.), Greenwood Press, Westport, 1993, pp. 237-241. (b) "The Academy of Sciences of Lisbon", Academia das Ciências de Lisboa, 1994 (booklet). (c) "Comemorações do II Centenário da Academia das Ciências de Lisboa", Academia das Ciências de Lisboa, 1995. (d) Iria, A. "A Fundação da Academia das Ciências de Lisboa" in "História e Desenvolvimento da Ciência em Portugal", vol. II, Academia das Ciências de Lisboa (1986). (e) Dias Agudo, F.R. "Contribuição da Academia das Ciências de Lisboa para o Desenvolvimento da Ciência", ibid. (f) Peixoto, J.P. "A Revolução Cultural e Científica dos Séculos XVII e XVIII e a Génese das Academias", ibid. (g) Baião, A. "A Infância da Academia (1788-1794)" (1934). (h) Ayres, C. "Para a História da Academia das Sciências de Lisboa", Academia das Ciências de Lisboa, Separata do Boletim da Segunda Classe, vol. XII (1927). (i) "III Jubileu da Academia das Ciências de Lisboa" (1931).

M. Toscano Rico, President of the Academy
F. Dias Agudo, Coordinator of International Affairs of the Academy
Armando J.L. Pombeiro, The Editor, Secretary General of the Academy

Part 1
Redox and Other Molecular Properties

Chapter 1

Homoleptic, Mononuclear Transition Metal Complexes of 1,2-Dithiolenes: Updating their Electrochemical-to-Structural Properties

Piero Zanello and Emanuela Grigiotti

Dipartimento di Chimica dell'Università di Siena, Via Aldo Moro, 53100 Siena, Italy

1   INTRODUCTION

Metal complexes of 1,2-dithiolenes appeared in literature about seventy years ago as analytical reagents [1]. The subsequent elucidation of their molecular and electronic structure at first and then either their relationships with some enzyme-mimicking activity, or the continuous discovery of their novel applications in material science (related to their uncommon conducting, magnetic, non-linear optic, sensing and dyeing properties) made them one of the most promising and persisting topic in inorganic chemistry. This is reflected on the numerous review papers from time to time devoted to the different aspects of their chemistry [2]. The 1968 McCleverty's pioneering work [2b] focussed on the structural and redox properties of such compounds, both the aspects being crucial to most innovative applications. The present review wishes to update just such a matter.

The different subclasses of ligands used in metal complexes discussed here are illustrated in Schemes 1-4. They have been subdivided depending upon if the 1,2-dithiolate substituents give rise to lateral ring(s), as well as on the type of lateral ring(s).

Since, from the structural viewpoint, we wish to give evidence only to the most significant reorganizations following electron transfer processes, for sake of simplicity bond lengths will be approximated to the hundredth of Å (neglecting any esd value).

It must be preliminarily taken into account that transition metal complexes with *dithiolene* ligands constitute an intriguing class of complexes as far as their

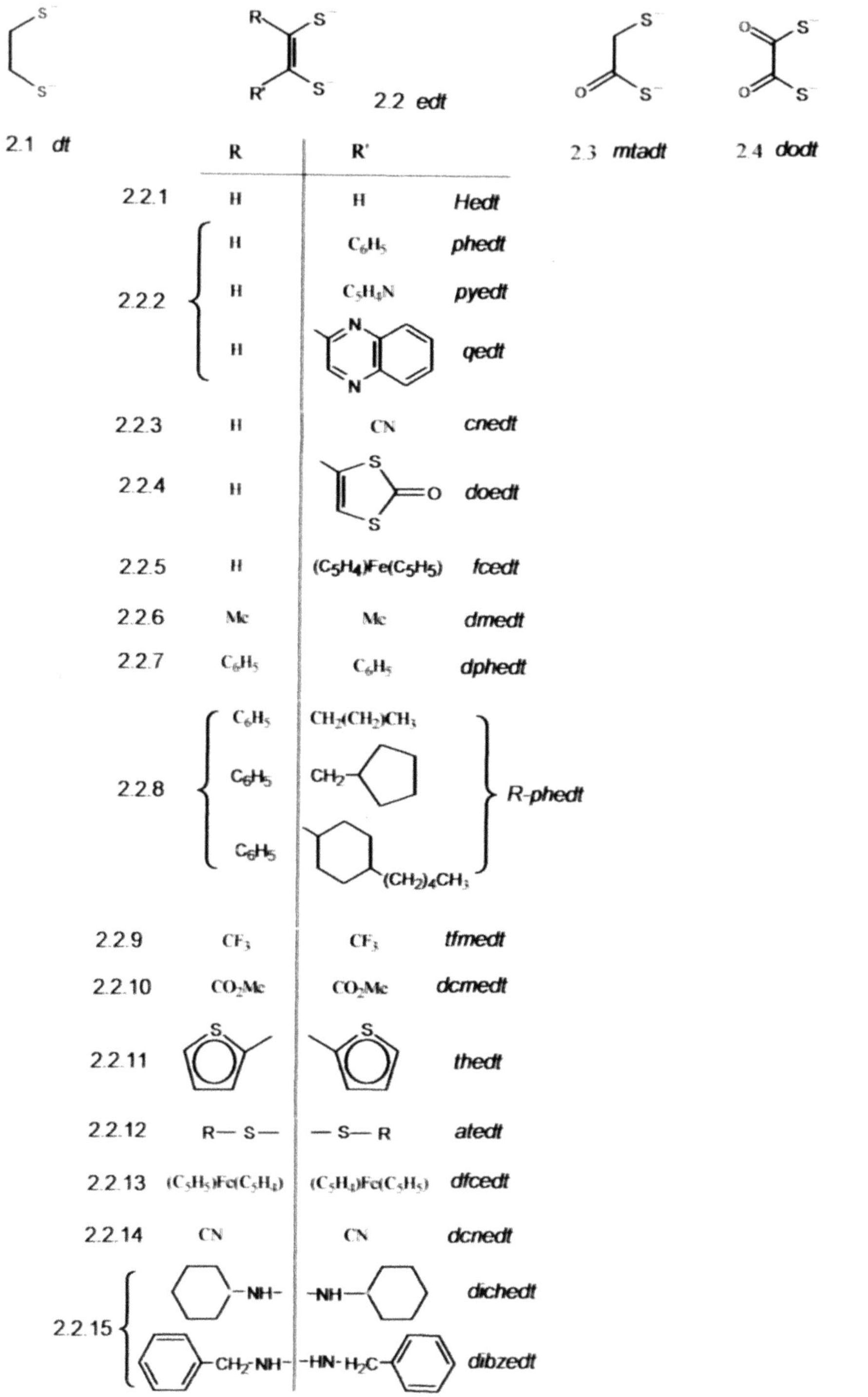

2.1 *dt*   2.2 *edt*   2.3 *mtadt*   2.4 *dodt*

	R	R'	
2.2.1	H	H	*Hedt*
2.2.2	H	C_6H_5	*phedt*
	H	C_5H_4N	*pyedt*
	H	(quinoxaline)	*qedt*
2.2.3	H	CN	*cnedt*
2.2.4	H	(1,3-dithiol-2-one, methyl)	*doedt*
2.2.5	H	$(C_5H_4)Fe(C_5H_5)$	*fcedt*
2.2.6	Me	Me	*dmedt*
2.2.7	C_6H_5	C_6H_5	*dphedt*
2.2.8	C_6H_5	$CH_2(CH_2)CH_3$	*R-phedt*
	C_6H_5	CH_2(cyclopentyl)	
	C_6H_5	(cyclohexyl)$(CH_2)_4CH_3$	
2.2.9	CF_3	CF_3	*tfmedt*
2.2.10	CO_2Me	CO_2Me	*dcmedt*
2.2.11	(2-methylthiophene)	(2-methylthiophene)	*thedt*
2.2.12	R—S—	—S—R	*atedt*
2.2.13	$(C_5H_5)Fe(C_5H_4)$	$(C_5H_4)Fe(C_5H_5)$	*dfcedt*
2.2.14	CN	CN	*dcnedt*
2.2.15	(cyclohexyl)—NH—	—NH—(cyclohexyl)	*dichedt*
	(phenyl)—CH_2-NH—	—HN-H_2C—(phenyl)	*dibzedt*

Scheme 1

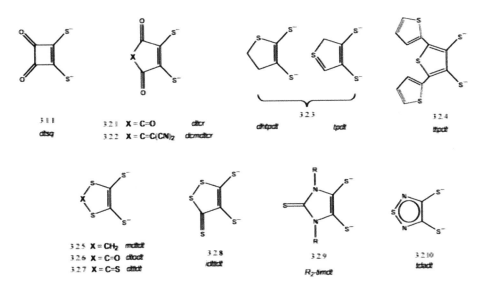

3.1.1 *dtsq*	3.2.1 **X = C=O** *dtcr* 3.2.2 **X = C=C(CN)₂** *dcrmdtcr*	3.2.3 *dhtpdt* *tpdt*	3.2.4 *ttpdt*

3.2.5 **X = CH₂** *mdtdt*
3.2.6 **X = C=O** *dtodt*
3.2.7 **X = C=S** *dttdt*

3.2.8
idtdt

3.2.9
R₂-bimdt

3.2.10
tdadt

Scheme 2

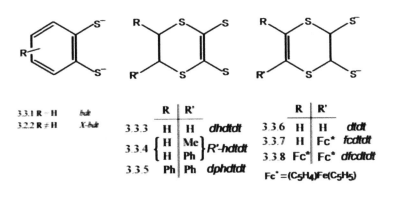

3.3.1 R = H   *bdt*
3.2.2 R ≠ H   *X-bdt*

	R	R'	
3.3.3	H	H	*dhdtdt*
3.3.4	H H	Me Ph	*R'-hdtdt*
3.3.5	Ph	Ph	*dphdtdt*

	R	R'	
3.3.6	H	H	*dtdt*
3.3.7	H	Fc*	*fcdtdt*
3.3.8	Fc*	Fc*	*dfcdtdt*

Fc* = (C₅H₄)Fe(C₅H₅)

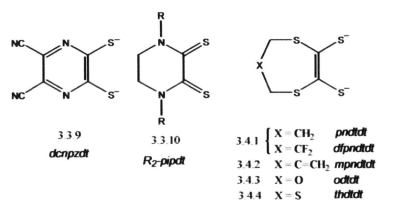

3.3.9
dcnpzdt

3.3.10
R₂-pipdt

3.4.1 { **X = CH₂**   *pndtdt*
{ **X = CF₂**   *dfpndtdt*
3.4.2 **X = C=CH₂** *mpndtdt*
3.4.3 **X = O**   *odtdt*
3.4.4 **X = S**   *thdtdt*

Scheme 3

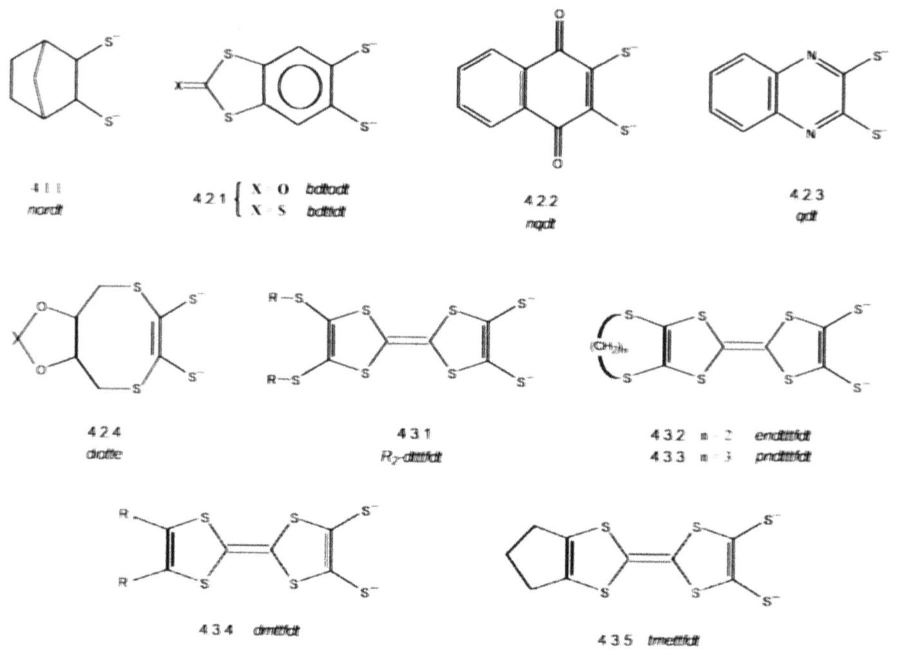

Scheme 4

electrochemical behaviour is concerned. In fact, considering that such ligands are able to shuttle reversibly the sequence *1,2-dithione/1,2-dithione monoanion/ene-1,2-dithiolate* illustrated in Scheme 5, their metal complexes should display a notably extended redox activity.

For instance, the most common homoleptic bis-dithiolene metal complexes $[M^{II}(S_2C_2R_2)_2]^{n-}$ (M = Ni, Pd, Pt) should potentially display the redox sequence illustrated in Scheme 6.

Obviously, the different steps can proceed at different or at coincident potential values depending upon the extent of intramolecular communication existing inside the molecule.

In reality, this has to be considered as a simplified sequence, in that it assumes that the central metal ion maintains its oxidation state unchanged, but the concomitant occurrence of metal centred redox processes cannot be disregarded, and out the subsequent occurrence of internal metal-to-ligand electron transfer reorganization can not be ruled out.

In fact, in those cases in which metal-based and ligand-based frontier orbitals are comparable in energy, the ligands under subject are defined as redox "non-innocent" [3].

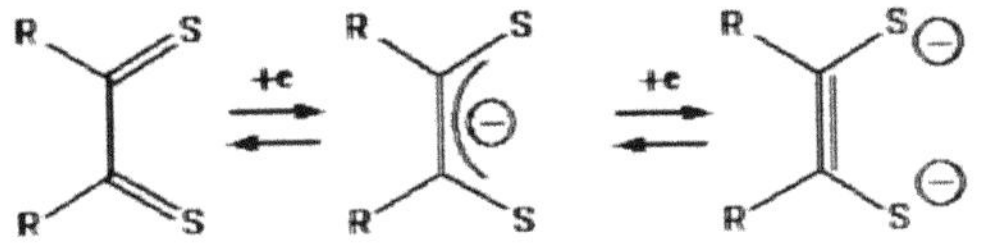

Scheme 5

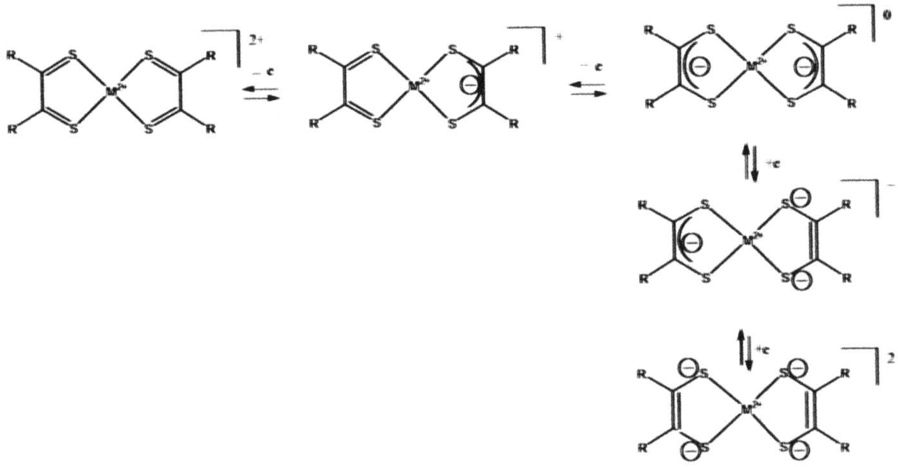

Scheme 6

2   1,2-DITHIOLATES NOT FORMING LATERAL RINGS

2.1   dt Complexes

In literature, metal complexes of the 1,2-ethanedithiolate dianion [SCH_2CH_2S]$^{2-}$ (here abbreviated as *dt*) and those of (unsubstituted) 1,2-ethenedithiolate dianion [$SCH=CHS$]$^{2-}$ (hereinafter abbreviated as *Hedt*) are commonly undifferentiated, at least as far as their nomenclature is concerned, both being often designated as "edt" complexes.

2.1.1 Bis(dt) complexes

The crystal structures of a number of homoleptic bis-dithiolenes of general formula [$M^{II}(dt)_2$]$^{2-}$ (M = Cr, Mn, Co, Ni, Zn, Pd, Cd) are available [4]. Excluding [$Ni(dt)_2$]$^{2-}$ [4d], the electrochemical behaviour of these complexes has not been reported, even if the molecular structure of the redox couple [$Co(dt)_2$]$^{2-/-}$ is known, Table 1 [4c].

Table 1. Selected structural parameters (average bond distances, Å; dihedral angles, °) in the redox couple $[Co(dt)_2]^{2-/-}$

Complex	Co-S	S-C	C-C	α^a	Cation
$[Co(dt)_2]^{2-}$	2.28	1.80	1.40	88.5	$[NMe_4]^+$
$[Co(dt)_2]^-$	2.17	1.79	1.33	6.8	$[NMe_4]^+$

^{a} Dihedral angle between the two CoS_2 planes

It is not only interesting to note that the dianion has a tetrahedral geometry, whereas the monoanion assumes a (nearly) square-planar geometry, but also the significant variation of the Co-S distance, which allows one to assume a mainly metal-centred electron transfer.

$[Ni(dt)_2]^{2-}$ in MeCN solution gives rise to a quasireversible one-electron oxidation (E°' = –0.68 V, *vs*. SCE), which is accompanied by relatively fast decomposition of the electrogenerated monoanion [4d].

The vanadyl and thiovanadyl complexes $[V^{IV}E(dt)_2]^{2-}$ (E = O, S) have been structurally characterised [5]. They possess a square-pyramidal geometry, with the chalcogenyl atom apically coordinated and the metal ion placed out (towards the apex) of the thiolene-sulfur basal plane. From the electrochemical viewpoint, these V(IV) complexes in MeCN solution only put in evidence irreversible electron–transfer processes [5b].

Brief electrochemical data have been also reported for the square pyramidal Re(V) and Tc(V) complexes $[ME(dt)_2]^-$ (E = O, S) [6a,b]. In MeCN solution, metal-centered reductions occur at very negative potential values (E°$_{-/2-}$: ReO, < –2.1 V; ReS = –1.83 V; TcO = –1.86 V, *vs*. SCE), whereas no ligand-centred reversible electron transfer process seem to be present [6a,c].

In conclusion, the absence of reversible electron transfer processes in the actual complexes is likely due to the absence of the carbon/carbon unsaturation in the dithiolene backbone. In fact, just the π-delocalization along the whole molecular frame arising from unsaturation commonly triggers electron mobility. In this light, the partial chemical reversibility of the one-electron oxidation of $[Ni(dt)_2]^{2-}$ can be attributed to the instability of the electrogenerated Ni(III) monoanion $[Ni(dt)_2]^-$.

2.1.2 Tris(dt) complexes

The solid-state structures of the tris-chelates $[M(dt)_3]^{n-}$ (M = Ti(IV), n = 2; M = Nb(V), Ta(V), n = 1) are known [4b,c,7]. As exemplified in Figure 1, which

(a)   (b)

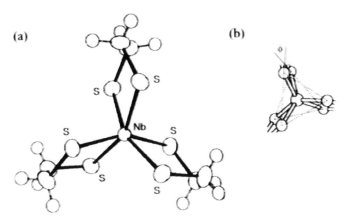

Figure 1. (a) X-Ray structure of $[Nb(dt)_3]^-$ ($[NEt_4]^+$ cation). Average Nb-S distance = 2.43 Å. Twist angle (ϕ) between the equilateral triangles formed by the three upper and the three lower sulfur atoms = about 30°. (b) Meaning of "twist" angle.

refers to $[Nb(dt)_3]^-$ [7], they possess a geometry, which, based on the value of the twist angle ϕ, is intermediate beween trigonal-prismatic and octahedral (trigonal prism: $\phi = 0$; octahedron: $\phi = 60°$).

Further supporting the redox inactivity of the saturated dt ligand, dmso solutions of $[M(dt)_3]^-$ (M = Nb, Ta) only display a metal-centred, quasireversible, one-electron reduction (Nb: $E^{o'}_{-/2-} = -1.20$ V; Ta: $E^{o'}_{-/2-} = -1.56$ V, *vs.* SCE) [7].

2.2 edt Complexes

2.2.1 Hedt Complexes (Better known as "edt")

2.2.1.1 Bis(Hedt) complexes

The square planar geometry of the monoanion $[Ni(Hedt)_2]^-$ is shown in Figure 2 [8].

In dmso solution it undergoes reversibly either a one-electron oxidation or a one-electron reduction, Table 2 [2b, 9].

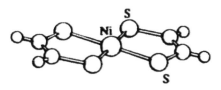

Figure 2. X-Ray structure of $[Ni(Hedt)_2]^-$ ($[Fe(C_5Me_5)_2]^+$ cation). Bond lengths: Ni-S = 2.14 Å; S-C = 1.71 Å; C=C = 1.32 Å.

The neutral complex [Ni(Hedt)$_2$] maintains the square-planar geometry of the monoanion, with a slight decrease of the Ni-S and S-C distances (by about 0.04 Å and 0.02 Å, respectively) and a slight increase of the C=C distance (by about 0.04 Å) [10].

Such structural change does not allow the precise determination of the oxidation state of the central Ni atom inside the redox couple [Ni(Hedt)$_2$]$^{-/0}$. In spite of the Ni(III)/Ni(IV) assignment [8], we note that [Pd(Hedt)$_2$]$^-$, which is difficulty assignable as Pd(III) compound, displays a similar electrochemical behaviour, Table 2 [2b]. Hence, it cannot be ruled out that the [Ni(Hedt)$_2$]$^{-/0}$ complexes might be seen as oxidation products of the Ni(II) precursor [Ni^{II}(Hedt)$_2$]$^{2-}$, with the extra electrons stepwise removed from the dithiolene ligands. The same likely holds for the other [M(Hedt)$_2$]$^-$ complexes listed in Table 2. In this connection, in agreement with the easy oxidation exhibited by [Pd(Hedt)$_2$]$^-$, the crystal structure of the neutral [Pd(Hedt)$_2$] has been reported [11]. As expected, the PdS$_4$C$_4$H$_4$ frame is planar, but in the solid state the neutral complex looks like a dimer because of the presence of a direct intermolecular metal-metal bond (2.79 Å).

Table 2. Formal electrode potentials (V, *vs.* SCE) for the redox changes exhibited by a few [M(Hedt)$_2$]$^{n-}$ complexes

M	$E^{o'}_{-/2-}$	$E^{o'}_{-/0}$	Solvent	Reference
Ni	−0.84	+0.16	dmso	2b,9
	−0.97	+0.07	dmf	2b
Pd	−0.64	-	dmso	2b,9
	−0.77	+0.11	dmf	2b
Co	−0.93	-	dmso	2b,9
Cu	−0.74	-	dmso	2b,9

Figure 3 illustrates the square-pyramidal structure of the oxo-Mo(IV) dianion [MoO(Hedt)$_2$]$^{2-}$ [12].

In MeCN solution it undergoes a reversible one-electron oxidation ($E^{o'} = -0.61$ V, *vs.* SCE), followed by a further irreversible oxidation [12]. The crystal structure of the corresponding monoanion has been solved [12]. It maintains the square pyramidal geometry of the parent dianion with the following structural variations: (a) the average Mo-S distance shortens by about 0.02 Å; (b) the Mo=O distance shortens by about 0.06 Å; (c) both the S-C and C=C distances

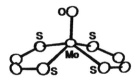

Figure 3. X-Ray structure of $[MoO(Hedt)_2]^{2-}$ ($[NEt_4]^+$ cation). Average bond lengths: Mo–S = 2.40 Å; Mo–O = 1.74 Å; S–C = 1.75 Å; C=C = 1.33 Å.

shorten by about 0.02 Å. Because of the significant variation of the Mo=O distance, it has been assumed that the one-electron removal is due to the $Mo^{IV}O/Mo^VO$ oxidation.

2.2.1.2 *Tris(Hedt) complexes*

To the best of our knowledge, the only crystallographically characterized tris(Hedt) metal complex still remains $[Mo(Hedt)_3]$, whose structural and voltammetric data have been already reported [2b].

2.2.2 *phedt, pyedt, qedt* Complexes

2.2.2.1 *Bis(phedt), bis(X-pyedt) (X = 2, 3, 4), and bis(qedt) complexes*

Like $[MoO(Hedt)_2]^{2-}$, the square pyramidal $[MoO(phedt)_2]^{2-}$ undergoes a reversible one-electron oxidation, which has been assigned to the $Mo^{IV}O/Mo^VO$ process [13]. Analogous assignment has been made to the one-electron oxidation exhibited by the pyridin-2-yl, -3-yl, 4-yl $[MoO(X-pyedt)_2]^{2-}$ and quinoxalin-2-yl $[MoO(qedt)_2]^{2-}$ dianions [13]. The pertinent redox potentials are compiled in Table 3.

Table 3. Formal electrode potentials (V, *vs.* SCE) for the one-electron oxidation of $[MoO(phedt)_2]^{2-}$, $[MoO(X-pyedt)_2]^{2-}$ and $[MoO(qedt)_2]^{2-}$ complexes in dmf solution

Complex	$E^{o\prime}_{2-/-}$
$[MoO(phedt)_2]^{2-}$	−0.48
$[MoO(2-pyedt)_2]^{2-}$	−0.42
$[MoO(3-pyedt)_2]^{2-}$	−0.38
$[MoO(4-pyedt)_2]^{2-}$	−0.35
$[MoO(qedt)_2]^{2-}$	−0.28

As seen, the different pendant groups of the dithiolene units exert appreciable inductive effects, the phedt ligand favouring the one-electron removal with respect to pyedt and qedt, respectively.

2.2.3   cnedt Complexes

2.2.3.1 Bis(cnedt) complexes

The square planar dianion $[Ni(cnedt)_2]^{2-}$ exhibits in MeCN solution a first, reversible one-electron oxidation ($E^{o\prime} = -0.31$ V, *vs.* SCE), followed by a further irreversible process ($E_p = +0.65$ V) [14]. As a consequence of the one-electron removal, the corresponding monoanion maintains the square planar geometry, the Ni-S and S-C distances shorten by 0.04 Å and 0.02 Å, respectively, whereas the other bond lengths remain unaltered [14]. Thus, in spite of theoretical calculation on the HOMO composition of $[Ni(cnedt)_2]^{2-}$, which suggests that it is essentially contributed by the ligand [14], it seems almost certain that the appreciable variation of the Ni-S distance implies that the HOMO level is contributed also by the metal.

2.2.4 doedt Complexes

2.2.4.1 Bis(doedt) complexes

Nothing is known about the redox properties of the (almost) square planar complex $[Ni(doedt)_2]^-$ [15].

2.2.5 fcedt Complexes

2.2.5.1 Bis(fcedt) complexes

It has been briefly reported that $[Ni(fcedt)_2]$ undergoes two separate, ligand-centred, reversible processes ($[Ni(fcedt)_2]^{0/-}$: $E^{o\prime} = +0.51$ V; $[Ni(fcedt)_2]^{-/2-}$: $E^{o\prime} = -0.20$ V, *vs.* SCE) and a single two-electron oxidation ($[Ni(fcedt)_2]^{0/2+}$: $E^{o\prime} = +0.76$ V) which has been attributed to the two non-communicating ferrocenyl subunits [16]. In reality, such an interpretation would deserve revision in favour of two separate ferrocene-centred oxidations, also in the light of the behaviour of the related tetraferrocenyl complex which will be discussed below (see Section 2.2.13).

2.2.6 dmedt Complexes

2.2.6.1 Bis(dmedt) complexes

The electrochemical behaviour of the dianions $[M(dmedt)_2]^{2-}$ (M = Ni, Pd, Pt) has been reviewed [2b]. As a typical example, Figure 4 shows the voltammetric response of the square planar $[Ni(dmedt)_2]^{2-}$ [17].

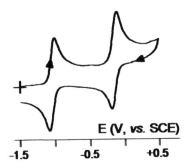

Figure 4. Cyclic voltammogram recorded at a platinum electrode in MeCN solution of [Ni(dmedt)$_2$]$^{2-}$. Scan rate 0.1 V s^{-1}.

It undergoes two separate one-electron oxidations with features of chemical reversibility ($E^{o'}_{2-/-}=-1.05$ V; $E^{o'}_{-/0}=-0.15$ V, *vs.* SCE). As a matter of fact, all the members of the reversible sequence 2–/–/0 have been structurally characterized, Table 4 [17].

Table 4. Selected bond distances (average values, Å) in the redox sequence [Ni(dmedt)$_2$]$^{2-/-/0}$

Complex	Ni–S	S–C	C=C	C-C$_{(Me)}$	Cation
[Ni(dmedt)$_2$]$^{2-}$	2.18	1.76	1.34	1.51	[NEt$_4$]$^+$
[Ni(dmedt)$_2$]$^-$	2.14	1.74	1.34	1.51	[NEt$_4$]$^+$
[Ni(dmedt)$_2$]^0	2.13	1.71	1.36	1.51	-

On passing gradually from the neutral 1,2-dithionate to the dianionic 1,2-dithiolate, the Ni-S and C-S distances progressively increase, whereas the C=C distance slightly decreases. The minor but significant variation of the Ni-S distance with the electron addition/removal processes suggests that the redox processes are centered on molecular orbitals significantly contributed not only by the dithiolene ligand but also by the metal.

As far as the oxo–dianions [MO(dmedt)$_2$]$^{2-}$ (M = Mo, W) are concerned, they also undergo two reversible one-electron oxidations according to the sequence [MoO(dmedt)$_2$]$^{2-/-/0}$ [18]. The crystal structures of two of the three-membered series are available, namely [MoO(dmedt)$_2$]$^{2-/-}$ [18a] (also available is the crystal structure of [WO(dmedt)$_2$]$^{2-}$ [18b]). Apart from their usual square-pyramidal geometry, the structural consequences of the one-electron removal are reported in Table 5, together with the formal electrode potentials of the electron transfer sequences [18].

Table 5. Formal electrode potentials (V, *vs.* SCE) for the redox sequence $[MO(dmedt)_2]^{2-/-/0}$ (MeCN solution) and selected interatomic distances (Å) in the couple $[MO(dmedt)_2]^{2-/-}$

Complex	$E^{o\prime}_{2-/-}$	$E^{o\prime}_{-/0}$	M-S	C-S	C=C	M-O	Cation
$[MoO(dmedt)_2]^-$	-	-	2.37	1.75	1.33	1.68	$[NEt_4]^+$
$[MoO(dmedt)_2]^{2-}$	−0.62	+0.14	2.39	1.78	1.33	1.71	$[NEt_4]^+$
$[WO(dmedt)_2]^{2-}$	−0.91	−0.05	2.38	1.79	1.33	1.74	$[NEt_4]^+$

Based on the variation of the Mo=O length, the overall reversible sequence has been assigned as: $M^{IV}O/M^{V}O/M^{VI}O$ (M = Mo, W) [18], even if the participation of the ligand to the electron transfer series cannot be ruled out.

2.2.6.2 Tris(dmedt) complexes

The trigonal prismatic derivatives $[M(dmedt)_3]$ (Mo, W) undergo reversibly the sequence $[M(dmedt)_3]^{0/-/2-}$ (M = Mo: $E^{o\prime}_{0/-} = -0.34$ V; $E^{o\prime}_{-/2-} = -0.86$ V. M = W: $E^{o\prime}_{0/-} = -0.36$ V; $E^{o\prime}_{-/2-} = -0.92$ V; MeCN solution, V *vs.* SCE) [19, 18a]. Stated that all the members of the two families $[M(dmedt)_3]^{0/-/2-}$ (M = Mo, W) are isostructural, the most significant structural changes accompanying the redox sequences are compiled in Table 6 [18a, 19].

Table 6. Selected bond distances (Å) and twist angles (°) in the three-membered series $[M(dmedt)_3]^{0/-/2-}$ (M = Mo, W)

Complex	M-S	S-C	C=C	ϕ	Cation
$[Mo(dmedt)_3]^0$	2.36	1.71	1.36	3.5	-
$[Mo(dmedt)_3]^-$	2.37	1.72	1.35	1.6	$[NEt_4]^+$
$[Mo(dmedt)_3]^{2-}$	2.40	1.75	1.33	2.6	$[NEt_4]^+$
$[W(dmedt)_3]^-$	2.38	1.73	1.36	2.8	$[NEt_4]^+$
$[W(dmedt)_3]^{2-}$	2.39	1.76	1.33	2.4	$[NEt_4]^+$

The slight variations of both the metal-to-ligand and the intraligand distances support that the electrons are transferred from orbitals contributed from either the metal or the ligand.

2.2.7 dphedt Complexes

2.2.7.1 Bis(dphedt) complexes

Metal complexes of 1,2-diphenylethene-1,2-dithiolate are long known and widely investigated either from the structural or redox viewpoints [2b]. Our discussion will be hence limited to updating previous data.

As illustrated in Figure 5, the square planar complex [Ni(dphedt)$_2$] undergoes two chemically reversible one-electron reductions and a partially chemically reversible one-electron oxidation [20a].

The same behaviour is exhibited by both [Pd(dphedt)$_2$] and [Pt(dphedt)$_2$], Table 7.

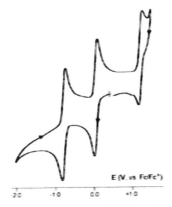

Figure 5. Cyclic voltammogram recorded in CH$_2$Cl$_2$ solution of [Ni(dphedt)$_2$]. Scan rate 0.2 V s^{-1}.

Table 7. Formal electrode potentials (V, *vs.* Fc/Fc$^+$) for the redox processes exhibited by [M(dphedt)$_2$] (M = Ni, Pd, Pt) and [Ni(dph-p-X-edt)$_2$] in CH$_2$Cl$_2$ solution

Complex	$E^{o\prime}_{0/+}{}^{a}$	$E^{o\prime}_{0/-}$	$E^{o\prime}_{-/2-}$	Reference
[Ni(dphedt)$_2$]	+1.15	+0.02	−0.80	20a
	–	+0.01	−0.81	20b
	–	−0.04	−0.87	20c
[Ni(dph-p-Br-edt)$_2$]	–	+0.07	−0.74	20c
[Ni(dph-p-F-edt)$_2$]	–	+0.03	−0.79	20c
[Ni(dph-p-Me-edt)$_2$]	–	−0.10	−0.92	20c
[Ni(dph-p-OMe-edt)$_2$]	–	−0.11	−0.93	20c
[Pd(dphedt)$_2$]	+1.17	+0.02	−0.61	20a
[Pt(dphedt)$_2$]	+1.20	−0.06	−0.80	20a

^{a} Partial chemical reversibility

The monoanions [M(dphedt)$_2$]$^-$ (M = Ni [21], Pt [22]) have been structurally characterised. Concerned with the Ni monoanion, the most significant variations in bond lengths with respect to the neutral species are summarized in Table 8 (together with those of a few asymetrically substituted complexes which will be discussed in the next Section 2.2.8.1).

Once again, both the metal and the dithiolene ligands look like involved in the electron transfer process.

A series of para-phenyl substituted Ni complexes have been also prepared and their electrochemical behaviour studied, Table 7. The inductive effects exerted by the para-substituents strictly obey the Hammett equation [20c].

Like the related [MoO(Hedt)$_2$]$^{2-}$ (and at variance with the dmedt analogues), the square pyramidal dianions [WE(dphedt)$_2$]$^{2-}$ (E = O, S) only exhibit a chemically reversible one-electron oxidation to the corresponding monoanions [WE(dphedt)$_2$]$^-$ (E = O, S: E$^{\circ\prime}$ = –0.62 V, *vs.* SCE; MeCN solution [18b]). No crystallographic data are available for the likely isostructural monoanions. A second irreversible oxidation is present for E = O (MeCN: E$_p$ = –0.21 V).

Table 8. Selected bond distances (average values, Å) in the redox couple [Ni(dphedt)$_2$]$^{0/-}$, together with those of the asymmetrically substituted [Ni(R-phedt)$_2$]^0

Complex	M-S	S-C	C=C	Cation	Reference
[Ni(dphedt)$_2$]^0	2.10	1.71	1.37	-	2b
	2.12	1.70	1.41	-	23
[Ni(dphedt)$_2$]$^-$	2.14	1.75	1.34	[NBu$_4$]$^+$	21a
	2.14	1.77	1.34	[NEt$_4$]$^+$	21b
[Pt(dphedt)$_2$]$^-$	2.26	1.73	1.38	[NBu$_4$]$^+$	22
[Ni(Bu-phedt)$_2$]^0	2.12	1.71	1.36	-	20b
[Ni(C$_5$H$_9$CH$_2$-phedt)$_2$]^0	2.12	1.71	1.37	-	20b

2.2.7.2 Tris(dphedt) complexes

Structural data for the neutral complexes [M(dphedt)$_3$] (M = V, Re) have been reviewed, together with the electron transfer ability of the series M = V, Mo, W, Re [2b]. They usually display the sequence [M(dphedt)$_3$]$^{3-/2-/-/0}$. In particular, [W(dphedt)$_3$] exhibits in dmf solution two reversible one-electron reductions (E$^{\circ\prime}_{0/-}$ = –0.54 V; E$^{\circ\prime}_{-/2-}$ = –1.13 V, *vs.* SCE) [2b]. Both the neutral derivative and its monoanion possess a trigonal prismatic geometry, and the passage from $[W(dphedt)_3]^-$ to $[W(dphedt)_3]^0$ involves an increase of the W-S, S-C, and C=C distances (by 0.03 Å, 0.05 Å, and 0.02 Å, respectively) [18b]. It seems evident that also in this case either the metal or the ligand is involved in the electron transfer process.

2.2.8 R-phedt Complexes

2.2.8.1 Bis(R-phedt) complexes

A series of asymmetrically substituted Ni(II)-dithiolenes bearing a phenyl group and either a butyl, or a cyclopentylmethyl, or a 4-pentylcyclohexyl group have been characterised [20b]. The crystal structures of the complexes $[Ni(Bu\text{-}phedt)_2]$ and $[Ni(C_5H_9CH_2\text{-}phedt)_2]$ have been solved and the pertinent data are collected in Table 8.

As a consequence of the electron-donating effect of the alkyl substituent, the present complexes in CH_2Cl_2 solution undergo reversibly the 2–/–/0 sequence at slightly more negative potential values than $[Ni(dphedt)_2]$ (R = Bu: $E^{\circ'}_{0/-}$ = –0.13 V, $E^{\circ'}_{-/2-}$ = –0.96 V; R = $CH_2C_5H_9$: $E^{\circ'}_{0/-}$ = –0.13 V, $E^{\circ'}_{-/2-}$ = –0.97 V; R = $C_5H_{11}C_6H_{10}$: $E^{\circ'}_{0/-}$ = –0.14 V, $E^{\circ'}_{-/2-}$ = –0.98 V; V, *vs*. Fc/Fc$^+$).

2.2.9 tfmedt Complexes (better known as "tdf")

2.2.9.1 Bis(tfmedt) complexes

Among complexes $[M(tfmedt)_2]^{n-}$, the Ni, Pd, Pt derivatives usually exhibit the reversible sequence 2–/–/0, the Cu, Co, Fe complexes only undergo the 2–/– process, and the Au complex is stable only as monoanion [2b].

Wishing to point out the structural variations following the electron transfer processes, we dwell upon the square planar couple $[Ni(tfmedt)_2]^{0/-}$, a few selected structural parameters of which are collected in Table 9.

Table 9. Selected bond distances (average values, Å) in the couple $[Ni(tfmedt)_2]^{0/-}$

Complex	Ni-S	S-C	C=C	C-C$_{(CF3)}$	Cation	Reference
$[Ni(tfmedt)_2]^0$	2.12	1.71	1.38	1.48	-	25
$[Ni(tfmedt)_2]^-$	2.14	1.70	1.40	1.47	$[C_7H_7]^+$	24a
	2.13	1.72	1.36	1.50	$[PTZ]^{+\,a}$	24b
	2.14	1.73	1.35	1.51	$[Fe(C_5Me_5)_2]^+$	24c

^{a} PTZ = phenothiazine

The slight structural variations occurring at the ligand frame suggest that the electron transfer involves orbitals which are substantially contributed by the ligand. Further support to such a proposal comes from the electrochemical finding, that in MeCN solution the ligand-centred reversible reductions $[Ni(tfmedt)_2]^{0/-/2-}$ are followed by a further reversible one-electron step at very negative potential values ($E^{\circ\prime} = -2.34$ V, *vs.* SCE), which proved to generate the Ni(I) trianion $[Ni(tfmedt)_2]^{3-}$ [26].

It has been briefly reported that the square-pyramidal [27] dianion $[MoO(tfmedt)_2]^{2-}$ undergoes in MeCN solution the (common) sequence $[MoO(tfmedt)_2]^{2-/-/0}$ [12], but no structural data are available for any of the oxidised members of the family. The crystal structure of $[WO(tfmedt)_2]^{2-}$ is also known [27].

2.2.9.2 Tris(tfmedt)complexes

The trigonal prismatic dianions $[M(tfmedt)_3]^{2-}$ (M = Mo, W) undergo, in MeCN solution, two reversible one-electron oxidations [2b].

The crystal structures of all the members of the redox family $[Mo(tfmedt)_3]^{2-/-/0}$ are available and selected structural parameters are reported in Table 10 [27, 28].

Table 10. Selected bond distances (average values, Å) in the redox family $[Mo(tfmedt)_3]^{2-/-/0}$

Complex	M-S	S-C	C=C	Cation	Reference
$[Mo(tfmedt)_3]^0$	2.35	1.70	1.36	-	27
$[Mo(tfmedt)_3]^-$	2.37	1.72	1.37	$[Fe(C_5Me_5)_2]^+$	28
$[Mo(tfmedt)_3]^{2-}$	2.37	1.74	1.33	$[NEt_4]^+$	27
	2.38	1.74	1.36	$Fe(C_5Me_5)_2]^+$	28
$[W(tfmedt)_3]^{2-}$	2.37	1.74	1.34	$[NEt_4]^+$	27

Once again, the (slight) variation of the interatomic distances with the overall charge appears like extended along the whole metallacycle fragments. Infact, the most significant structural differences inside the three members reside in the extent of trigonal prismatic geometry. The neutral complex adopts a perfect trigonal prismatic structure; the monoanion also possesses a nearly perfect trigonal prismatic structure; the dianion is significantly distorted ($\phi \approx 16°$) [27, 28].

2.2.10   dcmedt Complexes

2.2.10.1   Bis(dcmedt)complexes

The crystal structures of the square planar anions [Ni(dcmedt)$_2$]$^-$ and [Ni(dcmedt)$_2$]$^{2-}$ are known [29], even if the pertinent electrochemical data are not available. Passing from the dianion to the monoanion, The Ni-S and S-C distances undergo a slight decrease (by 0.04 Å and 0.02 Å, respectively), whereas the C=C and the peripheral C=O distances remain substantially unaltered [29b]. These data suggest that the extra electron is removed from a molecular orbital contributed by both the metal and the ligand.

The square-pyramidal dianion [MoO(dcmedt)$_2$]$^{2-}$ undergoes in MeCN solution a reversible one-electron oxidation (E°' = –0.03 V, *vs.* Ag/AgCl), followed by a further irreversible oxidation (E$_p$ = +0.82 V) [30a]. Nothing is known about the redox activity of the square-pyramidal monoanion [TcO(dcmedt)$_2$]$^-$ [30b].

2.2.10.2   Tris(dcmedt)complexes

The (slightly) distorted trigonal prismatic geometry of the dianion [Mo(dcmedt)$_3$]$^{2-}$ is shown in Figure 6, together with its cyclic voltammetric profile in CH$_2$Cl$_2$ solution [30a,c].

It undergoes two reversible one-electron oxidations (E$^{o'}_{2-/-}$ = +0.02 V; E$^{o'}_{-/0}$ = +0.42 V, *vs.* SCE), which support the stability of the corresponding monoanion and neutral derivatives [30a,c].

Even if not strictly pertinent to the present review, we would mention that crystal the structure and electrochemistry of the spectacular hexanuclear complex [Pd(dcmedt)$_2$]$_6$ are available [31].

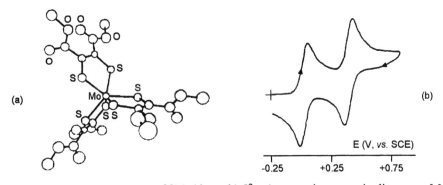

(a)   (b)

E (V, vs. SCE)

-0.25   +0.25   +0.75

Figure 6. (a) X-Ray structure of [Mo(dcmedt)$_3$]$^{2-}$. Average interatomic distances: Mo-S = 2.39 Å; C-S = 1.74 Å; C=C = 1.35 Å; C-C$_{(COOMe)}$ = 1.49 Å. Twist angle $\phi \approx 15°$. (b) Cyclic voltammetric behaviour of the dianion in CH$_2$Cl$_2$ solution.

2.2.11 thedt Complexes

2.2.11.1 Bis(thedt) complexes

[Ni(thedt)$_2$] gives rises in CH$_2$Cl$_2$ solution to two one-electron reductions ($E^{\circ'}_{0/-}$ = +0.13 V; $E^{\circ'}_{-/2-}$ = −0.66 V, *vs.* SSCE) and a one-electron oxidation ($E^{\circ'}_{0/+}$ = +1.08 V), all having features of chemical reversibility [32].

2.2.12 atedt Complexes

2.2.12.1 Bis(atedt) complexes

A few [Ni(atedt)$_2$]$^{n-}$ complexes can display up to three reversible one-electron processes, which have been assigned to the sequence [Ni(atedt)$_2$]$^{+/0/-/2-}$. Nevertheless, since the cyclic voltammogram of [Ni(C$_{11}$H$_{23}$-tedt)$_2$] [34a] could be better interpreted as due to the sequence [Ni(C$_{11}$H$_{23}$-tedt)$_2$]$^{2+/+/0/-}$ and in view of important discrepancies with the electrochemistry of [Ni(CH$_3$-tedt)$_2$]$^-$ [33b], we prefer to set the redox sequences exhibited by [M(atedt)$_2$]$^{n-}$ (M = Ni, Pd, Pt) within the series [Ni(atedt)$_2$]$^{2+/+/0/-/2-}$, Table 11.

Table 11. Formal electrode potentials (V, *vs.* SCE) for the redox processes exhibited by [M(atedt)$_2$]$^{n-}$ complexes

M	alkyl group	$E^{\circ'}_{2+/+}$	$E^{\circ'}_{+/0}$	$E^{\circ'}_{0/-}$	$E^{\circ'}_{-/2-}$	Solvent	Reference
Ni	CH$_3$	-	+0.78	−0.09	-	CH$_2$Cl$_2$	33a
	CH$_3$	-	-	+0.05	−0.72	dmf	33b
	C$_4$H$_9$	+1.12	+0.77	−0.08	-	CH$_2$Cl$_2$	34a
	C$_6$H$_{13}$	+1.11	+0.76	−0.08	-	CH$_2$Cl$_2$	34a
	C$_{11}$H$_{23}$	+1.12	+0.76	−0.11	-	CH$_2$Cl$_2$	34a
Pd	CH$_3$	-	-	+0.13	−0.50	dmf	33b
Pt	CH$_3$	-	-	+0.14	−0.64	dmf	33b

2.2.13 dfcedt Complexes

2.2.13.1 Bis(dfcedt) complexes

As illustrated in Figure 7, the tetraferrocenyl-substituted dithiolate [Ni(dfcedt)$_2$] [35] undergoes in CH$_2$Cl$_2$ solution, four separate one-electron oxidations ($E^{\circ'}_{0/+}$ = −0.02 V; $E^{\circ'}_{+/2+}$ = +0.15 V; $E^{\circ'}_{2+/3+}$ = +0.28 V; $E^{\circ'}_{3+/4+}$ = +0.49 V, *vs.* Fc/Fc$^+$), which are attributed to the gradual oxidation of the four ferrocenyl subunits, and two one-electron reductions ($E^{o\prime}_{0/-}$ = −0.67 V; $E^{o\prime}_{-/2-}$ = −1.43 V) centred on the Ni-dithiolene core [36].

The result is significant in that proves that the central NiS_4C_4 group does not create a barrier to the mutual electronic interaction among the four ferrocene subunits.

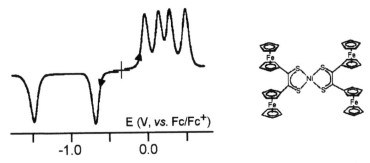

Figure 7. Differential pulse voltammogram recorded in CH_2Cl_2 solution of [Ni(dfcedt)$_2$]. Anodic scan: [NBu$_4$][B(C$_6$F$_5$)$_4$] supporting electrolyte; cathodic scan: [NBu$_4$][PF$_6$] supporting electrolyte.

2.2.14 dcnedt Complexes (better known as "mnt")

2.2.14.1 Bis(dcnedt) complexes

The square planar complexes [M(dcnedt)$_2$]$^{n-}$ (M = Ni, Cu, Pd, Pt, Au) undergo partially or completely the electron transfer sequence [M(dcnedt)$_2$]$^{2-/-/0}$, Table 12.

As a consequence of the reversibility of these processes, the crystal structures of different [M(dcnedt)$_2$]$^{2-/-}$ couples are available. In particular, a lot of structural data exist for [Ni(dcnedt)$_2$]$^{2-/-}$ as a function of the nature of the counter-cations. Figure 8 just exemplifies that in both oxidation states the square planar geometry

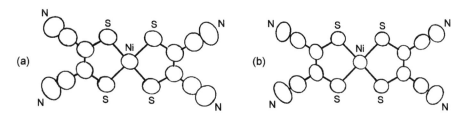

Figure 8. X-Ray structures of: (a) [Ni(dcnedt)$_2$]$^{2-}$; (b) [Ni(dcnedt)$_2$]$^{-}$.

Table 12. Formal electrode potentials (V, *vs.* SCE) for the redox processes exhibited by $[M(dcnedt)_2]^{2-}$.

M	$E^{o\prime}_{0/-}$	$E^{o\prime}_{-/2-}$	$E^{o\prime}_{2-/3-}$	Solvent	Reference
Ni	+1.02	+0.23	−1.68	MeCN	2b, 26
	-	+0.49	-	MeCN	37a
	+1.04[a]	+0.09	-	MeCN	37b
	-	+0.23	−1.74	dme[b]	26
	-	+0.21	-	dmf	2b
	-	+0.12	-	CH_2Cl_2	2b
Pd	-	+0.46	-	MeCN	2b
	+0.70	+0.26	-	MeCN	37b
	-	+0.44	−1.90	MeCN	37c
	+1.16	-	-	CH_2Cl_2	37c
Pt	-	+0.21	-	MeCN	2b
	+0.94	+0.06	-	MeCN	37b
	+1.07	+0.22	−2.41	MeCN	37c
Cu	-	+0.34	-	MeCN	2b
	+1.18[a]	+0.19	−0.85	MeCN	37b
Au	-	−0.42	-	MeCN	2b
	+1.10	−0.93	-	CH_2Cl_2	37d

[a] Peak-potential value for irreversible processes; [b] 1,2-dimethoxyethane

is maintained [38f] and Table 13 summarizes how some interatomic distances vary with the electron addition/removal.

As in most previous cases, the variations of the bond lengths are minimal and extended along the whole molecular frame, thus once again supporting the substantially ligand centred nature of the electron transfer. Also in this case, such an assumption is supported by the electrochemical detection of the $[Ni(dcnedt)_2]^{2-/3-}$ reduction step and the concomitant EPR characterization of the trianion as a Ni(I) species [40].

Crystallographic data for the square planar couples $[Pd(dcnedt)_2]^{2-/-}$ and $[Pt(dcnedt)_2]^{2-/-}$ are reported in Table 14.

Also in this case, the M(II)/M(I) process $[M(dcnedt)_2]^{2-/3-}$ occurs at very negative potential values [40b].

In agreement with the preferred square planar coordination of Cu(III) ion, the geometry of the monoanion $[Cu(dcnedt)_2]^-$ is square planar [2b]. The easy access

Table 13. Selected bond distances (average values; Å) in the redox couple [Ni(dcnedt)$_2$]$^{2-/-}$.

Complex	Ni-S	S-C	C=C	C-C$_{(CN)}$	C≡N	Cation	Ref.
[Ni(dcnedt)$_2$]$^{2-}$	2.16	1.75	1.30	1.43	1.13	[NMe$_4$]$^+$	38a
	2.17	1.75	1.33	1.41	1.13	[NMe$_4$]$^+$	38b
	2.18	1.73	1.38	1.42	1.14	[TMPD]$^{+\,a}$	38c
	2.17	1.74	1.35	1.44	1.14	[KOS]$^{+\,b}$	38d
	2.18	1.73	1.39	1.44	1.13	[Ag(PPh$_3$)$_2$]$^+$	38e
	2.17	1.73	1.36	1.43	1.14	[NBu$_4$]$^+$	38f
	2.17	1.73	1.36	1.43	1.14	[NMP]$^{+\,c}$	38g
	2.17	1.74	1.37	1.43	1.13	[MV]$^{2+\,d}$	38h
	2.18	1.73	1.38	1.43	1.14	[A]$^{2+\,e}$	38i
	2.19	1.71	1.36	1.50	1.12	[DPD-Me]$^{2+\,f}$	38l
	2.17	1.74	1.36	1.43	1.15	[N-MeA]$^{+\,g}$	38m
[Ni(dcnedt)$_2$]$^{-}$	2.15	1.71	1.36	1.43	1.14	[PMePh$_3$]$^+$	39a
	2.15	1.72	1.37	1.43	1.13	[NEt$_4$]$^+$	38f
	2.15	1.72	1.37	1.43	1.14	[C$_7$H$_7$]$^+$	39b
	2.15	1.71	1.31	1.43	1.15	[TMPD]$^{+\,a}$	39c
	2.14	1.72	1.34	1.44	1.12	[NMe$_3$Ph]$^+$	39d
	2.15	1.72	1.37	1.44	1.13	[TTM-TTF]$^{+\,h}$	39e
	2.15	1.70	1.40	1.43	1.12	[Fc$^{\#}$]$^{+\,i}$	39f
	2.14	1.71	1.37	1.42	1.15	[Fc†]$^{+\,l}$	39g
	2.14	1.71	1.37	1.44	1.14	[Fc‡]$^{+\,m}$	39g
	2.15	1.71	1.40	1.40	1.17	[Etpy]$^+$	39h
	2.15	1.71	1.39	1.44	1.13	[(DT-TTF)$_2$]$^{+\,n}$	39i
	2.15	1.71	1.34	1.45	1.14	[Na(15-crown-5)]$^+$	39l
	2.15	1.71	1.37	1.43	1.14	[NO$_2$ql]$^{+\,o}$	39m
	2.15	1.71	1.36	1.43	1.14	[C$_{13}$H$_{13}$-N$_2$O$_2$]$^{+\,p}$	39n
	2.14	1.70	1.38	1.46	1.11	[BrFBzPy]$^{+\,q}$	39o
	2.15	1.71	1.36	1.44	1.14	[FbzPy]$^{+\,r}$	39p
	2.14	1.72	1.36	1.43	1.14	[AMPY]$^{+\,s}$	39q
	2.13	1.71	1.35	1.43	1.15	[FbzPyNH$_2$]$^{+\,t}$	39r

^{a} TMPD = N,N,N',N'-tetramethyl-*p*-phenylendiamine; ^{b} KOS = 1-ethyl-4-carbomethoxypyridine; ^{c} NMP = N-methylphenazine; ^{d} MV = methyl viologen = 1,1'-dimethyl-4,4'-bipyridine; ^{e} A = trans-4,4'-(1,2-ethendiyl)bis[1-(3-cyano-propyl)pyridine]-dibromide; ^{f} DPD-Me = trans-4,4-azobis(1-methyl-pyridine); ^{g} N-MeA = N-methylacridine; ^{h} TTM-TTF = tetra(methylthio) tetrathiafulvalene; ^{i} Fc$^{\#}$ = 1,1'-bis[2-(4-methylthio)phenylethenyl]ferrocene; ^{l} Fc† = 2,2', 3,3', 4,4', 5,5'-octamethyl-1,1'-bis(methylthio) ferrocene; ^{m} [Fc‡] = 2,2', 3,3', 4,4', 5,5'-octamethyl-1,1'-bis(*tert*-butylthio)ferrocene; ^{n} DT-TTF = dithiopheno-tetrathiafulvalene; ^{o} NO$_2$ql = 1-(4-nitrobenzyl)quinoline; ^{p} C$_{13}$H$_{13}$-N$_2$O$_2$ = 2-methyl-1-(4-nitrobenzyl)pyridine; ^{q} BrFBzPy = 1-(4'-bromo-2'-fluorobenzyl)pyridine;^{r}FBzPy=1-(4'-fluorobenzyl)pyridine;^{s}AMPY=4-aminopyridine; ^{t} FbzPyNH$_2$ = [1-(4'-fluorobenzyl)-4-aminopyridine]

Table 14. Selected bond distances (average values; Å) in the redox couples
$[M(dcnedt)_2]^{2-/-}$ (M = Pd, Pt)

Complex	M-S	S-C	C=C	C-C$_{(CN)}$	C≡N	Cation	Ref.
$[Pd(dcnedt)_2]^{2-}$	2.30	1.73	1.38	1.43	1.14	$[MV]^{2+\,a}$	41a
	2.29	1.72	1.39	1.45	1.14	$[ITTF]^{+\,b}$	41b
	2.30	1.74	1.34	1.44	1.14	$[Pt(CNMe)_4]^{2+}$	41c
$[Pd(dcnedt)_2]^{-}$	2.26	1.71	1.34	1.46	1.10	$[(Perylene)_2]^{+}$	42a
	2.29	1.73	1.39	1.45	1.14	$[K]^{+}$	42b
	2.28	1.71	1.40	1.43	1.13	$[NH_4]^{+}$	42b
$[Pt(dcnedt)_2]^{2-}$	2.28	1.72	1.36	1.40	1.16	$[NBu_4]^{+}$	43a
	2.28	1.71	1.42	1.43	1.15	$[PhCNSSN]^{+\,c}$	43b
	2.30	1.73	1.36	1.44	1.15	$[Pt(Me_2pipdt)_2]^{2+\,d}$	43c
	2.31	1.75	1.35	1.41	1.14	$[Pt(CNMe)_4]^{2+}$	41c
	2.29	*e*	*e*	*e*	*e*	$[Pt(NH_2Oc)_4]^{2+\,f}$	43d
$[Pt(dcnedt)_2]^{-}$	2.26	1.70	1.39	1.43	1.14	$[Rb]^{+}$	44a
	2.27	1.70	1.39	1.42	1.14	$[H_3O]^{+}$	44b
	2.27	1.71	1.38	1.43	1.15	$[NEt_4]^{+}$	44b
	2.26	1.71	1.39	1.42	1.16	$[(p\text{-}ClC_6H_4CNSSN)_2Cl]^{+\,g}$	43a
	2.27	1.71	1.37	1.43	1.13	$[TTM\text{-}TTF]^{+\,h}$	39e
	2.26	1.70	1.38	1.42	1.15	$[Fc^{\ddagger}]^{+\,i}$	39h
	2.27	1.72	1.37	1.43	1.14	$[(DT\text{-}TTF)_2]^{+\,j}$	39l
	2.26	*e*	1.36	*e*	1.13	$[Pt(NH_2Oc)_4]^{2+\,f}$	44c

^{a}MV=methylviologen=1,1'-dimethyl-4,4'-bipyridine; ^{b}ITTF=4-iodo-tetrathioful-valene; ^{c}PhCNSSN = 4-Phenyl-1,2-dithia-3,5-diazole; $^d[Pt(Me_2pipdt)_2]^{2+}$ = N,N'-dimethyl-piperazine-2,3-dithione-Pt(II); ^{e}not reported; $^f[Pt(NH_2Oc)_4]^{2+}$ = octylamine-Pt(IV); $^g(p\text{-}ClC_6H_4CNSSN)_2Cl$ = bis(4-(4-chlorophenyl)- 1,2-dithia-3,5-diazole; ^{h}TTM-TTF = tetra(methylthio)tetrathiafulvalene; $^i[Fc^{\ddagger}]$ = 2,2', 3,3', 4,4', 5,5'-octamethyl-1,1'-bis(*tert*-butylthio)ferrocene; ^{j}DT-TTF = dithiopheno-tetrathiafulvalene;

to $[Cu(dcnedt)_2]^{2-}$ allowed its isolation and crystallographic characterisation, Table 15.

The significant elongation of the Cu-S bond upon one-electron reduction seems to support its Cu(III)/Cu(II) nature. Further support arises from the fact that, in some cases, the dianion $[Cu(dcnedt)_2]^{2-}$ displays a tetrahedrally distorted geometry. In this connection, Table 15 also reports the interatomic distances of the tetrahedral dianion $[Zn(dcnedt)_2]^{2-}$.

Table 15. Selected bond distances (average values; Å) and dihedral angles (°) in the redox couple $[Cu(dcnedt)_2]^{2-/-}$ and $[Zn(dcnedt)_2]^{2-}$

Complex	M-S	S-C	C=C	C-C$_{(CN)}$	C≡N	α	Cation	Ref.
$[Cu(dcnedt)_2]^-$	2.17	1.72	1.31	1.43	1.14	-	$[NBu_4]^+$	45
$[Cu(dcnedt)_2]^{2-}$	2.27	1.73	1.36	1.43	1.14	-	$[NEt_4]^+$	46a
	2.28	1.73	1.36	1.43	1.14	-	$[NBu_4]^+$	46b
	2.25	1.73	1.36	1.43	1.13	47.4	$[MB]^{+\ a}$	46c
	2.28	1.74	1.34	1.44	1.13	-	$[NMP]^{+\ b}$	46d
	2.26	1.74	1.34	1.45	1.13	41.1	$[NMe_4]^+$	46e
	2.27	1.73	1.36	1.42	1.15	-	$[Fe(C_5Me_5)_2]^+$	46f
$[Zn(dcnedt)_2]^{2-}$	2.33	1.74	1.36	1.43	1.14	83.9	$[AsPh_4]^+$	47a
	2.33	1.73	1.38	1.43	1.14	81.8	$[MV]^{2+\ c}$	47b
	2.32	1.73	1.36	1.42	1.15	90	$[N\text{-}MeA]^{+\ d}$	38m

^{a} MB = methylene blue = 3,9-bis(dimethylamino)phenazothione; ^{b} NMP = N-methyl-phenazine; ^{c} MV = methyl viologen = 1,1'-dimethyl-4,4'-bipyridine; ^{d} N-MeA = N-methylacridine

As far as the gold complex is concerned, the only structural data are concerned with the square planar Au(III) monoanion $[Au(dcnedt)_2]^-$ [37d, 39i, 42a, 48].

The square planar Co(II) dianion $[Co(dcnedt)_2]^{2-}$ undergoes in thf solution either a reversible one-electron reduction ($E^{\circ'}_{2-/3-} = -1.83$ V, *vs.* SCE), or a one-electron oxidation ($E^{\circ'}_{2-/-} = -0.02$ V), which is followed by slow dimerisation to $[Co(dcnedt)_2]_2^{2-}$ [49]. In agreement with the slowness of the dimerization reaction and the electrochemical reversibility of the $[Co(dcnedt)_2]^{2-/-}$ process, the isolated monoanion maintains a square planar geometry [39h]. The interatomic distances for the couple $[Co(dcnedt)_2]^{2-/-}$ are compiled in Table 16, and, also in this case, they support the substantial ligand-centered nature of the unpaired electron.

Concerned with the electrochemical behaviour of $[Zn(dcnedt)_2]^{2-}$, it has been briefly reported that in MeCN solution it undergoes an (obviously ligand-centred) irreversible oxidation ($E_{p(2-/-)} = +1.00$ V, *vs.* SCE) [47a, 38i].

The dianions $[MoO(dcnedt)_2]^{2-}$ [12, 51] and $[WO(dcnedt)_2]^{2-}$ [52] exhibit the square pyramidal geometry typical of $[MO(dithiolene)_2]^{n-}$. $[MoO(dcnedt)_2]^{2-}$ undergoes a one-electron oxidation, which displays features of chemical reversibility in the cyclic voltammetric time scale ($E^{\circ'}_{2-/-} = +0.35$ V, *vs.* SCE, in CH_2Cl_2 solution [53a]; $E^{\circ'}_{2-/-} = +0.40$ V, *vs.* SCE, in MeCN solution [53b]). Macroelectrolysis tests prove however that the electrogenerated monoanion slowly disproportionates to $[Mo(dcnedt)_3]^{2-}$ (see below) [53b].

Table 16. Selected bond distances (average values; Å) in the redox couple [Co(dcnedt)$_2$]$^{2-/-}$

Complex	Co-S	S-C	C=C	C-C$_{(CN)}$	C≡N	Cation	Reference
[Co(dcnedt)$_2$]$^{2-}$	2.16	1.72	1.34	1.40	1.15	[NBu$_4$]$^+$	50a
	2.16	1.72	1.37	1.44	1.13	[BQ]$^{2+}$ [a]	50b
	2.25	1.73	1.36	1.43	1.13	[Cu(tim)]$^{2+}$ [b]	50c
	2.19	1.73	1.37	1.42	1.14	[Ni(tim)]$^{2+}$ [b]	50c
[Co(dcnedt)$_2$]$^-$	2.19	1.71	1.36	1.43	1.14	[Fc†]$^+$ [c]	39h
	2.18	1.72	1.34	1.43	1.14	[Fc‡]$^+$ [d]	39h

[a] [BQ] = 6,7,8,9-tetrahydrodipyrido[1,4]diazocene; [b] tim = 2,3,9,10-tetramethyl-1,4,8,1-tetraaza-cyclo-tetradeca-1,3,8,10-tetraene; [c] Fc† = 2,2', 3,3', 4,4', 5,5'-octamethyl-1,1'-bis(methylthio)-ferrocene; [d] [Fc‡] = 2,2', 3,3', 4,4', 5,5'-octamethyl-1,1'-bis(*tert*-butylthio)ferrocene

Finally, the (likely square pyramidal) anions [MO(dcnedt)$_2$]$^-$ (M = Tc, Re), in MeCN solution display a reversible one-electron reduction (M = Tc: $E^{o'}_{-/2-}$ = −0.64 V; M = Re: $E^{o'}_{-/2-}$ = −1.05 V, *vs.* SCE), and a reversible one-electron oxidation (M = Tc: $E^{o'}_{-/0}$ = +1.73 V; M = Re: $E^{o'}_{-/0}$ = +1.76 V) [6a], whereas [VO(dcnedt)$_2$]$^{2-}$ only exhibits, in CH$_2$Cl$_2$ solution, a reversible one-electron oxidation ($E^{o'}_{2-/-}$ = +0.40 V, *vs.* SCE) [53a] .

2.2.14.2 Tris(dcnedt) complexes

The redox ability of the tris-chelates [M(dcnedt)$_3$]$^{n-}$ (M = V, Cr, Fe, Co, Mo, W, Re) is long known [2b, 54]. From the structural viewpoint, they usually display geometrical arrangements ranging from octahedral (M = Fe [55a]) to distorted trigonal prismatic (M = V, Mo, W [55 b,c]). For example, [Cr(dcnedt)$_3$]$^{3-}$ undergoes two reversible one-electron oxidations (MeCN:

Table 17. Selected bond distances (average values; Å) in the redox couples [Cr(dcnedt)$_3$]$^{3-/2-}$ and [Ru(dcnedt)$_3$]$^{3-/2-}$

Complex	M-S	S-C	C=C	C-C$_{(CN)}$	C≡N	φ	Cation
[Cr(dcnedt)$_3$]$^{3-}$	2.39	1.73	1.35	1.43	1.14	a	[PPh$_4$]$^+$
[Cr(dcnedt)$_3$]$^{2-}$	2.34	1.73	1.33	1.45	1.13	a	[PPh$_4$]$^+$
[Ru(dcnedt)$_3$]$^{3-}$	2.35	1.71	1.39	1.43	1.13	50.2	[NEt$_4$]$^+$
[Ru(dcnedt)$_3$]$^{2-}$	2.34	1.72	1.36	1.43	1.13	47.1	[AsPh$_4$]$^+$

[a] Not reported

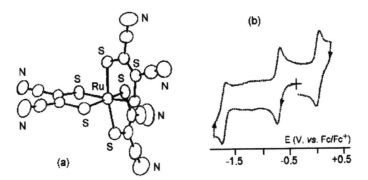

Figure 9. (a) X-Ray structure and (b) cyclic voltammogram (glassy carbon electrode; CH_2Cl_2 solution; scan rate 0.1 Vs^{-1}) of $[Ru(dcnedt)_3]^{2-}$ ($[AsPh_4]^+$ cation).

$E^{\circ\prime}_{3-/2-} = +0.16$ V; $E^{\circ\prime}_{2-/-} = +0.76$ V, *vs.* SCE) [2b], and the crystal structures of the redox couple $[Cr(dcnedt)_3]^{3-/2-}$ are available, Table 17 [56].

Both the complexes are described as octahedral. The significant variations of the Cr-S distance in this case seem to support a chromium centered [Cr(IV)/Cr(V)] process.

In this connection, however, complex $[Ru(dcnedt)_3]^{2-}$ probably better accounts for either the redox and structural properties of this class of compounds. In fact, as shown in Figure 9, it undergoes in CH_2Cl_2 solution two one-electron reductions ($E^{\circ\prime}_{2-/3-} = -0.70$ V; $E^{\circ\prime}_{3-/4-} = -1.71$ V, *vs.* Fc/Fc$^+$) and a one-electron oxidation ($E^{\circ\prime}_{2-/-} = +0.03$ V), all displaying features of chemical reversibility in the cyclic voltammetric time scale [57].

In the longer times of exhaustive electrolysis, only the trianion proved to be quite stable. The crystallographic data of the couple $[Ru(dcnedt)_3]^{3-/2-}$ are compiled in Table 17 [57]. The variations in the interatomic distances are minimal. The most significant structural change probably resides in the twist angle between the two S3 triangles. In fact it passes from 50.2° in the trianion to 47.1° in the dianion. Taking into account that for a perfect octahedron $\phi = 60°$, whereas for a perfect trigonal prism $\phi = 0°$, it ensues that the electron removal causes the molecular frame to pass from an (almost perfect) octahedron to a trigonally distorted octahedron.

2.2.15 dichedt, dibzedt Complexes

2.2.15.1 Bis(dichedt) and bis(dibzedt) complexes

The square planar complexes $[Pd(dichedt)_2]$ and $[Cu(dibzedt)_2]^{2+}$ have been crystallographically characterised [58], but no electrochemical data are available.

2.3 mtadt Complexes

2.3.1 Bis(mtadt) complexes

In MeCN solution, the square pyramidal monoanion $[TcO(mtadt)_2]^-$ [59] exhibits a reversible one-electron reduction ($E^{o\prime}_{-/2-} = -1.35$ V, *vs.* SCE) [6a]. The isoelectronic $[ReO(mtadt)_2]^-$ is harder to reduce ($E^{o\prime}_{-/2-} = -1.84$ V), but, as a consequence, easier to oxidise ($E^{o\prime}_{-/0} = +1.23$ V) [6a].

2.4 dodt Complexes

2.4.1 Bis(dodt) complexes

The dithioxalato ligand dodt constitutes a special case of dithiolene ligand, in that in its bis- or tris-chelates the α-diketone group can simultaneously coordinate more than one metal ion [60a]. Even in $[K_2Ni(dodt)_2]$, the dianion $[Ni(dodt)_2]^{2-}$ strongly interacts with the K^+ ion *via* the dioxo group such to form a true adduct (K-O = 2.77 Å) [60a].

Structural data exist for complexes $[M(dodt)_2]^{n-}$ (M = Ni, Pd: n = 2, square planar [60a,b]; M = Pt: n = 3, square planar [60b]; M = Zn, n = 2, tetrahedral [60c]).

It has been recently mentioned that the Ni and Pd dianions (unexpectedly) do not show reversible redox processes [60a]. In contrast, the dianion $[Cu(dodt)_2]^{2-}$ undergoes in CH_2Cl_2 solution a one-electron oxidation ($E^{o\prime}_{2-/-} = +0.13$ V, *vs.* SCE) [60d]. Even if such a process is accompanied by slow chemical complications (at slow scan rate the i_{pc}/i_{pa} ratio is significantly lower than 1), fast precipitation of $[Cu(dodt)_2]^-$ blocks such decomposition. In fact, the crystal structure of the redox couple $[Cu(dodt)_2]^{2-/-}$ is known [60d,e]. Both the complexes have a square planar geometry; the passage from the dianion to the monoanion causes a significant shortening of the Cu-S bond length (by 0.09 Å), whereas the intraligand distances undergo minor variations (of the order of 0.02 Å). These results support the formal assignment as $[Cu^{II}(dodt)_2]^{2-}$ and $[Cu^{III}(dodt)_2]^-$, respectively.

As far as the square pyramidal oxoanions are concerned, the crystal structures of $[MoO(dodt)_2]^{2-}$ [60f] and $[ReO(dodt)_2]^-$ [60g] are available. The monoanions $[MO(dodt)_2]^-$ (M = Tc, Re) exhibit in MeCN solution two reversible one-electron reductions (Tc: $E^{o\prime}_{-/2-} = -0.75$ V, $E^{o\prime}_{2-/3-} = -1.56$ V; Re: $E^{o\prime}_{-/2-} = -0.94$ V, $E^{o\prime}_{2-/3-} = -1.35$ V, *vs.* SCE) and two reversible one-electron oxidations (Tc: $E^{o\prime}_{-/0} = +1.91$ V, $E^{o\prime}_{0/+} = +2.28$ V; Re: $E^{o\prime}_{-/0} = +1.64$ V, $E^{o\prime}_{0/+} = +2.17$) [6a].

2.4.2 Tris(dodt) complexes

As far as we know, the distorted octahedral geometry of the trianion $[Co(dodt)_3]^{3-}$ is available [60h], but nothing is known about its redox activity.

3.0   1,2-DITHIOLATES FORMING LATERAL SINGLE RINGS

3.1   1,2-Dithiolates forming lateral four-membered rings

3.1.1   dtsq Complexes

3.1.1.1 Bis(dtsq) complexes

The square planar geometry of the dianion $[Ni(dtsq)_2]^{2-}$ in $[K_2Ni(dtsq)_2]$ is crystallographically ascertained [61a,b]. As it happens in $[K_2Ni(dodt)_2]$, the two oxo groups of each ligand coordinate one K^+ counterion (K-O = 2.91 Å). As exemplified in Figure 10, complexes $[M(dtsq)_2]^{2-}$ (M = Pt [61c], Pd [61d,e], Cu [61f]) possess a square planar geometry.

The Pt dianion exhibits a partially chemically reversible one-electron oxidation ($E^{o'}_{2-/-}$ = +0.58 V, *vs.* Ag/AgCl) [61b,c], whereas $[Au(dtsq)_2]^-$ exhibits a partially chemically reversible one-electron reduction [61b].

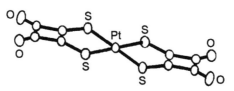

Figure 10. X-Ray structure of $[Pt(dtsq)_2]^{2-}$ ([BEDT-TTF]$^+$ cation).

3.2 1,2-Dithiolates forming lateral five-membered rings

3.2.1 dtcr Complexes

3.2.1.1 Bis(dtcr) complexes

We are not aware of crystal structures of bis(dtcr) metal complexes. $[Pt(dtcr)_2]^{2-}$ undergoes two one-electron reductions and a one-electron oxida-

Table 18. Formal electrode potentials (V, *vs.* SCE) for the redox processes exhibited by complexes $[M(dtcr)_2]^{2-}$ in MeCN solution

M	$E^{o'}_{2-/-}$	$E^{o'}_{2-/3-}$	$E^{o'}_{3-/4-}$
Ni	+0.47	−1.26	−1.44
Pd	+0.66	−1.26	−1.5^a
Pt	+0.40	−1.30	−1.50

^{a} Peak-potential for irreversible processes

tion, all exhibiting features of chemical reversibility in the cyclic voltammetric time scale [62a]. Similar behaviour is exhibited by the analogues [Ni(dtcr)$_2$]$^{2-}$ and [Pd(dtcr)$_2$]$^{2-}$. The relative formal electrode potentials are summarized in Table 18 [62a].

EPR spectroscopy supports the substantial ligand centred nature of the [M(dtcr)$_2$]$^{2-/-}$ process [62a].

3.2.1.2 Tris(dtcr) complexes

A few [M(dtcr)$_3$]$^{3-}$ trianions (M = Cr, Fe, Co) have been characterised. [Co(dtcr)$_3$]$^{3-}$ displays a (almost perfect) octahedral geometry [62b].

All these complexes display in MeCN solution a reversible one-electron oxidation (M = Cr: $E^{\circ\prime}_{3-/2-} = +0.47$ V; M = Fe: $E^{\circ\prime}_{3-/2-} = +0.15$ V; M = Co: $E^{\circ\prime}_{3-/2-} = +0.48$ V, *vs.* SCE) and a one-electron reduction, which possesses features of chemical reversibility only in the case of the Fe(III) complex ($E^{\circ\prime}_{3-/4-} = -0.59$ V) [62b].

3.2.2 dcmdtcr Complexes

3.2.2.1 Bis(dcmdtcr) complexes

Figure 11 shows the crystal structure of the dianion [Pd(dcmdtcr)$_2$]$^{2-}$ present in the adduct [NBu$_4$]$_2$[Pd(dcmdtcr)$_2$]·I$_2$ [62a].

The PdS$_4$ core is planar, but the ligands are slightly inclined (about 4°) with respect to the central plane.

[Cu(dcmdtcr)$_2$]$^{2-}$ exhibits a tetrahedrally distorted geometry (dihedral angle 36.8°) [63].

Like the dianions [M(dtcr)$_2$]$^{2-}$, complexes [M(dcmdtcr)$_2$]$^{2-}$ (M = Ni, Pd, Pt) exhibit two reversible one-electron reductions and a one-electron oxidation, Table 19 [62a].

It is interesting to note that, with respect to the related [M(dtcr)$_2$]$^{2-}$ derivatives, the cyano-substituted complexes are more difficult to oxidize by about 0.2 V, but

Table 19. Formal electrode potentials (V, *vs.* SCE) for the redox processes exhibited by complexes [M(dcmdtcr)$_2$]$^{2-}$ in MeCN solution

M	$E^{\circ\prime}_{2-/-}$	$E^{\circ\prime}_{2-/3-}$	$E^{\circ\prime}_{3-/4-}$
Ni	+0.66[a]	−0.70	−0.94
Pd	+0.74[a]	−0.66	−0.85
Pt	+0.58[a]	−0.69	−0.94

[a] Peak-potential for oxidation processes affected by electrode adsorption

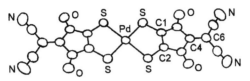

Figure 11. X-Ray structure of $[Pd(dcmdtcr)_2]^{2-}$ ($[NBu_4]^+$ cation). Average bond distances: Pd-S = 2.30 Å; S-C = 1.70 Å; C1-C2 = 1.38 Å; $C-C_{(ring)}$ = 1.47 Å; C=O = 1.22 Å; C4-C6 = 1.35 Å; $C-C_{(CN)}$ = 1.43 Å.

notably easier to reduce by about 0.5 V. Also in this case, EPR spectroscopy confirms the substantial ligand centered nature of the $[M(dcmdtcr)_2]^{2-/-}$ process [62a].

3.2.2.2 Tris(dcmdtcr) complexes

Like the trianions $[M(dtcr)_3]^{3-}$ (M = Cr, Fe, Co), the corresponding complexes $[M(dcmdtcr)_3]^{3-}$ (M = Cr, Fe, Co) have been characterised [62b]. $[Fe(dcmdtcr)_3]^{3-}$ displays either a reversible one-electron oxidation or a reversible one-electron reduction, both the processes being anodically shifted by about 0.3 V with respect to $[Fe(dtcr)_3]^{3-}$ ($E^{o'}_{3-/2-}$ = +0.30 V; $E^{o'}_{3-/4-}$ = −0.31 V, *vs.* SCE) [62b]. A similar shift in redox potentials occurs for the oxidation $[M(dcmdtcr)_3]^{3-/2-}$ (M = Cr, Co) [62b].

3.2.3 dhtpdt, tpdt Complexes

3.2.3.1 Bis(dhtpdt) and bis(tpdt) complexes

The unsaturated square planar monoanion $[Au(tpdt)_2]^-$ and its isomer $[Au(\alpha$-tpdt)_2]^-$ undergo in CH_2Cl_2 solution a reversible one-electron reduction ($E^{o'}$= −0.90 V and −1.16 V, *vs.* Ag/AgCl, respectively) and a one-electron oxidation affected by electrode adsorption ($E^{o'}$ = +0.75 V and +0.46 V, respectively) [64].

In contrast, the (almost) planar saturated monoanion $[Au(dhtpdt)_2]^-$ only displays in CH_2Cl_2 solution the one-electron oxidation process affected by electrode adsorption ($E^{o'}_{-/0}$ = +0.20 V, *vs.* Ag/AgCl) [64].

3.2.4 ttpdt Complexes

3.2.4.1 Bis(ttpdt) complexes

The square planar dianion $[Ni(ttpdt)_2]^{2-}$ [65] undergoes at first, chemically reversible, dithiolene-based, one electron oxidation ($E^{o'}_{2-/-}$ = −0.13 V, *vs.* Ag/AgCl), followed by a second irreversible process ($E_{p\ -/0}$ = +0.46 V). A reduction process is also present at negative potential values which is centred on the terthiophene ligand [65].

A roughly similar behaviour is displayed by [Pd(ttpdt)$_2$]$^{2-}$ and [Au(ttpdt)$_2$]$^-$ [65].

3.2.5 mdtdt Complexes (better known as "mdt")

3.2.5.1 Bis(mdtdt) complexes

The monoanion [Ni(mdtdt)$_2$]$^-$ possesses a tetrahedrally distorted planar geometry (dihedral angle, 14.77°) [66]. In nitrobenzene solution it exhibits a chemically reversible one-electron reduction ($E^{o'}_{-/2-}$ = –0.55 V, *vs.* SCE) and a chemically reversible one-electron oxidation ($E^{o'}_{-/0}$ = +0.08 V). A further oxidation process (E_p = +0.74 V) is affected by electrode adsorption phenomena [66].

3.2.6 dtodt Complexes (better known as "dmid")

3.2.6.1 Bis(dtodt) complexes

The tetrahedrally distorted [Ni(dtodt)$_2$]$^{2-}$ [67a] undergoes a chemically reversible oxidation to [Ni(dtodt)$_2$]$^-$ ($E^{o'}_{2-/-}$ = –0.33 V, in MeCN; [67b]; $E^{o'}_{2-/-}$ = –0.25 V, in dmf [67c,d]; V, *vs.* SCE). The monoanion is less distorted from the planar geometry than the dianion [67b,e], but, as usual, the variations of the interatomic distances along the whole molecular frame are minimal, Table 20.

Supported also by EPR spectroscopic data, it follows that the electron removal is substantially centred on the dithiolene ligand [67e].

[Pd(dtodt)$_2$]$^{2-}$ undergoes a reversible one-electron oxidation ($E^{o'}_{2-/-}$ = –0.03 V, *vs.* SCE) [67b].

Table 20. Selected bond distances (average values; Å) and dihedral angles (°) in the couple [Ni(dtodt)$_2$]$^{2-/-}$.

Complex	Ni-S	S-C	C=C	C-S$_{(outer ring)}$	C=O	α	Cation	Ref.
[Ni(dtodt)$_2$]$^{2-}$	2.19	1.73	1.34	1.76	1.21	17.4	[NEt$_4$]$^+$	67a
[Ni(dtodt)$_2$]$^-$	2.16	1.72	1.34	1.75	1.23	≈2	[NBu$_4$]$^+$	67b
	2.15	1.72	1.35	1.74	1.20	10.7	[AsPh$_4$]$^+$	67e
	2.14	1.71	1.31	1.73	1.21	-	[Fe(C$_5$Me$_5$)$_2$]$^+$	67f
	2.16	1.72	1.36	1.74	1.21	-	[Mn(C$_5$Me$_5$)$_2$]$^+$	67g

3.2.6.2 Tris(dtodt) complexes

[W(dtodt)$_3$]$^{2-}$ possesses a trigonally distorted octahedral geometry [68]. Like the Mo analogue, it undergoes two reversible one-electron oxidations, Table 21 [68].

Table 21. Formal electrode potentials (V, *vs.* Ag/AgCl) for the redox processes exhibited by complexes $[M(dtodt)_3]^{2-}$ in CH_2Cl_2 solution

M	$E^{o'}_{2-/-}$	$E^{o'}_{-/0}$	$E^{o'}_{2-/3-}$
Mo	−0.02	+0.33	−1.44
W	−0.07	+0.27	-

3.2.7 dttdt Complexes (better known as "dmit")

3.2.7.1 Bis(dttdt) complexes

Dttdt metal complexes appeared in literature in the mid 1970's and they still constitute the most studied class of metallo-dithiolenes [2i,n].

The crystal structure of the Fe(III) anion $[Fe(dttdt)_2]^-$ shows that it is dimerised, thus giving rise to monoanion units that are not perfectly planar [69]. It is, however, thought that in strong coordinating solvents (such as dmf, dmso, py) dimerisation might be prevented. As a matter of fact, in dmf solution, the monoanion exhibits a reversible one-electron reduction (E°' = −0.43 V, *vs.* SCE), and an oxidation process affected by adsorption phenomena at the platinum electrode (E_p = +0.14 V). The cathodic process is attributed to the Fe(III)/Fe(II) reduction, whereas the anodic process is attributed to a ligand-centred electron transfer [68].

The dianion $[Co(dttdt)_2]^{2-}$ possesses a tetrahedral geometry; in fact, the two dithiolene ligands around the Co(II) centre form a dihedral angle of 98.1° [70]. The cyclic voltammetric behaviour of the dianion in CH_2Cl_2 solution has been reported, but the interpretation is doubted [70].

Let us now deal with the abundant series of Ni complexes. The monoanion $[Ni(dttdt)_2]^-$ undergoes a chemically reversible one-electron reduction and a one-electron oxidation which is complicated by electrode adsorption phenomena. Table 22 summarizes the formal electrode potentials of such electron transfer processes in different solvents.

The dianion $[Ni(dttdt)_2]^{2-}$ shows the typical planar geometry of this class of complexes [72a], and in agreement with the electrochemical reversibility of the 2−/−/0 changes, the same geometry is maintained by $[Ni(dttdt)_2]^-$ and $[Ni(dttdt)_2]^0$, Table 23.

In confirmation of the extended delocalization of the electron transfers, most distances inside the ligand remain essentially unaltered. Nevertheless, the progressive shortening of the Ni–S bond suggests a significant contribution of the metal *d* orbitals to the frontier orbitals.

At variance with $[Ni(dttdt)_2]^{2-}$ and $[Pt(dttdt)_2]^{2-}$, $[Pd(dttdt)_2]^{2-}$ only exhibits

Table 22. Formal electrode potentials (V, *vs.* SCE) for the redox processes exhibited by [Ni(dttdt)$_2$]$^{n-}$

$E^{o'}_{2-/-}$	$E^{o'}_{-/0}$	Solvent	Reference
−0.23	+0.42	Me$_2$CO	71°
−0.20	+0.19^a	MeCN	71°
−0.13	+0.22^a	MeCN	71b
−0.17	≈+0.2^a	MeCN	71c
−0.14	+0.18	MeCN	71d
−0.25	-	dmf	71e
−0.18	≈+0.3^a	dmf/MeCN (3:2)	71e

^{a} Affected by electrode adsorption

a single oxidation. Literature reports are not unequivocal, describing such a process either as a two-electron process complicated by electrode adsorption [72c] or as a reversible one-electron process [75, 71e, 73m], Table 24.

The square planar geometry of the monoanion [Pt(dttdt)$_2$]$^-$ is known [73e].

Like [Co(dttdt)$_2$]$^{2-}$, the dianion [Cu(dttdt)$_2$]$^{2-}$ possesses a tetrahedrally distorted geometry (the dihedral angle between the two ligand planes is 57.3°) [76]. In MeCN solution, it displays two close one-electron oxidations affected by electrode adsorption (E$_{p\,2-/-}$ = +0.02 V; E$_{p\,-/0}$ = +0.08 V, *vs.* SCE) [76]. In view of the preference of Cu(III) ion for planar geometry, it cannot be ruled out that one of the two processes might be metal based.

The square planar monoanion [Au(dttdt)$_2$]$^-$ [67g, 77a-d] undergoes a one-electron reduction having features of chemical reversibility in the cyclic voltammetric time scale (E$^{o'}_{-/2-}$ = −0.62 V, dmf solution [77a]; −0.60 V, MeCN solution [77e]; V, *vs.* SCE) and a one-electron oxidation {E$_{p\,-/0}$ = +0.72 V (coupled to electrode adsorption, in dmf solution) [77a]; E$^{o'}_{-/0}$ = +0.35 V, MeCN solution [77e]}.

Like the dianions [M(dttdt)$_2$]$^{2-}$ (M = Co, Cu), [Zn(dttdt)$_2$]$^{2-}$ also has a tetrahedrally distorted geometry (the dihedral angle between the two ligand planes is around 95°) [78].

The square pyramidal dianion [MoO(dttdt)$_2$]$^{2-}$ [79] exhibits a first, chemically reversible, one-electron oxidation (E$^{o'}_{2-/-}$ = +0.12 V, *vs.* SCE) followed by a second one-electron oxidation affected by electrode adsorption (E$^{o'}_{-/0}$ = +0.52 V) [79].

The isostructural monoanion [ReO(dttdt)$_2$]$^-$ [80], undergoes in CH$_2$Cl$_2$ solution a partially chemically reversible one-electron reduction (E$_{p\,-/2-}$ = −1.7 V, *vs.* SCE) and an irreversible one-electron oxidation (E$_{p\,-/0}$ = +0.9 V) [80].

Table 23. Selected bond distances (average values; Å) in the redox family [Ni(dttdt)$_2$]$^{2-/-/0}$

Complex	Ni-S	S-C^a	C=C	C-S^b	S-C^b	C=S	Cation	Ref.
[Ni(dttdt)$_2$]$^{2-}$	2.22	1.75	1.39	1.71	1.73	1.68	[NBu$_4$]$^+$	72a
	2.20	1.73	1.35	1.74	1.72	1.65	[NBu$_4$]$^+$	72b
	2.19	1.74	1.33	1.74	1.72	1.65	[NHBu$_3$]$^+$	72c
	2.19	1.74	1.34	1.75	1.72	1.66	[Oct-py]$^{+\,c}$	72d
	2.19	1.73	1.35	1.74	1.72	1.65	[Rb(dchyl-18c6)]$^{+\,d}$	72e
	2.19	1.73	1.34	1.74	1.73	1.65	[Cs(dchyl-18c6)]$^{+\,d}$	72e
[Ni(dttdt)$_2$]$^-$	2.16	1.72	1.35	1.73	1.73	1.63	[NBu$_4$]$^+$	73a
	2.16	1.70	1.42	1.70	1.73	1.63	[DB-TTF]$^{+\,e}$	73b
	2.16	1.70	1.32	1.76	1.71	1.66	[BEDT-TTF]$^{+\,f}$	73c
	2.16	1.72	1.35	1.74	1.74	1.62	[NEt$_4$]$^+$	73d
	2.16	1.72	1.34	1.75	1.72	1.65	[NBu$_4$]$^+$	73e
	2.16	1.71	1.36	1.74	1.72	1.65	[NMe$_4$]$^+$	73f
	2.16	1.71	1.35	1.75	1.73	1.63	[NPr$_4$]$^+$	73f
	2.16	1.71	1.36	1.74	1.73	1.66	[Fe(C$_5$Me$_5$)$_2$]$^+$	73g
	2.17	1.70	1.38	1.73	1.73	1.63	[tmiz]$^{+\,g}$	73h
	2.15	1.71	1.36	1.74	1.72	1.64	[DIPSPh$_4$]$^{+\,h}$	73i
	2.16	1.72	1.33	1.74	1.72	1.64	[Co(C$_5$H$_5$)$_2$]$^+$	73l
	2.16	1.71	1.36	1.74	1.73	1.64	[PPN]$^{+\,i}$	71a
	2.16	1.71	1.36	1.74	1.73	1.64	[SmeEt$_2$]$^+$	71a
	2.16	1.71	1.35	1.74	1.74	1.62	[PPN]$^{+\,i}$	73m
	2.16	1.71	1.36	1.73	1.72	1.65	[EDA]$^{+\,l}$	73n
	2.17	1.72	1.36	1.74	1.73	1.64	[DMMP]$^{+\,m}$	73o
	2.16	1.71	1.37	1.74	1.72	1.65	[DMPPEMP]$^{+\,n}$	73o
	2.16	1.73	1.33	1.74	1.73	1.65	[DPD-Me]$^{+\,o}$	38l
	2.16	1.72	1.35	1.74	1.73	1.64	[K(DA-18-crown-6)]$^{+\,p}$	73p
	2.16	1.71	1.35	1.74	1.73	1.64	[Rb(DA-18-crown-6)]$^{+\,p}$	73p
	2.15	1.73	1.35	1.74	1.72	1.65	[Rb(dchyl-18c6)]$^{+\,d}$	72e
	2.16	1.71	1.35	1.75	1.73	1.63	[V]$^{+\,q}$	73q
	2.16	1.71	1.34	1.75	1.72	1.64	[NBu$_4$]$^+$	73r
	2.16	1.71	1.40	1.73	1.73	1.65	[Mn(C$_5$Me$_5$)$_2$]$^+$	67g
[Ni(dttdt)$_2$]^0	2.15	1.70	1.39	1.73	1.75	1.62	-	74
	2.14	1.70	1.39	1.73	1.74	1.63	-	72c

^{a}Inner ring; ^{b}outer ring; ^{c}Oct-py = N-octadecylpyridine; ^{d}dchyl-18c6 = dicyclohexyl-18-crown-6; ^{e}DB-TTF = dibenzotetrathiafulvalene; ^{f}BEDT-TTF = bis(ethylenedithio)tetrathiafulvalene; ^{g} tmiz = 1,2,3-trimethylimidazole; ^{h}DIPSPh$_4$ = tetraphenyldithiapyranylidene; ^{i}PPN = bis(triphenyl-phosphoranylidene); ^{l} EDA = 2-diethylamino-1,3-dithiolane; ^{m}DMMP = 4-(dimethylamino)-1-methylpyridine; ^{n}DMPPEMP = 4-[2-(4-(dimethylamino)phenyl)ethenyl]-1-methylpyridine; ^{o}DPD-Me = trans-4,4'-azobis(1-methylpyridine); ^{p}DA-18-crown-6 = 4,13-diaza-18-crown-6; ^{q}V = 3-[4-(diethyl methylamino)phenyl-1,5-diphenyl-6-oxoverdazole

Table 24. Formal electrode potentials (V, *vs.* SCE) for the redox processes exhibited by [M(dttdt)$_2$]$^{n-}$ (M = Pd, Pt) in MeCN solution

M	$E^{o'}_{2-/-}$	$E^{o'}_{-/0}$	Reference
Pd	+0.07[a,b]	-	74b
	+0.05	-	75
	+0.11	-	73m
	−0.09	-	71e
Pt	−0.08	+0.11	75
	−0.27	-	71e

[a] Peak potential value for a two-electron process; [b] affected by electrode adsorption

An even more limited redox aptitude is exhibited by [VO(dttdt)$_2$]$^{2-}$ (an irreversible oxidation at E_p = +0.19 V, *vs.* SCE, in MeCN solution) [81].

3.2.7.2 Tris(dttdt) complexes

A few tris-chelate complexes [M(dttdt)$_3$]$^{n-}$ (M = V, Mo, W, Re) have been electrochemically characterised. The V and Re complexes exhibit the whole sequence 3−/2−/−/0, whereas the Mo and W complexes only exhibit the sequence 2−/−/0, Table 25.

As a consequence of the chemical reversibility of the 2−/− electron transfer, the structural data for the (almost) trigonal prismatic redox couples [W(dttdt)$_3$]$^{2-/-}$ and [V(dttdt)$_3$]$^{2-/-}$ are available (in the last couple the integral oxidation states could be uncertain), Table 26.

As seen, the structural variations are minimal, thus once again supporting the ligand centred nature of the respective anodic processes.

Table 25. Formal electrode potentials (V, *vs.* SCE) for the redox processes exhibited by [M(dttdt)$_3$]$^{n-}$ (M = V, Mo, W, Re)

M	$E^{o'}_{3-/2-}$	$E^{o'}_{2-/-}$	$E^{o'}_{-/0}$	Solvent	Reference
V	−0.78	+0.15	+0.65[a]	MeCN	81a
Re	−1.28	−0.63	+0.76[a]	CH$_2$Cl$_2$	80
Mo		+0.18	+0.38[a]	dmf	82b
		−0.01	+0.26	CH$_2$Cl$_2$	68
W		+0.14	+0.32[a]	dmf	82b

[a] Affected by electrode adsorption

Table 26. Selected bond distances (average values, Å) in the couples $[M(dttdt)_3]^{n-}$.

Complex	M-S	S-C[a]	C=C	C-S[b]	S-C[b]	C=S	Cation	Reference
$[V(dttdt)_3]^-$	2.36	1.70	1.38	1.74	1.73	1.63	$[BEDT\text{-}TTF]^{+\ c}$	81b
$[V(dttdt)_3]^{2-}$	2.38	1.73	1.34	1.75	1.72	1.66	$[NMP]^{+\ d}$	81a
$[W(dttdt)_3]^-$	2.38	1.71	1.34	1.74	1.72	1.65	$[Fe(C_5Me_5)_2]^+$	82b
$[W(dttdt)_3]^{2-}$	2.39	1.73	1.34	1.73	1.72	1.65	$[NBu_4]^+$	82b
$[Mo(dttdt)_3]^{2-}$	2.40	1.73	1.35	1.72	1.71	1.66	$[NBu_4]^+$	82a,b

[a] Inner ring; [b] outer ring; [c] BEDT-TTF = bis(ethylenedithio)tetrathiafulvalene; [d] NMP = N-methylphenazine

3.2.8 idttdt Complexes (better known as "dmt")

3.2.8.1 Bis(idttdt) complexes

In sharp contrast with the huge number of metal complexes of the dttdt ligand, a limited number of complexes of the isomeric ligand idttdt have been characterised.

A series of dianions $[M(idttdt)_2]^{2-}$ (M = Ni, Pd, Pt, Cu) have been prepared, but the only crystal structure we are aware of is that of the square planar $[Ni(idttdt)_2]^{2-}$ [83a,b].

All these complexes only display the reversible oxidation to the corresponding monoanions, Table 27.

Comparison with the corresponding process for the $[M(dttdt)_2]^{2-}$ analogues shows that the actual complexes are slightly more difficult to oxidize by about 0.2 V.

Table 27. Formal electrode potentials (V, *vs.* SCE) for the one-electron oxidation of $[M(idttdt)_2]^{2-}$ (M = Ni, Pd, Pt, Cu) in MeCN solution

M	$E^{o\prime}_{2-/-}$	Reference
Ni	−0.07	83a
	+0.20	83c
	+0.06	71d
Pd	+0.18	83a
Pt	−0.08	83a
Cu	+0.09	83a

3.2.8.2 *Tris(idttdt) complexes*

To the best of our knowledge, the only crystal structure of tris-idttdt complexes is concerned with [V(idttdt)$_3$]$^{2-}$ [84], which is substantially similar to that of [V(dttdt)$_3$]$^-$ [81]. No electrochemical data are in this case available.

3.2.9 *R$_2$-timdt complexes*

3.2.9.1 *Bis(R$_2$-timdt) complexes*

Figure 12 shows the square planar geometry of the neutral [Ni(Pr$^i_2$-timdt)$_2$], together with its cyclic voltammetric behaviour [85a,b].

In addition to the reversible sequence 0/–/2– exhibited by most [Ni(dithiolene)$_2$]$^{n-}$ complexes, it also undergoes a single two-electron oxidation with features of partial chemical reversibility.

It is interesting to note that, as illustrated in Figure 13, the square planar [Pd(Et$_2$-timdt)$_2$] [85c] displays that five-membered sequence [Pd(Et$_2$-timdt)$_2$]$^{2+/}$

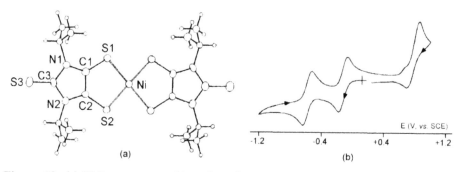

Figure 12. (a) X-Ray structure (a) and cyclic voltammogram (Pt working electrode; CH$_2$Cl$_2$ solution; scan rate 0.1 V s^{-1}) (b) of [Ni^{II}(Pr$^i_2$-timdt)$_2$].

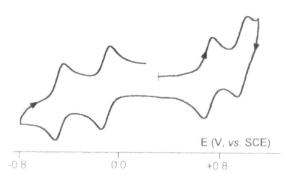

Figure 13. Cyclic voltammogram recorded at a platinum electrode in CH$_2$Cl$_2$ solution of [Pd(Et$_2$-timdt)$_2$]. Scan rate 0.2 V s^{-1}.

Table 28. Formal electrode potentials (V, *vs.* SCE) for the redox processes exhibited by [M(R$_2$-timdt)$_2$] (M = Ni, Pd) in CH$_2$Cl$_2$ solution

M	R	$E^{o\prime}_{2+/+}$	$E^{o\prime}_{+/0}$	$E^{o\prime}_{0/-}$	$E^{o\prime}_{-/2-}$	Reference
Ni	Me	+0.63^a	+0.63^a	−0.13	−0.59	85d
	Et	+0.78^a	+0.78^a	−0.11	−0.56	85a
		+0.65	+0.65	−0.14	−0.59	85d
	Pr^i	+0.75^a	+0.75^a	−0.17	−0.65	85a
	Bu	+0.77^a	+0.77^a	−0.12	−0.59	85a
	Ph	+0.73	+0.73	−0.04	−0.53	85d
Pd	Et	+0.72^a	+0.72^a	0.00	−0.36	85c
		+0.97	+0.71	−0.12	−0.49	85d
	Ph	+1.00	+0.74	+0.01	−0.38	85d
Pt	Me	+0.91^a	+0.69	−0.06	−0.43	85d
	Et	+1.03	+0.74	−0.13	−0.59	85d
	Ph	+1.2^b	+0.74	−0.01	−0.51	85d

^{a} Peak potential value for partially chemically reversible processes; ^{b} overlapped by a further oxidation

[Pd(Et$_2$-timdt)$_2$]$^+$/[Pd(Et$_2$-timdt)$_2$]^0/[Pd(Et$_2$-timdt)$_2$]$^-$/[Pd(Et$_2$-timdt)$_2$]$^{2-}$ theoretically expected for bis-dithiolenes (see Scheme 6) [85d].

As a matter of fact, concerned with the anodic path, inside the class [M(R$_2$-timdt)$_2$] (M = Ni, Pd, Pt), Ni complexes afford a single two electron removal, whereas Pd and Pt complexes give rise two separate one-electron removals. Table 28 compiles the formal electrode potentials for the whole redox changes exhibited by the present derivatives.

In agreement with the electrochemical reversibility of the 0/− reduction, the monoanion [Ni(Pr$^i_2$-timdt)$_2$]$^-$ maintains the square planar geometry of the neutral precursor [85e]. The pertinent structural parameters show quite minimal changes in bond lengths (the largest variation, 0.03 Å, is concerned with the C=C bond), which also in this case support the ligand centred nature of the electron transfer process.

3.2.10 tdadt Complexes (better known as "tdas")

3.2.10.1 Bis(tdadt) complexes

The square planar dianion [Ni(dtadt)$_2$]$^{2-}$ [86a] undergoes in MeCN solution a quasireversible oxidation to the corresponding monoanion [Ni(dtadt)$_2$]$^-$ ($E^{o\prime}_{2-/-}$

= +0.95 V, *vs.* SCE) [86b]. Crystallographic data on the monoanion seem to support a slight departure from planarity because of its tendence to dimerise, but structural data are not accessible [86c]. Crystal data are available for the monoanion [Fe(dtadt)$_2$]$^-$ [86c], but no electrochemical data are known.

3.3 1,2-Dithiolates forming six-membered lateral rings

3.3.1 bdt Complexes

3.3.1.1 Bis(bdt) complexes

As seen in most of the preceding cases, transition metal bis(bdt) complexes commonly exhibit the 0/–/2– redox processes [2b]. For example, in dmf solution, the formal electrode potential for the couple [Ni(bdt)$_2$]$^{2-/-}$ is –1.05 V, *vs.* Ag/AgClO$_4$ [2b]. Both the members of the couple adopt a square planar geometry, and also in this case the pertinent bond lengths show quite minimal variations. In fact, on passing from the dianion to the monoanion, the Ni-S distance shortens by 0.02 Å, the S-C length elongates by 0.02 Å, and the inner ring C=C distance remains unvaried [87]. These data once again suggest that the electron transfer is delocalised along the whole molecular frame.

A square planar geometry is also displayed by [Au(bdt)$_2$]^0 obtained by electrochemical oxidation of the monoanion [Au(bdt)$_2$]$^-$ [88].

The redox couples of the square pyramidal oxo-complexes [MO(bdt)$_2$]$^{2-/-}$ (M = Mo, W) have been crystallographically characterised (Mo: $E^{o'}_{2-/-}$ = –0.35 V, *vs.* SCE, in dmf solution [89a]; W: $E^{o'}_{2-/-}$ = –0.63 V, *vs.* SCE, in MeCN solution [18b]). The related structural data are compiled in Table 29.

Once again, the relative increase of the M=O distance upon one-electron reduction with respect to the metal-ligand and intraligand distances is diagnostic of metal-centred processes.

Table 29. Selected bond distances (average values; Å) in the redox couples [MO(bdt)$_2$]$^{2-/-}$ (M = Mo, W)

Complex	M-S	S-C	C=C^a	M=O	Cation	Reference
[MoO(bdt)$_2$]$^{2-}$	2.39	1.77	1.39	1.70	[NEt$_4$]$^+$	89a
[MoO(bdt)$_2$]$^-$	2.38	1.76	1.40	1.67	[PPh$_4$]$^+$	89a
[WO(bdt)$_2$]$^{2-}$	2.37	1.76	1.38	1.73	[NEt$_4$]$^+$	89c
[WO(bdt)$_2$]$^-$	2.37	1.76	1.39	1.69	[PPh$_4$]$^+$	89c

^{a} Inner ring

3.3.1.2 Tris(bdt) complexes

Structural data are available for the (distorted) octahedral anions $[Zr(bdt)_3]^{2-}$ [90a], $[Nb(bdt)_3]^-$ [90b], and $[Ta(bdt)_3]^-$ [90c], and $[Tc(bdt)_3]^-$ [90d]. The electrochemical The electrochemical irreversibility of the 2–/– redox change in these complexes testifies that they are the only stable members of their potential redox families [91].

As far as we know, the only stable redox couples of the $[M(bdt)_3]^{n-}$ complexes are concerned with the M = Mo, W families. The neutral $[Mo(bdt)_3]$ undergoes in CH_2Cl_2 solution two sequential reversible reductions to the corresponding mono- and di-anions, respectively ($E^{o\prime}_{0/-} = +0.20$ V; $E^{o\prime}_{-/2-} = -0.39$ V, *vs.* SCE) [92a]. Table 30 compares the bond distances in the neutral and monoanion complexes.

With respect to the trigonal prismatic geometry of the neutral complex, the one-electron addition causes a geometrical reorganization intermediate between trigonal prismatic and octahedral.

In turn, the dianion $[W(bdt)_3]^{2-}$ undergoes in MeCN solution two reversible oxidations to the corresponding monoanion and neutral complexes [93c]. The crystallographic data of the couple $[W(bdt)_3]^{2-/-}$ are reported in Table 30.

In both the Mo and W redox couples, the frontier orbitals look like metal and ligand based.

Table 30. Selected bond distances (average values; Å) and twist angles (°) in the redox couples $[M(bdt)_3]^{n-}$ (M = Mo, W)

Complex	M-S	S-C	C=C[a]	ϕ	Cation	Reference
$[Mo(bdt)_3]^0$	2.37	1.73	1.41	0	-	92b
$[Mo(bdt)_3]^-$	2.38	1.72	1.38	33.5	$[PPh_4]^+$	92a
$[W(bdt)_3]^-$	2.37	1.71	1.39	33	$[PPh_4]^+$	93a
	2.39	1.78	1.42	b	$[NMe_4]^+$	93b
$[W(bdt)_3]^{2-}$	2.39	1.75	1.40	3.5	$[NEt_4]^+$	93c
	2.39	1.72	1.39	b	$[AsPh_4]^+$	93d

[a] Inner ring; [b] not reported

3.3.2 X-bdt Complexes

3.3.2.1.1 Bis(4-Me-bdt) Complexes (better known as "tdt")

The electrochemical behaviour of a number of $[M^{II}(4\text{-Me-bdt})_2]^{2-}$ (M = Mn, Fe, Co, Ni, Cu, Zn) complexes has been investigated [94]. But for the Zn(II) complex, all the dianions undergo the chemically reversible oxidation to the corresponding monoanions, Table 31.

Table 31. Formal electrode potentials (V, *vs.* SCE) for the redox processes exhibited by [M^{II}(4-Me-bdt)$_2$]$^{2-}$

M	E°'$_{2-/-}$	E°'$_{-/0}$	Solvent	Reference
Mn	−0.63	+0.22^a	MeCN	94a
	−0.81	-	dmf	94b,c
	−0.68	-	CH$_2$Cl$_2$	94b,c
	−0.67	-	C$_2$H$_4$Cl$_2$	94b,c
Fe	−0.83^b	-	MeCN	94a
Co	−0.73	+0.20	MeCN	94a
Ni	−0.47	+0.44	MeCN	94a
Cu	−0.53	+0.62	MeCN	94a

^{a} Peak potential value for irreversible processes; ^{b} the anion is dimeric

It is interesting to note that, as illustrated in Figure 14, as a consequence of the chemically reversible ($i_{pa}/i_{pc} \approx 1$), but electrochemically quasireversible (ΔE_p = 290 mV, at 0.1 Vs^{-1}), one electron reduction, the square planar monoanion [Mn(4-Me-bdt)$_2$]$^-$ reorganizes to the tetrahedral dianion [Mn(4-Me-bdt)$_2$]$^{2-}$ [94b,c].

The pertinent structural data indicate that the dianion/monoanion redox change is characterised by a shortening of the Mn-S distance by 0.14 Å, whereas the S-C and C=C distances do not undergo variations. It seems hence evident that the extra electron is essentially delocalised on the manganese atom.

Structural data are also available for the square planar monoanions [M(4-Me-bdt)$_2$]$^-$ (M = Co [94d]; M = Cu [94a]; M = Au [94e]) and the tetrahedral dianion [Cd(4-Me-bdt)$_2$]$^{2-}$ [94f].

The square pyramidal oxo-complex [MoO(4-Me-bdt)$_2$]$^{2-}$ [89b] undergoes in dmf solution two reversible one-electron oxidations (E$^{o'}_{2-/-}$ = −0.46 V; E$^{o'}_{-/0}$ = +0.52 V, *vs.* SCE) [89b]. For the isostructural monoanions [MO(4-Me-bdt)$_2$]$^-$

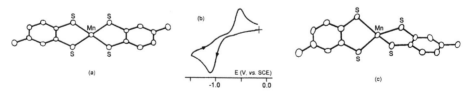

Figure 14. (a) X-Ray structure of [Mn(4-Me-bdt)$_2$]$^-$ ([PPh$_4$]$^+$ cation); (b) cyclic voltammogram recorded at a glassy carbon electrode in dmf solution of [Mn(4-Me-bdt)$_2$]$^-$ (scan rate 0.1 V s^{-1}); (c) X-ray structure of [Mn(4-Me-bdt)$_2$]$^{2-}$ ([PPh$_4$]$^+$ cation).

(M = Re, Tc), the 2–/–/0 redox processes occur at notably different potential values (Re: $E^{\circ'}_{2-/-} = -2.00$ V, $E^{\circ'}_{-/0} = +1.00$ V; Tc: $E^{\circ'}_{2-/-} = -1.52$ V, $E^{\circ'}_{-/0} = +0.95$ V, *vs.* SCE) [6a].

3.3.2.1.2 Tris(4-Me-bdt) complexes

The electrochemical behaviour of the tris-chelates $[M(4\text{-Me-bdt})_3]^{n-}$ (M = V, Mo, W, Re) has been reviewed [2b]. $[Zr(4\text{-Me-bdt})_3]^{2-}$ only exhibits an irreversible reduction ($E_p = -1.71$ V, *vs.* Ag/AgClO$_4$) [91].

3.3.2.2.1 Bis(3-SiPh₃-bdt) and bis (3-SiPh₃-5-Me-bdt) complexes

As expected on the basis of the inductive effects, the square pyramidal dianions $[MoO(3\text{-SiPh}_3\text{-bdt})_2]^{2-}$ [89b] and $[MoO(3\text{-SiPh}_3\text{-5-Me-bdt})_2]^{2-}$ are slightly more difficult to oxidise than $[MoO(4\text{-Me-bdt})_2]^{2-}$ ($E^{\circ'}_{2-/-} = -0.41$ V and -0.45 V, respectively; V, *vs.* SCE) [89b].

3.3.2.3.1 Bis(3,5-Buᵗ-bdt) complexes

The square planar $[Ni(3,5\text{-Bu}^t\text{-bdt})_2]^-$ undergoes in thf solution either a one electron oxidation ($E^{\circ'}_{-/0} = +0.35$ V, *vs.* NHE) or a one electron reduction ($E^{\circ'}_{-/2-} = -0.61$ V), both processes being chemically reversible [95]. As a matter of fact, the square planar three-membered series has been structurally characterised and the pertinent data are compiled in Table 32 [95].

Slight variations along the whole molecular frame occur, suggesting that the frontier orbitals have both metallic and ligand character.

Table 32. Selected bond distances (average values; Å) in the redox family $[Ni(3,5\text{-But-bdt})_2]^n$

Complex	Ni-S	S-C	C=C[a]	Cation
Ni(3,5-Buᵗ-bdt)₂]²⁻	2.17	1.76	1.40	[AsPh₄]⁺
Ni(3,5-Buᵗ-bdt)₂]⁻	2.14	1.75	1.41	[AsPh₄]⁺
Ni(3,5-Buᵗ-bdt)₂]⁰	2.13	1.73	1.42	-

[a] Inner ring

3.3.3 dhdtdt Complexes (better known as "dddt")

3.3.3.1 Bis(dhdtdt) complexes

The square planar monoanion $[Ni(dhdtdt)_2]^-$ in MeCN solution undergoes a chemically reversible one-electron reduction and a one-electron oxidation

Table 33. Formal electrode potentials (V, *vs.* SCE) for the redox processes exhibited by $[M(dhdtdt)_2]^{n-}$ complexes

M	$E^{o\prime}_{2-/-}$	$E^{o\prime}_{-/0}$	$E^{o\prime}_{0/+}$	Solvent	Reference
Ni	−0.85	−0.03[a]	-	dmf	33b, 96b
	−0.78	0.00	+0.80[b]	dmf	96c
	−0.69	+0.06[a]	-	MeCN	71b
	−0.68	+0.06[a]	-	MeCN	96d
	−0.59	+0.17	-	MeCN	96e
	−0.73	+0.05	-	thf	96f
Cu	−0.54	+0.33[a]	-	dmf	96g
Pd	−0.47	+0.16	-	dmf	33b
	−0.43	+0.14	-	Me_2CO	71a
Pt	−0.64	−0.02	-	dmf	33b
	−0.74	−0.11	+0.80[c]	PhCN	96c
Au	−1.32	+0.41	+0.82[c]	CH_2Cl_2	96h

[a] Coupled to chemical reactions; [b] peak-potential value for irreversible processes; [c] coupled to electrode adsorption

Table 34. Selected bond distances (average values; Å) in the square planar redox couple $[Ni(dhdtdt)_2]^{-/0}$

Complex	Ni-S	S-C	C=C[a]	C-S[b]	S-C[b]	C-C[b]	Cation	Reference
$[Ni(dhdtdt)_2]^-$	2.15	1.73	1.34	1.77	1.80	1.51	$[NEt_4]^+$	96b
	2.14	1.72	1.34	1.75	1.82	1.28	$[NBu_4]^+$	96h
	2.15	1.72	1.35	1.75	1.78	1.39	$[NMe_4]^+$	97a
$[Ni(dhdtdt)_2]^0$	2.12	1.69	1.37	1.75	1.82	1.47	-	97b

[a] Inner ring; [b] outer ring

Table 35. Selected bond distances (average values; Å) in the square planar redox couples $[M(dhdtdt)_2]^{-/0}$ (M = Pt, Au)

Complex	M-S	S-C[a]	C=C[a]	C-S[b]	S-C[b]	C-C[b]	Cation	Reference
$[Pt(dhdtdt)_2]^-$	2.27	1.74	1.34	1.77	1.80	1.50	$[NEt_4]^+$	98a
$[Pt(dhdtdt)_2]^0$	2.24	1.70	1.40	1.75	1.83	1.52	-	98b
$[Au(dhdtdt)_2]^-$	2.31	1.76	1.33	1.77	1.78	1.46	$[TTF]^+$	98c
$[Au(dhdtdt)_2]^0$	2.30	1.70	1.38	1.77	1.83	1.39	-	96h

[a] Inner ring; [b] outer ring; [c] TTF = tetrathiofulvalene

coupled to slow degradation of the electrogenerated neutral complex [71b]. Table 33 compiles the relative formal electrode potentials, together with those of the related copper, palladium and platinum complexes.

In spite of the relative instability, the neutral complex [Ni(dhdtdt)$_2$] has been isolated and crystallographically characterised. Selected interatomic distances relative to the redox couple [Ni(dhdtdt)$_2$]$^{-/0}$ are reported in Table 34. Once again, they suggest that the extra electron is delocalised along the whole molecular frame.

Surprisingly, to the best of our knowledge, no structural data are available for the stable dianion [Ni(dhdtdt)$_2$]$^{2-}$.

As far as the monoanion [Cu(dhdtdt)$_2$]$^-$ is concerned, the central CuS$_4$ core is planar in [NMe$_3$H][Ni(dhdtdt)$_2$] and tetrahedrally distorted (dihedral angle = 29°) in [NBu$_4$][Ni(dhdtdt)$_2$] [96g]. This points out the role played by countercations in modifying those crystal packing forces which are crucial to tune conducting properties triggered by stacking effects in metallodithiolenes.

Crystal data for the square planar couples [Pt(dhdtdt)$_2$]$^-$/[Pt(dhdtdt)$_2$]^0 and [Au(dhdtdt)$_2$]$^-$/[Au(dhdtdt)$_2$]^0 are reported in Table 35.

As usual, significant variations of the bond distances occur both at the metal-sulfur cores and at the ligand itself. Fractionally positive [Pt(dhdtdt)$_2$]$^{\delta+}$ cations have been also crystallographically characterised. No significant structural variation seems be present with respect to [Pt(dhdtdt)$_2$]^0 [71a,b].

3.3.3.2 Tris(dhdtdt) complexes

The electrochemical behaviour of a number of [M(dhdtdt)$_3$]$^{n-}$ complexes (M = Ti, V, Mn, Nb, Ta) has been investigated. As shown in Table 36, they can accede the 3–/2–/–/0 redox changes, only part of which are chemically reversible.

The crystal structures of the redox couple [V(dhdtdt)$_3$]$^{-/0}$, Table 37, indicate minimal variations in bonding distances, but, as previously noted for other

Table 36. Formal electrode potentials (V, *vs.* SCE) for the redox processes exhibited by [M(dhdtdt)$_3$]$^{n-}$ complexes in dmf solution

Complex	E$^{\circ\prime}_{3-/2-}$	E$^{\circ\prime}_{2-/-}$	E$^{\circ\prime}_{-/0}$	Reference
[Ti(dhdtdt)$_3$]$^{2-}$	−1.47	+0.03[a]	+0.50[a]	99a
[V(dhdtdt)$_3$]$^-$	−1.44	−0.51	+0.39	99b
[Mn(dhdtdt)$_3$]$^{2-}$	−0.62[b]	+0.08[a]	-	99a
[Nb(dhdtdt)$_3$]$^{2-}$	−1.71	−0.68	+0.48[a]	99b

[a] Peak-potential value for irreversible processes; [b] coupled to slow chemical complications

thris-dithiolenes, the main structural change following the electron transfer is the geometrical reorganization from the (almost) perfect trigonal prism of the monoanion to the octahedrally distorted trigonal prism of the neutral complex.

The crystal structure of the dianion $[Ti(dhdtdt)_3]^{2-}$ is also known [99c].

Table 37. Selected bond distances (average values, Å) and twist angles (°) in the couple $[V(dhdhdt)_3]^{-/0}$

Complex	V-S	S-C^a	C=C^a	C-S^b	S-C^b	C-C^b	φ	Cation	Reference
$[V(dhdtdt)_3]^-$	2.34	1.72	1.35	1.76	1.74	1.34	3.7	$[NBu_4]^+$	99a
	2.34	1.72	1.36	1.75	1.78	1.47	5.0	$[TTF]^{+c}$	99b
$[V(dhdtdt)_3]^0$	2.35	1.70	1.38	1.74	1.76	1.28	15.7	-	99b

^{a} Inner ring; ^{b} outer ring; ^{c} TTF = tetrathiofulvalene

3.3.4 R'- hdtdt Complexes

3.3.4.1 Bis(mehdtdt) and bis(phhdtdt) complexes (better known as "medt" and as "phdt", respectively)

As far as we know, the only mehdtdt and phhdtdt metallo-complexes up-to-now structurally and electrochemically studied are the square-planar nickel complexes. The monoanions $[Ni(mehdtdt)_2]^-$ and $[Ni(phhdtdt)_2]^-$ in MeCN solution reversibly undergo either a one-electron reduction or a one-electron oxidation, Table 38.

The crystal structures of the redox couple $[Ni(mehdtdt)_2]^{-/0}$ are available, Table 39.

The most significant structural change looks like centred on the peripheral regions of the ligand.

By way of comparison, Table 39 also reports the structural data of $[Ni(phhdtdt)_2]^0$, and those of the related diphenyl substituted $[Ni(dphdtdt)_2]^-$, which will be discussed in the next Section.

Table 38. Formal electrode potentials (V, *vs.* SCE) for the redox processes exhibited by $[Ni(R'-hdtdt)_2]^{n-}$ complexes

Complex	$E^{o'}_{2-/-}$	$E^{o'}_{-/0}$	Solvent	Reference
$[Ni(mehdtdt)_2]^-$	−0.70	+0.05	MeCN	96d
$[Ni(phhdtdt)_2]^-$	−0.65	+0.06	MeCN	100a
	−0.36	+0.37	MeCN	100b
	−0.71	+0.08	thf	96f

Table 39. Selected bond distances (average values; Å) in the square planar complexes [Ni(R'-hdtdt)$_2$]$^{-/0}$ and related complexes

Complex	Ni-S	S-C^a	C=C^a	C-S^b	S-C^b	C-C^b	C-C$_{(R')}$	Cation	Reference
[Ni(mehdtdt)$_2$]$^-$	2.13	1.71	1.32	1.76	1.74	1.24	1.16	[NBu$_4$]$^+$	98a
[Ni(mehdtdt)$_2$]^0	2.13	1.70	1.40	1.74	1.78	1.32	1.52	-	98b
[Ni(phhdtdt)$_2$]$^-$	2.13	1.70	1.38	1.74	1.76	1.28	1.51	-	98c
[Ni(dphdtdt)$_2$]$^-$	2.15	1.75	1.34	1.75	1.86	1.70	1.57	[PPh$_4$]$^+$	96h

^{a} Inner ring; ^{b} outer ring

3.3.5   dphdtdt Complexes

3.3.5.1 Bis(dphdtdt) complexes

Like the monosubstituted R'-hdtdt complexes, the diphenyl substituted [Ni(dphdtdt)$_2$]$^-$ exhibits in thf solution either a reversible one-electron reduction ($E^{o'}_{-/2-} = -0.71$ V, *vs.* SCE) or a reversible one-electron oxidation ($E^{o'}_{-/0} = +0.09$ V) [96f]. Structural data for the monoanion are reported in Table 39.

3.3.6   dtdt Complexes (better known as "ddt")

3.3.6.1 Bis(dtdt) complexes

Electrochemical data for a few [M(dtdt)$_2$]$^-$ monoanions (M = Ni, Pd) are available, Table 40.

Like the saturated dhdtdt derivatives, they undergo reversibly the 2−/−/0 redox changes. It is evident that the unsaturation of the peripheral ring makes the −/2− reduction of the nickel complex easier with respect to the saturated analogue.

Based on the crystallographic data for [Ni(dtdt)$_2$]$^-$ [101b] it seems likely that in the formally represented S−C=C−S fragment of the outer ring, the unsaturation is in reality delocalised over the whole SCCS frame.

Table 40. Formal electrode potentials (V, *vs.* SCE) for the redox processes exhibited by [M(dtdt)$_2$]$^{n-}$ complexes in MeCN solution and related complexes

Complex	$E^{o'}_{2-/-}$	$E^{o'}_{-/0}$	Reference
[Pd(dtdt)$_2$]$^{n-}$	−0.37	+0.11	101a
[Ni(dtdt)$_2$]$^{n-}$	−0.51	+0.06	101a
[Ni(dhdtdt)$_2$]$^{n-}$	−0.69	+0.06^a	71b

^{a} Coupled to chemical reactions

3.3.7 fcdtdt Complexes (better known as "dphdt")

3.3.7.1 Bis(fcdtdt) complexes

Figure 15 shows the structural and electrochemical properties of the NiS_4 square planar monoanion $[Ni(fcdtdt)_2]^-$ [102a].

It undergoes a quasireversible one-electron reduction and a quasireversible one-electron oxidation, both centred on the Ni-dithiolene moiety ($E^{o'}_{-/2-}$ = -0.95 V, ΔE_p = 190 mV; $E^{o'}_{-/0}$ = -0.18 V, ΔE_p = 180 mV; V, *vs.* Fc/Fc⁺), as well as a single two-electron oxidation (affected by electrode adsorption; $E_{p(0/2+)}$ = $+0.48$ V), which is centred on the two not-communicating ferrocenyl subunits.

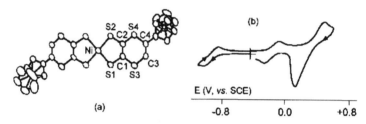

Figure 15. (a) X-Ray structure (average bonding distances: Ni-S = 2.15 Å; S1-C1 = 1.72 Å; C1-C2 = 1.37 Å; C1-S3 = 1.76 Å; S3-C3 = 1.75 Å; S4-C4 = 1.77 Å; C3-C4 = 1.34 Å) and (b) cyclic voltammogram (glassy carbon electrode; thf solution; scan rate 0.05 V s⁻¹) of $[Ni(fcdtdt)_2]^-$ ($[NBu_4]^+$ cation).

3.3.8 dfcdtdt Complexes

3.3.8.1 Bis(dfcdtdt) complexes

The monoanion $[Ni(dfcdtdt)_2]^-$ possesses structural (average bond distances: Ni-S = 2.14 Å; S1-C1 = 1.68 Å; C1-C2 = 1.39 Å; C1-S3 = 1.78 Å; S3-C3 = 1.73 Å; Å; C3-C4 = 1.33 Å) and redox properties (thf solution: $E^{o'}_{-/2-}$ = -0.98 V, ΔE_p = 170 mV; $E^{o'}_{-/0}$ = -0.20 V, ΔE_p = 170 mV; $E_{p(0/4+)}$ = $+0.54$V, *vs.* Fc/Fc⁺) rather similar to those of $[Ni(fcdtdt)_2]^-$ [102a,b]. Comparison with the related $[Ni(dfcedt)_2]$ (see Section 2.2.13.1) suggests that the interposition of the dithiin ring between the ferrocenyl subunits and the dithiolate function rises a barrier to the intramolecular communication of the ferrocene groups.

3.3.9 dcnpzdt Complexes

3.3.9.1 Bis(dcnpzdt) complexes

The dianions $[M(dcnpzdt)_2]^{2-}$ (M = Ni, Pd) have been characterised [103]. In MeCN solution, $[Ni(dcnpzdt)_2]^{2-}$ displays two reversible one-electron oxidations ($E^{o'}_{2-/-}$ = +0.52 V; $E^{o'}_{-/0}$ = +1.23 V, *vs.* SCE), whereas $[Pd(dcnpzdt)_2]^{2-}$ only exhibits an irreversible oxidation (E_p = +0.77 V). The crystal structure of $[Pd(dcnpzdt)_2]^{n-}$ present in $(TTF)_5[Pd(dcnpzdt)_2]_2$ has been reported [103].

3.3.10 R₂-pipdt Complexes

3.3.10.1 Bis(R₂-pipdt) complexes

A number of square planar $[M(R_2\text{-pipdt})_2]^{2+}$ dications have been characterised (M = Ni, Pd, Pt) [43c, 104a,b]. Figure 16,which refers to $[Pt(Me_2\text{-pipdt})_2]^{2+}$, shows that, as happens for the related R_2-timdt complexes (see Section 3.2.9), the present complexes are able to shuttle reversibly the sequence 2+/+/0/−/2−, Table 41.

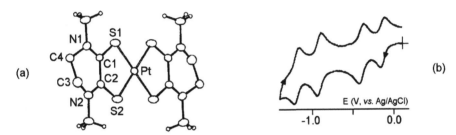

Figure 16. (a) X-Ray structure (average bonding distances: Pt-S = 2.28 Å; S1-C1 = 1.68 Å; C1-C2 = 1.53 Å; C1-N1 = 1.32 Å; N1-C4 = 1.45 Å; C4-C3 = 1.46 Å; N1-C(Me) = 1.45 Å) and (b) cyclic voltammogram (platinum electrode; MeCN solution; scan rate 0.1 V s⁻¹) of $[Pt(Me_2\text{-pipdt})_2]^{2+}$ ($[Pt(dcnedt)_2]^{2-}$ anion).

The fact that the different processes occur at essentially the same potential values, independently from the metal, suggests that the electron transfers are mainly ligand centered. The structural parameters for $[Ni(Me_2\text{-pipdt})_2]^{2+}$ are available [104a].

Table 41. Formal electrode potentials (V, *vs.* Ag/AgCl) for the redox processes exhibited by complexes $[M(Me_2\text{-pipdt})_2]^{2+}$ in MeCN solution [43c]

M	$E^{o'}_{2+/+}$	$E^{o'}_{+/0}$	$E^{o'}_{0/-}$	$E^{o'}_{-/2-}$
Ni	−0.16	−0.41	−0.96	−1.26
Pd	−0.18	−0.42	−0.87	−1.13
Pt	−0.15	−0.41	−0.92	−1.21

3.4   1,2-Dithiolates forming seven-membered lateral rings

3.4.1   pndtdt (better known as "pddt" or "ddtdt") and dfpndtdt (first named "F$_2$pdt") Complexes

3.4.1.1 Bis(pndtdt) and bis(dfpndtdt) complexes

Like the related dhdtdt complexes (Section 3.3.3), the monoanions [M(pndtdt)$_2$]$^-$ (M = Ni, Pt, Cu) and [Ni(dfpndtdt)$_2$]$^-$ undergo reversibly either the oxidation to the neutral species or the reduction to the corresponding dianions, Table 42.

The electron-withdrawing effect of the fluoride atoms obviously makes the reduction easier and the oxidation more difficult in [Ni(dfpndtdt)$_2$]$^-$ with respect to [Ni(pndtdt)$_2$]$^-$.

Crystallographic data for the square planar couple [Ni(pndtdt)$_2$]$^{-/0}$ are summarized in Table 43.

Once again, the electron transfer looks like delocalised along the whole molecular frame.

The crystal structures of [Cu(pndtdt)$_2$]$^-$ [105a] and [Ni(dfpndtdt)$_2$]$^-$ [105b] are also available.

Table 42. Formal electrode potentials (V, *vs.* SCE) for the redox processes exhibited by [M(pndtdt)$_2$]$^-$ and [M(dfpndtdt)$_2$]$^-$ complexes

Complex	$E^{\circ\prime}_{2-/-}$	$E^{\circ\prime}_{-/0}$	Solvent	Reference
[Ni(pndtdt)$_2$]$^-$	−0.71	+0.16	MeCN	71b
	−0.69	+0.17	MeCN	96d
	−0.65	+0.25	dmf	105a
[Ni(dfpndtdt)$_2$]$^-$	−0.55	+0.30	MeCN	105b
[Pt(pndtdt)$_2$]$^-$	−0.64	+0.15	MeCN	71b
[Cu(pndtdt)$_2$]$^-$	−0.61	-	dmf	105a

Table 43. Selected bond distances (average values; Å) in the square planar couple [Ni(pndtdt)$_2$]$^{-/0}$

Complex	Ni-S	S-C^a	C=C^a	C-S^b	S-C^b	C-C^b	Cation	Reference
[Ni(pndtdt)$_2$]$^-$	2.15	1.73	1.36	1.76	1.81	1.51	[NEt$_4$]$^+$	105a
	2.18	1.75	1.33	1.76	1.81	1.52	[NBu$_4$]$^+$	105c
[Ni(pndtdt)$_2$]^0	2.13	1.71	1.39	1.76	1.81	1.53	-	105d

^{a} Inner ring; ^{b} outer ring

3.4.2 mpndtdt Complexes (better known as "dpdt")

3.4.2.1 Bis(mpndtdt) complexes

A series of square planar monoanions $[M(mpndtdt)_2]^-$ (M = Ni, Au, Cu) have been structurally characterised [106a]. The Ni complex exhibits the usual reversible sequence 2–/–/0, the Cu complex reversibly undergoes the –/2– reduction, and the gold complexes is stable only as monoanion, Table 44.

Table 44. Formal electrode potentials (V, *vs.* SCE) for the redox processes exhibited by $[M(mpndtdt)_2]^-$ complexes in MeCN solution [106a]

Complex	$E^{\circ'}_{2-/-}$	$E^{\circ'}_{-/0}$
$[Ni(mpndtdt)_2]^-$	–0.62	+0.24
$[Cu(mpndtdt)_2]^-$	–0.67	+0.47^a
$[Au(mpndtdt)_2]^-$	+0.63^a	+1.28^a

^{a} Peak-potential value for irreversible processes

The neutral complex $[Ni(mpndtdt)_2]$ has been crystallographically solved [106b]. The major variations resulting from the monoanion-to-neutral change lie in the shortening of the Ni-S distance by 0.02 Å and the elongation of the inner ring S-C and C=C distances by 0.03 Å and 0.04 Å, respectively, which support also in this case that the extra electron is removed from a metal-ligand contributed frontier orbital.

3.4.3 odtdt Complexes (better known as "diod")

3.4.3.1 Bis(odtdt) complexes

The square planar monoanions $[M(odtdt)_2]^-$ (M = Ni, Cu) have been structurally characterised [107a,b]. As illustrated in Table 45, the Ni, Cu, Pd,

Table 45. Formal electrode potentials (V, *vs.* SCE) for the redox processes exhibited by $[M(odtdt)_2]^-$ complexes in MeCN solution

Complex	$E^{\circ'}_{2-/-}$	$E^{\circ'}_{-/0}$
$[Ni(odtdt)_2]^-$	–0.65	+0.27
$[Pd(odtdt)_2]^-$	–1.24^a	+0.26
$[Cu(odtdt)_2]^-$	–1.40^b	+0.56^a
$[Au(odtdt)_2]^-$	-	+0.64

^{a} Peak-potential value for irreversible processes; ^{b} coupled to slow chemical reactions

Au monoanions display electrochemical features qualitatively similar to those illustrated in previous complexes forming outer seven-membered rings [107].

3.4.4 thdtdt Complexes (better known as "dtdt" or "ttdt")

3.4.4.1 Bis(thdtdt) complexes

A series of monoanions [M(thdtdt)$_2$]$^-$ (M = Ni, Pd, Pt, Au, Cu) have been characterised [71a,b, 77e]. As a typical example, Figure 17 shows the structural and electrochemical features of [Ni(thdtdt)$_2$]$^-$ [71b].

Within the overall molecular chair conformation, the central NiS$_4$ core is square planar. The monoanion undergoes the reversible reduction to the corresponding dianion, whereas the one-electron oxidation is complicated by electrode adsorption phenomena.

Table 46. Formal electrode potentials (V, *vs.* SCE) for the redox processes exhibited by [M(thdtdt)$_2$]$^-$ complexes

Complex	$E^{\circ'}_{2-/-}$	$E^{\circ'}_{-/0}$	Solvent	Reference
[Ni(thdtdt)$_2$]$^-$	−0.61	+0.27[a]	MeCN	71a
	−0.59	+0.28[a]	MeCN	71b
	−0.59	+0.21	MeCN	77e
	−0.64	+0.23	PhCN	71a
[Pt(thdtdt)$_2$]$^-$	−0.53	+0.31[a]	MeCN	71a
	−0.59	+0.20	PhCN	71a
[Cu(thdtdt)$_2$]$^-$	−0.63	+0.45[b]	MeCN	77e
[Au(pndtdt)$_2$]$^-$	−0.91	+0.46	MeCN	77e

[a] Peak-potential value for processes affected by adsorption; [b] peak-potential value for irreversible processes

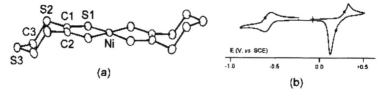

(a)

(b)

Figure 17. (a) X-Ray structure (average bonding distances: Ni-S = 2.14 Å; S1-C1 = 1.72 Å; C1-C2 = 1.38 Å; C1-S2 = 1.76 Å; S2-C3 = 1.81 Å; C3-S3 = 1.81) and cyclic voltammogram (glassy carbon electrode; MeCN solution; scan rate 0.1 V s^{-1}) (b) of [Ni(thdtdt)$_2$]$^-$ ([NBu$_4$]$^+$ cation).

A similar behaviour is displayed by the other metal complexes, Table 46.

The crystal structures of [Pt(thdtdt)$_2$]$^-$ [71a], [Cu(thdtdt)$_2$]$^-$ [77e] and [Au(thdtdt)$_2$]$^-$ [77e] are known.

4   1,2-DITHIOLATES FORMING LATERAL MULTIPLE RINGS

4.1   *1,2-Dithiolates forming bicyclic lateral rings*

4.1.1   *nordt Complexes*

4.1.1.1   *Bis(nordt) complexes*

Figure 18 illustrates the square planar geometry of the NiS$_4$ core of the dianion [Ni(nordt)$_2$]$^{2-}$ and its electrochemical behaviour [108].

It reversibly undergoes oxidation to the corresponding monoanion ($E^{0'}_{2-/-}$ = −0.76 V, *vs.* SCE). Given the low redox activity of the saturated dithiolene carbon-carbon bond (see Section 2.1.1), it seems plausible to assume that the electron removal involves the Ni(II)/Ni(III) process.

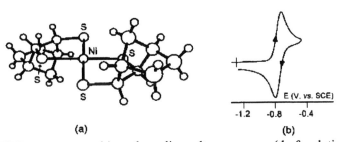

(a)   (b)

Figure 18. X-Ray structure (a) and cyclic voltammogram (dmf solution) (b) of [Ni(nordt)$_2$]$^{2-}$ ([Li]$^+$ cation). Selected bond lengths: Ni-S = 2.18 Å; S-C = 1.83 Å; C-C$_{(inner\ ring)}$ = 1.55 Å.

4.2   *1,2-Dithiolates forming lateral fused rings*

4.2.1   *bdtodt, bdttdt Complexes*

4.2.1.1   *Bis(bdtodt) and bis(bdttdt) complexes*

Brief electrochemical data are available for [Ni(bdtodt)$_2$]$^-$ and [Ni(bdttdt)$_2$]$^-$, Table 47, but the pertinent crystallographic data are lacking.

As rather common, the monoanions reversibly reduce to the corresponding dianions, whereas the stability of the neutral congeners is doubted.

Table 47. Formal electrode potentials (V, *vs.* SCE) for the redox processes exhibited by [Ni(bdtodt)$_2$]$^-$ and [Ni(bdtttdt)$_2$]$^-$

Complex	$E^{o'}_{2-/-}$	$E^{o'}_{-/0}$	Solvent	Reference
[Ni(bdtodt)$_2$]$^-$	−0.37	+0.32[a]	CH_2Cl_2	109a
	−0.36	0.0[b]	MeCN	109b
[Ni(bdtttdt)$_2$]$^-$	−0.30	+0.26[a]	CH_2Cl_2	109a

[a] Peak-potential value for processes affected by adsorption; [b] peak-potential value for irreversible processes

4.2.2 nqdt Complexes

4.2.2.1 Bis(nqdt) complexes

A structural and electrochemical study on a wide series of [M(nqdt)$_2$]$^{n-}$ complexes (M = Mn, Fe, Ni, Cu, Zn, Pd, Pt) has been reported [110]. The concomitant presence of the two redox-active quinone and dithiolate centers in the present ligand increases the difficulty to assign the nature of the pertinent redox changes. In fact, most complexes exhibit four redox processes, which, even if not clearly attributed, are thought to be ligand centered, Table 48.

Table 48. Formal electrode potentials (V, *vs.* Ag/AgCl) for the redox processes exhibited by [M(nqdt)$_2$]$^{n-}$ in MeCN solution

Complex	$E^{o'}_{1st}$	$E^{o'}_{2nd}$	$E^{o'}_{3rd}$	$E^{o'}_{4th}$
[Mn(nqdt)$_2$]$^{2-}$	−0.28	−0.68	−0.95	−1.47
[Fe(nqdt)$_2$]$^{2-}$	−0.30	−0.69	-	−1.49
[Ni(nqdt)$_2$]$^{2-}$	−0.28	−0.65	−0.95	−1.44
[Cu(nqdt)$_2$]$^-$	−0.29	−0.68	-	−1.47
[Zn(nqdt)$_2$]$^{2-}$	−0.27	−0.67	−0.94	−1.44
[Pd(nqdt)$_2$]$^{2-}$	−0.28	−0.54	−1.22	-
[Pt(nqdt)$_2$]$^{2-}$	−0.29	−0.67	-	−1.41

The structural parameters of the couple [Ni(nqdt)$_2$]$^{2-/-}$ point out that also in this case the extra electron looks like delocalised on the metal-ligand frame, in that passing from the dianion to the monoanion the Ni-S and C=C distances shorten (by 0.03 Å and 0.02 Å, respectively), whereas the S-C distance remains unvaried. It is however noted that, unexpectedly, the C-O length of the quinoidal function does not change appreciably (from 1.21 Å to 1.22 Å) [110].

X-Ray crystal data are also available for $[Cu(nqdt)_2]^-$ and $[Pd(nqdt)_2]^{2-}$ [110].

4.2.3 qdt Complexes

4.2.3.1 Bis(qdt) complexes

The electrochemical behaviour of the dianions $[M(qdt)_2]^{2-}$ (M = Co, Ni, Cu) shows that they are able to undergo reversibly the oxidation to the corresponding monoanions, Table 49.

The crystallographic data for the square planar redox couple $[Cu(qdt)_2]^{2-/-}$ show that, passing from the dianion to the monoanion, the Cu-S and C=C bond lengths shorten by 0.07 Å and 0.02 Å, respectively, whereas the S-C and C-N distances are substantially unaltered [111c]. The significant shortening of the Cu-S distance upon electron removal suggests that the electron transfer is essentially metal centred (*i.e.* it corresponds to the Cu(II)/Cu(III) process).

The X-ray structure of $[Ni(qdt)_2]^{2-}$ is available [111d].

Table 49. Formal electrode potentials (V, *vs.* SCE) for the redox processes exhibited by $[M(qdt)_2]^{2-}$

Complex	$E^{\circ\prime}_{2-/-}$	$E^{\circ\prime}_{2-/3-}$	Solvent	Reference
$[Co(qdt)_2]^{2-}$	+0.12	-	MeCN	111a
$[Ni(qdt)_2]^{2-}$	+0.12	-	dmf	111b
$[Cu(qdt)_2]^{2-}$	−0.18	−1.28	dmf	111c

4.2.3.2 Tris(qdt) complexes

Apart from minor information on the redox aptitude of $[M(qdt)_3]^{n-}$ (M = Co, n = 3; M = Fe, n = 2) [111a], the study most pertinent to the present review is that

Table 50. Selected bond distances (average values; Å) and twista angle (°) in the couple $[Mo(qdt)_3]^{2-/-}$

Complex	Mo-S	S-C	C=C	C-N	φ	Cation	Reference
$[Mo(qdt)_3]^-$	2.39	1.74	1.43	1.32	14.6	$[PPh_4]^+$	112c
$[Mo(qdt)_3]^{2-}$	2.39	1.74	1.44	1.32	4.5	$[PPh_4]^+$	112b

devoted to $[Mo(qdt)_3]^{n-}$ [112]. The dianion $[Mo(qdt)_3]^{2-}$ undergoes, in MeCN solution, either a reversible one-electron oxidation ($E^{o'}_{2-/-} = -0.19$ V, *vs.* Fc/Fc$^+$), or a reversible one-elecron reduction ($E^{o'}_{2-/3-} = -1.70$ V) [112a,b].

The crystal structures of the couple $[Mo(qdt)_3]^{2-/-}$ substantially show that, upon electron removal, the MoS$_6$ core passes from a almost perfect to a octahedrally distorted trigonal prism, Table 50.

4.2.4 diotte Complexes

4.2.4.1 Bis(diotte) complexes

Complex $[Ni(diotte)_2]^-$ has been recently prepared [113]. In MeCN solution, it displays a one-electron oxidation, which is affected by electrode adsorption ($E_p = +0.42$ V, *vs.* AgAgCl), and a reversible one-electron reduction ($E^{o'} = -0.56$ V). Chemical oxidation (by I_2) affords the neutral $[Ni(diotte)_2]$ [113]. No crystal structures are available.

4.3 1,2-Dithiolates forming lateral tetrathiafulvalene rings

4.3.1 R$_2$-dtttfdt Complexes

4.3.1.1 Bis(R$_2$-dtttfdt) complexes

The redox activity of a few complexes of general formula $[M(R_2\text{-}dtttfdt)_2]^{n-}$ (M = Co, R = Bu, C_6H_{13}, C_8H_{17}, $C_{10}H_{21}$, n = 1; M = Ni, R = Et, n = 2; M = Cu,

Table 51. Formal electrode potentials (V, *vs.* SCE) for the redox processes exhibited by $[M(R_2\text{-}dtttfdt)_2]^{n-}$

Complex	$E^{o'}_{2-/-}$	$E^{o'}_{-/0}$	Solvent	Ref.
$[Co(Bu_2\text{-}dtttfdt)_2]^-$	−0.61	−0.32	thf	114a
$[Co\{(C_6H_{13})_2\text{-}dtttfdt\}_2]^-$	−0.62	−0.26	thf	114a
$[Co\{(C_8H_{17})_2\text{-}dtttfdt\}_2]^-$	−0.56	−0.27	thf	114a
$[Co\{(C_{10}H_{21})_2\text{-}dtttfdt\}_2]^-$	−0.60	−0.27	thf	114a
$[Ni(Et_2\text{-}dtttfdt)_2]^{2-}$	−0.45	−0.07[a]	CH$_2$Cl$_2$	114b
$[Cu(Me_2\text{-}dtttfdt)_2]^-$	+0.25	+0.64	dmf	114c
$[Zn(Me_2\text{-}dtttfdt)_2]^{2-}$	+0.15	+0.18	dmf	114c
$[Hg(Et_2\text{-}dtttfdt)_2]^{2-}$	−0.04	+0.07[b]	CH$_2$Cl$_2$	114d
$[Pt\{(C_{10}H_{21})_2\text{-}dtttfdt\}_2]^{2-}$	−0.03	+0.38[a]	CH$_2$Cl$_2$	114e
$[Pt\{(C_{14}H_{29})_2\text{-}dtttfdt\}_2]^{2-}$	−0.25	+0.25	thf	114e
$[Pt\{(C_{14}H_{29})_2\text{-}dtttfdt\}_2]^{2-}$	−0.25	+0.05	thf	114e

[a] Peak-potential value for processes affected by adsorption; [b] followed by further oxidation processes

R = Me, n = 1; M = Zn, R = Me, n = 2; M = Hg, R = Et, n = 2; M = Pt, R = $C_{10}H_{21}$, $C_{14}H_{29}$, $C_{18}H_{37}$, n = 2) has been examined. As usual, they undergo the 2–/–/0 sequence, Table 51.

[Ni(Et$_2$-dtttfdt)$_2$] (obtained by chemical oxidation of the corresponding dianion) [114a] and [Pt{($C_{10}H_{21}$)$_2$.dtttfdt}$_2$]$^{2-}$ [114d] have a planar geometry, whereas [Hg(Et$_2$-dtttfdt)$_2$]$^{2-}$ has a tetrahedral geometry [114c].

4.3.2   endtttfdt Complexes

4.3.2.1 Bis(endtttfdt) complexes

[Au(endtttfdt)$_2$]$^-$ shows in dmf solution a quasireversible one-electron oxidation ($E^{o'}_{-/0}$ = +0.12 V, *vs.* SCE) [115], whereas [Co(endtttfdt)$_2$]$^-$ in thf solution exhibits either a quasireversible reduction ($E^{o'}_{-/2-}$ = –0.58 V, *vs.* SCE) or a quasireversible oxidation ($E^{o'}_{-/0}$ = –0.24 V) [114a].

4.3.3   pndtttfdt Complexes

4.3.3.1 Bis(pndtttfdt) complexes

Figure 19 shows the crystal structure of [Ni(pndtttfdt)$_2$]$^-$ [116a].

In MeCN solution, it exhibits the (usual) reversible one-electron reduction ($E^{o'}_{-/2-}$ = –0.45 V, *vs.* Ag/Ag$^+$) and the reversible one-electron oxidation ($E^{o'}_{-/0}$ = –0.15 V). As a matter of fact, the one-electron oxidised complex maintains the original planar geometry [116b]. Table 52 compares the bond lengths of the neutral complex with those of the corresponding monoanion.

It is evident that the electron transfer is essentially centred on the ligand.

The dianion [Cu(pndtttfdt)$_2$]$^{2-}$ exhibits a redox behaviour qualitatively similar to that of [Ni(pndtttfdt)$_2$]$^-$ ($E^{o'}_{-/2-}$ = –0.73 V, $E^{o'}_{-/0}$ = –0.08 V) [116a], but, at variance with the square planar geometry of this latter, has a tetrahedrally distorted geometry [116a,c].

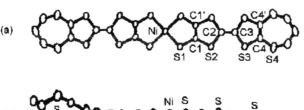

Figure 19. (a) ORTEP drawing and (b) side view of [Ni(pndtttfdt)$_2$]$^-$ ([PPh$_4$]$^+$ cation).

Table 52. Selected bond distances (average values; Å) in the square planar couple [Ni(pndtttfdt)$_2$]$^{-/0}$ (atomic labellings are referred to Figure 19)

Complex	Ni-S1	S1-C1	C1=C1'	C2=C3	C4=C4'	Cation	Reference
[Ni(pndtttfdt)$_2$]$^-$	2.16	1.72	1.34	1.32	1.36	[PPh$_4$]$^+$	116a
	2.17	1.71	1.36	1.35	1.34	[NMe$_4$]$^+$	116a,c
[Ni(pndtttfdt)$_2$]^0	2.18	1.73	1.38	1.38	1.34	-	116b

4.3.4 dmttfdt Complexes (originally named "dmdt")

4.3.4.1 Bis(dmttfdt) Complexes

[Ni(dmttfdt)$_2$]$^{2-}$ undergoes, in MeCN solution, two reversible one-electron oxidations (E$^\circ{}'_{2-/-}$ = −0.79 V; E$^\circ{}'_{-/0}$ = −0.48 V, *vs.* Ag/Ag$^+$) followed by a further irreversible anodic process (E$_p$ = +0.40 V) [117]. Both the neutral and the dianionic derivatives have been chemically, but not yet crystallographically, characterised [117].

4.3.5 tmettfdt Complexes (originally named "tmdt")

4.3.5.1 Bis(tmettfdt) Complexes

Like the related [Ni(dmttfdt)$_2$]$^{2-}$, the dianion [Ni(tmettfdt)$_2$]$^{2-}$ undergoes, in MeCN solution, two reversible one-electron oxidations (E$^\circ{}'_{2-/-}$ = −0.80 V; E$^\circ{}'_{-/0}$ = −0.46 V, *vs.* Ag/Ag$^+$) followed by a further irreversible anodic process (E$_p$ = +0.55 V) [117].

The neutral congener [Ni(tmettfdt)$_2$] possesses a square planar geometry [118].

5   FINAL REMARKS

A systematic examination of the structural consequences of the redox changes exhibited by the homoleptic, mononuclear metal complexes of a wide variety of dithiolene ligands points out that they do not constitute a homogeneous class of redox active derivatives. In fact, depending upon the overall frame of each dithiolene ligand (in terms of geometrical and inductive effects), the relative electrochemical responses cover the whole range of classical electron transfer processes, *i.e.* essentially metal centered, essentially ligand centered, or metal-ligand centered. This means that too often the multiple redox processes exhibited by metallodithiolenes are quickly solved as ligand (or metal) centred. In reality, the precise and accurate attribution of their nature (if wished) would always need structural and theoretical supports.

ACKNOWLEDGEMENTS

Piero Zanello gratefully acknowledges the financial support of the University of Siena (PAR), which since a long time allows him to carry out multiple research investigations.

ABBREVIATIONS

atedt = 1,2-di(alkylthio)-1,2-ethenedithiolate
bdt = benzene-1,2-dithiolate
bdtodt = 2-oxobenzo[d]-1,3-dithiole-5,6-dithiolate
bdttdt = 2-thionebenzo[d]-1,3-dithiole-5,6-dithiolate
cnedt = 2-cyano-1,2-ethenedithiolate
dcmdtcr = 2-dicyanomethylene-4,5-disulfanylcyclopent-4-ene-1,3-dionate
dcmedt = 1,2-dicarbomethoxy-1,2-ethenedithiolate
dcnedt = 1,2-dicyano-1,2-ethenedithiolate
dcnpzdt = 2,3-dicyano-pyrazine-5,6-dithiolate
dfcdtdt = 2,3-diferrocenyl-1,4-dithiin-5,6-dithiolate
dfcedt = 1,2-diferrocenyl-1,2-ethenedithiolate
dfpndtdt = 2,2-difluoro-1,3-propanediyldithioethylene-1,2-dithiolate
dhdtdt = 5,6-dihydro-1,4-dithiin-2,3-dithiolate
dhtpdt = 2,3-dihydro-4,5-thiophenedithiolate
dibzedt = 1,2-diiminobenzyl-1,2-ethenedithiolate
dichedt = 1,2-diiminocyclohexyl-1,2-ethenedithiolate
diotte = 1,3-dioxolane-tetrathiaethylene
dmedt = 1,2-dimethyl-1,2-ethenedithiolate
dmttfdt = dimethyltetrathiafulvalenedithiolate
dodt = 1,2-dioxo-1,2-dithiolate
doedt = 2-(1',3'-dithiole-2'-one)-1,2-ethenedithiolate
dphdtdt = 5,6-diphenyl-1,4-dithiin-2,3-dithiolate
dphedt = 1,2-diphenyl-1,2-ethenedithiolate
dt = ethane-1,2-dithiolate
dtcr = dithiocroconate = 4,5-disulfanylcyclopent-4-ene-1,2,3-trionate
dtdt = 1,4-dithiin-2,3-dithiolate
dtodt = 1,3-dithiole-2-one-4,5-dithiolate
dtsq = dithiosquarate

dttdt = 1,3-dithiole-2-thione-4,5-dithiolate

edt = ethene-1,2-dithiolate

endtttfdt = ethylenedithiotetrathiafulvalenedithiolate

fcdtdt = 2-ferrocenyl-1,4-dithiin-5,6-dithiolate

fcedt = ferrocenyl-1,2-ethenedithiolate

Hedt = unsubstituted ethene-1,2-dithiolate

idttdt = 1,2-dithiole-3-thione-4,5-dithiolate

mdtdt = 1,3-dithiole-4,5-dithiolate

mehdtdt = 5-methyl-6-hydro-1,4-dithiin-2,3-dithiolate

mpndtdt = 6,7-dihydro-6-methylene-5H-1,4-dithiepine-2,3-dithiolate

mtadt = mercaptothioacetate-1,2-dithiolate

nordt = norbornane-1,2-dithiolate

nqdt = naphthoquinonedithiolate

odtdt = 1,4-dithia-6-oxa-2,3-dithiolate = 1,2,6-oxadithiepin-4,5-dithiolate

phedt = 2-phenyl-1,2-ethenedithiolate

phhdtdt = 5-phenyl-6-hydro-1,4-dithiin-2,3-dithiolate

pndtdt = 1,3-propanediyldithioethylene-1,2-dithiolate = 6,7-dihydro-5H-1,4-dithiepine-2,3-dithiolate

pndtttfdt = propylenedithiotetrathiafulvalene-dithiolate

pyedt = pyridin-X-yl-1,2-ethenedithiolate (X = 2, 3, 4)

qdt = quinoxaline-2,3-dithiolate

qedt = quinoxalin-2-yl-1,2-ethenedithiolate

R_2-dtttfdt = 2,3-dialkylthiotetrathiafulvalene-6,7-dithiolate

R'-hdtdt = 5-substituted-6-hydro-1,4-dithiin-2,3-dithiolate

R-phedt = 1-substituted-2-phenyl-1,2-ethenedithiolate

R_2-pipdt = N,N'-dialkyl-piperazine-2,3-dithione

R_2-timdt = 1,3-dialkylimidazoline-2,4,5-trithionate

tfmedt = 1,2-di(trifluoromethyl)-1,2-ethenedithiolate

thdtdt = 5,7-dihydro-1,4,6-trithiin-2,3-dithiolate

thedt = 1,2-di(2-thienyl)-1,2-ethenedithiolate

tmettfdt = trimethylenetetrathiafulvalenedithiolate

tpdt = 3,4-thiophenedithiolate

ttpdt = 3,4-terthiophenedithiolate

X-bdt = benzene-substituted-1,2-dithiolate

REFERENCES

1. Clark, R.E.D. *Analyst*, *60*, 242 (1936).
2. (a) Gray, H.B. *Transition Met.Chem.*, *1*, 240 (1965); (b) McCleverty, J.A. *Prog.Inorg.Chem.*, *10*, 49 (1968); (c) Schrauzer, G.N. *Acc.Chem.Res.*, *2*, 72 (1969); (d) Hoyer, E.; Dietzsch, W.; Schroth, W. *Z.Chem. 11*, 41 (1971); (e) Burns, R.P.; McAuliffe, C.A. *Adv.Inorg.Chem.*, *22*, 303 (1979); (f) Mueller-Westerhoff, U.T.; Vance, B. in "Comprehensive Coordination Chemistry", Pergamon Press, Oxford, Vol. 2, 1987, 595-631; (g) Clemenson, P.I. *Coord.Chem.Rev.*, *106*, 171 (1990); (h) Mueller-Westerhoff, U.T.; Vance, B.; Yoon, D.I. *Tetrahedron*, *47*, 909 (1991); (i) Cassoux, P.; Valade, L.; Kobayashi, H.; Kobayashi, A.; Clark, R.A.; Underhill, A.E. *Coord.Chem.Rev.*, *110*, 115 (1991); (l) Olk, R.-M.; Olk, B.; Dietzsch, W.; Kirmse, R.; Hoyer, E. *Coord.Chem.Rev.*, *117*, 99 (1992); (m) Fourmigué, M. *Coord.Chem.Rev.*, *178-180*, 823 (1998); (n) Pullen, A.E.; Olk, R.-M. *Coord.Chem.Rev.*, *188*, 211 (1999); (o) Pilato, R.S.; Stiefel, E.I.; in "Bioinorganic Catalysis", Reedijk, J.; Bouwman, E.; Eds.; Marcel Dekker, New York, 1999, 81-152; (p) Robertson, N.; Cronin, L. *Coord.Chem.Rev.*, *227*, 93 (2002).
3. Ward M.D.; McCleverty, J.A. *J.Chem.Soc., Dalton Trans.*, 275 (2002).
4. (a) Costa, T; Dorfman, J.R.; Hagen, K.S.; Holm, R.H. *Inorg.Chem.*, *22*, 4091 (1983); (b) Dorfman, J.R.; Rao, Ch.P.; Holm, R.H. *Inorg.Chem.*, *24*, 453 (1985); (c) Rao, Ch.P.; Dorfman, J.R.; Holm, R.H. *Inorg.Chem.*, *25*, 428 (1986); (d) Snyder, B.S.; Rao, Ch.P.; Holm, R.H. *Aust. J.Chem.* *39*, 963 (1986).
5. (a) Szeymies, D.; Krebs, B.; Henkel, G. *Angew.Chem.Int.Ed.Engl.*, *23*, 804 (1984); (b) Money, J.A.; Huffman, J.C.; Christou, G. *Inorg.Chem.*, *24*, 3297 (1985); (c) Wiggins, R.W.; Huffman, J.C.; Christou, G. *J.Chem.Soc., Chem.Commun.*, 1313 (1983).
6. (a) Davison, A.; Orvig, C.; Trop, H.S.; Sohn, M.; DePamphilis, B.V.; Jones, A.G. *Inorg.Chem.*, *19*, 1988 (1980); (b) Blower, P.J.; Dilworth, J.R.; Hutchinson, J.P.; Zubieta, J.A. *Inorg.Chim.Acta*, *65*, L225 (1982); (c) Smith, J.E.; Byrne, E.F.; Cotton, F.A.; Sekutowski, J.C. *J.Am.Chem.Soc.*, *100*, 5571 (1978);.
7. Tatsumi, K.; Sekiguchi, Y.; Nakamura, A.; Cramer, R.E.; Rupp, J.J. *Angew.Chem.Int.Ed.Engl.*, *25*, 86 (1986).
8. Gama, V; Belo, D; Rabaça S.; Santos, I.C.; Alves, H.; Waerenborgh, J.C.; Duarte, M.T.; Henriques, R.T. *Eur.J.Inorg.Chem.*, 2101 (2000).
9. Hoyer, E.; Dietzsch, W.; Hennig, H.; Schroth, W. *Chem.Ber.*, *102*, 603 (1969).
10. Herman,Z.S.; Kirchner,R.F.; Loew,G.H.; Mueller-Westerhoff, U.T.; Nazzal, A; Zerner, M.C. *Inorg.Chem. 21*, 46 (1982).
11. Browall, K.W.; Bursch, T.; Interrante, L.V.; Kasper, J.S. *Inorg.Chem.*, *11*, 1800 (1972).
12. Donahue, J.P.; Goldsmith, C.R.; Nadiminti, U.; Holm, R.H. *J.Am.Chem.Soc.*, *120*, 12869 (1998).
13. Davies, E.S.; Beddoes, R.L.; Collison, D.; Dinsmore, A.; Docrat, A.; Joule, J.A.; Wilson, C.R.; Garner, C.D. *J.Chem.Soc., Dalton Trans.*, 3985 (1997)
14. Fourmigué, M.; Bertran, J.N. *Chem.Commun.*, 2111 (2000).
15. Keefer, C.E.; Purrington, S.T.; Bereman, R.D.; Boyle, P.D. *Inorg.Chem.*, *38*, 5437 (1999).

16. Wilkes, S.B.; Butler, I.R.; Underhill, A.E.; Kobayashi, A.; Kobayashi, H. *J.Chem.Soc., Chem.Commun.*, 53 (1994).

17. B.S.Lim, D.V.Fomitchev, and R.H.Holm, *Inorg.Chem.*, *40*, 4257 (2001).

18. (a) Lim, B.S.; Donahue, J.P.; Holm, R.H. *Inorg.Chem.*, *39*, 263 (2000); (b) Goddard, C.A.; Holm, R.H. *Inorg.Chem.*, *38*, 5389 (1999).

19. Fomitchev, D.V., Lim, B.S.; Holm, R.H. *Inorg.Chem.*, *40*, 645 (2001).

20. (a) Bowmaker, G.A.; Boyd, P.D.W.; Campbell, G.K. *Inorg.Chem.*, *22*, 1208 (1983); (b) Hill, C.A.S.; Charlton, A.; Underhill, A.E.; Malik, K.M.A.; Hursthouse, M.B.; Karaulov, A.I.; Oliver, S.N.; Kershaw, S.V. *J.Chem.Soc., Dalton.Trans.*, 587 (1995); (c) Sung, K.-M.; Holm, R.H. *J.Am.Chem.Soc.*, *124*, 4312 (2002).

21. (a) Mahadevan, C.; Seshasayee, M.; Kuppusamy, P.; Manoharan, P.T. *J.Cryst.Spectros.Res.*, *14*, 179 (1984); (b) Rajalakshmi, A.; Radha, A.; Seshasayee, M.; Kuppusamy, P.; Manoharan, P.T. *Z.Kristallogr.*, 159 (1988).

22. Koz'min, P.A.; Larina, T.B. *Koord.Khim. (Engl.Trans.)*, *5*, 464 (1979).

23. Megnamisi-Belombe, M.; Nuber, B. *Bull.Chem.Soc.Jpn.*, *62*, 4092 (1989).

24. (a) Wing, R.M.; Schlupp, R.L. *Inorg.Chem.*, *9*, 471 (1970); (b) Singhabhandhu, A.; Robinson, P.D.; Fang, J.H.; Geiger, W.E. *Inorg.Chem.*, *14*, 318 (1975); (c) Miller, J.S.; Calabrese, J.C.; Epstein, A.J. *Inorg.Chem.*, *28*, 4230 (1989).

25. Schmitt, R.D.; Wing, R.M.; Maki, A.H. *J.Am.Chem.Soc.*, *91*, 4394 (1969).

26. Geiger, W.E.; Mines, T.E.; Senftleber, F.C. *Inorg.Chem.*, *14*, 2141 (1975).

27. Wang, K.; McConnachie, J.M.; Stiefel, E.I. *Inorg.Chem.*, *38*, 4334 (1999).

28. Heuer, W.B.; Mountford, P.; Green, M.L.H.; Bott, S.G.; O'Hare, D.; Miller, S.J. *Chem.Mater.*, *2*, 764 (1990).

29. (a) Brown, R.K.; Bergendahl, T.J.; Wood, J.S.; Waters, J.H. *Inorg.Chim.Acta*, *68*, 79 (1983); (b) Baudron, S.A.; Batail, P. *Acta Cryst.*, *C58*, m575 (2002).

30. (a) Coucouvanis, D.; Hadjikyriacou, A.; Toupadakis, A.; Koo, S.-M.; Ileperuma, O.; Draganjac, M.; Salifoglou, A. *Inorg.Chem.*, *30*, 754 (1991); (b) Bandoli, G.; Nicolini, M; Mazzi, U.; Spies, H.; Munze, R. *Transition Met.Chem.*, *9*, 127 (1984); (c) Draganjac, M.; Coucouvanis, D. *J.Am.Chem.Soc.*, *105*, 139 (1983).

31. Beswick, C.L.; Terroba, R.; Greaney, M.A.; Stiefel, E.I. *J.Am.Chem.Soc.*, *124*, 9664 (2002).

32. Kean, C.L.; Pickup. P.G. *Chem.Commun.*, 815 (2001).

33. (a) Keller, C.; Walther, D.; Reinhold, J.; Hoyer, E. *Z.Chem.*, *11*, 410 (1988); (b) Vance,T.; Bereman, R.D. *Inorg.Chim.Acta*, *149*, 229 (1988).

34. (a) Charlton, A.; Hill, C.A.S.; Underhill, A.E.; Malik, K.M.A.; Hursthouse, M.B.; Karaulov, A.I.; Møller, J. *J.Mater.Chem.*, *4*, 1861 (1994); (b) Charlton, A.; Kilburn, J.D.; Underhill, A.E.; Webster, M. *Acta Cryst. C52*, 2441 (1996).

35. Mueller-Westerhoff, U.T.; Yoon, D.I.; Plourde, K. *Mol.Cryst.Liq.Cryst.*, *183*, 291 (1990)

36. Barrière, F.; Camire, N.; Geiger, W.E.; Mueller-Westerhoff, U.T.; Sanders, R. *J.Am.Chem.Soc.*, *124*, 7262 (2002).

37. (a) Kisch, H.; Nüsslein, F.; Zenn, I. *Z.Anorg.Allg.Chem.*, *600*, 67 (1991); (b) Persaud, L.; Langford, C.H. *Inorg.Chem.*, *24*, 3562 (1985); (c) Senftleber, F.C.; Geiger, W.E. *J.Am.Chem.Soc.*, *97*, 5018 (1975); (d) Fitzmaurice, J.C.; Slawin, A.M.Z.; Williams, D.J.; Woolins, J.D.; Lindsay, A.J. *Polyhedron*, *9*, 1561 (1990);

38. (a) Eisenberg, R.; Ibers, J.A. *Inorg.Chem.*, *4*, 605 (1965); (b) Eisenberg, R.; Ibers, J.A.; Clark, R.J.H.; Gray, H.B. *J.Am.Chem.Soc.*, *86*, 113 (1964); (c) Hove, M.J.; Hoffman, B.M.; Ibers, J.A. *J.Chem.Phys.*, *56*, 3490 (1972); (d) Dance, I.G.; Solstad, P.J.; Calabrese, J.C. *Inorg.Chem.*, *12*, 2161 (1973); (e) Coucouvanis, D.; Baenziger, N.C.; Johnson, S.M. *Inorg.Chem.*, *13*, 1191 (1974); (f) Kobayashi, A.; Sasaki, Y. *Bull.Chem.Soc. Jpn.*, *50*, 2650 (1977); (g) Endres, H.; Keller, H.J.; Moroni, W.; Nöthe, D. *Acta Cryst.*, *B35*, 353 (1979); (h) Kisch, H.; Fernández, A.; Wakatsuki, Y.; Yamazaki, H. *Z.Naturforsch.*, *40B*, 292 (1985); (i) Schmauch, G.; Knoch, F.; Kisch, H. *Chem.Ber.*, *127*, 287 (1994); (l) Handrosch, C.; Dinnebier, R.; Bondarenko, G.; Bothe, E.; Heineman, F.; Kisch, H. *Eur.J.Inorg.Chem.*, 1259 (1999); (m) Yang, L.-F.; Peng, Z.-H.; Cheng, G.-Z.; Peng, S.-M. *Polyhedron*, *22*, 3547 (2003).

39. (a) Fritchie, C.J. *Acta Cryst.*, *20*, 107 (1966); (b) Manoharan, P.T.; Noordik, J.H.; de Boer, E.; Keijzers, C.P. *J.Chem.Phys.*, *74*, 1980 (1981); (c) Ramakrishna, B.L.; Manoharan, P.T. *Inorg.Chem.*, *22*, 2113 (1983); (d) Mahadevan, C.; Seshasayee, M.; Murthy, B.V.R.; Kuppusamy, P.; Manoharan, P.T. *Acta Cryst.*, *C39*, 1335 (1983); (e) Brunn, K.; Endres, H.; Weiss, J. *Z.Naturforsch.*, *42B*, 1222 (1987); (f) Hobi, M.; Zürcher, S.; Gramlich, V.; Burckhardt, U.; Mensing, C.; Spahr, M.; Togni, A. *Organometallics*, *15*, 5342 (1996); (g) Zürcher, S.; Gramlich, V.; von Arx, D.; Togni, A *Inorg.Chem.*, *37*, 4015 (1998); (h) Robertson, N.; Bergemann,C.; Becker, H.; Agarwal, P.; Julian, S.R.; Friend, R.H.; Hatton, N.J.; Underhill, A.E.; Kobayashi, A. *J.Mater.Chem.*, *9*, 1713 (1999); (i) Ribera, E.; Rovira, C.; Veciana, J.; Tarrés, J.; Canadell, E.; Rousseau, R.; Molins, E.; Mas, M.; Schoeffel, J.-P.; Pouget, J.-P.; Morgado, J.; Henriques, R.T.; Almeida, M. *Chem.Eur.J.*, *5*, 2025 (1999); (l) Robertson, N.; Roehrs, S.; Akutagawa, T.; Nakamura, T.; Underhill, A.E. *J.Chem.Res., (S)* 54 (1999); (m) Ren, X.; Lu, C.; Liu, Y.; Zhu, H.; Li, H.; Hu, C.; Meng, Q. *Transition Met.Chem.* *26*, 136 (2001); (n) Ren, X-M.; Li, H.-F.; Wu, P.-H.; Meng, Q.-J. *Acta Cryst.*, *C57*, 1022 (2001); (o) Xie, J.; Ren, X.; Song, Y.; Zou, Y.; Meng, Q. *J.Chem.Soc., Dalton Trans.*, 2868 (2002); (p) Xie, J.; Ren, X.; Song, Y.; Zhang, W.; Liu, W.; He, C.; Meng, Q. *Chem.Commun.*, 2346 (2002); (q) Ren, X.; Chen, Y.; He, C.; Gao, S. *J.Chem.Soc., Dalton Trans.*, 3915 (2002); (r) Xie, J.; Ren, X.; Gao, S.; Zang, W.; Liu, W.; Meng, Q.; Yao, Y. *Eur.J.Inorg.Chem.*, 2393 (2003).

40. (a) Mines, T.E.; Geiger, W.E. *Inorg.Chem.*, *12*, 1189 (1973); (b) Geiger, W.E.; Allen, C.S.; Mines, T.E.; Senftleber, F.C. *Inorg.Chem.*, *16*, 2003 (1977).

41. (a) Lemke, M.; Knoch, F.; Kisch, H. *Acta Cryst.*, *C49*, 1630 (1993); (b) Batsanov, A.S.; Moore, A.J.; Robertson, N.; Green, A.; Bryce, M.R.; Howard, J.A.K.; Underhill, A.E. *J.Mater.Chem.*, *7*, 387 (1997); (c) Bois, H.; Connelly, N.G.; Crossley, J.G.; Guillorit, J.-C.; Lewis, G.R.; Orpen, G.; Thornton, P. *J.Chem.Soc., Dalton Trans.*, 2833 (1998).

42. (a) Domingos, A.; Henriques, R.T.; Gama, V.; Almeida, M.; Vieira, A.L.; Alcacer, L. *Synth.Met.*, *27*, B411 (1988); (b) Hursthouse, M.B.; Short, R.L.; Clemenson, P.I.; Underhill, A.E. *J.Chem.Soc., Dalton Trans.*, 67 (1989).

43. (a) Güntner, W.; Gliemann, G.; Klement, U.; Zabel, M. *Inorg.Chim.Acta*, *165*, 51 (1989); (b) Clegg, W.; Birkby, S.L.; Banister, A.J.; Rawson, J.M.; Wait, S.T.; Rizkallah, P.; Harding, M.M.; Blake, A.J. *Acta Cryst.*, *C50*, 28 (1994); (c) Bigoli,

F.; Deplano, P.; Mercuri, M.L.; Pellinghelli, M.A.; Pilia, L.; Pintus, G.; Serpe, A.; Trogu, E.F. *Inorg.Chem.*, *41*, 5241 (2002); (d) Bremi, J.; Gramlich, V.; Caseri, W.; Smith, P. *Inorg.Chim.Acta*, *322*, 23 (2001).

44. (a) Ahmad, M.M.; Turner, D.J.; Underhill, A.E.; Kobayashi, A.; Sasaki, Y.; Kobayashi, H. *J.Chem.Soc., Dalton Trans.*, 1759 (1984); (b) Clemenson, P.I.; Underhill, A.E.; Hursthouse, M.B.; Short, R.L. *J.Chem.Soc., Dalton Trans.*, 61 (1989); (c) Bremi, J.; d'Agostino, E.; Gramlich, V.; Caseri, W.; Smith, P. *Inorg.Chim.Acta*, *335*, 15 (2002).

45. Forrester, J.D.; Zalkin, A.; Templeton, D.H. *Inorg.Chem.*, *3*, 1507 (1964).

46. (a) Plumlee, K.W.; Hoffman, B.M.; Ratjack, M.T.; Kannewurf, R.C. *Solid State Commun.*, *15*, 1651 (1974); (b) Plumlee, K.W.; Hoffman, B.M.; Ibers, J.A. *J.Chem.Phys.*, *63*, 1926 (1975); (c) Snaathorst, D.; Doesburg, H.M.; Perenboom, J.A.A.; Keijzers, C.P. *Inorg.Chem.*, *20*, 2526 (1981); (d) Kuppusamy, P.; Ramakrishna, B.L.; Manoharan, P.T. *Inorg.Chem.*, *23*, 3886 (1984); (e) Mahadevan, C.; Seshasayee, M. *J.Cryst.Spectrosc.Res.*, *14*, 215 (1984); (f) Fettouhi.M.; Ouahab, L.; Hagiwara, M.; Codjovi, E.; Kahn, O.; Constant-Machado, H.; Varret, F. *Inorg.Chem.*, *34*, 4152 (1995).

47. (a) Stach, J.; Kirmse, R.; Sieler, J.; Abram, U.; Dietzsch, W.; Böttcher, R.; Hansen, L.K.; Vergoossen, H.; Gribnau, M.C.M.; Keijzers, C.P. *Inorg.Chem.*, *25*, 1369 (1986); (b) Lahner, S.; Wakatsuki, Y.; Kish, H. *Chem.Ber.*, *120*, 1011 (1987).

48. Noordik, J.H.; Beurskens, P.T. *J.Cryst.Mol.Struct.*, *1*, 339 (1971).

49. Vlček, Jr.,A.; Vlček, A.,A. *J.Electroanal.Chem.*, *125*, 481 (1981)

50. (a) Forrester, J.D.; Zalkin, A.; Templeton, D.H. *Inorg.Chem.*, *3*, 1500 (1964); (b) Schmauch, G.; Knoch, F.; Kisch, H. *Chem.Ber.*, *128*, 303 (1995); (c) Schmauch, G.; Chihara, T.; Wakatsuki, Y.; Hagiwara, M.; Kisch, H. *Bull.Chem.Soc.Jpn.*, *69*, 2573 (1996).

51. Götz, B.; Knoch, F.; Kisch, H. *Chem.Ber.*, *129*, 33 (1996).

52. Das, S.K.; Biswas, D.; Maiti, R.; Sarkar, S. *J.Am.Chem.Soc.*, *118*, 1387 (1996).

53. (a) McCleverty, J.A.; Locke, J.; Ratcliff, B.; Wharton, E.J. *Inorg.Chim.Acta*, *3*, 283 (1969); (b) Das, S.K.; Chaudhury, P.K.; Biswas, D.; Sarkar, S. *J.Am.Chem.Soc.*, *116*, 9061 (1994).

54. Stiefel, E.I.; Bennett, L.E., Dori, Z.; Crawford, T.H.; Simo, C.; Gray, H.B. *Inorg.Chem.*, *9*, 281 (1970).

55. (a) Sequeira, A.; Bernal, I. *J.Cryst.Mol.Struct.*, *3*, 157 (1973); (b) Stiefel, E.I.; Dori, Z.; Gray, H.B. *J.Am.Chem.Soc.*, *89*, 3353 (1967); (c) Brown, G.F.; Stiefel, E.I. *Inorg.Chem.*, *12*, 2140 (1973).

56. Lewis, G.R.; Dance, I. *J.Chem.Soc., Dalton Trans.*, 3176 (2000).

57. (a) Maiti, R.; Shang, M.; Lappin, A.G. *Chem.Commun.*, 2349 (1999); (b) Maiti, R.; Shang, M.; Lappin, A.G. *J.Chem.Soc., Dalton Trans.*, 244 (2002).

58. Antolini, L.; Fabretti, A.C.; Franchini, G.; Menabue, L.; Pellacani, G.C.; Desseyen, H.O.; Dommisse, R.; Hofmans, H.C. *J.Chem.Soc., Dalton Trans.*, 1921 (1987).

59. DePamphilis, B.V., Jones, A.G., Davis, M.A.; Davison, A. *J.Am.Chem.Soc.*, *100*, 5570 (1978).

60. (a) Coucouvanis, D.; Baenziger, N.C.; Johnson, S.M. *J.Am.Chem.Soc.*, *95*, 3875 (1973); (b) Bellitto, C.; Bonamico, M.; Fares, V.; Imperatori, P.; Patrizio, S.

J.Chem.Soc., Dalton Trans., 719 (1989); (c) Golič, L.; Bulc, N.; Dietzsch, W. *Acta Cryst.*, *C42*, 811 (1986); (d) Imamura, T.; Ryan, M.; Gordon, G.; Coucouvanis, D. *J.Am.Chem.Soc.*, *106*, 984 (1984); (e) Knatzidis, M.G.; Baenziger, N.C.; Coucouvanis, D. *Inorg.Chem.*, *24*, 2680 (1985); (f) Mennemann, K.; Mattes, R. J.Chem.Res. (S), 102 (1979); (g) Mattes, R.; Weber, H. *Z.Anorg.Allg.Chem.*, *474*, 216 (1981); (h) Butler, K.R.; Snow, M.R. *Inorg.Nucl.Chem.Letters*, *8*, 541 (1972).

61. (a) Coucouvanis, D.; Hollander, F.J.; West, R.; Eggerding, D. *J.Am.Chem.Soc.*, *96*, 3006 (1974); (b) Coucouvanis, D.; Holah, D.G.; Hollander, F.J. *Inorg.Chem.*, *14*, 2657 (1975); (c) Bellitto, C.; Bonamico, M.; Fares, V.; Serino, P. *Inorg.Chem.*, *35*, 4070 (1996); (d) Bonnet, J.-J.; Cassoux, P.; Castan, P.; Laurent, J.-P.; Soules, R. *Mol.Cryst.Liq.Cryst.*, *142*, 113 (1987); (e) Arrizabalaga, Ph.; Bernardinelli, G.; Geoffroy, M.; Castan, P. *Inorg.Chim.Acta*, *154*, 35 (1988); (f) Arrizabalaga, Ph.; Bernardinelli, G.; Castan, P.; Geoffroy, M.; Soules, R. *C.R.Acad.Sc.Paris* Série II, 559 (1987).

62. (a) Heuer, W.B.; Pearson, W.H. *J.Chem.Soc., Dalton Trans.*, 3507 (1996); (b) Heuer, W.B.; Pearson, W.H. *Polyhedron*, 15, 2199 (1996).

63. Venkatalakshmi, N.; Varghese, B.; Lalitha, S.; Williams, R.F.X.; Manoharan, P.T. *J.Am.Chem.Soc.*, *111*, 5748 (1989).

64. Belo, D.; Alves, H.; Lopes, E.B.; Duarte, M.T.; Gama, V.; Henriques, R.T.; Almeida, M.; Pérez-Benítez, A.; Rovira, C.; Veciana, J. *Chem.Eur.J.*, *7*, 511 (2001).

65. Pozo-Gonzalo, C.; Berridge, R.; Skabara, P.J.; Cerrada, E.; Laguna, M.; Coles, S.J.; Hursthouse, M.B. *Chem.Commun.*, 2408 (2002).

66. Veldhuizen, Y.S.J.; Veldman, N.; Spek, A.L.; Cassoux, P.; Carlier, R.; Mulder, M.J.J.; Haasnoot, J.G.; Reedijk, J *J.Chem.Soc., Dalton Trans.*, 2989 (1998).

67. (a) Sun, S.Q.; Zhang, B.; Wu, P.J.; Zhu, D.B. *J.Chem.Soc., Dalton Trans.*, 227 (1997); (b) Olk, R.-M.; Dietzsch, W.; Köhler, K.; Kirmse, R.; Reinhold, J.; Hoyer, E.; Golič, L.; Olk, B. *Z.Anorg.Allg.Chem.* 567, 131 (1988); (c) Liu, S.-G.; Wu, P.-J.; Li, Y.-F.; Zhu, D.-B. *Phosphorus, Sulfur and Silicon*, *90*, 219 (1994); (d) Liu, S.-G.; Liu, Y.-Q.; Liu, S.-H.; Zhu, D.-B. *Synth. Met.*, *74*, 137 (1995); (e) Vicente, R.; Ribas, J.; Zanchini, C.; Gatteschi, D.; Legros, J.-P., Faulmann, C.; Cassoux, P. *Z.Naturforsch.*, *43b*, 1137 (1988); (f) Fettouhi, M.; Ouahab, L.; Codjovi, E.; Kahn, O. *Mol.Cryst.Liq.Cryst.*, *273*, 29 (1995); (g) Faulmann, G.; Rivière, E.; Dorbes, S.; Senocq, F.; Coronado, E.; Cassoux, P. *Eur.J.Inorg.Chem.*, 2393 (2003).

68. Yang, X.; Freeman, G.K.W.; Rauchfuss, T.B.; Wilson, S.R. *Inorg.Chem.*, *30*, 3034 (1991).

69. Tanaka, S.; Matsubayashi, G. *J.Chem.Soc., Dalton Trans.*, 2837 (1992).

70. Matsuda, F.; Tamura, H.; Matsubayashi, G. *Inorg.Chim.Acta*, *235*, 239 (1999).

71. (a) Faulmann, C.; Errami, A.; Donnadieu, B.; Malfant, I.; Legros, J.-P.; Bowlas, C.; Cassoux, P.; Rovira, C.; Canadell, E. *Inorg.Chem.*, *35*, 3856 (1996); (b) Kato, R.; Kobayashi, H.; Kobayashi, A.; Sasaki, Y. *Bull.Chem.Soc.Jpn.*, *59*, 627 (1986); (c) Tommasino, J.-B.; Pomarede, B.; Medus, D.; de Montauzon, D.; Cassoux, P. *Mol.Cryst.Liq.Cryst.*, *237*, 445 (1993); (d) Nunn, I.; Eisen, B.; Benedix, R.; Kish, H. *Inorg.Chem.*, *33*, 5079 (1994); (e) Olk, R.-M.; Röhr, A.; Sieler, J.; Köhler, K.; Kirmse, R.; Dietzsch, W.; Hoyer, E.; Olk, B. *Z.Anorg.Allg.Chem.*, *577*, 206 (1989);

72. (a) Lindqvist, O.; Sjölin, L.; Sieler, J.; Steimecke, G.; Hoyer, E. *Acta Chem.Scand.*, *A33*, 445 (1979); (b) Steimecke, G.; Sieler, H.-J.; Kirmse, R.; Hoyer, E. *Phosphorus Sulfur*, *7*, 49 (1979); (c) Kushch, N.; Faulmann, C.; Cassoux, P.; Valade, L.; Malfant, I.; Legros, J.-P.; Bowlas, C.; Errami, A.; Kobayashi, A.; Kobayashi, H. *Mol.Cryst.Liq.Cryst.*, *284*, 247 (1996); (d) Bryce, M.R.; Moore, A.J.; Batsanov, A.S.; Howard, J.A.K.; Goldenberg, L.M.; Pearson, C.; Petty, M.C.; Tanner, B.K. *Synth.Met.*, *86*, 1839 (1997); (e) Cronin, L.; Clark, S.J.; Parsons, S.; Nakamura, T.; Robertson, N. *J.Chem.Soc., Dalton Trans.*, 1347 (2001).

73. (a) Lindqvist, O.; Andersen, L.; Sieler, J.; Steimecke, G.; Hoyer, E. *Acta Chem.Scand.*, *A36*, 855 (1982); (b) Kato, R.; Kobayashi, H.; Kobayashi, A.; Sasaki, Y. *Chem.Lett.*, 131 (1985); (c) Kobayashi, H.; Kato, R.; Kobayashi, A.; Sasaki, Y. *Chem.Lett.*, 191 (1985); (d) Groeneveld, L.R.; Schuller, B.; Kramer, G.J.; Haasnoot, J.G.; Reedijk, J. *Recl.Trav.Chim.Pays-Bas.*, *105*, 507 (1986); (e) Mentzafos, D.; Hountas, A.; Terzis, A. *Acta Cryst.*, *C44*, 1550 (1988); (f) van Diemen, J.H.; Groeneveld, L.R.; Lind, A.; de Graaff, R.A.G.; Haasnoot, J.G.; Reedijk, J. *Acta Cryst.*, *C44*, 1898 (1988); (g) Broderick, W.E.; Thompson, J.A.; Godfrey, M.R.; Sabat, M.; Hoffman, B.M. *J.Am.Chem.Soc.*, *111*, 7656 (1989); (h) Reefman, D.; Cornelissen, J.P.; Haasnoot, J.G.; de Graaff, R.A.G.; Reedijk, J. *Inorg.Chem.*, *29*, 3933 (1990); (i) Strzelecka, H,; Vicente, R.; Ribas, J.; Legros, J.-P.; Cassoux, P.; Petit, P.; Andre, J.-J. *Polyhedron*, *10*, 687 (1991); (l) Qi, F.; Xiao-Zeng, Y.; Jin-Hua, C.; Mei-Yun, H. *Acta Cryst.*, *C49*, 1347 (1993); (m) Kochurani; Singh, H.B.; Jasinski, J.P.; Paight, E.S.; Butcher, R.J. *Polyhedron*, *16*, 3505 (1997); (n) Sun, S.; Wu, P.; Zhu, D.; Ma, Z.; Shi, N. *Inorg.Chim.Acta*, *268*, 103 (1998); (o) Malfant, I.; Andreu, R.; Lacroix, P.G.; Faulmann, C.; Cassoux, P. *Inorg.Chem.*, *37*, 3361 (1998); (p) Takamatsu, N.; Akutagawa, T.; Hasegawa, T.; Nakamura, T.; Inabe, T.; Fujita, W.; Awaga, K. *Inorg.Chem.*, *39*, 870 (2000); (q) Mukai, K.; Hatanaka, T.; Senba, N.; Nakayashiki, T.; Misaki, Y.; Tanaka, K.; Ueda, K.; Sugimoto, T.; Azuma, N. *Inorg.Chem.*, *41*, 5066 (2002); (r) Liu, G.; Xue, G.; Yu, W.; Xu, W. *Acta Cryst.*, *C58*, m436 (2002).

74. Valade, L.; Legros, J.-P.; Bousseau, M.; Cassoux, P.; Garbauskas, M.; Interrante, L.V. *J.Chem.Soc., Dalton Trans.*, 783 (1985).

75. Sakamoto, Y.; Matsubayashi, G.-E.; Tanaka, T. *Inorg.Chim.Acta*, *113*, 137 (1986).

76. Matsubayashi, G.-E.; Takahashi, K.; Tanaka, T. *J.Chem.Soc., Dalton Trans.*, 967 (1988).

77. (a) Matsubayashi, G.-E.; Yokozawa, A. *J.Chem.Soc., Dalton Trans.*, 3535 (1990); (b) Nagapetyan, S.S.; Arakelova, E.R.; Belousava, L.V.; Struchkov, Yu.T.; Ukhin, L.Yu.; Shklover, V.E. *Russ.J.Inorg.Chem.*, *33*, 88 (1988); (c) Matsubayashi, G.-E.; Yokozawa, A. *Inorg.Chim.Acta 193*, 137 (1992); (d) Swaminathan, K.; Carroll, P.J.; Kochurani; Singh, H.B.; Bhargava, H.D. *Acta Cryst. C49*, 1243 (1993); (e) Takahashi, M.; Robertson, N.; Kobayashi, A.; Becker, H.; Friend, R.H.; Underhill, A.E. *J.Mater.Chem.*, *8*, 319 (1998).

78. (a) Chohan, Z.H.; Howie, R.A.; Wardell, J.L.; Wilkens, R.; Doidge-Harrison, S.M.S.V. *Polyhedron*, *16*, 2689 (1997); (b) Harrison, W.T.A.; Howie, R.A.; Wardell, J.L.; Wardell, S.M.S.V.; Comerlato, N.M.; Costa, L.A.S.; Silvino, A.C.; de Oliveira, A.I.; Silva, R.M. *Polyhedron*, *19*, 821 (2000).

79. Matsubayashi, G.-E.; Nojo, T.; Tanaka, T. *Inorg.Chim.Acta*, *154*, 133 (1988).

80. Matsubayashi, G.-E.; Maikawa, T.; Nakano, M. *J.Chem.Soc., Dalton Trans.*, 2995 (1993).
81. (a) Matsubayashi, G.-E.; Akiba, K.; Tanaka, T. *Inorg.Chem.*, *27*, 4744 (1988); (b) Broderick, W.E.; McGhee, E.M.; Gofrey, M.R.; Hoffman, B.M. *Inorg.Chem.*, *28*, 2904 (1989).
82. (a) Matsubayashi, G.-E.; Douki, K.; Tamura, H. *Chem.Lett.*, 1251 (1992); (b) Matsubayashi, G.-E.; Douki, K.; Tamura, H., Nakano, M.; Mori, W. *Inorg.Chem.*, *32*, 5990 (1993).
83. (a) Steimecke, G.; Sieler, H.-J., Kirmse, R.; Dietzsch, W.; Hoyer, E., *Phosphorus and Sulfur*, *12*, 237 (1982); (b) Lindqvist, O.; Sjölin, L.; Sieler, J.; Steimecke, G.; Hoyer E. *Acta Chem.Scand.*, *A36*, 853 (1982); (c) Kisch, H.; Nüsslein, F.; Zenn, I *Z.Anorg.Allg.Chem.*, *600*, 67 (1991).
84. Olk, R.-M.; Dietzsch, W.; Kirmse, R.; Stach, J.; Hoyer, E.; Golič, L. *Inorg.Chim.Acta*, *128*, 251 (1987).
85. (a) Bigoli, F.; Deplano, P.; Devillanova, F.A.; Lippolis, V.; Lukes, P.J.; Mercuri, M.L.; Pellinghelli, M.A.; Trogu, E.F. *J.Chem.Soc., Chem.Commun.*, 371 (1995); (b) Bigoli, F; Deplano, P.; Devillanova, F.A.; Ferraro, J.R.; Lippolis, V.; Lukes, P.J.; Mercuri, M.L.; Pellinghelli, M.A.; Trogu, E.F.; Williams, J.M. *Inorg.Chem.*, *36*, 1218 (1997); (c) Arca, M.; Demartin, F.; Devillanova, F.A.; Garau, A.; Isaia, F.; Lelj, F.; Lippolis, V.; Pedraglio, S.; Verani, G. *J.Chem.Soc., Dalton Trans.*, 3731 (1998); (d) Grigiotti, E.; Laschi, F.; Zanello, P.; Arca, M.; Denotti, C.; Devillanova F.A. *Portugaliae Electrochimica Acta*, in press; (e) Bigoli, F.; Deplano, P.; Mercuri, M.L.; Pellinghelli, M.A.; Pintus, G.; Trogu, E.F.; Zonnedda, G.; Wang, H.H.; Williams, J.M. *Inorg.Chim.Acta*, *273*, 175 (1998).
86. (a) Dyachenko, O.A.; Konovalikhin, S.V.; Kotov, A.I.; Shilov, G.V.; Yagubskii, E.B.; Faulmann, C.; Cassoux, P. *J.Chem.Soc., Chem.Commun.*, 508 (1993); (b) Hawkins, I.; Underhill, A.E. *J.Chem.Soc., Chem.Commun.*, 1593 (1990); (c) Schenk, S.; Hawkins, I.; Wilkes, S.B.; Underhill, A.E.; Kobayashi, A.; Kobayashi, H. *J.Chem.Soc., Chem.Commun.*, 1648 (1993); (d) Awaga, K.; Okuno, T.; Maruyama, Y.; Kobayashi, A.; Kobayashi, H.; Schenk, S.; Underhill, A.E. *Inorg.Chem.*, *33*, 5598 (1994).
87. (a) Sellmann, D.; Fünfgelder, S.; Knoch, F.; Moll, M. *Z.Naturforsch. B46*, 1601 (1991); (b) Mahadevan, C.; Seshasayee, M.; Kuppusamy, P.; Manoharan, P.T. *J.Crys.Spectrosc.Res.*, *15*, 305 (1985); (c) Xie, J.-L.; Ren, X-M.; He, C.; Song, Y. Duan, C.-Y.; Gao, S.; Meng, Q-J. *Polyhedron*, *22*, 299 (2003).
88. Rindorf, G.; Thorup, N.; Bjørnholm, T.; Bechgaard, K. *Acta Cryst. C46*, 1437 (1990).
89. (a) Boyde, S.; Ellis, S.R.; Garner, C.D.; Clegg, W. *J.Chem.Soc., Chem.Commun.*, 1541 (1986); (b) Oku, H.; Ueyama, N.; Kondo, M.; Nakamura, A. *Inorg.Chem.*, *33*, 209 (1994); (c) Ueyama, N.; Oku, H.; Nakamura, A. *J.Am.Chem.Soc.*, *114*, 7310 (1992); (d) Colmanet, S.F.; Mackay, M.F. *Austr. J.Chem.*, 40, 1301 (1987).
90. (a) Cowie, M.; Bennett, M.J. *Inorg.Chem.*, *15*, 1595 (1976); (b) Cowie, M.; Bennett, M.J. *Inorg.Chem.*, *15*, 1589 (1976); (c) Martin, J.L.; Takats, J. *Inorg.Chem. 14*, 1358 (1975); (d) Colmanet, S.F.; Williams, G.A.; Mackay, M.F. *J.Chem.Soc., Dalton Trans.*, 2305 (1997).

91. Martin, J.L.; Takats, J. *Inorg.Chem. 14*, 73 (1975).

92. (a) Cervilla, A.; Llopis, E.; Marco, D.; Pérez, F. *Inorg.Chem.*, *40*, 6525 (2001); (b) Cowie, M.; Bennett, M.J. *Inorg.Chem.*, *15*, 1584 (1976).

93. (a) Burrow, T.E.; Morris, R.H.; Hills, A.; Hughes, D.L.; Richards, R.L. *Acta Cryst.*, *C49*, 1591 (1993); (b) Knock, F.; Sellmann, D.; Kern, W. *J.Kristallogr.*, *202*, 326 (1992); (c) Lorber, C.; Donahue, J.P.; Goddard, C.A.; Nordlander, E.; Holm, R.H. *J.Am.Chem.Soc.*, *120*, 8102 (1998); (d) Knock, F.; Sellmann, D.; Kern, W. *J.Kristallogr.*, *205*, 300 (1993).

94. (a) Sawyer, D.T.; Srivatsa, G.S.; Bodini, M.E.; Schaefer, W.P.; Wing, R.M. *J.Am.Chem.Soc.*, *108*, 936 (1986); (b) Henkel, G.; Greiwe, K.; Krebs, B. *Angew.Chem.Int.Ed.Engl.*, *24*, 117 (1985); (c) Greiwe, K.; Krebs, B.; Henkel, G. *Inorg.Chem.*, *28*, 3713 (1989); (d) Eisenberg, R.; Dori, Z.; Gray, H.B.; Ibers, J.A. *Inorg.Chem.*, *7*, 741 (1968); (e) Mazid, M.A.; Razi, M.T.; Sadler, P.J. *Inorg.Chem.*, *20*, 2872 (1981); (f) Bustos, L.; Khan, M.A.; Tuck, D.G. *Can.J.Chem.*, *61*, 1146 (1983).

95. Sellmann, D.; Binder, H.; Häußinger, D.; Heinemann, F.W.; Sutter, J. *Inorg. Chim.Acta*, *300-302*, 829 (2000).

96. (a) Welch, J.H.; Bereman, R.D.; Singh, P. *Inorg.Chem.*, *27*, 3680 (1988); (b) Vance, C.T.; Bereman, R.D.; Bordner, J.; Hatfield, W.E.; Helms, J.H. *Inorg.Chem.*, *24*, 2905 (1985); (c) Faulmann, C.; Cassoux, P.; Yagubskii, E.B.; Vetoshkina, L.V. *New J.Chem.*, *17*, 385 (1993); (d) Tian-Ming, Y.; Jing-Lin, Z.; Xiao-Zeng, Y.; Xiao-Ying, H. *Polyhedron*, *14*, 1487 (1995); (e) Noh, D.-Y.; Lee, H.-J.; Underhill, A.E. *Synth.Met.*, *86*, 1837 (1997); (f) Lee, H.J.; Noh, D.Y. *Polyhedron*, *19*, 425 (2000); (g) Vance, C.T.; Welch, J.H.; Bereman, R.D. *Inorg.Chim.Acta*, *164*, 191 (1989); (h) Schultz, A.J.; Wang, H.H.; Soderholm, L.C.; Sifter, T.L.; Williams, J.M.; Bechgaard, K.; Whangbo, M.-H. *Inorg.Chem.*, *26*, 3757 (1987).

97. (a) Welch, J.H.; Bereman, R.D.; Singh, P.; Haase, D.; Hatfield, W.E.; Kirk, M.L. *Inorg.Chim.Acta*, *162*, 89 (1989); (b) Kim, H.; Kobayashi, A.; Sasaki, Y.; Kato, R.; Kobayashi, H. *Bull.Chem.Soc.Jpn.*, *61*, 579 (1988).

98. (a) Welch, J.H.; Bereman, R.D.; Singh, P.; *Inorg.Chim.Acta*, *163*, 93 (1989); (b) Gritsenko, V.V.; Dyachenko, O.A.; Cassoux, P.; Kotov, A.L.; Laukhina, E.E.; Faulmann, C.; Yagubskii, E.B. *Izv.Akad.Nauk. SSR, Ser.Khim.*, 1207 (1993); (c) Geiser.U.; Schultz, A.J.; Wang, H.H.; Beno, M.A.; Williams, J.M. *Acta Cryst.*, *C44*, 259 (1988).

99. (a) Welch, J.H.; Bereman, R.D.; Singh, P. *Inorg.Chem.*, *27*, 2862 (1988); (b) Livage, C.; Formigué, M.; Batail, P.; Canadell, E.; Coulon, C. *Bull.Soc.Chim.Fr.*, *130*, 761 (1993); (c) Welch, J.H.; Bereman, R.D.; Singh, P. *Inorg.Chem.*, *29*, 68 (1990).

100. (a) Zuo, J.-L.; You, F.; Fun, H.-K. *Polyhedron*, *16*, 1465 (1997); (b) Zuo, J.-L.; Yao, T.-M.; You, F.; You, X.-Z.; Fun, H.-K.; Yip, B.-C. *J.Mater.Chem.*, *6*, 1633 (1996).

101. (a) Nakamura, T.; Nogami, T.; Shirota, Y. *Bull.Chem.Soc.Jpn.*, *60*, 3447 (1987); (b) Kim, H.; Kobayashi, A.; Sasaki, Y.; Kato, R.; Kobayashi, H.; Nakamura, T.; Nogami, T.; Shirota, Y. *Bull.Chem.Soc.Jpn. 61*, 2559 (1988).

102. (a) Lee, H.J.; Noh, D.Y. *J.Mater.Chem.*, *10*, 2169 (2000); (b) Lee, H.J.; Noh, D.Y. *Synth.Met.*, *102*, 1696 (1999).

103. (a) Tomura, M.; Tanaka, S.; Yamashita, Y. *Synth.Met.*, *64*, 197 (1994).

104. (a) Bigoli, F.; Chen, C.-T.; Wu, W.-C.; Deplano, P.; Mercuri, M.L.; Pellinghelli, M.A.; Pilia, L.; Pintus, L.; Serpe, A.; Trogu, E.F. *Chem.Commun.*, 2246 (2001); (b) Bigoli, F.; Deplano, P.; Mercuri, M.L.; Pellinghelli, M.A.; Pintus, L.; Serpe, A.; Trogu, E.F. *J.Am.Chem.Soc.*, *123*, 1788 (2001).

105. (a) Bereman, R.D.; Lu, H. *Inorg.Chim.Acta*, *204*, 53 (1993); (b) Dautel, O.J.; Fourmigué, M. *Inorg.Chem.*, *40*, 2083 (2001); (c) Yao, T.-M.; You, X.-Z.; Li, C.; Li, L.-F. *Acta Cryst.*, *C50*, 67 (1994); (d) Geiser, U.; Tytko, S.F.; Allen, T.J.; Wang, H.H.; Kini, A.M.; Williams, J.M. *Acta Cryst.*, *C47*, 1164 (1991).

106. (a) Charlton, A.; Underhill, A.E.; Kobayashi, A.; Kobayashi, H. *J.Chem.Soc., Dalton Trans.*, 1285 (1995); (b) Charlton, A.; Underhill, A.E.; Malik, K.M.A.; Hursthouse, M.B.; Jørgensen, T.; Becher, J. *Synth.Met.*, 68, 221 (1995).

107. (a) Robertson, N.; Parkin, D.L.; Underhill, A.E.; Hursthouse, M.B.; Hibbs, D.E.; Malik, K.M.A. *J.Mater.Chem.*, *5*, 1731 (1995); (b) Cleary, C.F.; Robertson, N.; Takahashi, M.; Underhill, A.E.; Hibbs, D.E.; Hursthouse, M.B.; Malik, K.M.A. *Polyhedron*, *16*, 1111 (1997); (c) Underhill, A.E.; Robertson, N.; Parkin, D.L. *Synth.Met.*, *71*, 1955 (1995).

108. Fox, S.; Wang, Y.; Silver, A.; Millar, M. *J.Am.Chem.Soc.*, *112*, 3218 (1990).

109. (a) Larsen, J.; Bechgaard, K. *J.Org.Chem.*, *52*, 3285 (1987); (b) Noh, D.-Y.; Mizuno, M.; Choy, J.-H. *Inorg.Chim.Acta*, *216*, 147 (1994).

110. Eduok, E.E.; Krawiec, M.; Buisson, Y.-S.L.; O'Conner, C.J.; Sun, D.; Watson, W.H. *J.Chem.Cryst.*, *26*, 621 (1996).

111. (a) Ganguli, K.K.; Carlisle, G.O.; Hu, H.J.; Theriot, L.J.; Bernal, I. *J.Inorg. Nucl.Chem.*, *33*, 3579 (1971); (b) Mukhopadhyay, S.; Ray, D. *J.Chem.Soc., Dalton Trans.*, 1159 (1993); (c) Boyde, S.; Garner, C.D.; Clegg, W. *J.Chem.Soc., Dalton Trans.*, 1083 (1987); (d) Pignedoli, A.; Peyronel, G.; Antolini, L. *Acta Cryst.*, *B30*, 2181 (1974).

112. (a) Boyde, S.; Garner, C.D.; Enemark, J.H.; Ortega, R.B. *Polyhedron*, *5*, 377 (1986); (b) Boyde, S.; Garner, C.D.; Enemark, J.H.; Ortega, R.B. *J.Chem.Soc., Dalton Trans.*, 297 (1987); (c) Boyde, S.; Garner, C.D.; Enemark, J.H.; Bruck, M.A.; Kristofzki, J.G. *J.Chem.Soc., Dalton Trans.*, 2267 (1987).

113. Kisch, H.; Eisen, B.; Dinnebier, R.; Shankland, K.; David, W.I.F.; Knoch, F. *Chem.Eur.J.*, *7*, 738 (2001).

114. (a) Kubo, K.; Nakano, M.; Hamaguchi, S.; Matsubayashi, G. *Inorg.Chim.Acta*, *346*, 43 (1993); (b) Le Narvor, N.; Robertson, N.; Weyland, T.; Kilburn, J.D.; Underhill, A.E.; Webster, M.; Svenstrup, N.; Becher, J. *Chem.Commun.*, 1363 (1996); (c) Ueda, K.; Goto, M.; Sugimoto, T.; Endo, S.; Toyota, N.; Yamamoto, K.; Fujita, H. *Synth.Met.*, *85*, 1679 (1997); (d) Le Narvor, N.; Robertson, N.; Wallace, E.; Kilburn, J.D.; Underhill, A.E.; Bartlett, P.N.; Webster, M. *J.Chem.Soc., Dalton Trans.*, 823 (1996); (e) Nakazono, T.; Nakano, M.; Tamura, H.; Matsubayashi, G. *J.Mater.Chem.*, *9*, 2413 (1999).

115. (a) Nakano, M.; Kuroda, A.; Maikawa, T.; Matsubayashi, G. *Mol.Cryst.Liq.Cryst.*, *284*, 301 (1996); Nakano, M.; Kuroda, T.; Matsubayashi, G. *Inorg.Chim.Acta*, *254*, 189 (1997).

116. (a) Kumasaki, M.; Tanaka, H.; Kobayashi, A. *J.Mater.Chem.*, *8*, 301 (1998); (b) Kobayashi, A.; Tanaka, H.; Kumasaki, M.; Torii, H.; Narymbetov, B.; Adachi, T.

J.Am.Chem.Soc., *121*, 10763 (1999); (c) Kobayashi, A.; Kumasaki, M.; Tanaka, H. *Synth.Met.*, *102*, 1768 (1999).

117. Kobayashi, A.; Tanaka, H.; Kobayashi, H. *J.Mater.Chem.*, *11*, 2078 (2001).
118. Tanaka, H.; Okano, Y.; Kobayashi, H.; Suzuki, W.; Kobayashi, A. *Science*, *291*, 285 (2001).

Chapter 2

Mono- and Dinuclear Molybdenum and Tungsten Complexes: Electrochemistry, Optics and Magnetics

Jon A. McCleverty and Michael D. Ward

School of Chemistry, University of Bristol, Cantock's Close, Bristol BS8 1TS, UK

1 INTRODUCTION

In polynuclear complexes, *electronic* interactions between redox-active metal fragments linked by bridging ligands is evidenced by a separation between metal-centred redox couples and the subsequent formation of stable mixed-valence states. Such behaviour is a function of the length, substitution pattern and conformation of the bridging ligand. *Magnetic exchange* interactions between metals in polynuclear complexes are also dependent on the nature of the pathway linking the metal ions [1]. If the metal-based magnetic orbitals are sufficiently close to overlap directly, then the nature of the magnetic interaction depends on their relative symmetry [2] and this has been exploited in the preparation of complexes with predictable magnetic properties [1,3]. If however the magnetic orbitals are too far apart to overlap directly, but require the participation of bridging ligand orbitals to mediate the interaction (a super-exchange process), then the properties of the bridging ligand become as important as they are in mediating electronic interactions. This principle has received relatively little systematic attention for metal complexes, in contrast to extensive work on organic polyradicals whose magnetic properties are a function of structure and topology [4].

Until recently magnetic and electronic interactions in bridged polynuclear compounds have been treated quite separately, despite the fact that both clearly depend on the nature of the ligand through which the interaction is transmitted.

This partly reflects the fact that the complexes most often used to probe electronic interactions are based on kinetically inert, diamagnetic fragments such as Ru^{II} [5] whereas studies on magnetic exchange interactions frequently involve relatively labile first-row transition metals and lanthanide(III) ions in coordination environments where they do not show reversible redox interconversions [6].

However, by development of the chemistry of the kinetically and thermo-dynamically stable dinuclear tris(3,5-dimethylpyrazolyl)borato molybdenum and tungsten nitrosyl and oxo complexes, it has been possible to perform *combined* studies of electronic and magnetic interactions [7]. A number of factors combine to make the two metal fragments {M(NO)Tp*Cl} and {M(O)Tp*Cl} particularly useful for studying electronic and magnetic interactions between metal centres across bridging ligands. First, they are quite easily attached to a range of difunctionalised ligands whose length, conformation and topology can be varied extensively and systematically. Second, both fragments are redox active, allowing the study of electronic interactions by various electrochemical methods (voltammetry, square wave and differential pulse voltammetry) and spectro-electrochemistry. Third, they can be made paramagnetic ($S = 1/2$), enabling the study of magnetic exchange interactions by epr spectroscopy and susceptibility measurements. Fourth, the M–NO and M=O groups provide easily identifiable and strong or moderately strong IR absorptions which are very convenient spectroscopic reporters for monitoring changes in electronic density at the metal centres; a feature which has also been exploited spectroelectrochemically.

2   BASIC STRUCTURAL UNITS

2.1   Mononuclear Complexes

The starting point for the studies of the dinuclear species discussed in this article has been the mononuclear species [M(NO)Tp*XY}], where M = Mo or W and X = Y = anionic ligand [8], [Mo(NO)Tp*ClL] where L = pyridine and related derivatives [8,9,10], [M(O)Tp*ClX] where M = Mo or W and X = phenolato ligand [11,12] and [Mo(O)Tp*ClL] where L = pyridine and related derivatives [13,14], Figure 1a.

The structures of these mononuclear species are well established. Both types of complexes have pseudo-octahedral geometries and there are no substantial distortions. In the nitrosyl species, [M(NO)Tp*XY] (X = Y = anionic ligand), the metal atom may be thought of as formally coordinatively unsaturated, having a 16 valence electron (ve) configuration. This, when combined with coordination of the strong π-acceptor NO, means that the metal is relatively "electron deficient". The Tp* ligand is a good σ-donor but an indifferent π-acceptor and the halides are electronegative (π-donation is possible). The overall effect of this

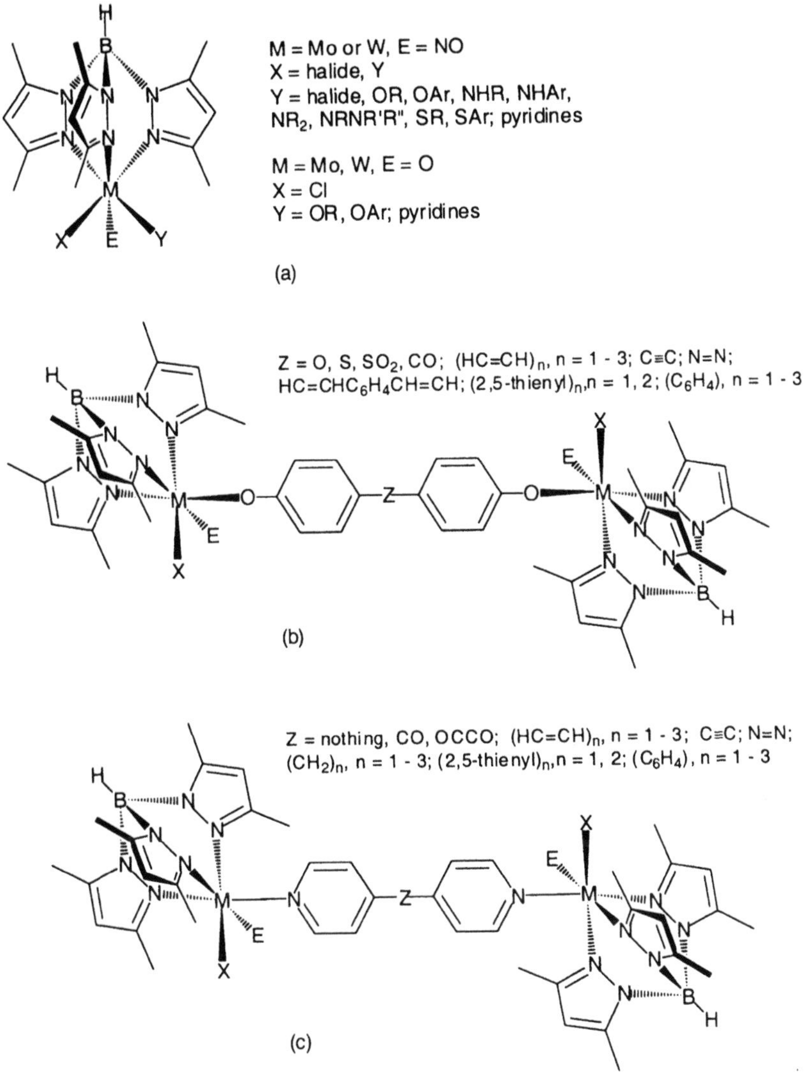

M = Mo or W, E = NO
X = halide, Y
Y = halide, OR, OAr, NHR, NHAr,
NR$_2$, NRNR'R'', SR, SAr; pyridines

M = Mo, W, E = O
X = Cl
Y = OR, OAr; pyridines

(a)

Z = O, S, SO$_2$, CO; (HC=CH)$_n$, n = 1 - 3; C≡C; N=N;
HC=CHC$_6$H$_4$CH=CH; (2,5-thienyl)$_n$,n = 1, 2; (C$_6$H$_4$), n = 1 - 3

(b)

Z = nothing, CO, OCCO; (HC=CH)$_n$, n = 1 - 3; C≡C; N=N;
(CH$_2$)$_n$, n = 1 - 3; (2,5-thienyl)$_n$,n = 1, 2; (C$_6$H$_4$), n = 1 - 3

(c)

Figure 1. Mono- and dinuclear complexes discussed in this chapter: (a) basic building block; (b) dinuclear arene-diolato species, M = Mo or W; E = NO or O; (c) dinuclear dipyridyl complexes, M = Mo, W, E = NO; M = Mo, E = O

particular ligand grouping around the molybdenum makes the {Mo(NO)Tp*}$^{2+}$ group strongly electronegative. If the co-ligands are polarisable, there should be significant $p_\pi \to d_\pi$ donation from the heteroatom E (O, N, S) to the metal in [Mo(NO)Tp*X(ER)] and, indeed, this has been confirmed crystallographically [8]. The Mo–E bond distances in 16 ve complexes are significantly shorter than those predicted for a single bond, even when taking into account the difficulties of calculating covalent radii for the $Mo(NO)^{3+}$ group. Typical Mo–E bond lengths are 1.90Å for Mo–O, 1.95Å for Mo–NHR, 2.31Å for Mo–SR and 2.20Å for Mo–py. The $p_\pi \rightarrow d_\pi$ donation also requires some rehybridisation of the E atom in anionic ligands, leading to an expansion of the Mo–E–C(R) bond angle, which was also observed.

Only a few oxo-molybdenum(V) and tungsten(V) complexes have been structurally characterised, viz. [Mo(O)Tp*Cl(OC$_6$H$_4$SC$_6$H$_4$OMe)] [15] and [W(O)Tp*Cl(OAr)] where Ar = OC$_6$H$_5$, OC$_6$H$_4$OMe–p and OC$_6$H$_4$N=NC$_6$H$_4$OH [12]. The Mo–O distance in the mononuclear species was 1.96Å, and the W–OAr distances averaged 1.95Å. The structure of the Mo(IV) species [Mo(O)Tp*Cl(py)] has also been reported [14].

In the nitrosyl complexes, the *formal* oxidation state of the metal in [M(NO)Tp*Cl(py)] is I (assuming coordination by NO^+, low-spin d^5 configuration), and in [M(NO)Tp*Cl(OAr)] it is II (low-spin d^4 configuration). In [M(NO)Tp*Cl(py)] and [M(NO)Tp*Cl(OAr)] the metal has 17 and 16 valence electrons, respectively [8]. One-electron reduction of [M(NO)Tp*Cl(py)] affords a diamagnetic monoanion formally containing M^0. In the oxo species [M(O)Tp*Cl(OPh)] the metal is in oxidation state V (d^1 configuration), and in [Mo(O)Tp*Cl(py)] is it IV (d^2 configuration).

2.2   Dinuclear Complexes

The types of complexes investigated are shown in Figure 1b and 1c. Although no structures containing metal nitrosyl termini bridged by arenediolato ligands could be determined, those of three complexes containing dipyridino bridging ligands have been reported. The crystallographic analyses were bedevilled by nitrosyl/halide group disorder and so accurate bond lengths and angles were not always accessible. However, it is the conformation of the bridging ligands which is the most interesting and informative structural feature, and this has always been clear.

The 4,4'-dipyridyl ligand in the oxo-molybdenum(IV) species [{Mo(O)Tp*Cl}$_2$(bipy)] [13] was essentially planar. Bridging ligand planarity also occurred in [{Mo(NO)Tp*Cl}$_2$(pyCH=CHpy)] [16] and [{Mo(NO)Tp*Cl}$_2$-{py(CH=CH)$_4$py}] [10]. This ligand conformation clearly facilitates overlap of metal d_{xy} and pyridine π-orbitals, and probably contributes significantly to the strong solvatochromism observed in the nitrosyl complexes, behaviour which is consistent with the highly polarising nature of the {Mo(NO)Tp*}$^+$ group [17].

The structures of several dinuclear oxo-metal arenediolates have been determined successfully. These include [M(O)Tp*Cl}$_2$(1,4-O$_2$C$_6$H$_4$)] (M =

Mo [18] or W [12]) and [{Mo(O)Tp*Cl}$_2$(1,3-O$_2$C$_6$H$_4$)] [18], [{M(O)Tp*Cl}$_2$ (OC$_6$H$_4$N=NC$_6$H$_4$O)] (M = Mo [19] or W [12]), [{Mo(O)Tp*Cl}$_2$(OC$_6$ H$_4$EC$_6$H$_4$O] where E = S [15] or CO [19]. The intermetallic distances in [Mo(O)Tp*Cl}$_2$(1,4-O$_2$C$_6$H$_4$)] and its 1,3-benzenediolate analogue were 8.74 and 7.57 Å, respectively. In the structure of [{Mo(O)Tp*Cl}$_2$(OC$_6$H$_4$SC$_6$H$_4$O], the C–S–C bond angle is 103° which, by comparison with the structure of the mononuclear [Mo(O)Tp*Cl(OC$_6$H$_4$SC$_6$H$_4$OH] [15], shows that there is no steric interaction between the two metal nitrosyl fragments. Furthermore, the two phenyl rings in the bridging ligand tend towards orthogonality with an angle between their mean planes of 77.4° which implies that it is not possible in this conformation for a π-symmetry orbital (*d* or *p*) on the S atom to interact with both phenyl π systems simultaneously. This is significant in terms of the electrochemical behaviour of this complex. In [{Mo(O)Tp*Cl}$_2${OC$_6$H$_4$(C O)C$_6$H$_4$O}] the bridging ligand is buckled, with an angle of 52° between the mean planes of the two aromatic rings, this type of conformation being typical of benzophenone groups in the solid state. In contrast, however, the bridging ligands in [{M(O)Tp*Cl}$_2$(OC$_6$H$_4$N=NC$_6$H$_4$O)] (M = Mo or W) is planar. The intermolecular Mo···Mo separations in [{Mo(O)Tp*Cl}$_2${OC$_6$H$_4$(CO)C $_6$H$_4$O}] and [{Mo(O)Tp*Cl}$_2$(OC$_6$H$_4$N=NC$_6$H$_4$O)] being 11.87 and 14.61°, respectively.

3   ELECTROCHEMICAL BEHAVIOUR

3.1   Mononuclear Complexes

The mononuclear nitrosyl complexes constitute a three-membered electron transfer chain (E = NO):

$$[M(E)Tp*X(L)]^- \quad \underset{+e^-}{\overset{-e^-}{\rightleftharpoons}} \quad [M(E)Tp*X(L)] \quad \underset{+e^-}{\overset{-e^-}{\rightleftharpoons}} \quad [M(E)Tp*X(L)]^+$$

where L may be anionic, (halide, OR, NHR, SR) or neutral (pyridines). The formation potentials for the monoanionic species are dependent on the metal, the donor atom of L and on the substituents attached to L, and to a limited extent on X (Cl, Br and I). However, the potentials for formation of [Mo(E)Tp*X(L)]$^+$ do not vary much when L = py [8].

The electrochemical properties of the Mo^V complexes [Mo(O)Tp*Cl(OR)] (R = alkyl or aryl), [Mo(O)Tp*Cl(SR)] and [Mo(O)Tp*{Q(CH$_2$)$_n$Q}] (Q = O or S) are similar to the nitrosyls, monoanionic Mo^{IV} and monocationic Mo^{VI} species being detected [20]. Their tungsten analogues, which are restricted to three phenolato complexes at present, behave similarly although the potentials are significantly more negative than their molybdenum analogues [12].

3.2　Dinuclear Complexes

The three systems [{Mo(NO)Tp*Cl}$_2$(Q)], where Q = arene-diolato or dipyridyl ligand, [{M(O)Tp*Cl}$_2$(arene-diolate)] (M = Mo or W) and [{Mo(O)Tp*Cl}$_2$(dipy)] exist, in principle, within five-membered electron transfer chains [9,10,13,15,18,19,21].

$$[\{M(E)Tp^*Cl\}_2Q]^{2-} \rightleftharpoons [\{M(E)Tp^*Cl\}_2Q]^- \rightleftharpoons [\{M(E)Tp^*Cl\}_2Q]$$
$$\rightleftharpoons [\{M(E)Tp^*Cl\}_2Q]^+ \rightleftharpoons [\{M(E)Tp^*Cl\}_2Q]^{2+}$$

Within the electron transfer chain involving [{Mo(NO)Tp*Cl}$_2$(dipy)]^z there are two distinct one-electron reduction processes [z = −1 and −2] in which the interaction between the metal centres, measured by the difference between the first and second formation potentials (ΔE_f), can be very large; *e.g.* up to *ca.* 800 mV corresponding to a comproportionation constant of *ca.* 10^{12} in CH_2Cl_2. However, there are two apparently coincident oxidations, due to formation of [{Mo(NO)Tp*Cl}$_2$(dipy)]$^{2+}$. The opposite situation exists for [{Mo(O)Tp*Cl}$_2$-(arenediolate)]^z where reduction seems to involve coincident one-electron processes, leading to a dianion, whereas the one-electron oxidations are clearly separate, the interactions again being significant (up to *ca.* 1000 mV in CH_2Cl_2).

In general, the electrochemical behaviour of the tungsten complexes [{W(E)-Tp*Cl}$_2$(arenediolate)] is identical to that of their molybdenum analogues, although the redox potentials are shifted anodically by 450–550 mV [8,12].

Electrochemical examination of the nitrosyl dipyridyl and oxo benzenediolato complexes revealed that the connecting group Z (Figure 1b, c) has a significant influence on the separation between the first and second oxidation potentials [10,18,19,22]. The efficiency of Z in facilitating delocalisation across the complexes is thienyl > ethenyl > phenyl, the thienyl units being particularly effective in maintaining large separations between the potentials over long distances (similar effects were detected in a more limited set of nitrosyl complexes [22]). An ethenyl spacer was almost as effective as an ethynyl spaces at mediating electronic interactions, but the azo spacer, N=N, is significantly worse. As with the dipyridyl nitrosyl complexes, interaction between the two redox centres diminishes significantly as the pathway between them lengthens.

The electrochemical behaviour of the molybdenum(IV) complexes [{Mo(O)Tp*Cl}$_2$(dipy)] is not so well-defined [13]. Both [{Mo(O)Tp*Cl}$_2$-(4,4-dipy)] and [{Mo(O)Tp*Cl}$_2$(pyCH=CHpy)] exhibited two coincident oxidation processes, corresponding to generation of [{Mo(O)Tp*Cl}$_2$(dipy)]$^{2+}$ which may be regarded as containing two Mo^V centres. Comparable processes were observed with mononuclear analogues [13,20]. While the reduction behaviour of [{Mo(O)Tp*Cl}$_2$(4,4-dipy)] is complicated, that of its bis(pyridino)ethene analogue is relatively clear, the species being reduced in two separate one-electron steps. It is tempting to assign these two electron transfers to the generation of Mo^{III}, but the CVs of related mononuclear species showed no evidence for Mo^{IV}/Mo^{III} couples. It is more reasonable to regard the reduction of [{Mo(O)Tp*Cl}$_2$(pyCH=CHpy)] as a bridging ligand centred process analogous to that observed in comparable nitrosyl species.

4   ELECTRONIC STRUCTURE

A knowledge of the frontier orbitals in the *mononuclear* [M(E)Tπ*ΞΛ] species (E = NO or O) is instructive in understanding the electrochemical behaviour of the nitrosyl and oxo species discussed above. The M–NO and M=O axis is defined as the z-axis and all other ligand donor atoms occupy the

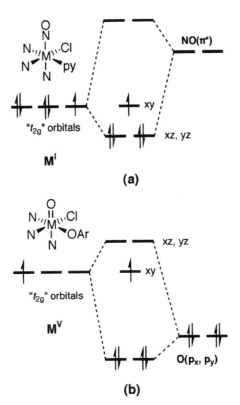

Figure 2. Ground-state electronic configurations for mononuclear complexes containing (a) {M^I(NO)Tp*Cl} bound to a pyridine ligand, and (b) {M^V(O)Tp*Cl}$^+$ bound to a phenolate ligand.

other co-ordinates, as shown in Figure 2 [7]. In [M(NO)Tp*XL] the NO ligand is a strong π-acceptor and so the two empty NO π^* orbitals overlap with the d_{xz} and d_{yz} orbitals, thereby lowering them but leaving the d_{xy} orbital unchanged. Therefore the electronic configuration of the metal centre in the nitrosyl complex is $d_{xz}^2 d_{yz}^2 d_{xy}^0$ when L = anionic ligand (*e.g.* phenolate) and $d_{xz}^2 d_{yz}^2 d_{xy}^1$ when L = pyridine. In the $\{M(O)Tp^*Cl\}^+$ fragment, the oxo ligand is a strong π-donor, the filled oxygen p_x and p_y orbitals overlapping with the metal d_{xz} and d_{yz} orbitals, raising them but again leaving the d_{xy} orbital unchanged. The M^V electronic configuration is therefore $d_{xy}^1 d_{xz}^0 d_{yz}^0$. In both cases, the d_{xy} orbital is of the correct symmetry for d_π–p_π overlap with the donor atom of the phenolato or pyridine ligand.

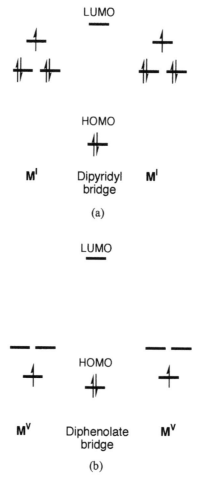

Figure 3. Qualitative MO scheme for (a) [{Mo(NO)Tp*Cl}$_2$(dipyridyl)] and (b) [Mo(O)Tp*Cl}$_2$(arenediolate)]

Using the frontier orbitals developed for the mononuclear core nitrosyl and oxo group, a simple qualitative molecular orbital scheme can be constructed for the dinuclear nitrosyl dipyridino and oxo arenediolato species, Figure 3 [7].

4.1   Nitrosyl Complexes

In the *nitrosyl* complexes, the relatively high-energy d_{xy} orbitals lie just below the low-lying π^* orbitals of the dipyridyl bridging ligand over which delocalisation can occur. Addition of a further electron to each d_{xy} orbital will raise them in energy even closer to the π^* level so that on reduction the added d_{xy} electrons are very effectively delocalised over the bridging group. This doubly reduced species can be represented by two extreme forms:$[\{M^0\}–L–\{M^0\}]$ and $[\{M^I\}–(L^{2-})–\{M^I\}]$, such an arrangement recognising the well-documented ability of oligo-pyridine ligands to reduce relatively easily [23]. In such a situation the two added electrons will be relatively close to each other, and the electrostatic repulsion will be strong, hence leading to relatively large values of ΔE_f, as observed. On the other hand, oxidation of the metals, with the concomitant formation of positively charged species, will cause the d_{xy} orbitals to drop significantly below the π^* levels, so that the d_{xy} electrons are much more metal-localised. Furthermore, the HOMO of the bridging dipyridyl is much lower in energy than the d_{xy} orbitals and therefore cannot participate in stabilising the oxidised species. The doubly oxidised species may be represented as $[\{M^{II}\}–L–\{M^{II}\}$ and there will be little contribution from the $[\{M^L–(L^{2+})–\{M^I\}]$ form. A consequence of this is that oxidation substantially reduces the interaction between the two metal centres and the formation potentials appear coincident. The value of ΔE_f is close to 36 mV (the statistical result for two redox centres with the same potentials) which is not usually resolved using cyclic voltammetry but can be extracted from differential pulse measurements [24].

4.2   Oxo Metal(V) Complexes

The situation in the dinuclear oxo diphenolato species is the opposite of that in the nitrosyls discussed above. There are significant separations between the two oxidation potentials in the CVs of dinuclear diphenolato-bridged complexes, and a near coincidence of the reduction processes. The d_{xy} orbital is thought to lie just above the HOMO of the bridging ligand, which is relatively high in energy because it carries a formal double negative charge. Oxidation will lower the energy of the d_{xy} orbital, and the positive charges will therefore be partly delocalised onto the bridging ligand. In contrast, reduction to Mo^{IV} will raise the d_{xy} orbitals above the bridging ligand HOMO so that delocalisation is decreased. The bridging ligand LUMO is too high in energy to participate, so the reductions are essentially metal-localised. This means that the reduced fragments interact minimally, and this is reflected in the coincident formation potentials ($\Delta E_f \approx 36$ mV). For the fully oxidised species [{Mo(O)Tp*Cl}$_2$(arenediolate)]$^{2+}$, two extreme canonical forms can be written: [{Mo^{VI}}–{arenediolate}$^{2-}$–{Mo^{VI}}] and [{Mo^V}–{quinone}^{0}–{Mo^V}]: *i.e.*, {arenediolate}$^{2-}$ is oxidised to a quinone. That oxidation of the arenediolato ligand could play a significant rôle in the description of these oxidised complexes is consistent with the known propensity of *para*-substituted arenediols to be oxidised to quinones and has important consequences for the behaviour of this class of complex, particularly their electrochromic properties [25]. Some differences in the electrochemical behaviour between molybdenum and tungsten oxo arenediolato analogues may be anticipated, since the tungsten orbitals will lie at energies significantly lower than their molybdenum counterparts, leading to less effective overlap with the bridging ligand π-orbitals.

4.3   Dinuclear Nitrosyl Arenediolates

The electronic structure of [{M(NO)Tp*X}$_2$(arenediolate) is currently less well understood. However, it is clear that the LUMO of the complex involves a mixture of d_{xy} and bridging ligand π-orbital character. On reduction this orbital is populated, and the strong dependence of E_f on the nature of the bridging group can be understood. Oxidation, however, is less clear. Both isomers of [{Mo(NO)Tp*Cl}$_2$(O$_2$C$_6$H$_4$)] exhibit two distinct oxidation processes but it is not yet clear from ZINDO and DFT calculations what happens to the HOMO of the dinuclear species on stepwise oxidation. The HOMO is a mixture of π*-NO, d_{xz}, d_{yz} and some bridging ligand π-character, and is not simply an orbital solely comprised of bridging ligand contributions [16].

5   SPECTROELECTROCHEMISTRY.

The simple electronic considerations outlined above show that the bridging ligands play a significant role in the electrochemistry of the dinuclear species discussed here. This poses the question as to what extent quinonoidal forms might be involved in the reduction of the nitrosyls and the oxidation of the arenediolates, respectively.

If oxidation led to electron loss from the metal centres, pathway (a) in Figure 4, the electronic spectra of the doubly oxidised species (containing two Mo^{VI} centres) should not be substantially different to those of the parent, except that the LMCT transitions should be stronger. The alternative to metal-based oxidation is ligand oxidation (pathway (b), Figure 4):

$$[M^V\text{–(arenediolate)–}M^V] \rightleftharpoons [M^V\text{–(semiquinone)–}M^V]^{1+}$$
$$\rightleftharpoons [M^V\text{–(quinone)–}M^V]^{2+}$$

If this occurred, semiquinone and then quinone-based $\pi \rightarrow \pi^*$ transitions should appear in addition to MLCT bands. The most effective way to probe this issue is by spectroelectrochemistry in the UV/VIS/NIR region.

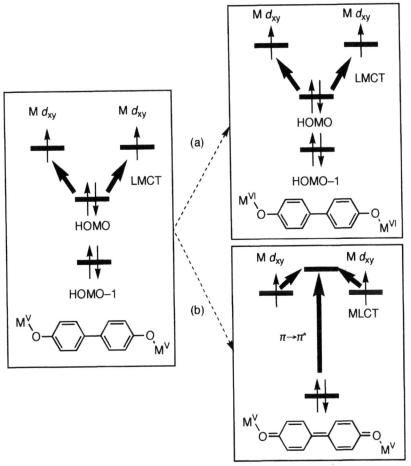

Figure 4. Generation of doubly oxidised forms of dinuclear oxometal complexes: metal-centred , (a), and ligand-centred oxidations, (b); and likely electronic transitions.

5.1   *Dinuclear Nitrosyl Complexes Containing Dipyridyl Ligands*

As described in 3.2, these complexes are oxidised in two simultaneous one-electron steps, but are reduced sequentially in two one-electron processes.

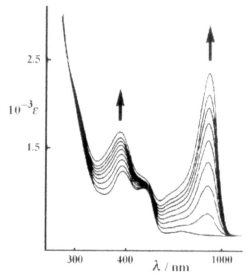

Figure 5. Spectroelectrochemistry of [Mo(NO)Tp*Cl(py)] (ε in dm^3 mol^{-1} cm^{-1})

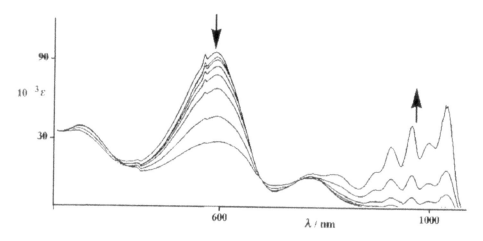

Figure 6. Spectroelectrochemistry of [{Mo(NO)Tp*Cl}$_2${py(CH=CH)$_3$py}] (ε in dm^3 mol^{-1} cm^{-1}).

As the mononuclear and dinuclear nitrosyl complexes are reduced to monoanions, strong bands in the NIR region evolved [10,26]. The behaviour of [Mo(NO)Tp*Cl(py)] on reduction in dichloromethane is illustrated in Figure 5. Of the original metal-to-ligand charge-transfer (MLCT) bands, the one at highest energy almost doubles in intensity but, more dramatically, a strong new band appears at 804 nm. This is also an MLCT absorption, occurring at lower energy because the energy of the metal-centred *dπ* orbitals increases on reduction bringing them closer to the ligand-centred acceptor LUMO orbitals, as described in 3.1. Even more remarkable events occur on reduction of the dinuclear species [{Mo(NO)Tp*Cl}$_2$(dipy)]{E = (CH=CH)$_3$}]. Very strong charge transfer bands evolve, but further into the NIR region than in the mononuclear species (Figure 6). The intensities and structure of these bands are actually characteristic of polyene chromophores, but the transitions are at lower energies than would normally be expected. This is probably because the LUMO of the neutral complex has substantial ligand character, and the shift to lower energy may occur because the separation of the HOMO and LUMO in the neutral molecule is much larger than the separation of the LUMO and the second lowest unoccupied orbital. These results show that reduction of the dinuclear dipyridyl nitrosyl species is a bridging ligand-dominated process.

5.2   *Dinuclear Oxo-Metal Diphenolato Complexes*

The species [{M(O)Tp*Cl}$_2$(arenediolate] (Figure 1b, Z = O) may be oxidised sequentially to a mono- and a dianion, and reduced, usually in two coincident one electron steps, to a dianion. Spectroelectrochemical studies provided unexpected information about the nature of the oxidised species, however [15,19]. The lowest energy electronic transition in neutral mononuclear species [Mo(O)Tp*Cl(OAr)] is a phenolate →Mo LMCT process, occurring in the region 500-700 nm (see Figure 7 for a typical example). On oxidation, this evolves into two stronger peaks, one at *ca.* 550 nm, and the other close to the near-IR region, between 700 and 900 nm, both being enhanced LMCT processes.

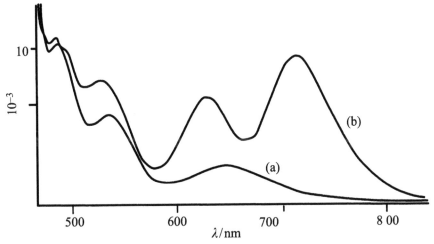

Figure 7. Electronic spectra of (a) [Mo(O)Tp*Cl(OC$_6$H$_4$OMe–*p*)] and (b) [Mo(O)Tp*Cl(OC$_6$H$_4$OMe–*p*)]$^+$. (ε in dm^3 mol^{-1} cm^{-1})

From a wealth of magnetic and EPR spectroscopic data, it is quite clear that in the dinuclear species [{M(O)Tp*Cl}$_2$(arenediolate)] the neutral precursors contain two oxometal(V) groups connected by an arenediolato bridge [15,19,27]. According to the MO scheme shown in Figure 4(b), the highest energy occupied orbitals are the singly occupied d_{xy} on each metal, the bridging ligand HOMO lying just below this. The first oxidation process would therefore be expected to be metal-based, although it would only require a small change in the relative energies of ligand frontier orbitals following the first oxidation to change this picture for subsequent oxidations. The spectral behaviour of [{Mo-(O)Tp*Cl}$_2${O(C$_6$H$_4$)$_n$O}] on oxidation, Figure 8, is very similar to that of the mononuclear species [Mo(O)Tp*Cl(OAr)], *i.e.* the strong near-IR transitions are typical of phenolate $\rightarrow$ Mo^{VI} LMCT processes, and this is supported by ZINDO calculations on the doubly oxidised dinuclear species [15]. It is also clear that the spectra of [{Mo(O)Tp*Cl}$_2${O(C$_6$H$_4$)$_n$O}]$^{n+}$ are quite different to those of related free polyphenylene quinones, whose $\pi \rightarrow \pi^*$ transitions are much higher in energy.

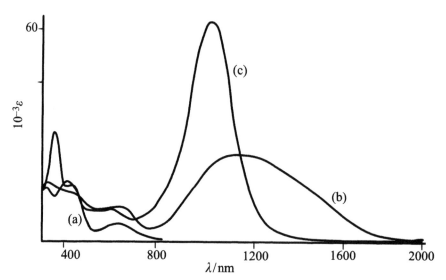

Figure 8. Electronic spectra of (a) [{Mo(O)Tp*Cl}$_2${O(C$_6$H$_4$)$_3$O}], (b) [{Mo(O)-Tp*Cl}$_2$O(C$_6$H$_4$)$_3$O}]$^+$ and (c) [{Mo(O)Tp*Cl}$_2${O(C$_6$H$_4$)$_3$O}]$^{2+}$ (ε in dm^3 mol^{-1} cm^{-1})

Furthermore, the absorption maxima for the free quinone series becomes progressively red-shifted as the quinone lengthens, which does not happen in the spectra of [{Mo(O)Tp*Cl}$_2${O(C$_6$H$_4$)$_n$O}]$^{2+}$, and while λ_{max} for the latter is similar in energy for both the mono- and dications, free semiquinones and related quinones behave quite differently. So it seems clear that oxidation of $[\{Mo(O)Tp^*Cl\}_2\{O(C_6H_4)_nO\}]$ is predominantly a metal-based process, *i.e.* as implied by pathway (a), Figure 4.

The spectroelectrochemical behaviour of $[\{Mo(O)Tp^*Cl\}_2(OC_6H_4N=NC_6H_4O]$ (Figure 9) is different to that of $[\{Mo(O)Tp^*Cl\}_2\{O(C_6H_4)_nO\}]$ [19]. The spectrum of $[\{Mo(O)Tp^*Cl\}_2(OC_6H_4N=NC_6H_4O]^+$ exhibits a principal absorption maximum at λ_{max} = 1268 nm, characteristic of a phenolato $\rightarrow Mo^{VI}$ LMCT process, whereas that of $[\{Mo(O)Tp^*Cl\}_2 (OC_6H_4N=NC_6H_4O]^{2+}$ has λ_{max} = 409 nm, suggesting that the second oxidation leads to the formation of a bridging quinone. The occurrence of this intense transition in the visible region in the dication is more like a $\pi \rightarrow \pi^*$ transition of a free quinone than a phenolate $\rightarrow Mo^{VI}$ transition which occurs in the near-IR region. ZINDO calculations on this and closely related species are broadly consistent with this view [15,19].

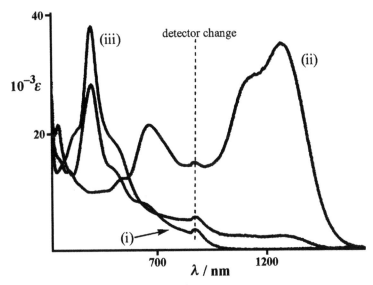

Figure 9. Electronic spectra of $[\{Mo(O)Tp^*Cl\}_2(OC_6H_4N=NC_6H_4O]^n$: (i) n = 0; (ii) n = +1; (iii) n = +2.

So the oxidative behaviour of $[\{Mo(O)Tp^*Cl\}_2\{O(C_6H_4)_nO\}]$ is significantly different to that of $[\{Mo(O)Tp^*Cl\}_2OC_6H_4N=NC_6H_4O]$ as a result of an internal charge redistribution associated with the second oxidation. In other words, one-electron oxidation of {Mo^{VI}–(arenediolate)–Mo^V} affords {Mo^{VI}–(arenediolate)–Mo^{VI}} with the former but {Mo^V–(quinone)–Mo^V} with the latter. This form of internal charge redistribution is not unprecedented, occurring in dithiolene, catecholato, dipyridyl and nitrosyl complexes. This behaviour is sometimes referred to as "non-innocence" and occurs in complexes in which the metal *and* at least one ligand is capable of redox behaviour, and when significant mixing of the HOMO (usually metal-based) and the LUMO (usually ligand-based), which are very close in energy, is possible [25].

The lowest energy transition in the electronic spectrum of [W(O)Tp*Cl(OPh)], the phenolate $\rightarrow W^V$ LMCT, occurs at 424 nm [12], at higher energy than that in the comparable Mo^V complexes. This is expected because of the relatively higher energy of the W $5d$ orbitals compared to molybdenum $4d$. On oxidation, the LMCT band red-shifted to 490 nm because the metal orbital is lowered in energy on conversion to W^{VI}. The intensity of this transition also increases significantly on oxidation because the receiving d orbital is now empty so that the transition dipole moment is increased. All of this is entirely consistent with the behaviour of [Mo(O)Tp*Cl(OPh)] on oxidation.

The electronic spectrum of [{W(O)Tp*Cl)$_2$(4,4'-OC$_6$H$_4$C$_6$H$_4$O)] is broadly similar to that of its molybdenum analogue [12]. The characteristic phenolate $\rightarrow W^V$ transition, which occurs at 491 nm (ε = 4900 M^{-1} cm^{-1}), red-shifts and gains in intensity on oxidation to [{W(O)Tp*Cl)$_2$(4,4'-OC$_6$H$_4$C$_6$H$_4$O)]$^+$ (λ_{max} = 738 nm, ε = 10000 M^{-1} cm^{-1}) and then blue-shifts and becomes even more intense on generation of [{W(O)Tp*Cl)$_2$(4,4'-OC$_6$H$_4$C$_6$H$_4$O)]$^{2+}$ (λ_{max} = 640 nm, ε = 26000 M^{-1} cm^{-1}). Again, this behaviour is entirely consistent with the molybdenum analogue, *viz.* that the oxidation processes are substantially metal-centred. The spectroelectrochemical oxidation behaviour of [{W(O)Tp*Cl}$_2$- (OC$_6$H$_4$N=NC$_6$H$_4$O)], however, is different to that of its molybdenum analogue. The electronic spectrum of the neutral species is complicated by the occurrence of bridging ligand absorptions which mask the phenolate $\rightarrow W^V$ LMCT band but, on successive oxidation to the mono- and dications, intense phenolate $\rightarrow W^{VI}$ transitions evolved at 664 and 615 nm, respectively. That there are no accompanying intense absorptions in the NIR region shows that the two oxidations are metal-based and that the bridging ligand is *not* oxidised to a quinonoidal form. In other words, [{W(O)Tp*Cl}$_2$(OC$_6$H$_4$N=NC$_6$H$_4$O)] exhibits "innocent" behaviour on oxidation. This behaviour reflects the less positive oxidation potentials of the tungsten complex when compared to its molybdenum analogue, and the ambiguity between metal- and ligand-centred oxidation is removed, with the metal centres clearly oxidising before the arenediolate ligand.

6 ELECTROCHROMIC NIR PROPERTIES OF DINUCLEAR MOLYBDENUM SPECIES

The unexpected evolution of strong near-IR absorption bands on oxidation of the dinuclear oxomolybdenum diphenolato complexes described above raised the possibility that some of these complexes might function as electrochromic near-IR dyes. Of particular interest was the discovery that a number of these oxidised complexes absorb in the region of the spectrum of interest for fibre-optic data transmission using silica fibres, Figure 10. [19,26]. These fibres have absorption minima at *ca.* 1300 and 1550 nm.

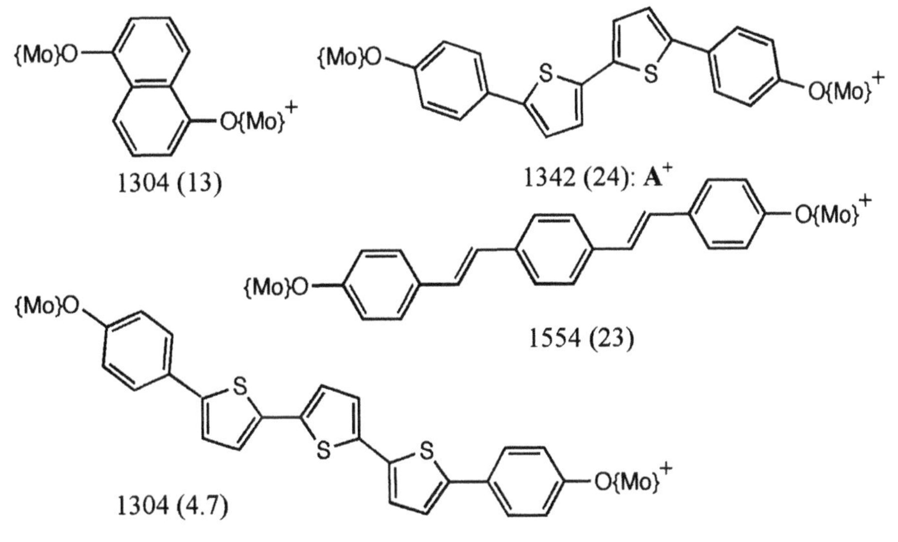

Figure 10. Near-IR absorption maxima for [{Mo(O)Tp*Cl}$_2$(bridge)]$^+$; wavelength in nm (e x 10^{-3} dm^3 M^{-1} cm^{-1}); {Mo} ≡ {Mo(O)Tp*Cl}

In order to test the viability of such species as potential electrochromic near-IR switches, it was necessary to ascertain whether the neutral species and the reduced (dianionic) species are optically transparent in the region around 1300 or 1550 nm, and also if there was a significant separation between the first and second oxidation potential. Spectroelectrochemical examination of these species showed that A$^+$ (Figure 10) was suitable for this testing. The electronic spectra of the mono-cationic, neutral and dianionic species are shown in Figure 11, and it is clear that both the neutral and dianionic species are transparent in the range 900 - 2000 nm [18].

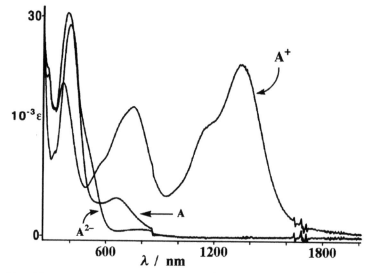

Figure 11. Electronic spectra of [{Mo(O)Tp*Cl}$_2${OC$_6$H$_4$(C$_4$H$_2$S)$_2$C$_6$H$_4$O}]^z, **A**^{z}, z = 0, +1, –2, in CH$_2$Cl$_2$ at –30°C.

The monocationic species has two strong absorptions, at 752 and 1342 nm which are due to phenolate → Mo^{VI} LMCT transitions. On further oxidation to the dication, the lower energy transition moves to *ca* 1200 nm, but this species gradually decomposes over *ca.* 30 min. Fortunately, the oxidation potentials for generation of the mono- and di-cation are sufficiently separate (+0.44 and +0.69 V) that it is possible to generate the former without significantly producing the latter. These optical and electrochemical properties indicated that it was possible to construct a near-IR optical switch operating by rapid and reversible switching between neutral **A** and monocationic **A**$^+$.

A two-electrode optically transparent thin-layer electrode (OTTLE) cell was constructed using two ITO-coated glass slides [28]. The absorption of the cell containing **A** dissolved in acetonitrile was monitored at 1160 nm while a stepped potential was applied to the electrodes. A growth in absorption was observed as the potential was increased from 0 to +1.0 V, but significant decomposition of the complex occurred at potentials greater than 1.6 V due to the formation of the unstable dication **A**$^{2+}$. Repeated stepping of the applied potential between 0 and 1.5 V showed reversible optical switching; a process which was repeatable over several thousand cycles before degradation set in. Furthermore, by varying the higher potential step during these experiments within the range of 1.0 - 1.5 V, the maximum absorbance of the material at a given wavelength could be varied. In other words, the extent of attenuation of the incident light beam could be controlled by adjusting the potential. This meant that the material had potential as a variable optical attenuator (VOA) device (in plain English: a dimmer switch!), and subsequent tests showed that the degree of optical power output attenuation was comparable to the best that was currently available from alternative technologies [29].

We had earlier noticed and have commented above that on reduction of the dinuclear nitrosyl dipyridyl complexes, strong near-IR charge transfer absorptions were generated [10]. Although this behaviour has not been investigated in detail, spectroelectrochemical studies of mononuclear [Mo(NO)Tp*Cl(pyR)] (pyR = 3- or 4-substituted pyridine) showed that some reduced (monoanionic) species also exhibited significant near-IR MLCT absorptions when R = 3-CN, 3-COMe and 3-COPh [26]. It therefore appears that some of these nitrosyl compounds also have potential as electrochromic dyes for use in electro-optic switching in the near-IR.

7   MAGNETIC PROPERTIES OF DINUCLEAR COMPLEXES

7.1   *EPR Spectral Characteristics*

The EPR spectra of dinuclear molybdenum nitrosyl complexes of the type shown in Figure 1c (E = NO) provided the first evidence that magnetic exchange was also an important feature of this class of compounds. The spectra frequently show significant and useful hyperfine coupling due to molybdenum nuclei with I (nuclear spin) $\gg 0$, and for species containing the $\{Mo(NO)\}^{2+}$ core, the hyperfine coupling constant A_{Mo} is *ca.* 5.0 mT [7]. However, when the unpaired spins are coupled to two molybdenum nuclei, the hyperfine coupling essential halved, to *ca.* 2.5 mT. This effect arises in a dinuclear complex containing two unpaired electrons via magnetic exchange interaction between the electron spins localised on individual metal centres, such that $|J| \gg A_{Mo}$, where J is the energy of the exchange interaction and A_{Mo} is the energy of the electron–nuclear hyperfine interactions [30]. Such "exchange coupled" spectra are well-documented in nitroxide di-radicals, where coupling of both electrons to both nitrogen nuclear spins occurs, even across short saturated spacers where there is no possibility of delocalisation of the electrons [30]. The energy of the hyperfine interaction is very small, the coupling between a single electron localised on one molybdenum centre (5.0 mT) being less than 0.01 cm^{-1}, which corresponds to an exchange interaction far too small to be measured by magnetic susceptibility methods. So even a very weak magnetic exchange interaction can give rise to an exchange-coupled spectrum. However, the sign of J has no influence on the spectrum at ambient temperatures: ferromagnetic and antiferromagnetic interactions will give the same spectrum as long as $|J|$ is above the very small lower limit.

Most of the dinuclear nitrosyl complexes shown in Figure 1c (E = NO) exhibited an exchange-coupled spectrum indicating that $|J| \gg A_{Mo}$ and $A_{Mo} \approx 2.5$ mT. If $A_{Mo} \approx 5.0$ mT in dinuclear species it was clear that the unpaired spins were isolated on each metal centre, and there was no magnetic exchange. However, intermediate situations could occur where $|J| \approx A_{Mo}$, the EPR spectra then having a second order appearance typical of weak exchange. Such situations arise when the bridging ligands are very long, or when there was a twist in the bridge, such as occurs when C_6H_4–C_6H_4 groups are incorporated [31]. These results show clearly that magnetic coupling is mediated by the bridging ligands.

The EPR spectra of mononuclear [Mo(O)Tp*Cl(OPh)] also consisted of a superimposed singlet and sextet with $A_{Mo} \approx 5.0$ mT, while the dinuclear species with relatively long bridging ligands (*e.g.* $(C_6H_4)_3$, –$(CH=CH)_3$) were like their nitrosyl counterparts with $A_{Mo} \approx 2.5$ mT [10,19,27]. With shorter ligands, the room-temperature spectra were broad and uninformative, except that a half-field $\Delta m_S = 2$ transition was detectable at 77 K, which provided further evidence of magnetic exchange.

EPR spectra of [{W(O)Tp*Cl}$_2$(1,4-OC$_6$H$_4$O)] and [{W(O)Tp*Cl}$_2$(4,4'-OC$_6$H$_4$C$_6$H$_4$O)] were similar to their molybdenum analogues, although the signals were much broader and the intensities of the $\Delta m_S = 2$ transitions were more intense with respect to the main signal [12].

7.2   *Magnetic Susceptibility Studies of Nitrosyl and Oxo Complexes*

While it is clear that magnetic exchange can occur, the EPR spectral evidence cannot provide information concerning the sign or magnitude of *J*. This may only be obtained by solid state magnetic susceptibility measurements, which have been made of a wide range of nitrosyl and oxo molybdenum complexes. Some data are given in Table 1.

The results obtained from [{Mo(NO)Tp*Cl}$_2$(dipyridyl)] show an alternation in the sign of *J* as the bridging ligand changes from the 4,4' through 4,3' to 3,3' geometry. The data indicate that the magnetic interaction switches from antiferromagnetic to ferromagnetic as the number of atoms in the pathway between the interacting spins changes from even to odd [7]. This suggested a spin-polarisation mechanism for propagation of the exchange interaction, as shown in Figure 12, a situation which is consistent with the spin exchange interaction detected by epr spectroscopy. The spin-polarisation mechanism arises from the Longuet-Higgins molecular orbital model for conjugated alternant hydrocarbons [32] which results in ferromagnetic coupling between two radicals separated by an *m*-phenylene bridge (odd atom pathway), and antiferromagnetic behaviour in *p*-phenylene diradicals (even atom pathway).

Table 1. Exchange coupling constants as a function of bridging ligand and metal.

Bridging ligand	M	J^a (cm^{-1})	Bridging ligand	M	J^a (cm^{-1})
(4,4′-dipyridyl structure)	Mo	−33	(1,4-phenylenedioxy structure)	Mo	−80
				W	−55
(pyridyl structure)	Mo	+0.8	(4,4′-biphenyldioxy structure)	Mo	−13.2
				W	−8.0
(bipyridyl structure)	Mo	−1.5	(methyl-biphenyldioxy structure)	Mo	−2.8
(methyl-bipyridyl structure)	Mo	−3.5	(1,3-phenylenedioxy structure)	Mo	+9.8
(pyridyl–E–pyridyl structure)			(O–phenyl–E–phenyl–O structure)		
E = HC=CH	Mo	−18	E = N=N	Mo	−12.8
				W	−10.7
E = (HC=CH)$_4$	Mo	−6.6	E = HC=CH	Mo	−3.6
			E = C≡C	Mo	−7.0
			E = (thiophene)	Mo	−12.8
			E = CO	Mo	−1.1

^{a} J determined using the exchange spin Hamiltonian in the form $H = -J(S_1S_2)$

The results in Table 1 also show that J is dependent on the conformation of the bridging ligand. Comparing the data obtained from the 4,4′-dipyridyl-bridged complex (where there is a twist of *ca.* 26° between the two pyridine rings) with that from the 2,2′-dimethyl-4,4-dipyridyl analogue (where the twist is 90°), it is clear that the magnitude of J decreases by about 90% as the two halves of the bridging ligand approach orthogonality. This is further evidence that the spin-polarisation is propagated via the delocalised π-system of the bridging ligand.

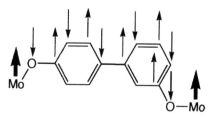

Even atom pathway ≡ antiferromagnetism   Odd atom pathway ≡ ferromagnetism

Even atom pathway ≡ antiferromagnetism   Odd atom pathway ≡ ferromagnetism

Figure 12. The spin-polarisation mechanism over dipyridyl and 4,4'-diphenolato ligands.

The magnetic behaviour of the oxomolybdenum complexes follows the same general pattern, as can also be seen from Table 1. Once again, if there is an even-atom pathway between the two spins, the exchange interaction is antiferromagnetic and if it is odd, the behaviour is ferromagnetic. Furthermore, when a twist is imposed between two halves of the bridge, as can be seen in comparing the complexes containing $OC_6H_4C_6H_4O$ and the 2,2'-dimethyl analogue $OC_6H_3(Me)C_6H_3(Me)O$, the exchange interaction is substantially diminished.

However, more detailed studies have revealed that the magnetic behaviour of $[\{Mo(O)Tp*Cl\}_2(1,4\text{-}OC_6H_4O)]$ is more satisfactorily accounted for by a super-exchange mechanism, but these same calculations confirmed that spin polarisation is an appropriate way in which to interpret the behaviour of the isomer $[\{Mo(O)Tp*Cl\}_2(1,3\text{-}OC_6H_4O)]$ [33].

The susceptibility data obtained from $[\{W(O)Tp*Cl\}_2(\text{arenediolate})]$ were comparable with their molybdenum analogues [12]. Only species having an even number of atoms in the bridging ligand systems were investigated. All coupling constants indicated intramolecular antiferromagnetic exchange, and that the value of *J* decreased as the bridging ligand increased and became twisted. However, the values of *J* for the tungsten complexes were lower than the values of their molybdenum analogues. This seems initially surprising since a comparison of the crystal structures of $[\{M(O)Tp*Cl\}_2(1,4\text{-}OC_6H_4O)]$

revealed that the geometric arrangement of metal fragments with respect to the bridging ligand is actually better optimised to promote magnetic exchange when M = W than when M = Mo [13]. The structural arrangements should therefore lead to more efficient super-exchange in the tungsten complex. However, J(W) < J(Mo), which presumably arises because there is less effective interaction between the W $5d_{xy}$ orbitals and the π-system of the bridging ligand than that involving the Mo $4d_{xy}$ orbitals. These results are consistent with the electrochemical data discussed in 3.2.

8   CONCLUSIONS

The results described above represent a major part of perhaps the single most coherent and extensive study of the effect of bridging ligands in metal complexes on the transmission of electronic and magnetic interactions. These interactions are exceptionally strong in the dinuclear nitrosyl and oxo molybdenum complexes mentioned above because of the near ideal matching in both symmetry and energy of the relevant metal orbitals with those of the bridging ligands.

Spectroelectrochemical studies of the first and second oxidation of [{Mo(O)Tp*Cl}$_2$(arenediolate)] revealed that the first charge transfer process is metal-based (giving {Mo^V(μ-arenediolate)Mo^{VI}} species), the second process being either metal-based ({Mo^{VI}(μ-arenediolate)Mo^{VI}}) or, because of internal charge redistribution, ligand-based giving quinonoidal species ({Mo^V(μ-quinone)Mo^V}). The tungsten analogues engaged only in metal-based oxidation behaviour.

The oxomolybdenum(V) complexes [{Mo(O)Tp*Cl}$_2$(μ-arenediolate)] exhibited strong switchable electrochromism in the near-IR region on oxidation, shown by the appearance of large MLCT bands whose position (λ_{max} 1200-1550 nm) depends on the arene-diolato link. This led to the development of an electrochromic molybdenum complex whose strong near-IR absorption could be reversibly switched between unoxidised and oxidised forms of the complex, the attenuation of incident light being dependent on the applied voltage. This voltage-dependent attenuation of an incident light beam forms a functional basis for a variable optical attenuator device.

The magnetic behaviour of the dinuclear species could be conveniently explained by a spin-polarisation mechanism. Magnetic exchange coupling in the molybdenum complexes was generally greater than that in comparable tungsten complexes, a manifestation of the relative effectiveness of overlap of the bridging ligand and metal orbitals.

ACKNOWLEDGEMENTS

The authors are grateful to the EPSRC (through postgraduate and postdoctoral support) and the European Commission (through Human Capital & Mobility and TMR Network grants) for providing the essential underpinning funding of the work described above. Without the skill, perseverance and dedication of our co-workers, whose names appear in the references below, most of the results could not have been obtained. We are extremely grateful to them, and also to Dr. C. J. Jones (University of Birmingham) who played a major role in much of the early work, Dr. John Maher (University of Bristol) for the EPR measurements, Dr. John Jeffery (University of Bristol) for ensuring crystallographic integrity, Professor Dante Gatteschi (University of Florence) and his colleagues for magnetic susceptibility measurements, and Professor Werner Blau (Trinity College, Dublin) who first drew our attention to the significance of electrochromism in the near-IR region.

REFERENCES:

1. Kahn, O. Molecular Magnetism, VCH publishers, Inc., New York, (1993); Bushby, R. J.; Paillaud, J.–L.; Molecular Magnets, in Petty, M. C.; Bryce, M. R.; Bloor, D. (eds.), Introduction to Molecular Electronics, Edward Arnold, London, 7, 2 (1995).
2. Goodenough, J. B. *Phys. Rev., 100,* 564 (1955); Goodenough, J. B. *J. Phys. Chem. Solids 6,* 287 (1958); Kanamori, J. *J. Phys. Chem. Solids, 10,* 87 (1959); Ginsberg, A. P. *Inorg. Chim. Acta Rev., 5,* 45 (1971).
3. Kahn, O. *Struct. Bonding, 68,* 89 (1971) Gordon–Wylie, S. W.; Bominaar, E. L.; Collins, T. J.; Workman, J. M.; Claus, B. L.; Patterson, R. E.; Williams, S. A.; Conklin, B. J.; Yee, G. T.; Weintraub, S. T. *Chem. Eur. J., 1,* 528 (1995).
4. Miller, J. S.; Epstein, A. J. *Angew. Chem., Int. Ed. Engl., 33,* 385 (1994); Iwamura, H. *Pure Appl. Chem., 65,* 57 (1993); Rajca, S. *Chem. Rev., 94,* 871 (1994); Rajca, S.; Rajca, A. *J. Am. Chem. Soc., 117,* 9172 (**1995**); Yoshizawa, K.; Hoffman, R. *Chem. Eur. J., 1,* 403 (1995); Iwamura, H.; Koga, N.; *Acc. Chem. Res., 26,* 346 (1993).
5. Ward, M. D. *Chem. Soc. Rev., 24,* 121 (1995); Ward, M. D. *Chem. Ind.,* 568 (1996); Reimers, J. R.; Hush, N. S. *Inorg. Chem., 29,* 3686 (1990); Ribou, A.-C.; Launay, J.-P.,;Sachtleben, M. L.; Li, H.; Spangler, C. W. *Inorg. Chem., 35,* 3735 (1996); Evans, C. E. B.; Naklicki, M. L.; Revzani, A. R.; White, C. A.; Kondratiev, V. V.; Crutchley, R. J. *J. Am. Chem. Soc., 120,* 13096 (1998); Kaim, W.; Klein, A.; Glockle, M. *Acc. Chem. Res., 33,* 755 (2000).
6. Brunold, T. C.; Gamelin, D. R.; Solomon, E. I.; *J. Am. Chem. Soc., 122,* 8511 (2000); Bernard, S; Yu, P.; Audiere, J.-P.; Rieviere, E.; Clement, R.; Guilhem, J.; Tchernatov, L.; Nakatani, K. *J. Am. Chen. Soc., 122,* 9444 (2000); Escuer, A.; Cano, J.; Goher, M. A. S.; Journaux, Y.; Lloret, F.; Mautner, F. A.; Vicente, R. *Inorg. Chem., 39,* 4688 (2000); Ruiz-Perez, C.; Hernandez-Molina, M.; Lorenzo-Luis, P.;

Lloret, F.; Cano, J.; Julve, M. *Inorg. Chem.*, *39*, 3845 (2000); Triki, S.; Berezovsky, F.; Pala, J. S.; Coronado, E.; Gomez-Garcia, C. J.; Clementa, J. M.; Riou, A.; Molinie, P *Inorg. Chem.*, *39*, 4165 (2000); Cui, Y.; Chen, G.; Ren, J.; Quan, Y. T.; Huang, J. S. *Inorg. Chem.*, *39*, 4165 (2000); Liu, Q. D.; Gao, S.; Li, J. R.; Zhon, Q. Z.; Yu, K. B.; Ma, B. Q.; Zhang, X. X.; Jin, T. Z. *Inorg. Chem.*, *39*, 2488 (2000).

7. McCleverty, J. A.; Ward, M. D., *Acc. Chem. Res.*, *31*, 842 (1998).

8. McCleverty, J. A.; *Chem. Soc. Rev.*, *12*, 331 (1983); McCleverty, J. A.; Ward, M. D.; Jones, C. J. *Comments Inorg.Chem.*, *22*, 293 (2001).

9. Das, A.; Maher, J. P.; McCleverty, J. A.; Navas, J. A.; Ward, M. D.; *J. Chem. Soc., Dalton Trans.*, 681 (1993).

10. McWhinnie, S. L. W.; Thomas, J. A.; Hamor, T. A.; .Jones, C. J.; McCleverty, J. A.; D.Collison, Mabbs, F. E.; .Harding, C. J.; Yellowlees, L. J.; Hutchings, M. G. *Inorg. Chem.*, *35*, 760 (1996).

11. Chang, C. S. J.; Enemark, J. H. *Inorg. Chem.*, *30*, 683 (1991); Basu, P.; Bruck, M. A.; Li, Z.; Dhawan, I. K.;Enemark, J. H. *Inorg. Chem.*, *34*, 405 (1995).

12. Stobie, K. M.; Bell, Z. R.; Munhoven, T. W.; Maher, J. P.; McCleverty, J. A.; Ward, M. D.; McInnes, E. J. L.;Totti, F.; Gatteschi, D. *Dalton Trans.*, 36 (2003).

13. Kassim, M. B.; Paul, R. L.; Jeffery, J. C.; McCleverty, J. A.; Ward, M. D.; *Inorg. Chim. Acta*, *327*, 160 (2002).

14. Roberts, S. A.; Young, C. G.; Kipke, C. A.; Cleland Jr, W. E.; Yamanouchi, K.; Carducci, M. D.; Enemark, J.H. *Inorg. Chem.*, *29*, 3650 (1990).

15. Harden, N. C.; Humphrey, E. R.; Jeffery, J. C.; Lee, S.-M.; Marcaccio, M.; McCleverty, J. A.; Rees, L. H.; Ward, M. D. *J.Chem.Soc., Dalton Trans.*, 2417 (1999).

16. Kassim, M., PhD Thesis, University of Bristol, 2003.

17. Thomas, J. A.; Hutchings, M. G.; Jones, C. J.; McCleverty, J. A. *Inorg. Chem.*, *35*, 289 (1996).

18. Ung, V. Â.; Bardwell, D. A.; .Jeffery, J. C.; Maher, J. P.; McCleverty, J. A.; Ward, M. D.; Williamson, A. *Inorg. Chem.*, *35*, 5290 (1996).

19. Bayly, S. R.; Humphrey, E. R.; de Chair, H.; Paredes, C. G.; Bell, Z. R.; Jeffery, J. C.; McCleverty, J. A.; Ward, M. D.; Totti, F.; Gatteschi, D.; Courric, S.; Steele, B. R.; Screttas, C. G. *J. Chem. Soc., Dalton Trans.*, 1401 (2001).

20. Chang, C. J. S.; Collison, D.; Mabbs, F. E.; Enemark, J. H. *Inorg. Chem.*, *29*, 2261 (1990); Chang, C. S. J.; Pecci, T. J.; Carducci, M. D.; Enemark,J. H. *Inorg. Chem.*, *32*, 4106 (1993); Basu, P.; Bruck, M. A.; Li, Z.; Dhawan, I. K.; Enemark, J. H. *Inorg. Chem.*, *34*, 405 (1995).

21. McDonagh, A. M.; Ward, M. D.; McCleverty, J. A.; *New J. Chem.*, *25*, 1236 (2001).

22. Hock, J.; Cargill Thompson, A. M. W.; McCleverty, J. A.; Ward, M. D. *J. Chem. Soc., Dalton Trans.*, 1257 (1996).

23. Kemp, W. In *Organic Spectroscopy;* Macmillan: London, 154.(1979).

24. Charsley, S. M.; .Jones, C. J.; McCleverty, J. A.; Neaves, B. D.; Reynolds, S. J. *J. Chem. Soc., Dalton Trans.*, 301 (1988).

25. Ward, M. D.; McCleverty, *J. Chem. Soc., Dalton Trans.*, 275 (2002), and references therein.

26. Kowallick, R.; Jones, A. N.; Reeves, Z. R.; Jeffery, J. C.; McCleverty, J. A.; Ward, M. D. *New J.Chem.*, *23*, 915 (1999).

27. Ung, V. Â.; Cargill Thompson, A. M. W.; Bardwell, D. A.; Gatteschi, D.; Jeffery, J. C.; McCleverty, J. A.; Totti, F.; Ward, *M. D. Inorg. Chem.*, *36*, 3447; Ung, V. Â.; Couchman, S.; Jeffery, J. C.; McCleverty, J. A.; Ward, M. D.; Totti, F.; Gatteschi, D. *Inorg. Chem.*, *38*, 365 (1999); Bayly., S. R.; McCleverty, J. A.; Ward, M. D.; Gatteschi, D.; Totti, F. *Inorg. Chem.*, *39*, 1288 (2000).

28. McDonagh, A. M.; Bayly, S. R.; Riley, D. J.; Ward, M. D.; McCleverty, J. A.; Cowan, M. A.; Morgan, C. N.; Varrazza, R.; Penty, R. V.; White, I. H.; *Mater. Chem.*, *12*, 2523 (2000).

29. Cowan, M. A.; Morgan, C. N.; Varrazza, R.; Penty, R. V.; White, I. H.; McDonagh, A. M.; Riley, D. J.; Ward, M. D.; McCleverty, J. A.; *FiberSystems International.*, 14 (2001).

30. Reitz, D. C., Weissman, S. I. *J. Chem. Phys.*, *33,* 700 (1960); Brière, R., Dupeyre, R.-M., Lemaire, H., Morat, C., Raassat, A., Rey, P., *Bull. Soc. Chim. Fr.*, *34*, 4828 (1965).

31. Shonfield, P. K. A., Behrendt, A., Jeffery, J. C., Maher, J. P., McCleverty, J. A., Psillakis, E., Ward, M. D., Western, C. *J. Chem. Soc., Dalton Trans.*, 4341 (1999).

32. Longuet-Higgins, J. C. *J. Chem. Phys.*, *18*, 265 (1950).

33. Bencini, A.; Gatteschi, D.; Totti, F.; Sanz, D. N.; McCleverty, J. A.; Ward, M. D. *J. Phys. Chem.*, *102*, 10545 (1998).

Part 2

Molecular Electroactivation and Electrocatalysis

Chapter 3

The Surface Electrochemistry of Metallophthalocyanines – Dioxygen Reduction

A. B. P. Lever and Yonglin Ma

Dept. of Chemistry, York University
CB 124, 4700 Keele St., Toronto, Ont. Canada M3J1P3

1   INTRODUCTION

The phthalocyanines [1, 2, 4, 32, 34, 60] are an important class of synthetic dyestuffs which became of great commercial significance in the early part of last century and continue to be extensively applied because of their intense color, high thermal and chemical stability and non-toxicity. Another very critical feature of these species is their redox activity, their ability to exist in a wide range of different oxidation states involving oxidation or reduction of the metal and/or phthalocyanine ring [4]. This combination of properties has spawned new applications and potential applications including CD-W disk dyes (information storage), photodynamic therapy, odor removal, removal of sulfur from oils, security applications, molecular metals, fuel cell catalysts, CHEMFET devices and chemical sensors [4, 29, 39, 60].

The solution electrochemical properties of metallophthalocyanines (MPc) have been extensively discussed [4] but these species have also been studied intensively as films on electrode surfaces, as so-called chemically modified electrodes (CMEs) [25,34,45, 68]. Such CMEs are of interest in the development of chemical sensors [60]. In general a redox active target species will interact with the MPc surface in one or more of the MPc oxidation states. We present here a non-comprehensive survey of this surface electrochemical behavior, illustrating it with work drawn mostly but not exclusively from our own laboratory and focusing mainly on dioxygen reduction.

Well defined surface films, with reproducible properties, can generally be obtained simply by dipping a clean graphite electrode surface into a solution of MPc, usually in an organic solvent. Depending on concentration, the time of dipping and the nature of the MPc, a monolayer or a small number of layers may be attached. While different types of graphite may be used, we will discuss here studies with ordinary pyrolytic graphite (OPG, deposited on the edge or basal plane), highly oriented pyrolytic graphite (HOPG) or glassy carbon. In the case of glassy carbon, the CME surface is usually formed by allowing a few microlitres of solvent containing MPc to dry on the surface. Other supports such as carbon paste[10], or MPc dispersed in various polymers [8, 19, 29, 31, 47, 59, 63], while quite commonly investigated are not covered here. While usually called 'chemically modified electrodes', the binding of the MPc to the surface, in most cases, is physical rather than chemical. Nevertheless the attachment is strong. The modified electrode is usually removed from the organic phase and introduced into a cell containing an aqueous electrolyte. Most MPc species are insoluble in water and the CME can be polarized and cycled many times without loss of material to the solution. Even when the MPc is soluble in water, its loss from the CME is usually very slow.

In solution electrochemical studies, species are oxidized or reduced by diffusing to the electrode. The product species will diffuse away from the electrode and, in particular, any reaction product formed from, say a reduced species and a target molecule at the surface of the electrode in the Helmholtz layer [12] will also diffuse away in time and will therefore be difficult to study. In surface electrochemical studies, however, oxidized or reduced products remain on the surface for study. Moreover any product of reaction with a target, providing it is non-gaseous, will also remain attached to the surface and can be probed. This is a very considerable advantage. For example a $M^{III}Pc(-2)$ species might be reduced to a $M^{I}Pc(-2)$ species and then stepwise to a $M^{I}Pc(-6)$ species, or oxidized stepwise to a $M^{IV}Pc(0)$ species, and all of these species will remain on the surface between successive steps, and can be independently assessed for example for chemical sensor, activity. The pH of the electrolyte solution into which the CME is placed, may also be varied, perhaps from 1 to 14, leading to the possible formation of hydroxo species in the basic regime and protonated or hydridic species in the acidic regime. All of these various products can be generated from a single deposition by variation of pH and polarization of the electrode.

A very large number of different target species have been explored with MPc CMEs, far too many to list here [39]; various aspects have been reviewed [11, 34, 60]– essentially any redox active species can, in principle, be assessed by a MPc CME.

Focusing specifically on graphite CMEs, there has been very considerable activity studying oxygen, hydrogen peroxide, many sulfur species, hydrazine, nitrogen oxides, species such as dopamine, and various other biological reagents etc [3, 11, 39]. In this contribution, we shall focus specifically on the reactivity of MPc/graphite CMEs towards molecular oxygen and hydrogen peroxide. The potential application here is for fuel cell anodes. This manuscript is an account of our work on these modified electrodes highlighting some of the unusual features that have been observed. We have explored several active metals known to be catalytically active, specifically iron, cobalt and ruthenium and have investigated the effect of various substituents, particularly sulfonic acid, neopentoxy, crown and perchloro derivatives. Further, the reactivity of dinuclear MPc species, of rhodium and ruthenium, have been explored to ascertain whether, mechanistically, the presence of a dinuclear unit can lead to a different reaction chemistry. Links to the relevant literature are also provided.

2   GENERAL BACKGROUND

Electrochemistry provides the means to probe the many oxidation states that can exist on the CME [4] and usefully to probe the reactivity of these various oxidation states.

Metallophthalocyanines can be sequentially oxidized or reduced at the phthalocyanine ring. The initial species is in the MPc(-2) oxidation state. It is generally facile to oxidize to the MPc(-1) and MPc(0) states and to reduce at least down to MPc(-6) provided one uses solvents with a wide operating window. In addition, reduction or oxidation of the metal will take place in many transition metal MPc species [4]. These processes can be carried out either in solution or on a surface. A typical surface voltammogram is shown in Figure 1, with its pH dependence in Figure 2.

If one exposes an electrocatalytically active surface bound MPc to a target such as dioxygen, an electrocatalytic current may be observed which has much greater magnitude than the redox waves of the surface species (Figure 3). This arises through a cycle which can be generically written as:

$$M^{II}Pc(-2) + e^- \rightarrow M^{I}Pc(-2) \qquad\qquad E^1_{1/2} \qquad\qquad (1)$$

$$M^{I}Pc(-2) + O_2 \leftrightarrow (O_2)M^{III}Pc(-2) \qquad\qquad\qquad (2)$$

$$(O_2)M^{III}Pc(-2) + 2H^+ + e^- \rightarrow H_2O_2 + M^{II}Pc(-2) \qquad E^3_{1/2} \qquad (3)$$

The oxygen is present as a saturated solution in the solvent used, typically about 10^{-3} M, and therefore in much greater concentration than the surface species; hence there is a greatly increased reduction current. A complex of the type shown in Eqn. (2) is assumed to be an intermediate though proof of its identity is difficult to obtain. The position of $E_{1/2}^3$ is often close to $E_{1/2}^1$. It is a catalytic process because the $M^{II}Pc(-2)$ which is formed in reaction (3) is sitting on the electrode at, or negative of the potential, $E_{1/2}^1$. where it will be spontaneously reduced, via equation (1), to $M^IPc(-2)$.

3   METHODOLOGY

The CME is usually prepared by soaking a carefully cleaned graphite surface in an organic phase solution of the phthalocyanine for a few seconds or minutes. Cycling the potential may help in some cases. Depending on the MPc and time of soaking, a sub-monolayer, monolayer, or multi-layer surface may be generated by this procedure. The active surface can be evaluated by measuring the charge under a wave of known character, i.e., usually known to be one-electron in nature, and, knowing the area of the MPc molecular species, evaluating the number of active molecules deposited on a known geometric area. A deposited multilayer surface may not all be active throughout the layers [57].

Once an active surface has been obtained, it is usual first to evaluate the redox behavior of that surface, i.e to explore which oxidation states of the MPc species are accessible and to identify them. Such information is critical before exposing the surface to a target of interest. Figure 1 portrays an example of an MPc/OPG surface (MPc = iron(III) tetrasulfonated phthalocyanine) exhibiting four redox processes [56, 58, 76]. It is not the intent in this article to discuss in depth how such redox processes are identified. Usually one obtains the corresponding solution phase voltammogram and employs spectroelectrochemistry to obtain the electronic spectrum of each oxidation state and thereby identify the redox processes, e.g. [69,77]. One then generally assumes that the corresponding processes on the graphite surface have the same origin. This is surely true in the majority of cases but there are known exceptions. Many of the surface processes have potentials which are pH dependent (Figure 2) (*vide infra*). Under such circumstances the solution redox potentials are normally comparable to the surface potentials at higher pH. Reviewing the data in Figure 1 recognize that the surface of the electrode changes at each redox process. Between 3 and 4, the surface is $[Fe^{III}TsPc(-2)]^+$ (displayed overall charge ignoring the sulfonate groups), beyond 4 it has oxidized to $[Fe^{III}TsPc(-1)]^{2+}$ or possibly $[Fe^{IV}TsPc(-2)]^{2+}$. Between 2 and 3 it is $[Fe^{II}TsPc(-2)]$ and between 1 and 2 it is $[Fe^ITsPc(-2)]^-$. Just negative of wave I, the surface will be $[Fe^ITsPc(-3)]^{2-}$. Each of these surfaces is a discrete chemical compound with its own chemical and catalytic reactivity.

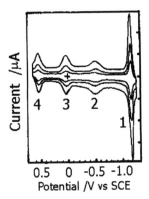

Figure 1. Scan rate dependence of $Fe^{II}TsPc/OPG$ in buffer at pH 10.7. Processes: 1. $[Fe^{I}TsPc(-3)]^{2-}/[Fe^{I}TsPc(-2)]^{-}$; 2. $Fe^{II}TsPc(-2)/[Fe^{I}TsPc(-2)]^{-}$; 3. $[Fe^{III}TsPc(-2)]^{+}/Fe^{II}TsPc(-2)$; 4. $[Fe^{III}Pc(-1)]^{2+}/[Fe^{III}TsPc(-2)]^{+}$. Adapted from [56].

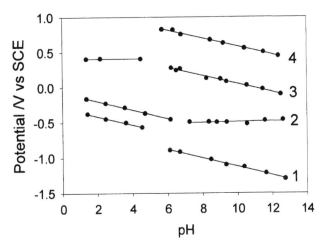

Figure 2. pH dependences of waves shown in Figure 1. Adapted from [56].

By exposing the electrode to a buffered electrolyte the behavior of each of these redox processes with variation of pH can readily be assessed (Figure 2). Horizontal lines in these Pourbaix diagrams indicate no change of the degree of protonation above or below the line while a slope of ca. –59 mV/pH unit implies that, for a one-electron process, there is a difference of one proton (or one hydroxide) between the species on either side of the line [12].

The next step will be to expose the surface to a target of interest and look for a large increase in current due to an electrocatalytic process. A nice example is shown in Figure 3 illustrating the oxygen activity of the aforementioned

$Fe^{III}TsPc/OPG$ surface under oxygen gas at low and high pH. It is obvious from this figure that the Fe(III)TsPc surface between waves 3 and 4 is not an active catalyst for oxygen reduction, while the Fe(II)TsPc surface formed at wave 3 (when scanning negatively) does electrocatalytically reduce molecular oxygen. No evidence is provided by this experiment as to whether the $Fe^{I}TsPc$ surface negative of wave 2 is also an active catalyst (*vide infra*). Under acidic conditions (Figure 3), the surface produced, scanning negatively, at wave 3 (also $Fe^{II}TsPc$) is evidently active for oxygen reduction. However note that the foot of the oxygen reduction wave begins at the base of wave 3 in the alkaline regime but more negatively than wave 3 in the acidic regime.

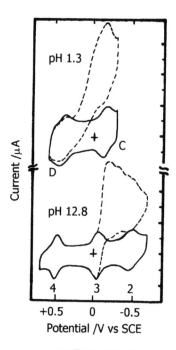

Figure 3. Cyclic voltammograms of $Fe^{II}TsPc/OPG$ in 0.05 M Na_2SO_4 buffered at pH 1.3 and 12.8 as shown. Solid lines under a nitrogen atmosphere and hatched lines saturated with oxygen (ca. 1 mM O_2). Adapted from [56].

On the positive side wave 3, at the foot of the wave, there is very little formation of Fe(II)TsPc so the appearance of the oxygen electrocatalytic current at so positive a potential implies that the surface is highly active and the rate of oxygen reduction is very fast [6, 7, 51]. On the other hand, the need to proceed to potentials negative of 3, implies the reverse, that on the $Fe^{II}TsPc$ surface in acidic medium, the oxygen reduction process is less facile and slow.

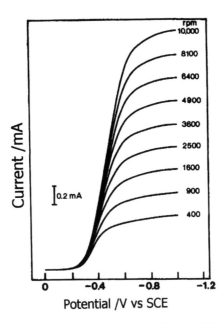

Figure 4. Reduction of molecular oxygen at a $Co^{II}TNPc/OPG$ electrode in oxygen saturated 0.10 M NaOH. Adapted from [69].

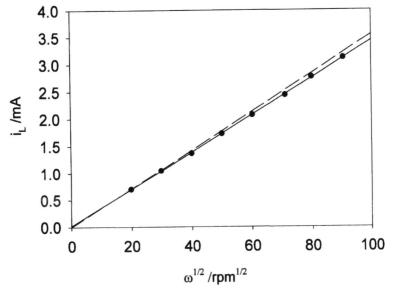

Figure 5. Levich plot for data in Figure 4. Solid line is regression line for data shown and hatched line is the theoretical line for n = 2. Adapted from [69].

The products of the electrocatalytic process should be identified. The first step to learn this is usually to carry out a rotating disk electrode (RDE) study measuring the electrocatalytic current at a series of rotation rates. A typical data set is shown in Figure 4 for tetraneopentoxyphthalocyaninatocobalt(II)/OPG. From the diffusion limiting current expression [46, 69] one generates a Levich plot of limiting current versus square root rotation rate (Figure 5) [40], the slope of which is directly related to the number of electrons involved in the electrocatalytic process.

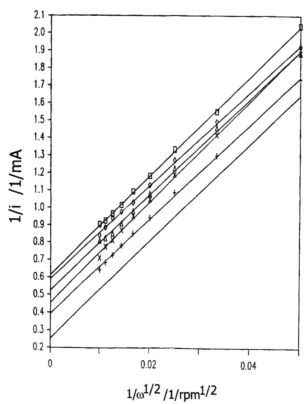

$$1/\omega^{1/2} /1/rpm^{1/2}$$

Figure 6. Koutecky-Levich plots for oxygen reduction at several different mononuclear and polynuclear cobalt complexes. The bottom line defines the slope for a theoretical n = 2 process. Adapted from [69].

Further, a plot of inverse current versus inverse square root of the rotation rate, known as a Koutecky-Levich plot [40] (Figure 6) yields, as intercept on the inverse current axis, a value for the limiting rate of target reduction when it is not controlled by diffusion, i.e. if the concentration at the electrode could be maintained at the bulk value; identified as the 'kinetic current' it is the predicted rate at an infinite rotation rate. It is a useful parameter to compare catalyst behavior.e.g. [69].

4   THE SURFACE CHEMISTRY OF CHLORO (PHTHALOCYANINE)RHODIUM(III) $ClRh^{III}Pc(-2)$ [66, 73]

4.1   Surface in the absence of oxygen.

We summarize here in some detail, the surface electrochemistry of chloro(p hthalocyanine)rhodium(III) and its reactivity towards molecular oxygen. More detailed supporting arguments can be found in the original publications. The chemistry is dominated by the instability of mononuclear $Rh^{II}Pc$ which has a strong tendency to dimerize by formation of a Rh-Rh bond to make $(Rh^{II}Pc)_2$. The bulk solution deposits mononuclear $ClRh^{III}Pc(-2)$ on the graphite surface but once this surface has been reduced to $Rh^{II}Pc$, the surface is then dominated by the chemistry of the dimeric rhodium(II) species. Comparison with the tetra-neopentoxyphthalocyanine analog, $ClRh^{III}TNPc$ was useful.

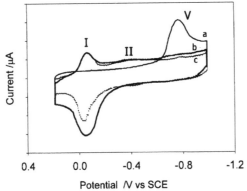

Figure 7. The cyclic voltammogram of a chlororhodium(III) tetraneopentoxyphthalo-cyanine/OPG electrode, $[ClRh^{III}TNPc(-2)/OPG]$ in 0.1 M KOH. a) first scan; b) second scan; c) equilibrium scan. 100 mV/s, drybox conditions. Adapted from [66].

The salient feature is that the chlororhodium(III) species has no redox activity between +0.2 and −0.6 V vs SCE but when scanned beyond −0.6 V, reduction of $ClRh^{III}Pc(-2)$ occurs at wave V (with loss of chloride) to yield $Rh^{II}Pc$ which rapidly dimerizes and yields a new reversible couple at ca. 0 V (wave I) which is a 2 electron/2 rhodium $[(Rh^{III}Pc)_2^{2+}/(Rh^{II}Pc(-2)]$ process. Proof of the assignment arises from two further observations, (i) wave V disappears in the second scan, i.e. all the active surface has been converted to a new species and (ii) the dimer species itself can be deposited on graphite from a pre-prepared solution of the dimer generated by reduction of a solution of ClRh^{III}Pc. This dimer surface exhibits wave I on the first scan, and no wave V.

There is a second and irreversible wave, wave II (Figure 8) which appears weakly or more strongly depending on scan rate, but the sum of the charges under waves I and II is a constant.

The species at wave I is unstable, and cycling around this wave causes its collapse; it is the oxidized component $[(Rh^{III}Pc)_2]^{2+}$ which breaks up to form mononuclear $[Rh^{III}Pc(-2)]^+$. If, after the collapse of wave I, the surface is scanned to about −0.4 V, the cathodic current of wave II increases. Apparently, wave II involves reduction to form Rh^{II}Pc(-2) which then dimerizes since upon scanning back to +0.2 V wave I is fully recovered. This confirms the identity of wave II as a mononuclear Rh^{III}/Rh^{II} process. Waves I and II probably do not involve chloride since if they fell apart to form ClRh^{III}Pc(-2), wave V should be recovered, and it is not. Addition of excess chloride in the form of electrolyte Cl$^-$ also does not lead to recovery of wave V.

4.2   Surface in the presence of oxygen.

The reactivity of the dimeric species towards dioxygen is demonstrated in Figure 8 a,b. The presence of oxygen (at very low concentration, 0.01 mM) while cycling over wave I, causes its partial collapse significantly more rapidly than under an argon atmosphere. If, after this partial collapse, scanning is continued beyond −0.4 V then a wave due to reduction of dioxygen is clearly seen (Figure 8b) at about the same potential as wave II. Subsequently returning to +0.2 V regenerates wave I. The collapse of wave I under oxygen must be due to the formation of an oxygen adduct, presumably monomeric, which is then reduced at about the same potential as wave II, to re-form Rh^{II}Pc(-2) which then rapidly dimerizes again.

Note that, in this experiment, during a positive going scan from about −0.2 to 0 V, positive of wave II and just negative of wave I, the surface is dimeric (Rh^{II}Pc(-2))$_2$. If oxygen were to react rapidly with this species, then wave I would not re-appear under a low oxygen concentration- yet it does reappear but with diminished intensity relative to the argon experiment. If oxygen were only to react with mononuclear Rh^{II}Pc(-2) then the intensity of the recovered wave I ought to be independent of whether argon or oxygen is present. Under the more usual 1 mM concentration of dioxygen, an electrocatalytic wave similar to that seen with other MPc species is observed (Figure 9) with the production of hydrogen peroxide.

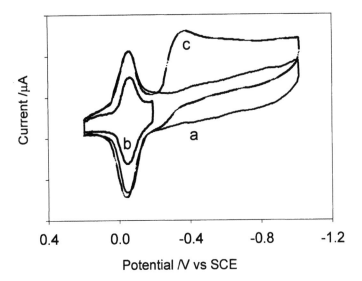

Figure 8. Cyclic voltammograms of chlororhodium(III) tetraneopentoxyphthalo-cyanine/OPG [ClRh^{III}TNPc(-2)/OPG] in 0.1 M KOH, argon degassed; scan rate100 mV/s:a) equilibrium scan; b) Scan from +0.2 to –0.2 about 60 s after addition of 0.01 M dioxygen; c) second scan after trace b) from +0.2 to –1.0 V. Adapted from [66].

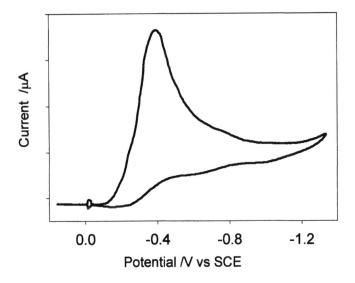

Figure 9. Cyclic voltammogram of chlororhodium(III) tetraneopentoxyphthalocyanin e/OPG [ClRh^{III}TNPc(-2)/OPG] in 0.1 M KOH saturated with dioxygen. Scan rate 100 mV/s. Adapted from [66].

Scheme I summarizes the chemistry of this species. The nature of the oxygen adduct is not proven; it may be a mononuclear $O_2RhPc(-2)$ species but it could also be dimeric $Pc(-2)Rh-O_2-RhPc(-2)$.

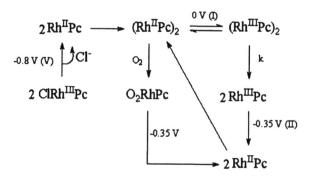

Scheme 1. Surface electrochemistry of chlororhodium phthalocyanine [66]

5   THE SURFACE ELECTROCHEMISTRY OF PERCHLORINATED PHTHALOCYANINES [26, 50, 78]

An interesting aspect of the surface electrochemistry of these perchlorinated species is the existence of redox processes that are hardly observable. Figure 10 displays the cyclic voltammetry of $Cl_{16}Fe^{II}Pc(-2)$, deposited on HOPG, at pH = 2. Wave A is the Fe^{III}/Fe^{II} process and the barely observable wave B is the Fe^{II}/Fe^{I} process. Wave C is $Fe^{I}Pc(-2)/Fe^{I}Pc(-3)$ which is of significantly greater intensity than wave A because some of the reduction current which should have occurred at wave B is actually being added to wave C.

The most likely explanation of this phenomenon relates to the mechanism of electron transfer from the electrode. One cannot assume that all the molecules deposited on the surface will be electroactive. This can be evaluated by, for example, spin coating the graphite surface with a known amount of material and then measuring the area under a wave, at slow scan rate to minimize problems due to kinetic slowness, to determine how much of the known quantity of material has actually been oxidized or reduced. Such studies generally yield only partial electroactivity for the surface [23, 24, 50, 57]. One can also expect that the actual degree of electroactivity will depend on the redox process being probed. In this case (Figure 10) the activity is small for the Fe^{II}/Fe^{I} process, perhaps limited to material actually in contact with the graphite surface, while possibly all the surface is active for the subsequent process, wave C, which would create

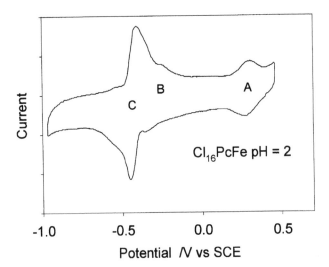

Figure 10. Surface electrochemical cyclic voltammogram of an iron(II) hexadecachloro-phthalocyanine/OPG [$Fe^{II}Cl_{16}Pc(-2)$]/OPG electrode in pH = 2 buffer; scan rate 100 mV/s. (A) $Fe^{III/II}$; (B) $Fe^{II/I}$; (C) $Fe^{I}Pc(-2)/Fe^{I}Pc(-3)$ Adapted from [26].

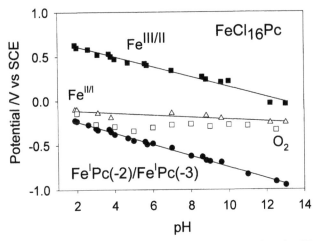

Figure 11. The pH dependences of the various redox processes for the [$Fe^{II}Cl_{16}Pc(-2)$]-/OPG electrode shown in Figure 10, as labeled. Oxygen reduction is also shown as open triangles. Adapted from [26].

a radical on the π^* orbital of the phthalocyanine which may be able to hop through the entire surface. However this cannot be the entire story since the M^{II}/M^{I} couple (M = Co, Fe) is also absent from the solution voltammetry of these species in DMF. The solutions are highly aggregated and the M^{II}/M^{I} couple is broadened greatly by a slow kinetic pathway. We note that if OPG is used instead of HOPG, a much better defined Fe^{II}/Fe^{I} redox process is observed [5].

The oxygen reduction process on $Cl_{16}FePc$ takes place at a potential corresponding roughly with the purported Fe^{II}/Fe^{I} process over the entire pH range (Figure 11) and thus reaction of oxygen with $[Cl_{16}Fe^{I}Pc(-2)]^-$ is probably the first step in the electrocatalysis. Both the Fe^{II}/Fe^{I} and oxygen reduction process show little dependence on pH except possibly at very low pH. The Co^{II}/Co^{I} process was not discerned and oxygen reduction takes place, also largely pH independent except at low pH, at a slightly more negative potential than for the iron catalyzed process.

The pH dependences are shown in Figure 11 for the various redox processes of $Cl_{16}PcFe$/HOPG. We note that the pH dependences of the Fe^{III}/Fe^{II} and $Fe^{I}Pc(-2)/Fe^{I}Pc(-3)$ couples on this electrode are very similar to those reported by Aguirre et al [5] on OPG. However the Fe^{II}/Fe^{I} couple lies at a somewhat more positive potential on OPG and has a greater pH dependence.

6   COBALT AND IRON CROWNED PHTHALOCYANINES [37, 38]

A series of crowned phthalocyanines (MCRPc) (Figure 12) was synthesized [37, 38]and their redox chemistry studied (also see [44, 52, 72]). Surface data for the Fe^{II} species are shown in Figure 13; the two waves are the Fe^{III}/Fe^{II} and Fe^{II}/Fe^{I} processes.

Figure 14 shows the pH dependence of these waves, while Figure 16 shows the oxygen reduction electrocatalytic current at pH 8.9. A single wave is seen and water is the product (4 electron reduction) rather than hydrogen peroxide. The pH dependence of this oxygen reduction follows a three step process shown in

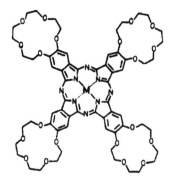

Figure 12. Iron and Cobalt crowned phthalocyanines

Figure 14. The oxygen wave occurs significantly negative of the Fe^{III}/Fe^{II} couple but also significantly positive of the Fe^{II}/Fe^{I} couple. Thus the active catalyst appears to be the $Fe^{II}CRPc(-2)$ species.

The reaction rate is fairly slow since the oxygen reduction wave lies somewhat negative (Figure 15) of the Fe^{III}/Fe^{II} couple [6, 7, 51]. However the pH profile for oxygen reduction is atypical since usually (Figure 11, and see further discussion

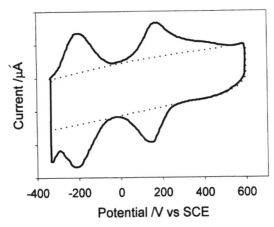

Figure 13. A cyclic voltammogram of the iron(II) crown phthalocyanine/HOPG $[Fe^{II}CRPc(-2)]$/HOPG electrode at pH 0.9, scan rate 100 mV/s. Adapted from [38].

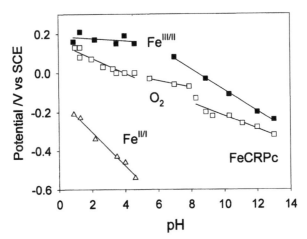

Figure 14. pH dependences of surface waves; (A) $Fe^{III}CRPc/Fe^{II}CRPc$; (B) Oxygen reduction at a $[Fe^{II}CRPc]$/HOPG electrode; C) $Fe^{II}CRPc(-2)/Fe^{I}CRPc(-2)$ Adapted from [38].

below) oxygen reduction becomes pH independent in the alkaline regime. Oxygen reduction at Co^{II}CRPc/HOPG is strikingly different (Figure 16) in that two successive reduction processes are observed, from oxygen to hydrogen peroxide (2 electrons) and then from hydrogen peroxide to water (2 electrons). This is a fairly common observation for oxygen electrocatalytic reduction at a R$_x$CoPc/graphite electrode [13, 23, 24, 67] although four electron reduction on cobalt phthalocyanine type surfaces has sometimes been observed [19,30,49, 53, 65] (see discussion below). The pH dependences of these waves are shown in Figure 17.

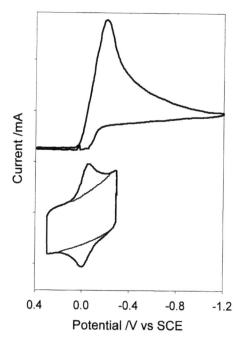

Figure 15. Cyclic voltammogram of an iron(II) crown phthalocyanine [Fe^{II}CRPc(-2)]/ HOPG electrode at pH 8.9. Lower: under argon; hatched line bare HOPG surface. Upper, buffer saturated with oxygen. Scan rate 100 mV/s. Adapted from [38].

Oxygen reduction occurs up to 300 mV positive of the Co^{II}/Co^{I} redox process with a very similar pH dependence. This might be interpreted as indicating that the active catalyst is Co^{II}CRPc or that Co^{I}CRPc is the catalyst in an extremely fast process. A Koutecky-Levich plot of the oxygen reduction data (Figure 16 C) shows a straight line that appears to go through the origin; this implies an exceptionally fast reaction rate and therefore favors the second explanation or that Co^{I}CRPc is the active catalyst. While there is evidence that simple Co^{II}Pc will bind oxygen at low temperatures [17, 18], Co^{II}CRPc does not react significantly in solution with oxygen at room temperature and therefore the formation constant for such an oxygen adduct must be very small. If this is the case, it is likely that should such a complex be the oxygen reduction intermediate its formation would probably be the rate determining step. This is inconsistent with the Koutecky-Levich data. We return to further consideration of these data below.

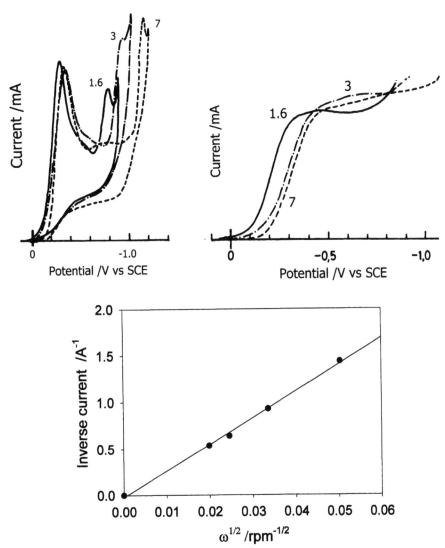

Figure 16. (A) Cyclic voltammogram of a cobalt(II) crown phthalocyanine [Co^{II}CRPc-(-2)]/HOPG in buffer saturated with oxygen at various pH as indicated; scan rate 100 mV/s. (B) Corresponding RDE responses for data in (A). (C) Koutecky Levich plot of data in (B). Adapted from [38].

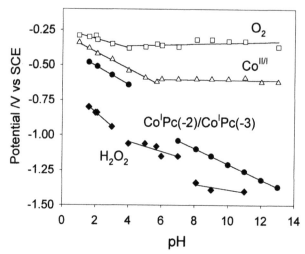

Figure 17. pH dependences of redox processes using the cobalt(II) crown phthalo-cya-nine [Co^{II}CRPc(-2)]/HOPG and cobalt(II) tetrasulfonatophthalocyanine electrodes. (A) Oxygen reduction at [CoCRPc(-2)]/OPG; (B) [Co^{II}CRPc(-2)/Co^{I}CRPc-(-2)]; (C) [Co^{I}TsPc(-2)/Co^{I}TsPc(-3)]; (D) hydrogen peroxide reduction on [Co^{II}CRPc-(-2)]/HOPG. Adapted from [38].

7   MONOMERIC AND POLYMERIC TETRAAMINO-PHTHALOCYANINATOCOBALT [65, 65]

A key feature of the chemistry of these species is that the polymerized surface is a four electron reductant of oxygen to water over a wide pH range from 2 to 13, while the un-polymerized species exhibits only a two electron reduction of oxygen to hydrogen peroxide. Polymerized tetraaminophthalocyaninato-cobalt(II) (p-CoTAPc) has been the subject of considerable study [9, 27, 28, 33, 35, 36, 43, 48, 54, 55, 61, 62, 64, 65] since early reports by Li and Guarr established that the coatings were conductive [41, 42, 70, 71]. The polymerized film is obtained by cycling the graphite electrode surface to + 1.0 V, in a DMF solution of Co^{II}TAPc [65]. The new p-CoTAPc surface exhibits three pH dependent waves, (I, II, III) the most negative of which (III) we associate with the Co^{II}/Co^{I} process on the basis of its pH dependence (below) and position with respect to a Hammett analysis [65, 75]. The two more positive waves are probably features of the polymer (see the original paper for details [65]). Indeed the appearance of waves I, II is very similar to those seen in the CV of p-Fe^{II}TAPc [55] although the authors in this latter work assign these waves as being iron localized. and as also seen in polyH$_2$TAPc (2 waves only) where they obviously cannot be metal localized [54].

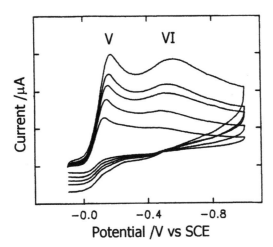

Figure 18. Cyclic voltammograms of cobalt(II) polytetraaminophthalocyanine [p-Co^{II}TAPc(-2)]/OPG electrodes in oxygen saturated buffer as a function of scan rate, pH = 10.0 Oxygen reduction waves V and VI are identified. Scan rates 12.5, 25, 50, 75 and 100 mV/s. Adapted from [65].

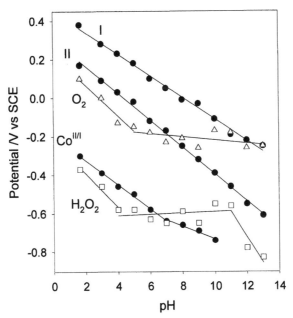

Figure 19 . pH dependences of redox processes using the cobalt(II) tetraaminophthalo-cyanine [p-Co^{II}TAPc(-2)]/OPG electrodes. (Solid circles) Waves I and II are associated with the polymer backbone. Wave III is believed to be [p-Co^{II}TAPc(-2)/p-Co^{I}TAPc-(-2)]; (Open squares) Oxygen and hydrogen peroxide reduction as identified. Adapted from [65].

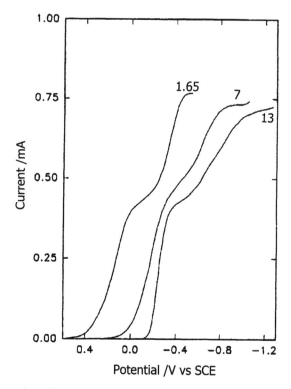

Figure 20. Rotating disk electrode data for oxygen reduction on p-CoTAPc/OPG electrodes at various pH as shown. Rotation rate 860 rpm, scan rate 5 mV/s. Adapted from [65].

Figure 18 shows the cyclic voltammogram of the reduction of oxygen exhibiting two successive waves in parallel with the study shown in Figure 16 for CoCRPc. Indeed the first wave is oxygen to hydrogen peroxide and second wave hydrogen peroxide to water. Their pH dependences are shown in Figure 19 yielding behavior very similar to that in Figure 17. However there is one major difference as noted by comparing the RDE data shown in Figure 20 with those shown in Figure 16. The p-CoTAPc surface shows two RDE waves corresponding with CV couples V and VI while the CoCRPc surface exhibits only a single wave (Figure 16).

The RDE experiment is a 'steady state' experiment and will generally only respond to the primary Faradaic processes. Thus any products, in this case hydrogen peroxide, will be spun off the active electrode before they can be electrolyzed. However if the product is oxidized or reduced at a very fast rate such that the redox process is complete before the material has time to be spun off, then there will be additional current corresponding to electroactivity of the product species. The CV experiment (Figure 18) reveals that hydrogen peroxide is formed when oxygen is reduced at wave V and that the hydrogen peroxide itself is reduced at the potential of wave VI. Unlike the data in Figure 16 for CoCRPc, we do observe a second wave in the RDE experiment. The Koutecky-Levich plot (for the top plateau in Figure 20) yields n = 4 for oxygen reduction at this wave VI potential. This could be interpreted in terms of oxygen being reduced to hydrogen peroxide and then this being very rapidly reduced to water before it could be spun off the electrode, i.e. a 2+2 reaction, or it could be a concerted 4 electron reduction of oxygen without the intermediacy of hydrogen peroxide. The lower plateau of the RDE dependence on rotation rate (Figure 20) is not sufficiently well defined to enable one to be sure that the second plateau always appears at twice the current of the first plateau irrespective of scan rate. If it is a 2+2 process, then at the faster scan rates one would expect that the top plateau current would diminish. There is no evidence for this in the KL plot (see Figure 11[65]). If this is a 2+2 process, the hydrogen peroxide must be reduced extremely rapidly before it can be spun off. It is relevant that for the CoCRPc/HOPG oxygen system, [38] the KL plot appears to go through the origin (Figure 16) implying an extremely fast reaction rate for oxygen reduction yet for this p-CoTAPC /OPG, the intercept is non-zero and therefore the reaction rate is not so fast. The RDE data for CoCRPc/ HOPG system (Figure 16) do not show much evidence for current arising from peroxide reduction and so the rate in this case must be sufficiently slow that the peroxide is spun off before it can be reduced. We do not have direct evidence for the relative rate of peroxide reduction in the CoCRPc/HOPG system compared to the p-CoTAPc/OPG system but we would surmise it is slower in the latter case. This may then be a true example of a 4 electron concerted reduction of oxygen.

Note that the unpolymerized CoTAPC surface catalyzes oxygen to hydrogen peroxide but the RDE experiment does not exhibit the second wave. We comment on the pH dependence of oxygen reduction below. Further, if the polymerized surface is obtained by cycling beyond +1.0 V, a different surface which does not show waves I-III is obtained. It exhibits a single wave (IV) and appears to be similar to the initial surface studied by Guarr and Li [41]. It does not display the 4 electron reduction of oxygen. Thus the generation of an appropriate polymeric surface for production of the 4-electron reduction process is crucial.

8   DINUCLEAR RUTHENIUM(II) PHTHALOCYANINE [15, 16, 21, 22, 39]

The electrochemical properties of both mononuclear $L_2Ru^{II}Pc$ and dinuclear $[LRuPc]_2$ have been described. The former is dominated by ring oxidation and reduction processes while the latter displays a series of metal localized processes.

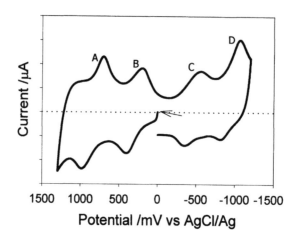

Figure 21. Surface electrochemistry in dichloromethane of $(Ru^{II}Pc)_2$/OPG; scan rate 100 mV/s. See text for assignments. Adapted from [22].

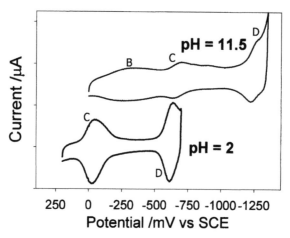

Figure 22. Surface electrochemistry of $(Ru^{II}Pc)_2$/OPG at pH 2.0 and 11.5; scan rate 100 mV/s. Adapted from [22].

The dinuclear species, which contains a metal-metal bond [14-16] shows (Figures 21,22) two reduction processes localized at the metal to yield, formally, $[Ru^IPc.Ru^{II}Pc]^-$ at wave C and $[Ru^IPc]_2$ at wave D. Two oxidation processes are observed to form $[Ru^{III}Pc.Ru^{II}Pc]^+$ at wave B and, probably, $[Ru^{III}Pc]_2$ at wave A. These same set of four waves can be observed in solution (not shown, see Figure 2 [22]) and on the surface exposed to dichloromethane (Figure 22) and exposed to aqueous electrolyte (Figure 23) and are assumed to have the same assignments.   Their pH dependence as shown in Figure 23. The slopes are consistent with protons being involved in the redox processes so the species noted above must actually have bound protons as noted in Figure 23.

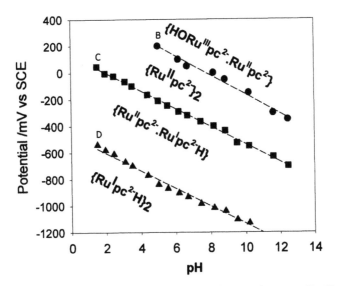

Figure 23. Pourbaix diagram showing pH dependence of waves B, C and D at a $(Ru^{II}pc)2/OPG$ electrode as shown in Figure 22. Adapted from [22].

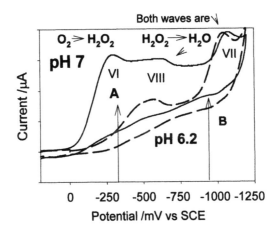

Figure 24. Oxygen reduction at $[Ru^{II}Pc]_2/OPG$ at pH 7 and hydrogen peroxide reduction at pH 6.2. Scan rate 100 mV/s. Adapted from [22].

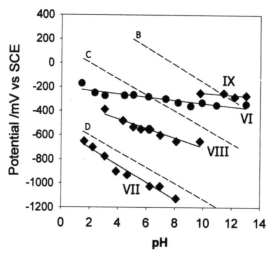

Figure 25. Pourbaix diagram showing pH dependence of oxygen and hydrogen peroxide reduction waves at $[Ru^{II}Pc]_2$/OPG. Adapted from [22].

The oxygen reduction chemistry at the $[RuPc]_2$/OPG electrode is very unusual, indeed unique. At low pH two waves similar to those shown in Figure 18 are observed [22]; these are the conventional steps from oxygen to hydrogen peroxide (wave VI) and then, at more negative potentials, hydrogen peroxide to water (wave VII). The pH dependence of wave VI is shown in Figure 25. It is almost pH independent but at high pH is actually a four electron reduction to water. The explanation lies in the fact that the reduction of hydrogen peroxide is highly pH dependent. At a pH higher than about 9, exposure of the $[RuPc]_2$/OPG electrode to oxygen-free hydrogen peroxide yields a reduction wave at a potential of about –200 mV (Figure 25, wave IX). This is slightly more positive than the oxygen reduction wave VI in this pH range. Thus as soon as oxygen is reduced to hydrogen peroxide, this product finds itself on the electrode at a potential where it is reduced immediately to water; therefore we observe a 2+2 reduction of oxygen to water.

Most surprisingly, in the intermediate pH range of ca. 3-10 there is a third pH dependent hydrogen peroxide reduction process (Wave VIII) lying at a potential intermediate between waves VII and IX. This is observed upon oxygen reduction and upon reduction of oxygen-free hydrogen peroxide (Figure 24). The appearance of two reduction waves from the bulk (H_2O_2) to the same product (H_2O) is generally not possible in a CV experiment. The observation here could result from the waves being two consecutive one-electron processes rather than a direct two-electron reduction to water, or it could be due to the nature of the surface changing such that the peak in wave VIII is due to a loss of surface activity between waves VIII and VII leaving some unreduced H_2O_2 at the surface for subsequent reduction at VII. Unfortunately the $[RuPc]_2$/OPG electrode is not very stable to bulk H_2O_2 so that a clear analysis of this problem was not possible. Current studies are looking at the chemistry of $[RuX_{16}Pc]_2$/OPG (X = Cl, F) which should be more stable and amenable to proper analysis [20]

9   FINAL COMMENTS

A detailed understanding of the mechanism by which oxygen or hydrogen peroxide is reduced at these surfaces is still elusive, despite studies going back several decades. While it is surely true that oxygen and hydroperoxide adducts of the surface must be involved, it is difficult to probe them. The pH dependences provide a clue. Oxygen reduction tends to be almost pH independent except at very low pH and generally mimicks the pH dependence of the M^{II}/M^{I} couple [24, 38, 56, 57, 65, 74]. Hydrogen peroxide reduction, on the other hand, frequently displays the stepped dependence shown in Figures 19 and 21. Four-electron oxygen reduction on FeCRPc in the alkaline regime (Figure 15) also shows a stepped dependence, suggesting that hydrogen peroxide is a key intermediate in the reduction process.

Oxygen reduction to peroxide is electrocatalyzed by $M^{I}(R_xPc)$ or $M^{II}(R_xPc)$ depending on M and R Usually, hydrogen peroxide reduction to water takes place at fairly negative potentials on a surface defined as $[M^{I}(R_xPc(-3))]^{2-}$ for M = Fe, Co (Figures 19,21), thus the radical anion is the effective catalyst. Clearly while the dinuclear $(Ru^{II}Pc)_2$ behaves towards oxygen in a fashion completely analogous to the other MPc species, and likely involves $Ru^{I}Pc(-2)$ as the active catalyst, it behaves in a startlingly different fashion towards hydrogen peroxide. This may be a consequence of its dinuclear character. Taken together with the unusual four-electron oxygen reduction chemistry of the p-CoTAPc polymer species it would seem that future research might well focus usefully on other dinuclear and polynuclear metallophthalocyanines.

Despite the quarter of century of research into phthalocyanine electrocatalysis there remain many questions to answer; it is clearly still a vibrant field of study.

ACKNOWLEDGEMENTS

We are indebted to the Natural Science and Engineering Council (Ottawa) for continuing support and to Office of Naval Research (Washington) for previous support of this research. ABPL thanks Canada Council for the Arts for a Killam Foundation Fellowship (200-2002). We thank Drs. Mehrdad Ebadi and Yu-Hong Tse for useful comment.

REFERENCE LIST

1. *The Phthalocyanines*, Properties and Applications. VCH, New York, Vols. 1-4, 1989-1996. (Eds. Leznoff, C. C.; Lever, A. B. P.)
2. Lever, A.B.P. *Adv. Inorg. Chem. and Radiochem. 7*, 27 (1965).
3. Lever, A.B.P. *Chemtech, 17*, 506 (1987).
4. Lever, A.B.P.; Milaeva , E.R.; Speier. G. *The Phthalocyanines*, Properties and Applications. Vol. 3 edition. VCH, New York. 1993. pp. 1-69.
5. Aguirre, M.J.; Isaacs, M.; Armijo, F.; Bocchi, N.; Zagal, J.H. *Electroanalysis 10*, 571 (1998).
6. Andrieux, C.P.; Dumas-Barchiat, J.M.; Saveant, J.M. *J. Electroanal,. Chem. 131*, 1 (1982b).
7. Andrieux, C.P.; Saveant, J.M. *J. Electroanal. Chem. 134*, 163 (1982).
8. Appel , G.; Mikaloa, R.; Henkela, K.; Opreaa, A.; Yfantisa, A.; Paloumpaa I.; Schmeißera, D. . *Solid-State Electronics, 44*, 855 (2000).
9. Ardiles, P.; Trollund, E.; Isaacs, M.; Armijo, F.; Canales, J.C.; Aguirre, M.J.,Canales, M.J. *J. Mol. Cat. A: Chemical 165*, 169 (2001).
10. Baldwin, R.P.; Ravichandran, K. *J. Electroanal. Chem 126*, 293 (1981).
11. Baldwin, R.P.; Thomsen, K.N. *Talanta, 38*, 1 (1991).
12. Bard, A.J.; Faulkner, L.R. "Electrochemical Methods: Fundamentals and Applications", Second Edition, 2001.
13. Behret, H. *Z. Physik. Chem 113*, 97 (1978).
14. Caminiti, R.; Donzello, M.P.; Ercolani; C.; Sadun, C. *Inorg. Chem. 38*, 3027 (1999).
15. Capobianchi, A.; Paoletti, A.M.; Pennesi, G.; Rossi, G.; Caminiti, R.; Ercolani, C. *Inorg. Chem. 33*, 4635 (1994).
16. Capobianchi, A.; Pennesi, G.; Paoletti, A.M.; Rossi, G.; Caminiti, R.; Sadun, C.; Ercolani, C. *Inorg. Chem. 35* , 4643 (1996).
17. Cariati, F.; Gallizzioli, D.; Morazzoni, F.; Busetto, C. *J. Chem. Soc. Dalton Trans.* 556 (1975).
18. Cariati, F.; Morazzoni, F; Busetto, C. *J. Chem. Soc. Dalton Trans* 496 (1976).
19. Coutanceau, C.; Crouigneau, P.; Leger, J.M.; Lamy, C. *J. Electroanal. Chem. 379*, 389 (1994).
20. Dharamdhat, C.; Lever, A.B.P.,Morin, S. to Be Published (2003).
21. Ebadi, M. Ph. D. Thesis, York University, (2002).
22. Ebadi, M.; Alexiou, C.; Lever, A.B.P. *Can. J. Chem. 79*, 992 (2001).
23. Elzing, A.; Van der Putten, A.; Visscher, W.; Barendrecht, E. *J. Electroanal. Chem. 200*, 313 (1986).
24. Elzing, A.; Van der Putten, A.; Visscher, W.; Barendrecht, E. *J. Electroanal. Chem. 233*, 99 (1987).
25. Gilmartin, M.A.T.; Hart, J.P. *Analyst 120*, 1029 (1995).
26. Golovin, M.N.; Seymour, P.; Jayaraj, K.; Fu, Y.S.; Lever, A.B.P. *Inorg. Chem. 29*, 1719 (1990).
27. Griveau, S.; Pavez, J.; Zagal , J.H.; Bedioui, F. *J. Electroanal. Chem. 497*, 75 (2001).

28. Griveau, S.; Albin, V.; Pauporte, T.; Zagal, J.H.; Bedioui, F. *J. Mat. Chem. 12*, 225 (2002).
29. Guillaud, G.; Simon, J.; Germain, J.P. *Coord. Chem. Rev. 178-180*, 1433 (1998).
30. Ikeda, O.; Itoh, S.; Yoneyama, H. *Bull. Chem. Soc. Japan 61*, 1428 (1988).
31. Riber, J.; de la Fuente, C.; Vazquez, M. D.; Tascón, M. L.; Bataner, P. S. *Talanta 52*, 241 (2000).
32. Janda, P.; Kobayashi, N.; Auburn, P.R.; Lam, H.; Leznoff, C.C.; Lever, A.B.P. *Can. J. Chem. 67*, 1109 (1989).
33. Jin, J.; Miwa, T.; Mao, L.; Tu, H.; Jin, L. *Talanta 48*, 1005 (1999).
34. Zagal, J. H. *Coord. Chem. Rev. 119*, 89 (1992).
35. Kang, T.F.; Shen, G.L.; Yu, R.Q. *Anal. Lett. 30*, 647 (1997).
36. Brown, K. L.; Shaw, J.; Ambrose, M.; Mottola, H. A. *Microchem. J. 72*, 285 (2002).
37. Kobayashi , N.; Lever, A.B.P. *J. Am. Chem. Soc., 109* , 7433 (1987).
38. Kobayashi, N.; Janda, P.; Lever, A.B.P. *Inorg. Chem. 31*, 5172 (1992).
39. Lever, A.B.P. *J. Porph. Phthalocyan. 3*, 488 (1999).
40. Levich, V.G. "Physicochemical Hydrodynamics", Prentice Hall, Englewood Cliffs, New Jersey. 1962.
41. Li, H.; Guarr, T.F. *Syn. Metals 38*, 243 (1990).
42. Li, H.; Guarr, T.F. *J. Electroanal. Chem. 317*, 189 (1991).
43. Magdesieva, T.V.; Zhukov, I.V.; Kravchuk, D.N.; Semenikhin, O.A.; Tomilova, L.G.,Butin, K.P. *Russ. Chem. Bull.* (Translation of *Izvestiya Akademii Nauk, Seriya Khimicheskaya*) *51*, 805 (2002).
44. Koçak, M.; Cihan, A.; Okur, A. I.; Gül, A.; Bekaroglu, O. *Dyes and Pigments 45*, 9. (2000).
45. Murray, R. W. *J. Electrochem. Soc. 13*, 201 (1984).
46. Newman, J. *J. Phys.Chem. 70*, 1327 (1966).
47. Nguyen Van, C.; Potje-Kamloth, K. *Thin Solid Films 392*, 113 (2001).
48. Oni, J.; Nyokong, T. *Anal. Chim. Acta 434*, 9 (2001).
49. Osaka, T.; Naoi, K.; Hirabayashi, T.; Nakamura, S. *Bull. Chem. Soc. Japan 59*, 2717 (1986).
50. Ouyang, J.; Shigehara, K.; Yamada, A.; Anson, F.C. *J. Electroanal. Chem. 297*, 489 (1991).
51. Oyama, N.; Oki, N.; Ohno, H.; Ohnuki, Y.; Matsuda, H.; Tsuchida, E. *J. Phys. Chem. 87*, 3642 (1983).
52. SaImagelam, O.; Gül, A. *Polyhedron 20*, 269 (2001).
53. Radyushkina, K.A.; Merenkova, M.V.; Tarasevich, M.R. *Elektrokhimiya 29*, 514 (1993).
54. Ramirez, G.; Trollund, E.; Canales, J.C.; Canales, M.J.; Armijo, F.; Aguirre, M. J. *Boletin De La Sociedad Chilena De Quimica 46*, 247 (2001).
55. Ramirez, G.; Trollund, E.; Isaacs, M.; Armijo, F.; Zagal, J.; Costamagna, J.; Aguirre, M.J. *Electroanalysis 14*, 540 (2002).
56. S. Zecevic; Simic-Glavaski, B.; Yeager, E.; Lever, A. B. P.; Minor, P.C. *J. Electroanal. Chem. 196*, 339 (1985).
57. Schlettwein, D.; Yoshida, T. *J. Electroanal. Chem. 441*, 139 (1998).

58. Sen, R.K.; Zagal, J.; Yeager, E. *Inorg. Chem. 16*, 3379 (1977).
59. Skarda, J.; Potje-Kamloth, K. *Proceedings – Electrochem. Soc. 97-19*, 696 (1997).
60. Snow, A.W.; Barger, W.R. in 'The Phthalocyanines; Properties and Applications". Vol. vol. 1. VCH, New York. 1989. pp. 341-392.
61. Somashekarappa, M.P.; Keshavayya, J.; Sampath, S. *Pure and Applied Chem. 74*, 1609 (2002).
62. Somashekarappa, M.P.; Sampath, S. *Chem. Commun.* 1262 (2002a).
63. Sun, Z.; Tachikawa, H. *Anal. Chem. 64*, 1112 (1992).
64. Trollund, E.; Ardiles, P.; Aguirre, M.J.; Biaggio, S.R.; Rocha-Filho, R.C. *Polyhedron 19*, 2303 (2000).
65. Tse, Y.-H.; Janda, P.; Lam, H.; Zhang, J.; Pietro, W.J.; Lever, A.B.P. *J. Porph. Phthalocyanines 1*, 3 (1997).
66. Tse, Y.H.; Seymour, P.; Kobayashi, N.; Lam, H.; Leznoff, C.C.; Lever, A.B.P. *Inorg. Chem. 30*, 4453 (1991).
67. Van den Brink, F.;Visscher, W.; Barendrecht, E. *J. Electroanal. Chem. 157*, 283 (1983).
68. Vasudevan, P.; Phougat, N.; Shukla, A.K. *Applied Organomet. Chem. 10*, 591 (1996).
69. Nevin, W.A.; Hempstead, M.R.; Liu, W.; Leznoff, C.C.;Lever, A. B. P. *Inorg. Chem., 26*, 570 (1987).
70. Xu, F.; Li, H.; Peng, Q.;Guarr, T.F. *Synthetic Metals 55*, 1668 (1993).
71. Xu, F.; Li, H.; Cross, S.J.,; Guarr, T.F. *J. Electroanal. Chem. 368*, 221 (1994).
72. Xu H.J.; Guo-Xiang, X. J. *Photochem. and Photobiology A, Chemistry 92*, 35 (1995).
73. Tse, Y.-H.; Manivannan, V.; Strelets, V.V.; Persaud, L.S.; Seymour, P.; Lever, A. B. P. *Inorg. Chem., 35*, 725 (1996).
74. Zagal, J.; Paez, M.; Tanaka, A.A.; Dos Santos Junior, J.R.;Linkous, C.A. *J. Electroanal. Chem. 339*, 13 (1992).
75. Zagal, J.H.; Gulppi, M.A.; Cardenas-Jiron, G. *Polyhedron 19*, 2255 (2000).
76. Zagal, J.; Paez, M.; Fierro, C. *Proc. Electrochem. Soc. 87-12*, 198 (1987).
77. Manivannan, V.; Nevin, W.A.; Leznoff, C.C.; Lever, A.B.P. *J. Coord. Chem., 19*, 139-158 (1988).
78. Zeng, Z.; Pan, Z. *Guangdong Gongxueyuan Xuebao 12*, 37 (1995).

Chapter 4

From Electron Transfer to Chemistry: Electrochemical Analysis of Organometallic Reaction Centers and of their Interaction Across Ligand Bridges

Wolfgang Kaim

Institut für Anorganische Chemie, Universität Stuttgart
Pfaffenwaldring 55, D-7550 Stuttgart, Germany

1   INTRODUCTION

The determination of electrochemical potentials has been a cornerstone of electron transfer research. In one of the elementary approaches the splitting of potentials E_1 and E_2 for metal-based one-electron transfer in symmetrically bridged dinuclear complexes such as the Creutz-Taube ion [1-4] has been used to confirm the existence of mixed-valent species as intermediates of degenerate inner-sphere electron transfer. According to (1) and (2) the difference $\Delta E = E_2 - E_1$ of redox potentials translates to the comproportionation constant K_c.

The Creutz-Taube-Ion:

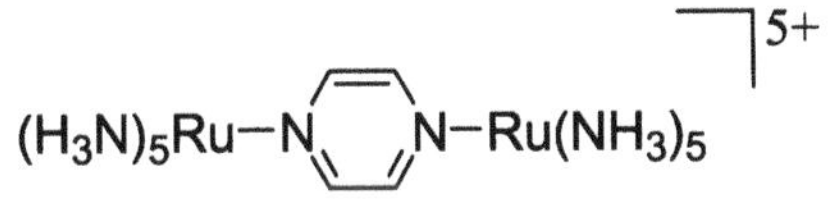

$$(H_3N)_5Ru-N\!\!\!\bigcirc\!\!\!N-Ru(NH_3)_5\;\Big]^{5+}$$

$$Ru^{II}/Ru^{III} \text{ or } Ru^{2.5}/Ru^{2.5} \text{ ?}$$

$$M^n\text{-}L\text{-}M^n \underset{+e^-}{\overset{-e^-}{\rightleftharpoons}} M^n\text{-}L\text{-}M^{n+1} \underset{+e^-}{\overset{-e^-}{\rightleftharpoons}} M^{n+1}\text{-}L\text{-}M^{n+1} \qquad (1)$$

Red (E_1) Int (E_2) Ox

$$\text{Comproportionation constant } K_c = \frac{[\text{Int}]^2}{[\text{Red}][\text{Ox}]} = 10^{\Delta E/59\text{mV}} \quad (\Delta E = E_2 - E_1) \quad (2)$$

In addition to the electronic interaction parameter H_{AB} and the spin-spin coupling constant J the K_c values have thus been used as one measure to quantify metal-metal interaction in bridged di- and oligonuclear compounds.

For the chemist, however, electron transfer processes are usually not isolated but part of more intricate reaction mechanisms which involve one or several atom transfer steps as well. Yet there are few systematic electrochemical studies available on the electronic coupling of metallic *reaction centers* instead of mere electron transfer centers across conjugated bridging ligands. As will be pointed out in this article, the coupling phenomena can indeed become rather elaborate even for rather modest reaction centers such as one-atom-dissociating/ -associating systems.

The reaction centers used in this article undergo the initial process (3) which can be exploited for hydride transfer catalysis in homogeneous phase. Mononuclear systems starting from $[(N^\wedge N)MCl(C_nR_n)]^+$ as well as dinuclear ligand-bridged species $\{(\mu\text{-BL})[MCl(C_nR_n)]_2\}^{2+}$ will be presented.

$$[(N^\wedge N)MCl(C_nR_n)]^+ + 2\,e^- \rightarrow [(N^\wedge N)M(C_nR_n)] + Cl^- \quad (3)$$

$N^\wedge N$: α-diimine ligand, BL: bis(α-diimine) bridging ligand;

$M = Rh$ or Ir: $n = 5$, $R = Me$;

$M = Ru$ or Os: $n = 6$, $R = H$, Me or other

2   THE TEST SYSTEM: ELECTROCATALYTIC HYDRIDE TRANSFER

Hydride transfer is an important step in many chemical and biochemical processes. For instance, the reactivity of the ubiquituous NAD(P)H/NAD(P)$^+$ coenzyme pair involves a formal hydride transfer as an essential biochemical reaction [5]. In organic and industrial chemistry the hydride transfer step can be part of hydrogenation reactions, yielding products such as H_2, hydrocarbons or alcohols.

Of special interest in this context are catalysts which facilitate hydride transfer. Whereas heterogeneous catalysts based on platinum or other metallic phases have long been known there are not many well-established systems for the homogeneous solution phase [6]. One such system developed for homogeneous H_2 production [7] and later applied to the important regeneration reaction of NAD(P)$^+$ to NAD(P)H [8] has been based on the rhodium(III) catalyst precursor $[(bpy)RhCl(C_5Me_5)]^+$. Together with its iridium(III) analogue $[(bpy)IrCl(C_5Me_5)]^+$ [9] these systems have been analyzed electrochemically, structurally, spectroscopically and theoretically [9-15]. The electrochemical and spectroscopic studies have then be extended to the analogous ruthenium(II) and osmium(II) complexes with the easier modified arene ligands [13] and to mononuclear complexes with α-diimine ligands other than 2,2'-bipyridine [11,12,16-25] (cf. Schemes 2-10). Scheme 1 summarizes the catalytic cycle as deduced from electrochemical and other experiments [7,9-13,26].

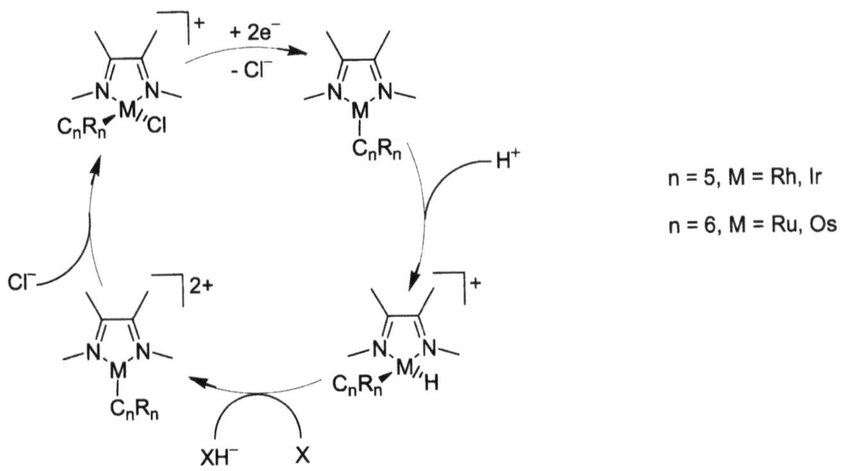

$n = 5, M = Rh, Ir$

$n = 6, M = Ru, Os$

Scheme 1

The activation step involves the chloride-dissociative two-electron reduction (ECE or EEC) of the precursors to yield electron-rich and coordinatively unsaturated neutral species $[(N^\wedge N)M(C_nR_n)]$. In the presence of protons an oxidative addition reaction can then produce the hydride complex $[(N^\wedge N)MH(C_nR_n)]^+$ which, in the case of the rhodium system, is an efficient hydride transfer agent for substrates such as $NAD(P)^+$ or H^+ [7,8]. The catalytic cycle is closed through rapid re-addition of chloride (or a solvent molecule) to achieve the stable d^6 18 valence electron (VE) configuration of the precursor.

3   MONONUCLEAR COMPOUNDS: AN ORGANOMETALLIC REACTION CENTER

A number of mononuclear complex ions $[(N^\wedge N)MCl(C_nR_n)]^+$, M = Co [14], Rh [7,10,11,13,14,16,20,24,25] or Ir [9,12-15,17-19,21,22,24,25] and n = 5, or M = Ru [13] or Os [13,18,23] and n = 6 has been studied with respect to the reactivity and the structure of intermediates within the reaction cycle of Scheme 1.

3.1   The Reductive Activation Step

It has been established for many of the above systems that the typical activation follows a chloride-dissociative, i.e. electrochemically irreversible two-electron process (ECE or EEC) [7-24].

$$[(N^\wedge N)MCl(C_nR_n)]^+ + 2\ e^- \xrightarrow{E^{pc}} [(N^\wedge N)M(C_nR_n)] + Cl^- \tag{3'}$$

$$[(N^\wedge N)M(C_nR_n)] - 2\ e^- + Cl^- \xrightarrow{E^{pa}} [(N^\wedge N)MCl(C_nR_n)]^+ \tag{4}$$

The differences between the peak potentials E^{pc} and E^{pa} are related to the properties of the starting complex (reduction) and to the stability of the neutral species formed (re-oxidation). Thus, the iridium and osmium complexes exhibit larger such differences than corresponding rhodium or ruthenium analogues [12-14] because of the stronger metal-to-ligand back donation in the former systems (cf. below). In the same vein, better π acceptor ligands such as bpz in the series of isomeric complexes $[(N^\wedge N)RhCl(C_5R_5)]^+$ [11] with the diazabipyridine = bidiazine [27] ligands (Scheme 2) give rise to higher differences E^{pc}–E^{pa}.

A typical cyclovoltammetric response for such a mononuclear species is shown in Figure 1 by example of the reduction and re-oxidation of the complex ion $[(bppz)OsCl(C_6Me_6)]^+$ (Scheme 3).

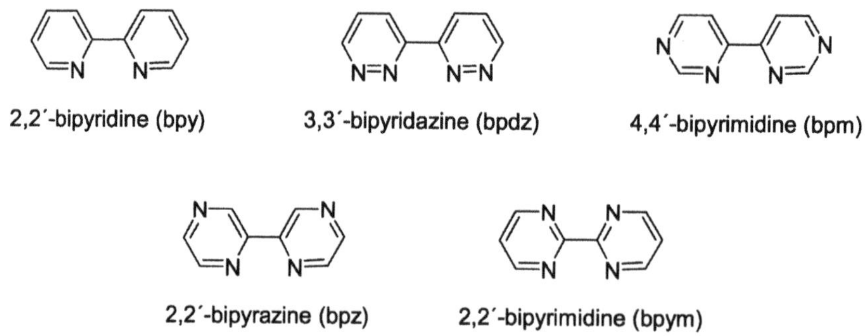

2,2´-bipyridine (bpy)   3,3´-bipyridazine (bpdz)   4,4´-bipyrimidine (bpm)

2,2´-bipyrazine (bpz)   2,2´-bipyrimidine (bpym)

Scheme 2

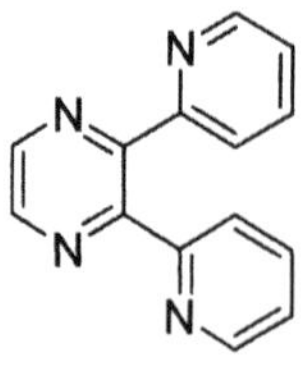

bis-2,3-(2-pyridyl)pyrazine (bppz)

Scheme 3

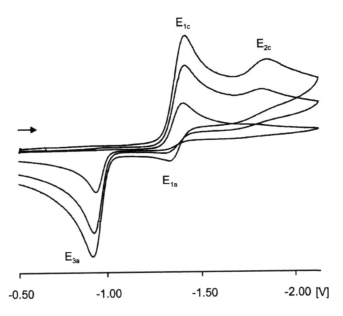

Figure 1. Cyclic voltammograms for the conversion [(bppz)OsCl(C_6Me_6)]$^+$/[(bppz)-Os(C_6Me_6)] in DMF/0.1 M Bu_4NPF_6 at 0.1, 0.5 and 1 V/s scan rates (potentials vs. $(C_5H_5)_2Fe^{+/0}$).

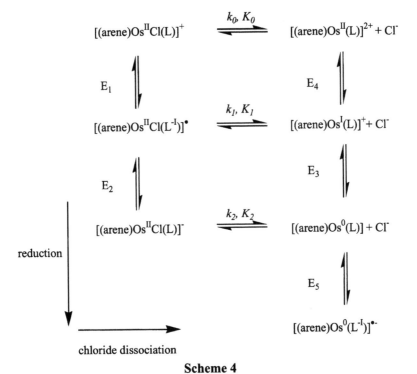

Scheme 4

Using a ladder scheme (Scheme 4) as a guideline, the graphical analysis of the cyclic voltammogram recorded at 1 V/s scan rate in DMF/0.1 M Bu_4NPF_6 yielded the parameters for E_i, k_j and K_j as given in Figure 2.

The good correspondence in Figure 2 and for the example with N^N = bpym [23] confirm the validity of the model and the usefulness of such an approach.

Similar results were obtained with heterocyclic N^N [16] and also N^O ligands [20-22] (Scheme 5).

With sterically shielded α-diimines such as 1,4-bis(2,6-dialkylphenyl)-1,4-diazabutadienes (Scheme 6) the re-oxidation can be separated from the slow re-coordination process according to the sequence (5) and (6) [16-18].

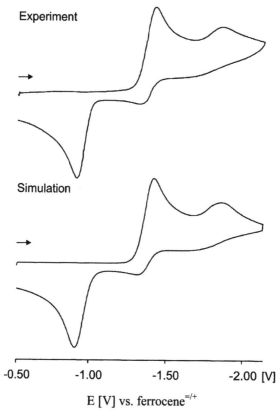

Experiment

Simulation

-0.50   -1.00   -1.50   -2.00 [V]

E [V] vs. ferrocene$^{=/+}$

Figure 2. Experimental and simulated cyclic voltammogram for the conversion [(bppz)OsCl(C_6Me_6)]$^+$/[(bppz)Os(C_6Me_6)] in DMF/0.1 M Bu_4NPF_6 at 1 V/s scan rate. Parameters according to Scheme 4: $E_1 = -1.42$ V, $E_2 = -1.80$ V, $E_3 = E_4 = -0.88$ V; k_0 =0.1 s^{-1} (fixed), $k_1 = 8.0$ s^{-1}, $k_2 > 1000$ s^{-1}; $K_0 = 2 \cdot 10^{-8}$ mol/l, $K_1 = 0.4$ mol/l, $K_2 = 8 \cdot 10^{12}$ mol/l (potentials vs. $(C_5H_5)_2Fe^{+/o}$).

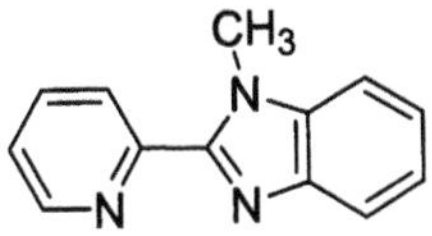

2-(2′-pyridyl)-N-methyl-
benzimidazole (mepybim)

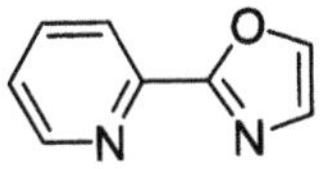

2-(2′-pyridyl)oxazole
(pyox)

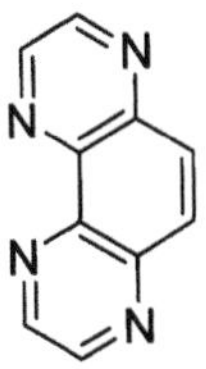

1,4,7,10-tetraaza-
phenanthrene (tap)

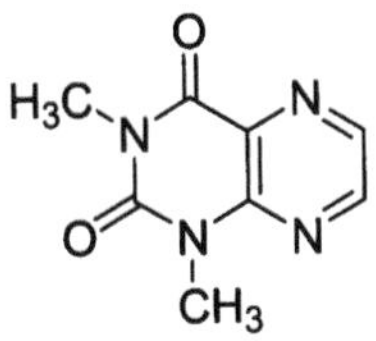

1,3-dimethyllumazine (dml)

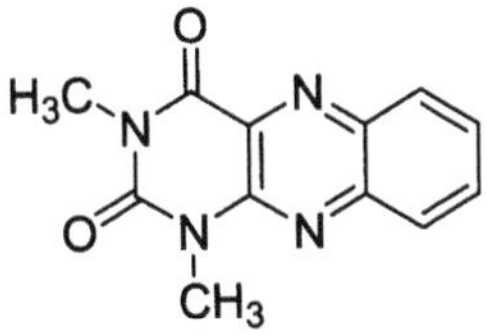

1,3-dimethylalloxazine (dma)

Scheme 5

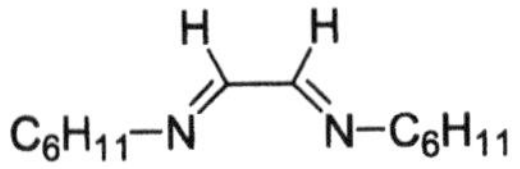

C_6H_{11}—N=CH—CH=N—C_6H_{11}

glyoxal-biscyclohexylimine
(c-Hex-DAB)

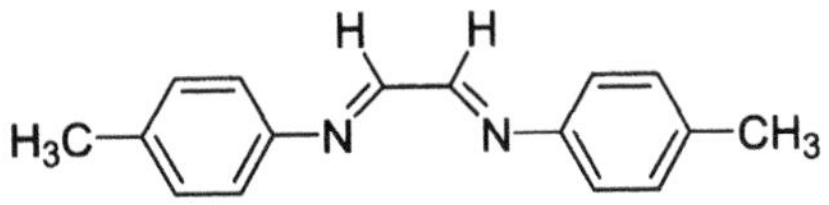

H_3C—C$_6$H$_4$—N=CH—CH=N—C$_6$H$_4$—CH_3

glyoxal-bis(4-methylphenylimine)
(p-Tol-DAB)

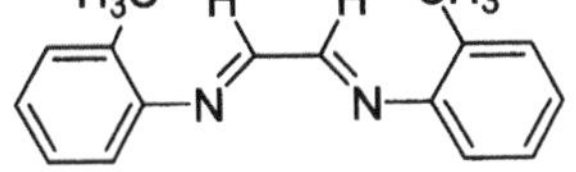

glyoxal-bis(2-methylphenylimine)
(o-Tol-DAB)

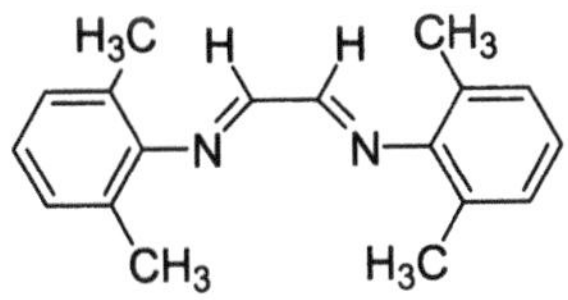

glyoxal-bis(2,6-dimethylphenylimine)
(2,6-Xyl-DAB)

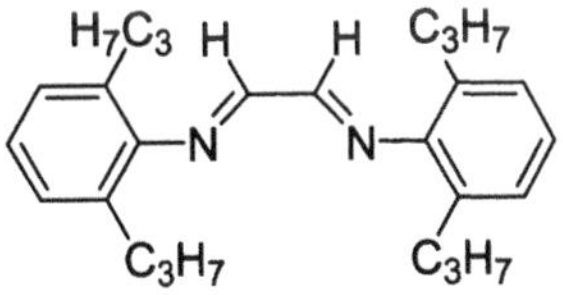

glyoxal-bis(2,6-diisopropyl-
phenylimine) (Dipp-DAB)

Scheme 6

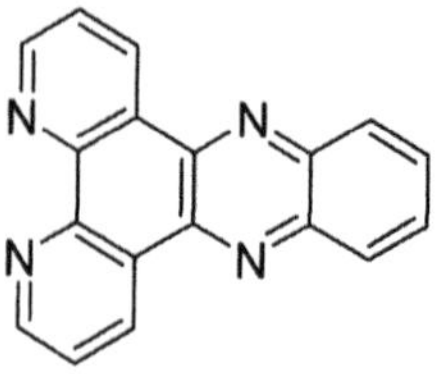

dipyrido[3,2-a:2′,3′-c]-
phenazine (dppz)

Scheme 7

$$[(N^\wedge N)M(C_nR_n)] - 2\,e^- \xrightleftharpoons{E_5} [(N^\wedge N)M(C_nR_n)]^{2+} \tag{5}$$

$$[(N^\wedge N)M(C_nR_n)]^{2+} + Cl^- \xrightarrow{\text{(slow)}} [(N^\wedge N)MCl(C_nR_n)]^+ \tag{6}$$

Even with $N^\wedge N$ = 1,4-disubstituted 1,4-diazabutadienes (R-DAB) [16-18] or dipyrido[3,2-a;2',3'-c]phenazine (dppz) [25] (Scheme 7) the cathodic two-electron process remains operating in spite of the less negative reduction potential (R-DAB) or the vanishing charge density at the coordination centers in the LUMO (dppz [28]) as compared to bpy.

However, there have been three instances reported where a separation of the two-electron reaction into one-electron processes occurs: Choosing very strongly π-accepting ligands such as *o*-quinones (e.g. pdo) [25] or azo compounds (e.g. abpy [29], Scheme 8) [24] one can observe the one-electron reduced form as an EPR or spectroelectrochemically identifiable labile intermediate involving the radical anion $(N^\wedge N)^{\cdot -}$ as a ligand. In other words, the activation splits into an {E + EC} reaction sequence (7, 8) where the intermediate exhibits enhanced tendency towards chloride dissociation [24,25].

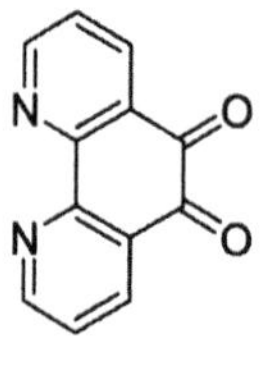

1,10-phenanthroline-
5,6-dione (pdo)

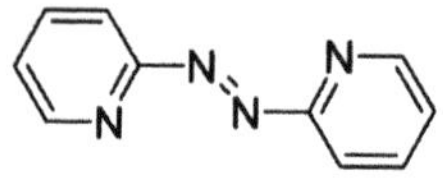

2,2′-azobispyridine
(abpy)

Scheme 8

$$[(N^\wedge N)MCl(C_nR_n)]^+ + e^- \stackrel{E_7}{\rightleftharpoons} [(N^\wedge N^{-I})MCl(C_nR_n)]^\bullet \tag{7}$$

$$[(N^\wedge N)MCl(C_nR_n)]^\bullet + e^- \stackrel{E_8^{pc}}{\longrightarrow} [(N^\wedge N)M(C_nR_n)] + Cl^- \tag{8}$$

After the second reduction there is a very rapid loss of chloride, indicating that the electron first stored in the ligand radical (electron reservoir function [30]) combines with the second electron to bring about the intended reaction. Apparently, the π^* orbitals of such very easily reduced ligands lie below the metal d orbital which is occupied on going from the d^6 to the d^8 configuration.

With the more conventional bpy ligand the splitting into an {EC + E} process was observed using the 3d element analogue, viz., $[(bpy)CoCl(C_5Me_5)]^+$ [14]. In contrast to the above, the intermediate here involves a *metal*-reduced species, $[(bpy)Co^{II}(C_5Me_5)]^+$, as evident most convincingly from a typical low-spin cobalt(II) EPR spectrum.

$$[(bpy)CoCl^{III}(C_5Me_5)]^+ + e^- \stackrel{E_9^{pc}}{\longrightarrow} [(bpy)Co^{II}(C_5Me_5)]^+ + Cl^- \tag{9}$$

$$[(bpy)Co^{II}(C_5Me_5)]^+ + e^- \stackrel{E_{10}}{\rightleftharpoons} [(bpy)Co(C_5Me_5)] \tag{10}$$

The loss of chloride after the first one-electron reduction step is obvious from the behavior of the isolated cobalt(I) form $[(bpy)Co(C_5Me_5)]$ in the absence of chloride, which is oxidized in two reversible steps ($E_{10} = E_{11}$):

$$[(bpy)Co(C_5Me_5)] - e^- \stackrel{E_{11}}{\rightleftharpoons} [(bpy)Co(C_5Me_5)]^+ \tag{11}$$

$$[(bpy)Co(C_5Me_5)]^+ - e^- \stackrel{E_{12}}{\rightleftharpoons} [(bpy)Co(C_5Me_5)]^{2+} \text{ (solv.)} \tag{12}$$

Like the rhodium and iridium compounds (cf. below) the neutral cobalt(I) complexes are further reduced in a reversible one-electron step:

$$[(bpy)Co(C_5Me_5)] + e^- \stackrel{E_{13}}{\rightleftharpoons} [(bpy)Co(C_5Me_5)]^- \tag{13}$$

In agreement with the (spectro)electrochemical analysis the long-wavelength absorptions in complexes $[(N^\wedge N)MCl(C_nR_n)]^+$ have been calculated as predominantly ligand-to-ligand charge transfer (LL'CT) transitions where L = Cl (involving the M-Cl bond) and L' = N^N [15]. Characteristically, the energy of that transition is lower for R-DAB complexes than for the bpy system because the π^* orbitals of the R-DAB ligands lie lower than that of bpy.

The neutral products [(N^N)M(C$_n$R$_n$)] as crucial, coordinatively unsaturated and very electron-rich intermediates in the catalytic cycle (Scheme 1) were identified through spectroscopy [11-19], electrochemistry (cf. below) [7-25], structure analysis [17] and quantum chemical calculations [15] as species with very strong metal d$_\pi$/ligand π^* orbital mixing. Accordingly, their long wavelength absorptions are due to transitions between two mixed π type orbitals and, remarkably, the energy of that transition is much higher for the R-DAB complexes because of the stronger π interaction with that ligand [15].

Structure analysis of (R-DAB)Ir(C$_5$Me$_5$), R = 2,6-dimethylphenyl [17], as supported by *ab initio* calculations has illustrated that the most valid formulation of the ground state is (R-DAB^{2-})Ir^{III}(C$_5$Me$_5$), i.e., the N^N ligand is the eventual site of a two-electron reduction. Formation of the enediamido form R-DAB^{2-} of the chelate ligand is evident from shortened C-C and lengthened C-N distances [17]. This result signifies a cooperativity between N^N (as electron storage site) and M (as center of the atom transfer reaction).

3.2 Further Electron Transfer Processes

Whereas the two-electron oxidized forms [(N^N)M(C$_n$R$_n$)]$^{2+}$ occur as solvated species within the reaction cycle (Scheme 1), it has been observed that most neutral compounds can be reduced in a largely reversible one-electron step (14) at very negative potentials [11-19] (Figure 3).

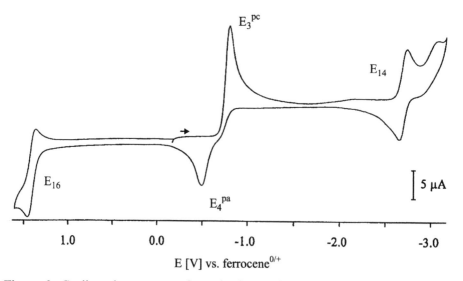

Figure 3. Cyclic voltammogram for reduction and oxidation of [(*p*-Tol-DAB)IrCl-(C$_5$Me$_5$)]$^+$ in CH$_3$CN/0.1 M Bu$_4$NPF$_6$ at 0.2 V/s scan rate.

$$[(N^N)M(C_nR_n)] + e^- \underset{}{\overset{E_{14}}{\rightleftharpoons}} [(N^N)M(C_nR_n)]^- \tag{14}$$

EPR studies in a few selected cases [24,25] have indicated that there is N^N/metal orbital mixing in the singly occupied MOs of $[(N^N)M(C_nR_n)]^-$. For instance, of the three relevant formulations (15)

$$[(N^N)M^o(C_5R_5)]^- \leftrightarrow [(N^N^{\cdot-})M^I(C_5R_5)]^- \leftrightarrow [(N^N^{2-})M^{II}(C_5R_5)]^- \tag{15}$$

the latter alternative with the fully reduced acceptor ligand and d^7 configuration at the metal appears most dominant on the basis of EPR and spectroscopic results for L = abpy and M = Rh,Ir [24]. The strong π acceptor nature of abpy [29] is expected to favor such a preference, there may be more contributions from the N^N ligand in other cases.

The abpy complexes [(abpy)M(C$_5$Me$_5$)] also exhibit an extreme degree of *cathodic* potential shift of up to 1 V (!) when comparing the reduction of free N^N and that of $(N^N)M(C_nR_n)$ [24]. Even for the prototypical bpy systems these cathodic shifts are about 0.1-0.2 V (Ru,Rh) and 0.25-0,45 V (Os,Ir) [11-13]. In contrast, all other metal complex fragments, including electron rich ones involving Mo^o or Re^I, induce *anodic* shifts for bound bpy due to net charge transfer from the ligand to the metal via the coordinative bonds [31,32]. Only strong orbital mixing or even a reversed orbital sequence (i.e. *d below* π^*(N^N)) would explain such a behavior, implying massive electron transfer in the ground state from the metal to the ligand [31].

Reversible one-electron oxidation of the precursor complexes at E_{16} to $[(N^N)MCl(C_nR_n)]^{2+}$ has been observed for a few $5d^6/5d^5$ species (Fig. 3) due to the stability of osmium(III) and iridium(IV) oxidation states [12,13,18,19,25].

$$[(N^N)MCl(C_nR_n)]^+ - e^- \underset{}{\overset{E_{16}}{\rightleftharpoons}} [(N^N)MCl(C_nR_n)]^{2+} \tag{16}$$

Although it will not be relevant for the following discussion of dinuclear species we should also like to mention the fact that the hydride intermediates $[(N^N)MH(C_nR_n)]^+$, generated either chemically or electrochemically (in protic solutions), can be reversibly reduced to corresponding radical anion complexes $[(N^N^{-I}MH(C_nR_n)]^\cdot$ as evident from EPR spectroscopy [12,13,32]. The potentials E_{17} are distinctly more negative than those of the chlorometal analogues.

$$[(N^N)MH(C_nR_n)]^+ + e^- \underset{}{\overset{E_{17}}{\rightleftharpoons}} [(N^N^{-I})MH(C_nR_n)]^\cdot \tag{17}$$

4   HOMODINUCLEAR COMPOUNDS: COUPLING OF REACTION CENTERS

Dinucleating bis(α-diimine)-type ligands, such as those shown in Schemes 9 and 10, have been used extensively in studies of metal-metal interaction. These include electrochemical investigations involving mixed-valent intermediates [2-4], electron transfer reactivity studies [33], spectroscopic measurements [34] and the determination of spin-spin interaction (magnetism) [35]. We thus set out to probe the effects of ligand-mediated coupling between two well understood reaction centers of the above kind in homodinuclear systems, starting from complexes $\{(\mu\text{-BL})[MCl(C_nR_n)]_2\}^{2+}$ [23,36-41].

Stability problems of some dinuclear systems (Rh,Ir [39]) with the widely used bpym ligand due to steric strain led us to study first complexes with the bptz, bpip and bxip ligands [36-38], later followed by compounds with bmtz, abpy and abcp [40,41]. Stability was achieved, however, for the diosmium complex ions $\{(\mu\text{-bpym})[OsCl(C_6R_6)]_2\}^{2+}$, to the extent that one of these precursors (C_6R_6 = *p*-cymene) could be crystallographically characterized as an *anti* (*trans*) isomer (Figure 4) [23].

In other instances [36-41], isomer mixtures of variable *syn/anti* (*cis/trans*) ratios were observed by NMR spectroscopy. However, variable compositions of these mixtures did not exhibit detectable differences of electrochemical potentials or absorption spectra.

The much facilitated reduction of dinuclear complexes of the above acceptor ligands [36-41] has allowed us to even isolate paramagnetic one-electron reduction products such as $\{(\mu\text{-BL})[MCl(C_5Me_5)]_2\}^{\bullet+}$, M = Rh or Ir with the azopyridine ligands BL = abpy or abcp [41] or $\{(\mu\text{-abcp})[OsCl(C_6R_6)]_2\}^{\bullet+}$ [42].

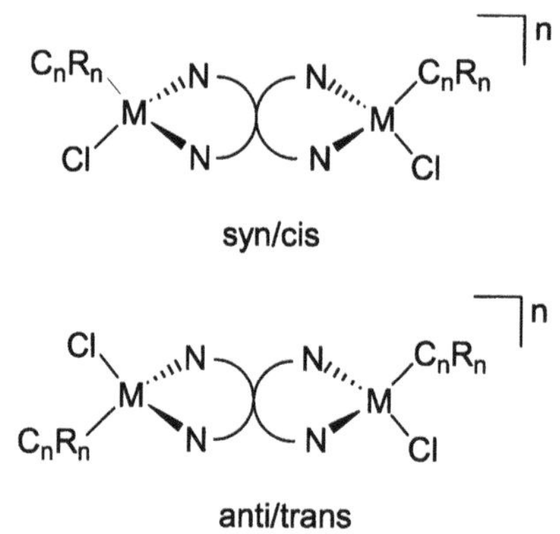

syn/cis

anti/trans

Figure 4. Positional isomerism in dinuclear complexes with planar bis-bidentate ligands.

In the following, the electrochemical behavior, the spectroelectrochemical identification and oxidation state assignment of the observed species will be described, illustrated by appropriate examples.

In most instances, i.e., except for the rhodium and iridium complexes of bpym [39], the first reduction step (18) involves a reversible one electron transfer.

$$\{(\mu\text{-BL})[MCl(C_nR_n)]_2\}^{2+} + e^- \xrightarrow{E_{18}} \{(\mu\text{-BL})[MCl(C_nR_n)]_2\}^{\cdot+} \qquad (18)$$

The complexes $\{(\mu\text{-BL})[MCl(C_nR_n)]_2\}^{\cdot+}$ are identified by UV-VIS-NIR spectroelectrochemistry (Fig. 5) and especially by EPR spectroscopy, including high-frequency EPR [41,42] of the stable, isolable azopyridyl compounds (BL = abpy, abcp), as complexes of two low-spin d^6 metal centers bridged by an anion radical ligand $BL^{\cdot-}$ [23,36-42]. Typically, the absorption spectra exhibit features of the ligand radical anion such as the presence or absence ($bptz^{\cdot-}$) of certain bands in the visible region. In the EPR spectra the sometimes resolved

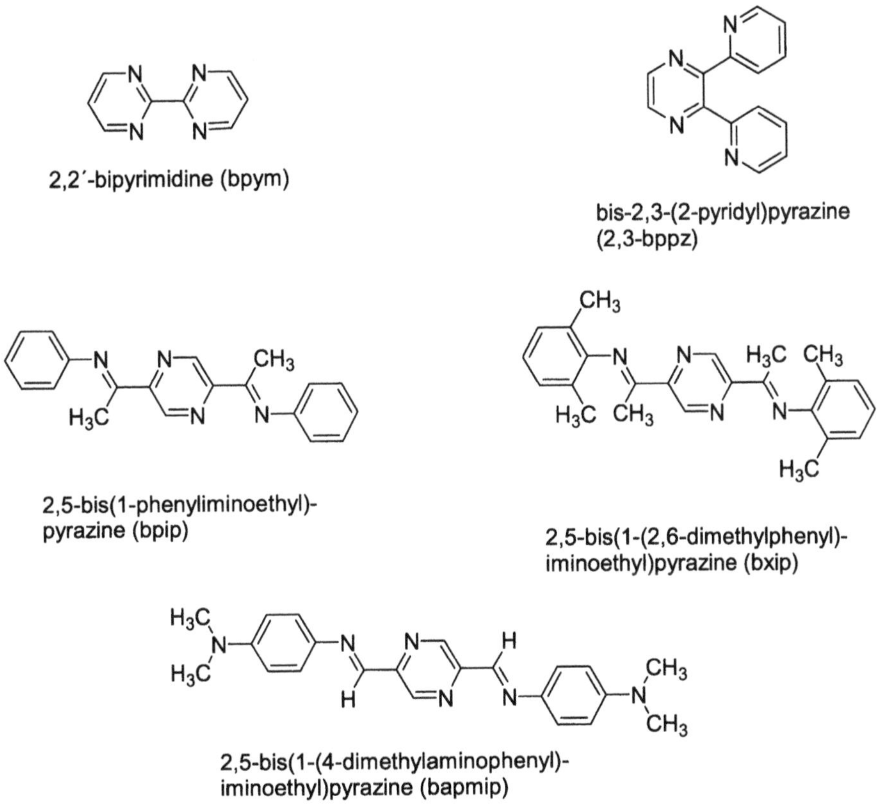

2,2′-bipyrimidine (bpym)

bis-2,3-(2-pyridyl)pyrazine
(2,3-bppz)

2,5-bis(1-phenyliminoethyl)-
pyrazine (bpip)

2,5-bis(1-(2,6-dimethylphenyl)-
iminoethyl)pyrazine (bxip)

2,5-bis(1-(4-dimethylaminophenyl)-
iminoethyl)pyrazine (bapmip)

Scheme 9

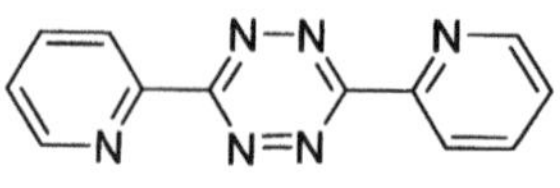

3,6-bis(2-pyridyl)-1,2,4,5-
tetrazine (bptz)

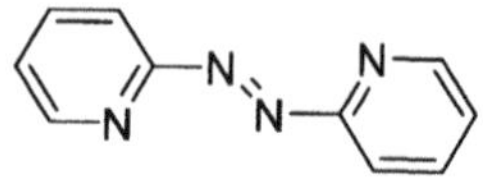

2,2′-azobispyridine
(abpy)

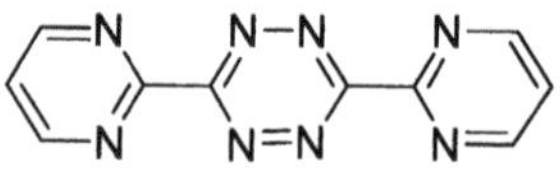

3,6-bis(2-pyrimidyl)-1,2,4,5-
tetrazine (bmtz)

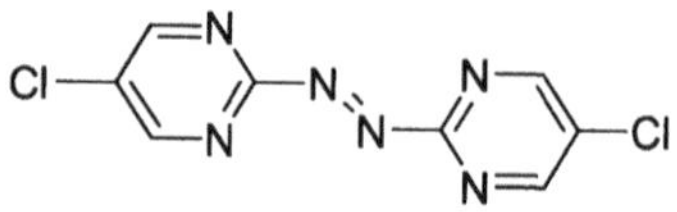

2,2 -azobis(5-chloro-
pyrimidine) (abcp)

Scheme 10

ligand hyperfine structure [37,38], the rather small metal coupling constants [23,42] and the small deviation of isotropic and anisotropic g values from the free electron value $g_e = 2.0023$ [23,36-42] indicate a ligand-centered reduction [43]. Conventional (Figure 6) and high-field EPR studies have illustrated, however, that the complexes of the 5d elements osmium and iridium have larger g anisotropies due to the effect of spin-orbit coupling [41,42].

The second electron added to the radical complex intermediate causes the familiar chloride dissociation in an irreversible process (19) at one of the two organometal sites to produce compounds $\{[(R_nC_n)M](\mu\text{-BL})[MCl(C_nR_n)]\}^+$. These are Class I [2-4,44] mixed-valent systems with the two metal centers in different chemical environments. Like the mononuclear complexes $(N^\wedge N)M(C_nR_n)$ they exhibit intense $\pi \rightarrow \pi^*$ transitions at low energies [23,36-42].

$$\{(\mu\text{-BL})[MCl(C_nR_n)]_2\}^{\cdot+} + e^- \xrightarrow{E_{19}^{pc}}$$
$$\{[(R_nC_n)M](\mu\text{-BL})[MCl(C_nR_n)]\}^+ + Cl^- \qquad (19)$$

Obviously, the two-electron (EEC or ECE) process as observed for most mononuclear complexes and for bpym-bridged dirhodium systems (20) [39] is split into an {E + EC} sequence with the first electron being stored in the ligand π system where it does not immediately labilize the metal-chloride bond. This electron reservoir behavior [30] is an essential part of any multielectron reactivity (including catalysis) where charge accumulation has to take place until the appropriate number of electrons allows for the intended chemical reaction at the most favourable potential.

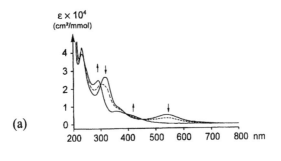

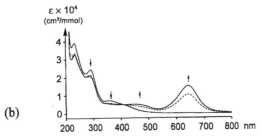

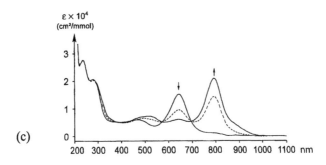

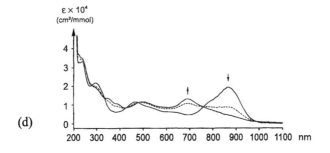

Figure 5. Spectroelectrochemical response on reduction:

(a) $\{(\mu\text{-bptz})[RhCl(C_5Me_5)]_2\}^{(2+)} \rightarrow {}^{(\bullet+)}$

(b) $\{(\mu\text{-bptz})[RhCl(C_5Me_5)]_2\}^{\bullet+} \rightarrow \{(Me_5C_5)Rh(\mu\text{-bptz})RhCl(C_5Me_5)\}^{+}$

(c) $\{(Me_5C_5) Rh(\mu\text{-bptz})RhCl(C_5Me_5)\}^{+} \rightarrow (\mu\text{-bptz})[Rh(C_5Me_5)]_2$

(d) $\{(\mu\text{-bptz})[Rh(C_5Me_5)]_2\}^{(-)} \rightarrow {}^{(2-)}$ (in $CH_3CN/0.1$ M Bu_4NPF_6).

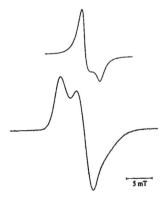

5 mT

Figure 6. EPR spectra of complex cations $\{(\mu\text{-bptz})[MCl(C_5Me_5)]_2\}^{\cdot+}$, M = Rh (top) and M = Ir (bottom) in frozen acetone solution at 3.4 K.

$$\{(\mu\text{-BL})[RhCl(C_5Me_5)]_2\}^{2+} + 2\ e^- \xrightarrow{E_{20}^{pc}}$$
$$\{[(Me_5C_5)Rh](\mu\text{-BL})[RhCl(C_5Me_5)]\}^+ + Cl \qquad (20)$$

The re-oxidation of (19) or (20) takes place in a two electron fashion (21):

$$\{[(R_nC_n)M](\mu\text{-BL})[MCl(C_nR_n)]\}^+ + Cl^- - 2\ e^- \xrightarrow{E_{21}^{pa}}$$
$$\{(\mu\text{-BL})[MCl(C_nR_n)]_2\}^{2+} \qquad (21)$$

Except for BL = bpip or bxip and M = Rh or Ir (n = 5) [36-38] the following reduction step is a chloride-dissociative two-electron EEC/ECE process (22) [23,36-42].

$$\{[(R_nC_n)M](\mu\text{-BL})[MCl(C_nR_n)]\}^+ + 2\ e^- \xrightarrow{E_{22}^{pc}}$$
$$\{(\mu\text{-BL})[M(C_nR_n)]_2\} + Cl^- \qquad (22)$$

re-oxidation:

$$\{(\mu\text{-BL})[M(C_nR_n)]_2\} + Cl^- - 2\ e^- \xrightarrow{E_{23}^{pa}}$$
$$\{[(R_nC_n)M](\mu\text{-BL})[MCl(C_nR_n)]\}^+ \qquad (23)$$

The symmetrical neutral compounds $\{(\mu\text{-BL})[M(C_nR_n)]_2\}$ have $\pi \rightarrow \pi^*$ transitions at longer wavelengths and with higher intensities than analogous complexes $\{[(R_nC_n)M](\mu\text{-BL})[MCl(C_nR_n)]\}^+$ or $(N^\wedge N)M(C_nR_n)$ [23,36-42] (Figure 5).

Especially for the stronger π-accepting bridging ligands containing azo functions the effective oxidation state distribution may be formulated according to B or C within resonance alternatives (24), as supported by theory and the very negatively shifted redox potentials for further reduction (cf. 29).

$$d^8(BL)d^8 \longleftrightarrow d^7(BL^{\cdot-})d^8 \longleftrightarrow d^7(BL^{2-})d^7 \tag{24}$$
$$\quad\ A \qquad\qquad\quad B \qquad\qquad\quad C$$

For BL = bpip or bxip and M = Rh or Ir (n = 5), however, this two-electron step of the second reductive elimination is also split into two one-electron processes {EC + E} (25, 27) [36-38].

$$\{[(R_nC_n)M](\mu\text{-BL})[MCl(C_nR_n)]\}^+ + e^- \xrightarrow{\ E_{25}^{pc}\ }$$
$$\{(\mu\text{-BL})[M(C_nR_n)]_2\}^{2+} + Cl^- \tag{25}$$

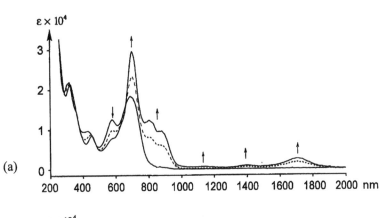

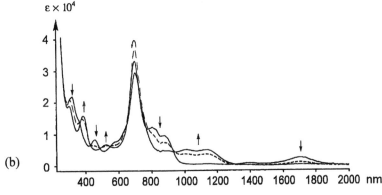

Figure 7. Spectroelectrochemical response on reduction: (a) {(Me$_5$C$_5$)Rh(μ-bpip)RhCl(C$_5$Me$_5$)}$^+$ → {(μ-bpip)[Rh(C$_5$Me$_5$)]$_2$}$^+$, and (b) {(μ-bpip)[Rh(C$_5$Me$_5$)]$_2$}$^{(+)\rightarrow(0)}$ (in CH$_3$CN/0.1 M Bu$_4$NPF$_6$).

$$\{(\mu\text{-BL})[M(C_nR_n)]_2\}^+ + Cl^- - e^- \xrightarrow{E_{26}^{pa}}$$
$$\{[(R_nC_n)M](\mu\text{-BL})[MCl(C_nR_n)]\}^+ \tag{26}$$

$$\{(\mu\text{-BL})[M(C_nR_n)]_2\}^+ + e^- \overset{E_{27}}{\rightleftharpoons} \{(\mu\text{-BL})[M(C_nR_n)]_2\} \tag{27}$$

The intermediate complexes $\{(\mu\text{-BL})[M(C_nR_n)]_2\}^+$ are mixed-valent species in an unusual acceptor-bridged d^7/d^8 combination (Rh^I/Rh^{II} or Ir^{I}Ir^{II}) as evident from typical near-infrared intervalence charge transfer (IVCT) bands at $\lambda_{max} >$ 1200 nm (Figure 7) and from large g anisotropies of the EPR signals [36-38]. Characteristically, the g anisotropy is again larger for the iridium analogues due to the high spin-orbit coupling constant of this 5d element [38].

According to the spectroscopic results, the contributions from a conceivable radical formulation B in (28) involving two rhodium(II) or iridium(II) centers are unlikely. The distinction between alternatives A and C, d^7/d^8 or d^6/d^7 configurations, is less straightforward, however, the former is strongly favored due to the unsaturated nature of both metal centers which would not be compatible with a d^6 configuration.

$$M^I(BL)M^{II} \leftrightarrow M^{II}(BL^{\cdot-})M^{II} \leftrightarrow M^{II}(BL^{2-})M^{III} \tag{28}$$
$$\quad A \qquad\qquad B \qquad\qquad C$$

Although direct evidence is not available, we consider these complexes as valence-averaged (Class III [2-4,44]) mixed-valent compounds despite the rather small comproportionation constant K_c according to Equ. 2 [4,38].

Apparently, the more negative potential for the reductive activation of the second chloroorganometal fragment causes dissociation of Cl$^-$ already after the uptake of one additional electron in an $\{EC + E\}$ fashion (24,26), whereas the first activation has occurred at less negative potentials with the chloride dissociation only after the second electron uptake ($\{E + EC\}$, (18,19)). As a consequence, the first paramagnetic intermediate is a radical complex whereas the second one is a mixed-valent species [37,38].

It is not clear at this point why only the rhodium and iridium complexes of the 2,5-diiminopyrazine ligands exhibit this stepwise second activation under formation of a mixed-valent intermediate, steric effects of the substituted ligands may play a role. Figure 8 illustrates that there is no such splitting of the second two-electron activation for areneosmium complexes of bpip. Complex ladder schemes can be constructed [23,42] as the basis of the graphical analysis of such cyclic voltammograms (Figure 8).

The next step is a reversible one-electron reduction (29), occurring at much more negative potentials than the corresponding values of the free ligands BL

[36-42]. This fact points to the rather small contribution from formulation A in alternatives (24).

$$\{(\mu\text{-BL})[M(C_nR_n)]_2\} + e^- \underset{}{\overset{E_{29}}{\rightleftharpoons}} \{(\mu\text{-BL})[M(C_nR_n)]_2\}^- \tag{29}$$

The odd-electron form $\{(\mu\text{-BL})[M(C_nR_n)]_2\}^-$ can be a radical with $d^8(\text{BL}^{\cdot-})d^8$ configuration or a metal-centered species with either $d^9(\text{BL}^0)d^8$ or $d^7(\text{BL}^{2-})d^8$

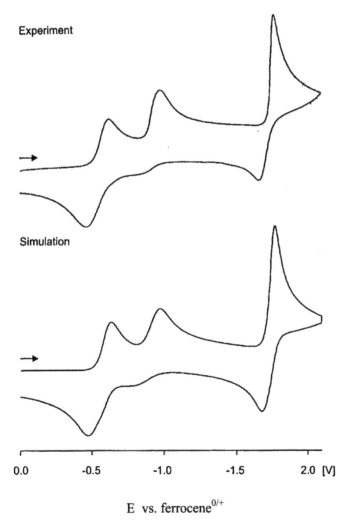

Figure 8. Experimental and simulated cyclic voltammogram for the conversion $\{(\mu\text{-bpip})$-$[\text{OsCl}(C_6Me_6)]_2\}^{2+}/\{(\mu\text{-bpip})[\text{Os}(C_6Me_6)]_2\}$ in DMF/0.1 M Bu$_4$NPF$_6$ at 1 V/s scan rate. Parameters: $E_{18} = -0.59$ V, $E_{19} = -0.97$ V, $E_{20} = -1.76$ V, $E_{21} = -0.48$ V, $E_{23} = -1.70$ V; $K_0 = 10^{-11}$ mol/l, $K_1 = 10^{-6}$ mol/l, $K_2 > 10^2$ mol/l (potentials vs. $(C_5H_5)_2\text{Fe}^{+/0}$).

formulation. On the basis of the above arguments for the neutral precursor we assume the latter alternative $d^7(BL^{2-})d^8$ which is supported by EPR studies for even mononuclear $[(abpy)Rh(C_5Me_5)]^-$ [24]. Unfortunately, the very negative reduction potentials E_{29} and the possibly rapid EPR relaxation rates of dinuclear complexes precluded EPR measurements in these cases, and UV-VIS-NIR spectroelectrochemical results were inconclusive (Figure 5).

Finally, the complexes $\{(\mu\text{-BL})[M(C_nR_n)]_2\}^{2-}$ observable in some instances [36,37] after one more reversible one-electron reduction (30) can only be formulated as fully reduced species with a $d^8(BL^{2-})d^8$ configuration.

$$\{(\mu\text{-BL})[M(C_nR_n)]_2\}^- + e^- \overset{E_{30}}{\rightleftharpoons} \{(\mu\text{-BL})[M(C_nR_n)]_2\}^{2-} \tag{30}$$

Figure 9 summarizes the electrochemical results for dinuclear rhodium species.

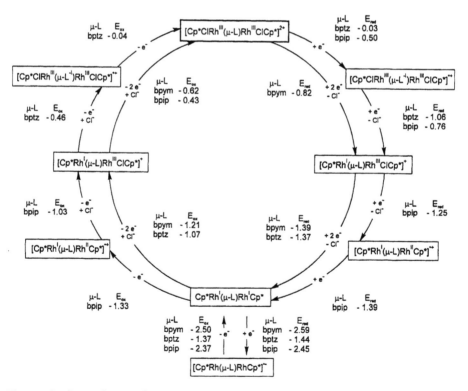

Figure 9. Synoptic reaction scheme for dinuclear rhodium complexes (potentials measured vs. $(C_5H_5)_2Fe^{0/+}$ in $CH_3CN/0.1$ M Bu_4NPF_6, 100 mV/s scan rate).

It has been stated in the Introduction that, for interacting *electron transfer* centers, the difference of redox potentials E_2-E_1, as convertible to K_c according to eq. (2) is a quantitative measure of the electrochemical coupling. Using one set of electron transfer centers such as $[Ru(bpy)_2]^{2+/3+}$ one can thus evaluate the coupling capability of different bridging ligands BL, increasing in the order of bpym < bpip < bptz < abpy [45] for the systems of Scheme 10.

Considering the cathodic peak potentials E^{pc} for the metal/chloride-dissociative processes of the ligand-coupled *reaction centers* one can use, in a first approach, the differences $\Delta^{pc} = E^{pc}(1) - E^{pc}(2) = E_{19}^{pc} - E_{22}^{pc}$ to estimate the corresponding coupling interaction. For the rhodium (iridium) complexes discussed here these values Δ_{pc} are [36-39,41]:

bpym: 0.51 V (0.36 V); bpip: 0.49 V (0.71 V); bptz: 0.31 V (0.51 V);

abpy: 0.73 V (1.01 V) (100 mV/s scan rate, CH_3CN/0.1 M Bu_4NPF_6)

Thus, the sequence only partially parallels that observed for pure electron transfer [45]. The differences are probably due to steric effects which are obviously more important for atom transfer involving reactions than for mere electron transfer processes.

For a more precise evaluation of that kind of electrochemical coupling between reaction centers it will be desirable to use $E_{1/2}$ values from simulated cyclic voltammograms in order to arrive at Δ values independent of scan rate [23,24].

5   HETERODINUCLEAR COMPOUNDS

In an extension of this work on homodinuclear systems we have recently prepared in a stepwise fahion new heterodinuclear compounds of the type $\{[(OC)_3 ClRe](\mu\text{-BL})[MCl(\eta^5\text{-}C_5Me_5)]\}^+$, M = Rh or Ir, and characterized them (spectro)electrochemically [46,47]. One such compound with BL = bptz and M = Rh was crystallographically identified as an *anti* structural isomer [46] (cf. Figure 4).

The purpose of these heterodinuclear systems is to combine hydride-transfer catalysis (at M) [7-10] with CO_2 activation (at Re [48,49]) at an *intramolecular* level to facilitate the formation of hydrogenated C_1 compounds from CO_2.

Studies in aprotic medium and in the absence of substrates [46,47] have shown the following (cf. Fig. 10):

Like in most of the homodinuclear cases, the first reduction (I) involves electron uptake [31] by the bridging acceptor ligand to form a radical species.

$$\{[(OC)_3ClRe](\mu\text{-BL})[MCl(\eta^5\text{-}C_5Me_5)]\}^+ + e^- \underset{\phantom{E_{31}}}{\overset{E_{31}}{\rightleftharpoons}}$$
$$\{[(OC)_3ClRe](\mu\text{-BL})[MCl(\eta^5\text{-}C_5Me_5)]\}^{\cdot} \qquad\qquad (31)$$

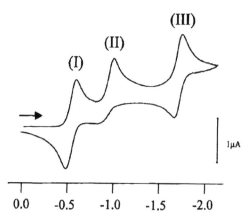

Figure 10. Cyclic voltammogram for the reduction (I) of [(OC)$_3$ClRe(μ-bpip)RhCl-(C$_5$Me$_5$)]+ in CH$_2$Cl$_2$/0.1 M Bu$_4$NPF$_6$ at 0.1 V/s scan rate.

On further one-electron reduction (II) there is chloride dissociation from M according to (32), producing the well established intense $\pi \rightarrow \pi^*$ transition of the (N^N)M(C$_n$R$_n$) chromophor at long-wavelengths.

$$\{[(OC)_3ClRe](\mu\text{-BL})[MCl(\eta^5\text{-C}_5Me_5)]\}^{\cdot} + e- \underset{}{\overset{E^{pc}_{32}}{\rightleftharpoons}}$$
$$\{[(OC)_3ClRe(\mu\text{-BL})[M(\eta^5 \text{ C}_5Me_5)]\} + Cl^- \tag{32}$$

$$\{[(OC)_3ClRe(\mu\text{-BL})[M(\eta^5\text{-C}_5Me_5)]\} + Cl^- - 2\,e^- \overset{E^{pa}_{33}}{\longrightarrow}$$
$$\{[(OC)_3ClRe](\mu\text{-BL})[MCl(\eta^5 C_5Me_5)]\}^+ \tag{33}$$

The third electron (III) is added in a quasi-reversible step (34), followed by slow dissociation of the rhenium-bound chloride [46]:

$$\{[(OC)_3ClRe(\mu\text{-BL})[M(\eta^5\text{-C}_5Me_5)]\} + e- \overset{E_{34}}{\rightleftharpoons}$$
$$\{[(OC)_3ClRe(\mu\text{-BL})[M(\eta^5\text{-C}_5Me_5)]\}^- \tag{34}$$

$$\{[(OC)_3ClRe(\mu\text{-BL})[M(\eta^5\text{-C}_5Me_5)]\}^- \overset{slow}{\longrightarrow}$$
$$\{[(OC)_3Re(\mu\text{-BL})[M(\eta^5\text{-C}_5Me_5)]\} + Cl^- \tag{35}$$

Even in the absence of the reducible substrates H$^+$ and CO$_2$ the system is thus capable of accepting at least three electrons to produce coordinatively unsaturated sites in close proximity and coupled by a conjugated bridging ligand. Current studies of substrate reduction are under way.

6   FINAL REMARKS

This article describes the electrochemical interaction of reversible electron transfer and bond-breaking processes between molecular coupled organometallic reaction centers through different conjugated bridging ligands. Not unexpectedly, this chemistry beyond electron transfer turns out to be rather complex in spite of the high symmetry of the systems investigated. One obvious result is that the electron transfer mediating properties of the bridging ligand are maintained to some extent for the coupling between the redox reaction centers, however, effects of steric interactions seem to play a greater role when atom transfer is involved in addition to electron transfer, causing variable patterns and sequences of E and EC steps. Analysis and rationalization of such effects are essential to better understand multielectron transfer-dependent processes in biological systems and synthetic chemistry, including energy-relevant processes [50].

ACKNOWLEDGEMENT

Support from the Deutsche Forschungsgemeinschaft, Volkswagenstiftung, EU (COST D14) and Fonds der Chemischen Industrie is gratefully acknowledged. I also thank the dedicated co-workers mentioned in those references which pertain to our own work. Special thanks are due to Dr. Brigitte Schwederski and Mrs. Angela Winkelmann for their contributions to preparing this article and to Drs. J. Fiedler and S. Zališ (J. Heyrovsky Institute, Prague) for continued cooperation.

REFERENCES

1. Creutz, C.; Taube, H. *J. Am. Chem. Soc.*, *95*, 1086 (1973).
2. Creutz, C. *Prog. Inorg.Chem.*, *30*, 1 (1983).
3. Kaim, W.; Bruns, W.; Poppe, J.; Kasack, V. *J. Mol. Struct.*, *292*, 221 (1993).
4. Kaim, W.; Klein, A.; Glöckle, M. *Acc. Chem. Res.*, *33*, 755 (2000).
5. (a) Stout, D.M.; Meyers, A.I. *Chem. Rev.*, *82*, 223 (1982). (b) Lee, I.-S. H.; Jeoung, E. H.; Kreevoy, M. M. *J. Am. Chem. Soc.*, *119*, 2722 (1997).
6. Kölle, U. *New J. Chem.*, *16*, 157 (1992).
7. (a) Kölle, U.; Grätzel, M. *Angew. Chem.*, *99*, 572 (1987); *Angew. Chem. Int. Ed. Engl.*, *26*, 568 (1987). (b) Kölle, U.; Kang, B.-S.; Infelta, P.; Compte, P.; Grätzel, M. *Chem. Ber.*, *112*, 1869 (1989).
8. (a) Westerhausen, D.; Hermann, S.; Hummel, W.; Steckhan, E. *Angew. Chem.*, *104*, 1496 (1992); *Angew. Chem. Int. Ed. Engl.*, *31*, 1529 (1992). (b) Steckhan, E.; Hermann, S.; Ruppert, R.; Thömmes, J.; Wandrey, C. *Angew. Chem.*, *102*, 445 (1990); *Angew. Chem. Int. Ed. Engl.*, *29*, 388 (1990). (c)
9. (a) Ziessel, R. *J. Am.Chem.Soc.*, *115*, 118 (1993). (b) Ziessel, R. *Angew. Chem.*, *103*, 863 (1991); *Angew. Chem. Int. Ed. Engl.*, *30*, 844 (1991). (c) Ziessel, R.;

Noblat-Chardon, S.; Deronzier, A.; Matt, D.; Toupet, L.; Balgroune, F.; Grandjean, D. *Acta Cryst., B49*, 515 (1992).

10. Cosnier, S.; Deronzier, A.; Vlachopoulos, N. *J. Chem. Soc., Chem. Commun.*, 1259 (1989). Rodriguez, M.; Romero, I.; Llobet, A., Deronzier, A.; Biner, M.; Parella, T.; Stoeckli-Evans, H. *Inorg. Chem.*, , *40*, 4150 (2001). Laguitton-Pasquier, H.; Martre, A.; Deronzier, A. *J. Phys. Chem. B*, *105*, 4801 (2001). Chardon-Noblat, S.; Deronzier, A.; Ziessel, R. *Collect. Czech. Chem. Commun.*, *66*, 207 (2001). Chardon-Noblat, S.; Deronzier, A.; Hartl, F.; Van Slageren, J.; Mahabiersing, T. *Eur. J. Inorg. Chem.*, 613 (2001). Chardon-Noblat, S.; Cripps, G. H.; Deronzier, A.; Field, J. S.; Gouws, S.; Haines, R.; Southway, F. *Organometallics*, *20*, 1668 (2001). Romero, I.; Collomb, M.-N.; Deronzier, A.; Llobet, A.; Perret, E.; Pecaut, J.; Le Pape, L.; Latour, J.-M. *Eur. J. Inorg. Chem.*, 69 (2001).

11. Ladwig,. M.; Kaim, W. *J. Organomet. Chem.*, *419*, 233 (1991).

12. Ladwig, M.; W. Kaim, W. *J. Organomet. Chem.*, *439*, 79 (1992).

13. Kaim, W.; Reinhardt, R.; Sieger, M. *Inorg. Chem.*, *33*, 4453 (1994).

14. Kaim, W.; Reinhardt, R.; Waldhör, E.; Fiedler, J. *J. Organomet. Chem.*, *524*, 195 (1996).

15. Zalis, S.; Sieger, M.; Greulich, S.; Stoll, H; Kaim, W. *Inorg. Chem.*, *42*, 5185 (2003).

16. Reinhardt, R.; Kaim, W. *Z. Anorg. Allg. Chem.*, *619*, 1998 (1993).

17. Greulich, S.; Kaim, W.; Stange, A.; Stoll, H.; Fiedler, J.; Zalis, S. *Inorg. Chem.*, *35*, 3998 (1996).

18. Berger, S.; Baumann, F.; Scheiring, T.; Kaim, W. *Z. Anorg. Allg. Chem.* **627** (2001) 620 (2001).

19. Greulich, S.; Klein, A.; Knoedler, A.; Kaim, W. *Organometallics*, *21*, 765 (2002).

20. Heilmann, O.; Hausen, H.-D.; Kaim, W. *Z. Naturforsch.*, *49b*, 1554 (1994).

21. Heilmann, O.; Hornung, F. M.; Kaim, W.; Fiedler, J. *J. Chem. Soc., Faraday Trans.*, *92*, 4233 (1996).

22. Heilmann, O; Hornung, F. M.; Fiedler, J.; Kaim, W. *J. Organomet. Chem.*, *589*, 2 (1999).

23. Baumann, F.; Stange, A.; Kaim, W. *Inorg. Chem. Commun.*, *1*, 305 (1998).

24. Kaim, W.; Reinhardt, R.; Greulich, S.; Fiedler, J. *Organometallics*, *22*, 2240 (2003).

25. Berger, S.; Scheiring, T.; Fiedler, J.; Kaim, W., manuscript in preparation.

26. Reinhardt, R.; Fees, J.; Klein, A.; Sieger, M.; Kaim, W. in *Wasserstoff als Energieträger*, VDI-Verlag,Düsseldorf, 1994, p. 133.

27. Ernst, S.; Kaim, W. *J. Am. Chem. Soc.*, *108*, 3578 (1986).

28. Fees, J.; Kaim, W.; Moscherosch, M.; Matheis, W.; Klima, J.; Krejcik, M.; Zalis, S. *Inorg. Chem.*, *32*, 166 (1993).

29. Kaim, W. *Coord. Chem. Rev.*, *219-221*, 463 (2001).

30. Astruc, D. *Electron Transfer and Radical Processes in Transition Metal Chemistry*, VCH, New York, 1995.

31. Bruns, W.; Kaim, W.; Ladwig, M.; Olbrich-Deussner, B.; Roth, T.; Schwederski, B. in *Molecular Electrochemistry of Inorganic, Bioinorganic and Organometallic*

Compounds, (Pombeiro, A.J.L.; McCleverty, J., eds.), Kluwer Academic Publishers, Dordrecht (Nato ASI Series C 385), 1993, p. 255.

32. Greulich, S.; Reinhardt, R.; Fiedler, J.; Kaim, W., unpublished results.
33. Haim, A. *Prog. Inorg. Chem., 30,* 273 (1983).
34. Kaim, W.; Kohlmann, S. *Inorg. Chem., 25,* 3306 (1986).
35. Kaim, W.; Kohlmann, S.; Jordanov, J.; Fenske, D. *Z. Anorg. Allg. Chem., 598/599,* 217 (1991).
36. Kaim, W.; Reinhardt, R.; Fiedler, J. *Angew. Chem., 109,* 2600 (1997); *Angew. Chem. Int. Ed. Engl., 36,* 2493 (1997).
37. Kaim, W.; Berger, S.; Greulich, S.; Reinhardt, R.; Fiedler, J. *J. Organomet. Chem., 582,* 153 (1999).
38. Berger, S.; Klein, A.; Wanner, M.; Fiedler, J.; Kaim, W. *Inorg. Chem., 39,* 2516 (2000).
39. Kaim, W.; Reinhardt, R.; Greulich, S.; Sieger, M.; Klein, A.; Fiedler, J. *Collect. Czech. Chem. Commun., 66,* 291 (2001).
40. Glöckle, M.; Fiedler, J.; Kaim, W. *Z. Anorg. Allg. Chem., 627,* 1441 (2001).
41. Frantz, S.; Reinhardt, R.; Greulich, S.; Wanner, M.; Kaim, W.; Fiedler, J.; Duboc-Toia, C., *Dalton Trans.,* 3370 (2003).
42. Baumann, F.; Kaim, W.; Sarkar, B.; Denninger, G.; Kümmerer, H.-J.; Fiedler, J., unpublished results.
43. Kaim, W. *Coord. Chem. Rev., 76,* 187 (1987).
44. Robin, M. B., Day, P. *Adv. Inorg. Chem. Radiochem., 10,* 247 (1967).
45. Ernst, S.; Kasack, V.; Kaim, W. *Inorg. Chem., 27,* 1146 (1988).
46. Scheiring, T.; Fiedler, J.; Kaim, W. *Organometallics, 20,* 1437 and 3209 (2001).
47. Frantz, S.; Weber, M.; Scheiring, T.; Fiedler, J.; Duboc, C.; Kaim, W., *Inorg. Chim. Acta,* in print.
48. Hawecker, J.; Lehn, J. M.; Ziessel, R. *Helv. Chim. Acta, 69,* 1900 (1986). Sullivan, B. P.; Meyer, T. J. *J. Chem. Soc., Chem. Commun.,* 1244 (1984).
49. Scheiring, T.; Klein, A.; Kaim, W. *J. Chem. Soc., Perkin Trans. 2,* 2569 (1997).
50. Tributsch, H. *J. Electroanal. Chem., 331,* 783 (1992).

Chapter 5

Bond and Structure Activation by Anodic Electron-Transfer: Metal-Hydrogen Bond Cleavage and cis/trans Isomerization in Coordination Compounds

Armando J.L. Pombeiro* and M. Fátima C. Guedes da Silva

Centro de Química Estrutural, Complexo I, Instituto Superior Técnico, Av. Rovisco Pais, 1049-001 Lisboa. Portugal. E-mail: pombeiro@ist.utl.pt

1   INTRODUCTION

The recognition that electrochemical methods can induce chemical reactivity, thus often promoting the formation and/or cleavage of chemical bonds in coordination compounds, has been reported since long. Following exploratory works by Wilkinson, Fischer and Vlcek, a pioneering and systematic study on organometallic complexes was undertaken by Dessy since the middle of the 60s. A growing interest on the field was triggered by the success of this approach and a number of reviews has appeared along the years (for those in the last decade see [1-5] and for the listing of earlier reviews and the discussion of the state-of-the-art in the early 80s see [6]). A couple of books [7,8] has also been published.

Nevertheless, the activation of a bond upon electron-transfer (ET), resulting in its weakening or even cleavage, what constitutes a fundamental "step" whose interpretation is of a great importance, is still the object of a considerable debate and the subject has been reviewed [9] for organic systems, namely organohalides. Coordination compounds have not yet attracted such an attention.

Among the bonds in coordination compounds with a high chemical and biological relevance, namely in processes involving redox reactions [10-15], one can consider the metal-hydrogen bond. However, the effect on such a bond of an ET to (or from) the complex, in particular the change of the metal oxidation state, has only been scarcely investigated.

Metal-ligand bond cleavage can also be involved in ET-induced structural changes of coordination compounds, namely in variations in ligand hapticity [2, 16, 17] and in ambient ligand linkage isomerizations [17d,18]. ET-promoted conformation changes *e.g.* in macrocyclic complexes [19] and other chelates [20], and geometrical isomerizations [3,5] of coordination compounds are processes related to the above which involve activation and weakening of chemical bonds (in particular metal-ligand bonds that can change their coordination positions upon ET), although their complete cleavage does not necessarily occur at least in the overall process. Such ET-generated structural modifications can lead to species with quite different redox and chemical properties, constituting a conceptually simple method for the control of reactivity with relevance *e.g.* in inorganic biology.

In this work we address the activation of metal-hydrogen bonds, as well as the geometrical isomerization (*cis/trans*) promoted by oxidation of a transition metal complex, focusing on systems whose electrochemical behaviours were mechanistically investigated in our Group, namely comprising ECEC schemes (E = electron-transfer step, C = chemical step) of square and non-square types.

The redox potential values are quoted relative to the saturated calomel electrode (SCE). They can be easily converted into the scale based on the ferrocene/ferricinium redox couple as the reference, by considering that, for the latter couple, $E° = 0.53, 0.55$ or 0.45 V *vs.* SCE (in 0.2 M $[NBu_4][BF_4]/CH_2Cl_2$, thf or NCMe, respectively).

2   SQUARE SCHEME (ECEC) FOR AN ELECTRON-TRANSFER INDUCED CONVERSION

The square ECEC (four-component) scheme was introduced over 30 years ago [21] for protonation reactions of organic molecules coupled to ET, and since then has been applied to various other situations [3-5, 22].

It can be represented by the general Scheme 1 which considers the effect of oxidation (for a reduction process a similar scheme, *mutatis mutantis* could be presented) on the interconversion (equilibrium) between species A and B (they can be *e.g.* an acid and its conjugate base or two isomers).

The equilibrium constants K_1 and K_2 (for the A/B and the A^+/B^+ equilibria, respectively) and the standard redox potentials $E°(A^{°/+})$ and $E°(B^{°/+})$ for the redox

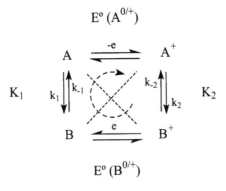

$$E° (A^{0/+})$$

$$E° (B^{0/+})$$

Scheme 1

couples are related by Equations 1 derived from the well-known relationship 2 between the standard Gibbs free energy change ($\Delta G°$) and the chemical equilibrium constant (K) or the standard potential for a redox process (E°), and also taking into account that $\Delta G° = 0$ for the thermochemical cycle.

$$\frac{K_2}{K_1} = \exp\left[\frac{E}{RT}\{E°(A^{o/+}) - E°(B^{o/+})\}\right] = \frac{[B^+]\cdot[A]}{[B]\cdot[A^+]} \qquad (1)$$

$$\Delta G° = -RT\ln K = -nFE° \qquad (2)$$

$$\Delta E° = E°(A^{o/+}) - E°(B^{o/+}) \qquad (3)$$

The ratio of the equilibrium constants K_2/K_1 is the equilibrium constant for the cross reaction (4) (dotted lines in Scheme 1) and the greater is the difference in the redox potentials, $\Delta E° = E°(A^{°/+}) - E°(B^{°/+})$ (Equation 3), the greater is the equilibrium shift towards the conversion (into B^+) upon oxidation (Figure 1). Hence, if that redox potential difference ($\Delta E°$) is 0.20 or 1.0 V, the K_2/K_1 ratio takes the values of 2.4×10^3 or 8.0×10^{16}, respectively, *i.e.* a high thermodynamic gain towards the A to B (in fact B^+) conversion results from a single electron oxidation of A. This commonly also corresponds to a kinetic gain ($k_2/k_{-1} > 1$).

However, when $E°(A^{°/+}) = E°(B^{°/+})$, there is no change in the equilibrium conversion ($K_2 = K_1$), and if $E°(A^{°/+}) < E°(B^{°/+})$, *i.e.* $\Delta E° < 0$, the conversion is hampered by oxidation ($K_2/K_1 < 1$).

$$A^+ + B \underset{}{\overset{K_2/K_1}{\rightleftharpoons}} A + B^+ \qquad (4)$$

The four related species (A, A^+, B and B^+) of a square scheme are not always detected, namely B if the A $\rightleftharpoons$ B equilibrium is very much shifted towards A

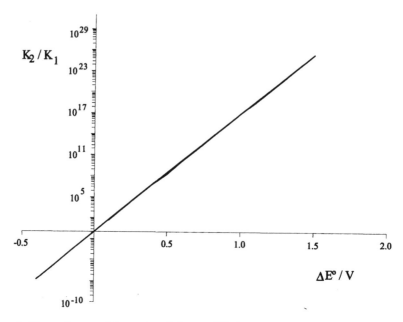

Figure 1. Dependence of the ratio of the equilibrium constants on the redox potential difference (see scheme 1 and equations 1 and 3).

(low K_1). This can occur *e.g.* when A is an hydride complex with a low acidity, but the hypothetical consideration of such a scheme remains conceptually valid. In the following section (3) we shall consider the heterolytic metal-H bond cleavage (H^+ loss) promoted by oxidation of an hydride complex [MH], *i.e* A = [MH], A^+ = $[MH]^+$, B^+ = [M] and B = $[M]^-$. K_1 and K_2 are then the acid constants (K_a) of the starting and the oxidized hydride complexes, [MH] and $[MH]^+$, respectively.

Section 4 deals with geometrical (*cis/trans*) isomerization of coordination compounds induced by oxidation, for which the general scheme 1 applies by considering that A and B denote those two isomers.

Square-type mechanisms can also be followed in processes involving ET-promoted reactions at ligands, as we have observed in interconvertible aminocarbyne (CNH_2), isocyanide (CNH) or cyanide (CN) complexes of iron or rhenium [23-27]. Anodically-induced ligand N-H bond cleavage then occurs as a result of the enhanced ligand acidity upon oxidation of the complex. However, these reactions (which follow single or double square processes whose mechanisms we have investigated in detail [25-27]) do not involve metal-hydrogen bond cleavage nor geometrical isomerization of the complexes and will not be discussed herein.

3   ANODICALLY INDUCED METAL-HYDROGEN BOND CLEAVAGE

3.1   *Heterolytic and Homolytic Cleavages*

Oxidation of an hydride complex, $M^{(n)}$-H, is normally an effective way of activating the metal-hydrogen bond. Thermochemical cycles for series of 18-electron hydride complexes of the type of the square Scheme 1 and related ones (incorporating the oxidation of those complexes and homolysis of the metal-H bond, *i.e.* $H^{\bullet}$ dissociation, as well as ET and H-atom transfer to the dehydro-genated forms) have been used [28-31] to probe the effect of the oxidation (i) on the acid constant (K_a) or (ii) on the metal-hydrogen bond dissociation energy (BDE), respectively. The $\Delta(pK_a)$ and $\Delta(BDE)$ values are determined by the differences of redox potentials $E°(M^{-/°})-E°(MH^{°/+})$ or $E°(M^{°/+}) - E°(MH^{°/+})$, respectively.

The acidity can increase dramatically upon oxidation (the pK_a decrease can reach 20-30 units in acetonitrile, corresponding to an activation that can be greater than 150 kJ mol^{-1} [28a, 29]) and therefore the oxidized $M^{(n+1)}$-$H^{+\bullet}$ radical complex then behaves as an extremely strong Brönsted acid. Deprotonation can then occur upon heterolytic M-H bond cleavage that corresponds to a H^+ reductive elimination leading to a reactive reduced radical, $M^{(n-1)\bullet}$, as shown by process (a), Scheme 2.

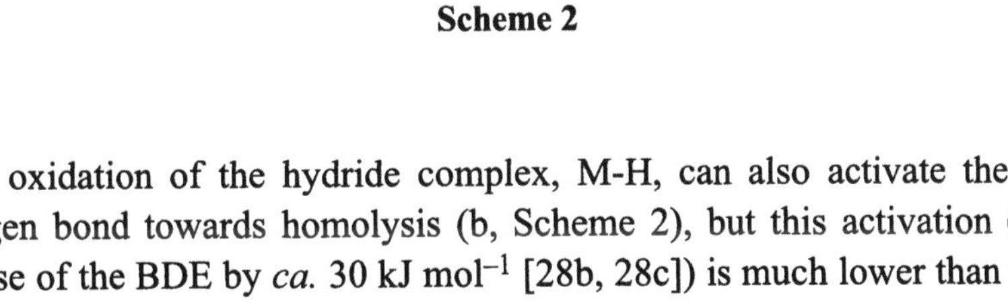

$$M^{(n)}-H \; \xrightleftharpoons{-e} \; M^{(n+1)}-H^{+\bullet} \quad \begin{cases} \xrightarrow{(a)} \; M^{(n-1)\bullet} \; + \; H^+ \\[2ex] \xrightarrow{(b)} \; M^{(n)+} \; + \; H^{\bullet} \end{cases}$$

Scheme 2

The oxidation of the hydride complex, M-H, can also activate the metal-hydrogen bond towards homolysis (b, Scheme 2), but this activation (typical decrease of the BDE by *ca.* 30 kJ mol^{-1} [28b, 28c]) is much lower than that for the heterolytic deprotonation which is the dominant one in most cases.

$$M^{(n)}-H \; \xrightleftharpoons[\;]{-e}^{(^{I}E°)} \; M^{(n+1)}-H^{+\bullet} \; \xrightarrow{-H^+} \; M^{(n-1)\bullet} \; \xrightleftharpoons[\;]{-e}^{(^{II}E°)} \; M^{(n)+} \; \xrightarrow{Nu} \; M-Nu^+$$

Scheme 3

3.2   ECEC and Other Processes

The reduced and coordinatively unsaturated radical $M^{(n-1)\cdot}$ formed by the oxidatively induced deprotonation can readily undergo further oxidation (Scheme 3) at a potential that is not higher than that of the parent complex ($^{II}E^{\circ} \leq {}^{I}E^{\circ}$), which can promote nucleophilic addition, according to an ECEC process.

Examples are given by the hydride complexes $[MoH(CO)_2(dppe)_2]^+$ [29] and possibly by *trans*-$[FeH(CN)(dppe)_2]$ [25], although the nucleophiles were not identified.

Other systems, which were studied in great detail [32, 33], are provided by the iron(II)-hydride isocyanide complexes *trans*-$[FeH(CNR)(dppe)_2]^+$. In particular, the CNMe member of the series, in thf/$[Bu_4N][BF_4]$, undergoes a complex ECEC-type reaction mechanism at the first oxidation wave (Scheme 4), as established by digital simulation of CV, that involves the oxidation of the complex and the reductive elimination of the hydride as a proton to form an unstable Fe(I) complex that is rapidly oxidized and then attacked by the tetrafluoroborate ion to yield the fluoro-complex *trans*-$[FeF(CNMe)(dppe)_2]^+$. Hence there occurred an overall replacement of hydride by fluoride, and all the $-2e/-H^+$ process occurred at the first oxidation wave ($^{I}E^{\circ} = 0.87$ V) of the initial complex. The fluoro-complex product of Fe(II) is oxidized at a slightly higher potential in the second oxidation wave ($E^{\circ} = 1.01$ V).

The rate constant (k) for the ET-induced proton loss was estimated and shown [32] to increase with the net electron acceptor minus donor character of the *trans* isocyanide ligand (CNR), as expressed by the electrochemical P_L ligand parameter and the Taft σ^* polar constant (for alkyl isocyanides) or the Hammett σ_p^+ constant (for the aromatic isocyanides). Hence, the variation of k is determined by polar effects or by direct conjugation between the phenyl substituent and the redox metal centre, respectively.

The study was recently extended [34] to the Fe(II) dinitrile complexes *trans*-$[FeH(LL)(dppe)_2]^+$ (LL = $N\equiv CCH=CHC\equiv N$, $N\equiv CC_6H_4C\equiv N$, $N\equiv CCH_2CH_2C\equiv N$) and to the corresponding dinuclear ones *trans*-$[\{FeH(dppe)_2\}_2(\mu\text{-}LL)]^{2+}$ with a bridging dinitrile, and has shown the occurrence of an overall $-2e^-/-H^+$ process for each iron site that follows an ECEC-type process.

The detailed mechanism (Scheme 5) was investigated by digital simulation for the fumaronitrile complex *trans*-$[\{FeH(dppe)_2\}_2(\mu\text{-}N\equiv CCH=CHC\equiv N)]^{2+}$ (in CH_2Cl_2) in which the unsaturated linking dinitrile allows an electronic communication between the two iron atoms as detected by CV [34]. The complex, onwards denoted by $[HFe^\frown FeH]^{2+}$, upon single-electron reversible oxidation ($E_1^{\circ} = 0.61$ V), leads to the mixed valent Fe(II)/Fe(III) form $[HFe^\frown FeH]^{3+}$ which undergoes either heterolytic Fe-H bond cleavage ($k_1 = 4.5$ s^{-1}) or further

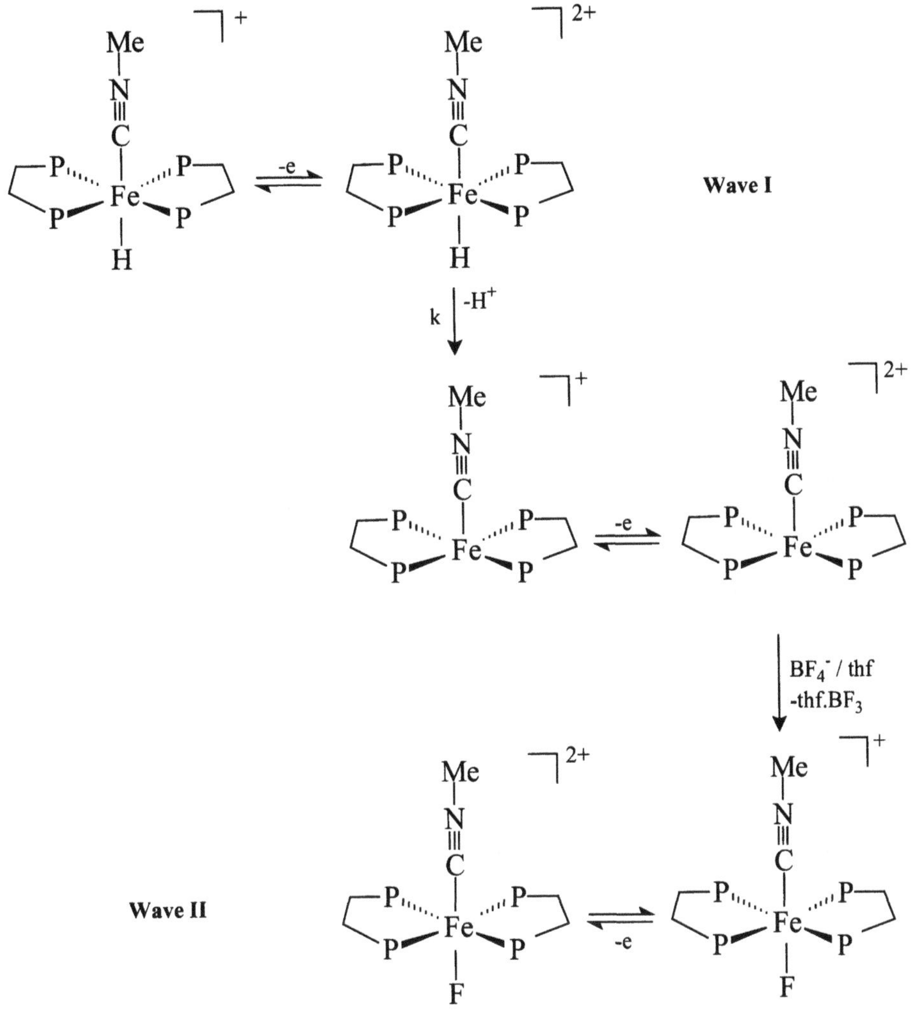

Scheme 4

oxidation, at a higher potential ($E_2^\circ = 0.80$ V), at the second Fe(II) site that had not yet been oxidized, followed by the two hydride-protons loss ($k_2 = 5.0$ s^{-1}).

The singly or doubly deprotonated compounds, $[Fe^\frown FeH]^{2+}$ or $[Fe^\frown Fe]^{2+}$, in which the deprotonated iron atom is in the reduced +1 oxidation state, undergo further oxidation and addition of a nucleophile, conceivably $[BF_4]^-$ [34].

The behaviour was rationalized by *ab initio* calculations on complex models (with PH$_3$ as the phosphine) which were used *inter aliae* to confirm or predict the order of the E$^\circ$ values, the nature of the ET-induced chemical steps, *i.e.* the heterolytic Fe-H bond cleavages, geometric isomerizations and

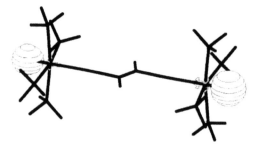

Figure 2. HOMO (one for the pair of degenerate ones) of *trans*-[{FeH(dppe)$_2$}$_2$-(μ-N≡CCH=CHC≡N)]$^{2+}$ (or [H*Fe⌢Fe*H]$^{2+}$).

nucleophilic [BF$_4$]$^-$ addition to the unsaturated iron centres [34]. Hence, *e.g.* the HOMOs (Figure 2) of the starting complex [H*Fe⌢Fe*H]$^{2+}$ represent the bonding combinations of hydride and iron orbitals, with predominance of the former, indicating that the first oxidation should be hydride centered, lead to the development of a positive charge at this atom and to a weakening of the Fe-H bond towards its heterolytic cleavage.

The theoretical studies [34] also predict (i) that the deprotonated Fe(I)/Fe(II) complex [*Fe⌢Fe*H]$^{2+}$ should be oxidized at a lower potential than that of the initial [H*Fe⌢Fe*H]$^{2+}$ complex (E$_5^\circ$ < E$_1^\circ$, in agreement with the experimental results) and (ii) that the oxidation of the former should not lead to Fe-H bond cleavage (also in accord with the mechanism of Scheme 5) since its HOMO is ligand bridge centered. Moreover, they indicate that structural arrangements can

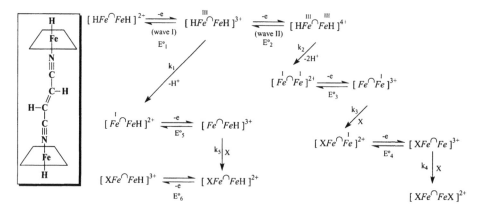

Scheme 5

occur in the oxidation process, *e.g.* in $[HFe^\frown FeH]^{2+}$ the unsaturated $\{Fe(PH_3)_4\}^+$ centre displays the vacant site (formed upon ET-induced hydride loss as H^+) in *cis* position to the bridging dinitrile (hence a *trans* → *cis* isomerization has occurred). They have also established [34] that the *E* isomeric form of the bridging fumaronitrile in the starting complex is more stable than the *Z* isomer.

In the hydride/isocyanide complex *trans*-$[FeH(CNH)(dppe)_2]^+$ there are two potential acidic centres, the metal and the CNH ligand, which could compete for ET-induced H^+ loss. The study of the oxidation process (which follows a square mechanism), assisted by digital simulation of cyclic voltammetry, has shown [25] that the single-electron oxidation of that complex, in thf, leads to the cyano complex $[FeH(CN)(dppe)_2]^+$ (upon H^+ liberation from the ligated isocyanide at a rate constant of 50 s^{-1}) which undergoes a further slower H^+ loss from the metal (at a rate constant of 0.25 s^{-1}). The acidic behaviour of the isocyanide ligand dominates that of the metal.

The oxidation of an hydride complex can, in principle, trigger a variety of other possible chemical/ET processes that can be different from the above $(EC)_n$-type. Hence, *e.g.* the 17-electron radical $M^\bullet$ formed upon $-e/-H^+$ from the initial 18-electron hydride M-H complex can undergo a nucleophilic addition [28a], a reductive elimination [35,36] or a dimerization [35]. In other processes, the oxidized hydride, $M\text{-}H^{+\bullet}$, can protonate its parent complex M-H which thus converts into a di-hydride MH_2^+ [25a, 36, 37] or a dihydrogen complex $M(\eta^2\text{-}H_2)^+$ [38-40].

Other oxidatively induced processes of hydride complexes involving M-H bond cleavage include disproportionation [41,42] and two-ligand reductive elimination [35,36] reactions. However, the anodic bimolecular disproportionation-type process reported [43] for the complexes initially formulated as the hydrides $[ReHCl(NCR)(dppe)_2]^+$ in fact does not involve metal-hydrogen bond cleavage since such compounds, which were obtained by protonation of the corresponding nitrile complexes $[ReCl(NCR)(dppe)_2]$, have been recently reformulated [44] as the azavinylidene species $[ReCl((N=CHR)(dppe)_2]^+$ derived upon proton addition to the nitrile ligand rather than to the metal. Therefore, the mechanism, which was correctly established, concerns an anodically induced C-H bond cleavage, at the N=CHR ligand, rather than a metal-H bond rupture.

In some cases, the proton extrusion from the hydride complex requires a second oxidation step [45-47] to occur.

4   ANODICALLY INDUCED *cis/trans* ISOMERIZATIONS

Electrochemical methods are particularly adequate to investigate the behaviours of coordination compounds that can exist in two geometrical forms,

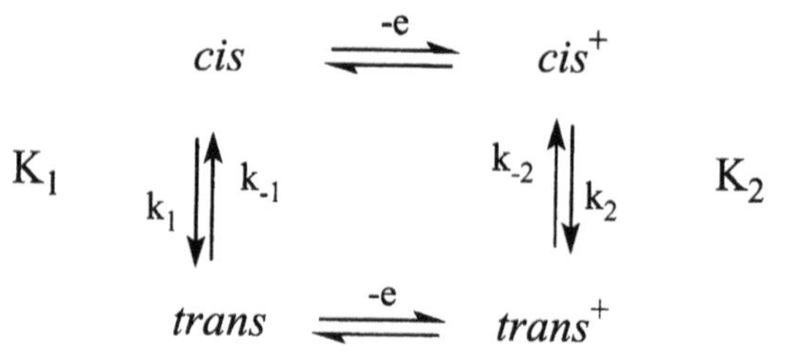

$$cis \quad \underset{}{\overset{-e}{\rightleftharpoons}} \quad cis^+$$

$$K_1 \quad k_1 \updownarrow k_{-1} \qquad k_{-2} \updownarrow k_2 \quad K_2$$

$$trans \quad \underset{}{\overset{-e}{\rightleftharpoons}} \quad trans^+$$

Scheme 6

namely *cis* and *trans* (or *fac* and *mer*), whose relative stability is dependent on the electron count, provided their redox processes occur at distinct potentials. Scheme 1 and Equation 1 then assume the forms of Scheme 6 and Equation 5, respectively.

$$\frac{K_2}{K_1} = \exp\left[\{E°(cis^{o/+}) - E°(trans^{o/+})\}\right] = \frac{[trans^+]\cdot[cis]}{[trans]\cdot[cis^+]} \tag{5}$$

As shown by this equation, the relative isomeric stability is determined by the difference in the redox potentials of the two isomeric redox couples and therefore depends on the factors (structural, electronic or steric) that influence such potentials, such as the electronic configuration, the metal oxidation state, the electron/donor acceptor properties of the ligands and even their bulkiness.

Hence, geometrical isomerizations can be induced by ET, in particular electrochemically, and the subject has been recently reviewed by us [5] for octahedral-type complexes, with emphasis on the processes that have been mechanistically investigated in detail. We now focus on the *cis/trans* systems that have been studied in our Laboratory, namely involving π-electron acceptor ligands (like nitriles, isocyanides and carbonyl), and related ones.

4.1 Single-electron Promoted Isomerizations

The 16-electron Mo(II) d^4 isocyanide complexes *cis*-[Mo(SR')$_2$(CNR)$_4$] (R' = C$_6$H$_2$Pr$_3^i$-2,4,6; R = Me or Bu^t), with two bulky thiolate ligands, undergo anodically-promoted *cis* to *trans* isomerization following a square ECEC-type mechanism (Scheme 6 assumes then the form of Scheme 7) for which (R = Bu^t) the rate and equilibrium constants were estimated by digital simulation of cyclic voltammetric data collected in CH$_2$Cl$_2$, at –50 °C to prevent decomposition [48].

The *cis* isomer displays an oxidation potential E°(*cis*$^{o/+}$) = –0.09 V *vs.* SCE, that is higher than that of the *trans* one, E°(*trans*$^{o/+}$) = –0.27 V *vs.* SCE [48], what

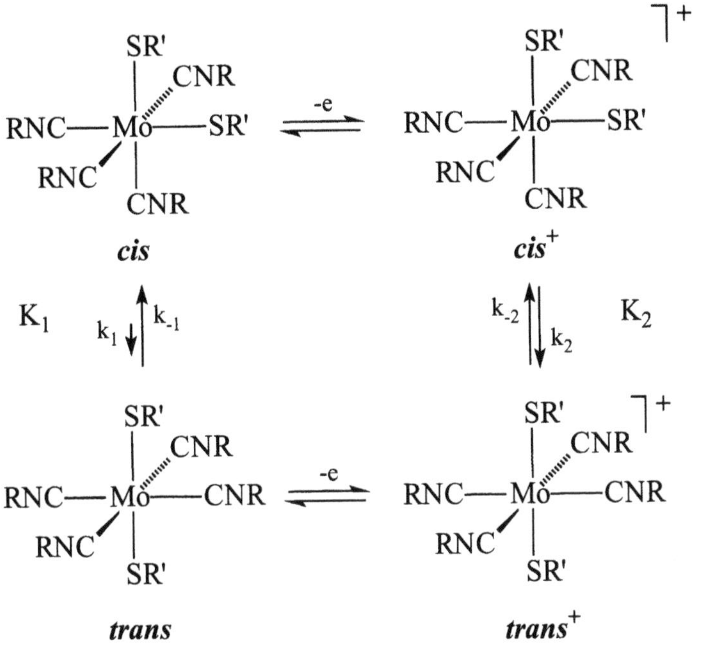

$$
\begin{array}{ccc}
\text{(cis complex)} & \overset{-e}{\rightleftharpoons} & \text{(cis}^+ \text{ complex)}^+
\end{array}
$$

cis   **cis$^+$**

$K_1 \quad k_1 \downarrow \uparrow k_{-1}$ $k_{-2} \uparrow \downarrow k_2 \quad K_2$

$$
\begin{array}{ccc}
\text{(trans complex)} & \overset{-e}{\rightleftharpoons} & \text{(trans}^+ \text{ complex)}^+
\end{array}
$$

trans   **trans$^+$**

Scheme 7

can be interpreted on the basis of a simplified d_π orbital-level splitting (Figure 3) for complexes with the general formula $[ML_4L_2']$ with L = better π-electron acceptor than L', *i.e.* L = CNR, L' = SR' in this particular case [48]. In this diagram, x is the number of L ligands that can interact with (π-acceptance from) a given metal d_π orbital, and the greater is x the more stabilized is this orbital. Hence, the HOMO is derived from the metal d_π orbital with the fewest interactions (x) with ligand L. For the diamagnetic *cis* isomer of the Mo d^4 complex, the HOMO is stabilized in relation to that of the *trans* compound in view of the higher number ($x = 3$ *vs.* 2) of π-acceptor isocyanide ligands interacting with the corresponding metal d_π orbital (d_{xy} or d_{yz} in the former, or d_{xz} or d_{yz} in the latter isomer).

The *cis* $\rightleftharpoons$ *trans* equilibrium (Scheme 7) strongly favours the *cis* isomer (K_1 = 1.7×10^{-4}) but single-electron oxidation shifts the equilibrium towards the *trans* isomeric form ($K_2 = 2$), corresponding to a substantial thermodynamic gain of *ca.* 10^4 (Figure 1 or Equation 5 for $\Delta E° = -0.09+0.27 = 0.18$ V) [48]. The kinetics are also accelerated by oxidation, the conversion into the more stable species being faster at the oxidized 15-electron compounds than at the neutral 16-electron level, *i.e.* k_2 (*cis$^+$* $\rightarrow$ *trans$^+$*) = 5 s^{-1} > k_{-1} (*trans* $\rightarrow$ *cis*) = 1 s^{-1} [48].

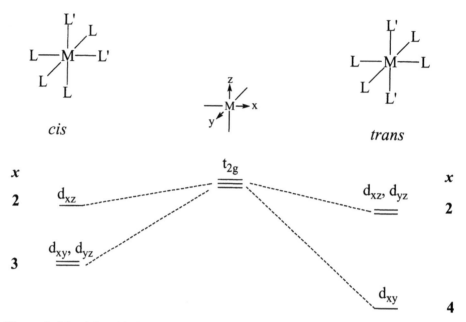

Figure 3. Metal d_π orbital level splitting for the *cis* and *trans* isomers of complexes with the general formula $[ML_4L'_2]$ (L = better π-electron acceptor than L'; x = number of stabilizing interactions of the L ligands with a given metal d_π orbital).

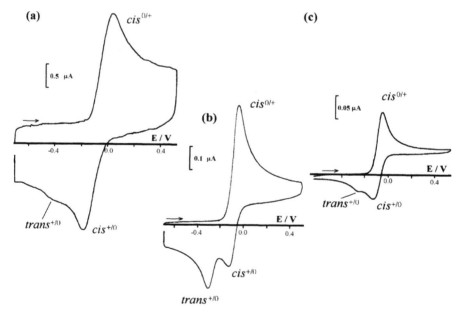

Figure 4. Cyclic voltammograms of *cis*-$[Mo(SR')_2(CNBu^t)_4]$ (R' = $C_6H_2Pr^i_3$-2,4,6) in CH_2Cl_2 with 0.2 M $[NBu_4][BF_4]$, at -50 °C and at a Pt-disc working electrode. Scan rates of 20 (a), 0.8 (b) and 0.05 (c) V.s^{-1}.

Typical cyclic voltammograms of *cis*-[Mo(SR')$_2$(CNBu^t)$_4$] are depicted in Figure 4, at different scan rates. At a sufficiently high scan rate, there is no time for an appreciable conversion of the oxidized *cis*$^+$ into *trans*$^+$ and therefore the oxidation wave (Figure 4a) approaches a reversible single-electron process. Decreasing the scan rate (Figure 4b) allows a partial and increasing *cis*$^+$ → *trans*$^+$ conversion which is recognized by the detection, on scan reversal, of increasing relative amounts of the latter, *i.e.* increasing $\rho = i_p(trans^{+/\circ})/i_p(cis^{\circ/+})$ (Figure 5a). However, ρ passes through a maximum and then starts to decrease (see also Figure 4c) upon decreasing the scan rate, what can be accounted for by the shift of the *trans*$^+$/*cis*$^+$ equilibrium towards *cis*$^+$ (during the slow reverse scan) due to the reduction of *cis*$^+$ before *trans*$^+$. As expected, the reversibility of the *cis*$^{\circ/+}$ wave follows the reverse trend of ρ (Figure 5b) [48].

The excellent fit of the simulated cyclic voltammogram, obtained for $k_5 = 5$, $k_{-2} = 2.5$, $k_{-1} = 1$ and $k_1 \approx 0$ s^{-1}, is illustrated in Figure 6 for the scan rate of 0.8 Vs^{-1} [48].

The general behaviour described above is also followed by a number of 18-electron octahedral-type carbonyl phosphine complexes, for which commonly $E^\circ(cis^{\circ/+}) > E^\circ(trans^{\circ/+})$, namely of the types [M(CO)$_2$(LL)$_2$] (M = Cr [49-

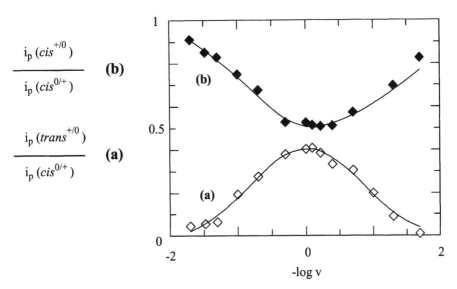

Figure 5. Experimental (symbols) (see conditions in Figure 4) and theoretical (lines) variations of the parameter $\rho = i_p(trans^{+/\circ})/i_p(cis^{\circ/+})$ (a) and of the reversibility of the *cis*$^{\circ/+}$ wave, $i_p(cis^{+/\circ})/i_p(cis^{\circ/+})$ (b) as a function of scan rate (v in Vs^{-1}). The lines correspond to the mechanism of Scheme 7 with $k_1 = 5$, $k_{-1} = 2.5$, $k_{-2} = 1$ and $k_2 = 0$ s^{-1}. Adapted from ref. [48].

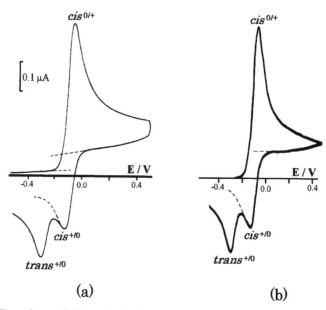

Figure 6. Experimental (a) and simulated (b) cyclic voltammograms at 0.8 V s^{-1} of a solution of *cis*-[Mo(SR')$_2$(CNBu^t)$_4$] (see conditions in Figures 4 and 5). Adapted from ref. [48].

51], Mo or W [50, 52, 53]; LL = chelating diphosphine), [ReBr(CO)(LL)$_2$] [54], [MX(CO)$_2$(L$_2$L')] (M = Mn or Re; X = Cl or Br; L$_2$L' = triphos) [55], [Mo(CO)$_2$(CNR)$_2$(PR'$_3$)$_2$] (R, R' = alkyl or aryl) [56] and the dinuclear cyanide-bridged complexes [L(LL)(CO)$_2$Mn-CN-Mn(CO)$_2$(LL)'L']$^+$ [L, L' = P(OPh)$_3$ or PEt$_3$; LL, (LL)' = diphosphine] [57,58] or [(η^5-C$_5$H$_5$)(LL)Fe-CN-Mn(CO)$_2$(LL')L] [59] in which the isomerization can be triggered by an intramolecular ET between the two metals via the cyanide bridge.

For the 18-electron dicarbonyl complexes of the general type [ML$_2$L'$_4$] (L = CO, a better π-electron acceptor than L'), with a metal d^6, the ordering E°(*cis*$^{°/+}$) > E°(*trans*$^{°/+}$) can be accounted for by considering the expected corresponding metal d$_\pi$ orbital level splitting (Figure 7) [5, 60], but such a simplified approach is not always successful for other types of complexes. Moreover, steric factors can also play a role [61-63] and *e.g.* for [M(CO)$_4$L$_2$] (M = Cr, Mo or W; L = organophosphine or phosphite ligand) [61] the *trans* isomers can become the thermodynamically more stable ones for bulky L ligands, although the *cis* isomers are favoured electronically.

The above systems were studied by a variety of electrochemical techniques (cyclic voltammetry, thin-layer cyclic voltammetry, diffusion linear potential sweep voltammetry, convolution and deconvolution potential sweep

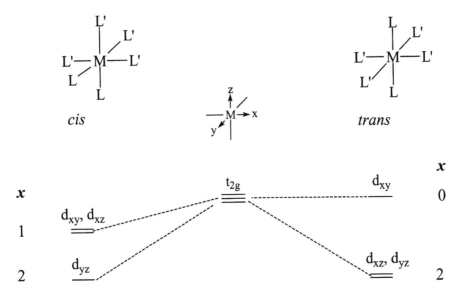

Figure 7. Metal d_π orbital level splitting for the *cis* and *trans* isomers of complexes of the general formula $[ML_2L'_4]$ (L = better π-electron acceptor than L' ; x = number of stabilizing interactions of the L ligands with a given metal d_π orbital) (adapted from reference [60]).

voltammetry, rotating-disc voltammetry, polarography or controlled potential electrolysis), in some cases associated with IR, ^{31}P or EPR spectroscopies and electrospray mass spectrometry.

The rate and equilibrium constants, when available for the square process (Scheme 6), indicate the following common features:

- Oxidation favours the relative *trans/cis* isomeric stability.

- Although isomerization can occur in both the 18-electron and the oxidized 17-electron complexes, the isomeric equilibrium favours the more stable form in the latter case ($K_2 > K_1^{-1}$). However, this feature is not observed for the above 16-/15-electron $[Mo(SR')_2(CNBu^t)_4]^{°/+}$ system ($K_2 < K_1^{-1}$) showing that generalizations have to be made cautiously.

- The isomerization is faster in both directions for the 17-electron complexes than for the 18-electron ones, the rate (k_2) towards the more stable oxidized isomer (*trans*$^+$) being faster than that (k_{-1}) towards the more stable non-oxidized one (*cis*), thus corresponding to a kinetic gain upon oxidation. Such a behaviour is also followed by the above $[Mo(SR')_2(CNBu^t)_4]^{°/+}$ couple.

Although the mechanism(s) of the isomerization itself has(have) not yet been ascertained, the low entropy of activation, the roughly non-dependence of the rates on the concentrations of the phosphine and of CO and on the solvent, in the cases these effects were investigated [50, 56] (see also below, section 4.3), support an intramolecular twist rather than a dissociative (metal-ligand bond rupture) mechanism.

4.2. Two - electron Promoted Isomerizations

The transfer of a single electron may not be sufficient to induce a geometrical isomerization which, however, can occur upon further transfer of a second electron, in particular according to a double-square mechanism (Scheme 8).

We have observed [64] such a behaviour for the series of dinitrile complexes *cis*-[Re(NCR)$_2$(dppe)$_2$][BF$_4$] (R = aryl or alkyl, *i.e.* C$_6$H$_4$NEt$_2$-4, C$_6$H$_4$OH-4, C$_6$H$_4$OMe-4, C$_6$H$_4$Me-4, C$_6$H$_5$, C$_6$H$_4$F-4, C$_6$H$_4$Cl-4, Pr^i, Bu^t, Me or CH$_2$C$_6$H$_4$Cl-4), for which a systematic study of the thermodynamic and kinetic effects of the R group was undertaken.

These complexes, as well as their corresponding trans isomers, heretofore denoted by *cis*$^+$ and *trans*$^+$, respectively, undergo two successive single-electron oxidations to form the corresponding 17-electron (*cis*$^{2+}$ and *trans*$^{2+}$) and 16-electron (*cis*$^{3+}$ and *trans*$^{3+}$) derivatives. The oxidation potentials are higher for the *cis* isomers what can be accounted for by the simplified d$_\pi$ orbital level splitting (Figure 7) expected for complexes of the general formula [ML$_2$L'$_4$] (in our case L = NCR, a better π-electron acceptor than the phosphine L' = ½dppe). Hence, E°(*cis*$^{+/2+}$) = 0.29 – 0.55 V > E°(*trans*$^{+/2+}$) = 0.11 – 0.46 V, and E°(*cis*$^{2+/3+}$) = 1.09 – 1.52 V > E°(*trans*$^{2+/3+}$) = 0.99 – 1.47 V *vs.* SCE, in CH$_2$Cl$_2$ [64].

The oxidation potential E° for the various redox couples depends on the electronic properties of the organic R group of the nitrile ligands, increasing linearly with the Hammett's σ$_p$ constant (electron-withdrawing character of

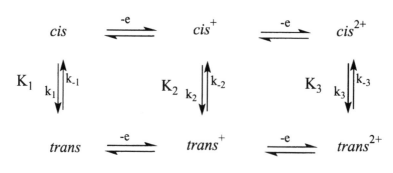

Scheme 8

the phenyl substituent) (Equations 6-9) in agreement with the expected HOMO stabilization. The sensitivity of the energy of this orbital to a change in the nitrile is higher (larger slopes of those linear plots) for the trans than for the cis isomers [64]. These observations, with IR spectroscopic [65] and X-ray structural [66, 67] data, indicate that the trans nitrile ligands are acting as more effective π-electron acceptors than the cis ones, thus being involved, to a greater extent, into the delocalised π-electron system comprising the metal and the aryl substituent which thus is able to transmit its electronic effect to the metal in a more effective way [64].

$$E°(trans^{+/2+}) = 0.340\ \sigma_p + 0.396 \tag{6}$$

$$E°(trans^{2+/3+}) = 0.452\ \sigma_p + 1.433 \tag{7}$$

$$E°(cis^{+/2+}) = 0.257\ \sigma_p + 0.503 \tag{8}$$

$$E°(cis^{2+/3+}) = 0.401\ \sigma_p + 1.463 \tag{9}$$

Steric effects can also play a role which becomes determinant in the case of the *alkyl* nitrile complexes for which $E°(cis^{+/2+})$ does not increase with the polar $\sigma*$ constant of the R group, but instead grows with its bulkiness, *i.e.* $E°(Me) <$ $E°(CH_2C_6H_4Cl\text{-}4) < E°(Pr^i) < E°(Bu^t)$ (for the other waves of both isomers this ordering is also roughly followed) [64].

Typical cyclic voltammograms for one member of the aromatic dinitrile series, *cis*-[Re(NCC$_6$H$_4$NEt$_2$-4)$_2$(dppe)$_2$]$^+$, are depicted in Figure 8 [64]. The first oxidation wave $(cis^{+/2+})$ is chemically reversible as observed upon scan reversal (*1*, in Figure 8a) and therefore no isomerization is detected. However, when the second oxidation $(cis^{2+/3+})$ wave is scanned, the $cis^{3+} \rightarrow trans^{3+}$ isomerization occurs (Scheme 9) and the $trans^{3+/2+}$ and $trans^{2+/+}$ reduction waves appear, as well as the corresponding anodic counterparts upon scan reversal (*e.g.*, *2*, in Figure 8a for the 2+/3+ level). The extent of the isomerization decreases with the increase of the scan rate (Figure 8b with lower relative current intensities for the trans isomer).

A further complication arises from the decomposition of $trans^{3+}$, but its rate (pseudo-first-order k_{dec}, Scheme 9) could be obtained by digital simulation of the cyclic voltammetric behaviour of the pure $trans^+$ species, by monitoring the current ratio $\rho = i_p(trans^{2+/+})/\ i_p(trans^{+/2+})$ *versus* the scan rate, *e.g.* Figure 9a for *trans*-[Re(NCC$_6$H$_4$F-4)$_2$(dppe)$_2$][BF$_4$]. This decomposition process is slower than the isomerization and does not occur significantly for sufficiently high scan rates, normally for log v above *ca.* 0.5 – 1.0, *e.g.* 1.0 in the case of the NCC$_6$H$_4$F-4 complex, when the $cis^{3+} \rightarrow trans^{3+}$ isomerization rate constant (k_3, Scheme 9) can be determined by simulation of the cyclic voltammetric behaviour

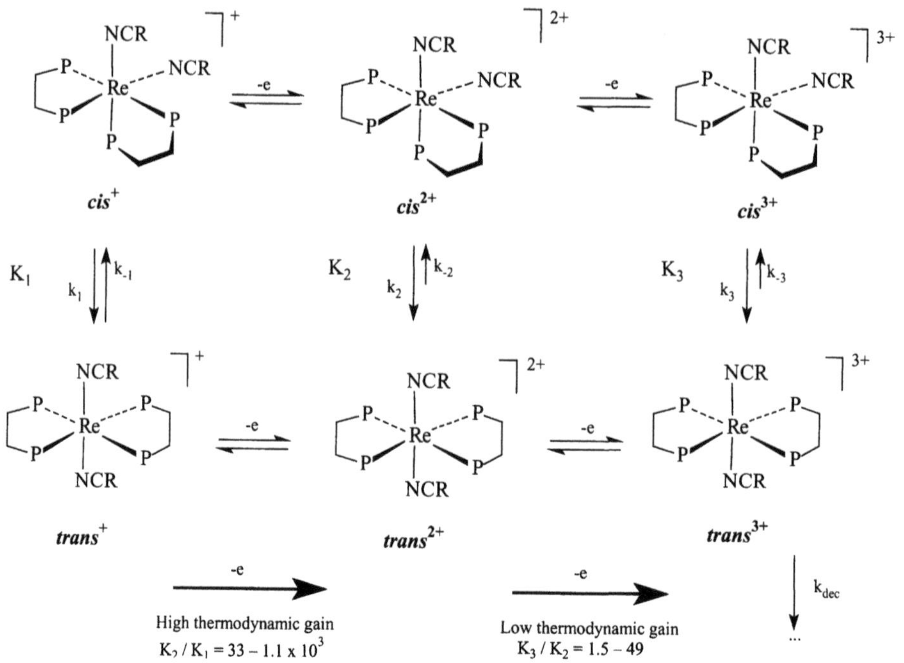

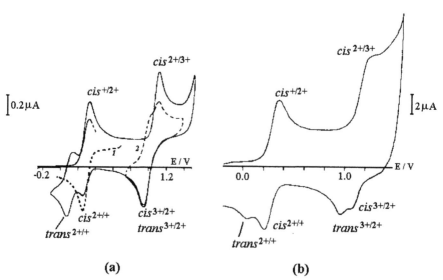

Scheme 9

Figure 8. Cyclic voltammograms for *cis*-[Re(NCC$_6$H$_4$NEt$_2$-4)$_2$(dppe)$_2$][BF$_4$] in CH$_2$Cl$_2$ with 0.2 M [NBu$_4$][BF$_4$], at 25 °C and at a Pt disc electrode at scan rates of 0.2 V s^{-1} (a) and of 20 V s^{-1}. Adapted from ref. [64].

of *cis$^+$* by monitoring the current ratio $\rho = i_p(trans^{2+/+})/ i_p(cis^{+/2+})$ as a function of the scan rate (Figure 9b illustrating the same complex). Such rate constants are dependent on electronic and steric effects as discussed below.

In contrast with the common situation encountered with the systems described above, for which the most stable non-oxidized isomeric form is the cis, in the aromatic dinitrile complexes [Re(NCR)$_2$(dppe)$_2$][BF$_4$] the trans isomers at the Re(I) 18-electron level (*trans$^+$*) appear to be thermodynamically more stable than the corresponding cis ones (the latter, *cis$^+$*, convert into the former, *trans$^+$*, upon prolonged heating) [64]. However, for the alkyl dinitrile complexes, this behaviour is only observed [65] for R = Bu^t, when steric effects favour the trans form, but not for R = Et, Pr or, as shown by theoretical studies [68], for R = Me.

Nevertheless, no isomerization was detected at ambient temperature either at the Re(I) or Re(II) oxidation state level, due to kinetic constraints. In all the cases, the relative stability of the trans isomer increases upon oxidation, normally in a pronounced way with the first oxidation ($K_2/K_1 = 33 - 1.1\times10^3$) but to a much lower extent with the second oxidation ($K_3/K_2 = 1.5 - 49$) [64]. The overall effect is given by $K_3/K_1 = 49 - 5.4\times10^4$ [64]. Theoretical studies performed on the diacetonitrile complexes confirm this behaviour: although both isomers exhibit a similar stability at the Re(I) level, the single-electron oxidation leads to an increase of the relative stability of the trans isomer at the Re(II) level, which, however, does not enhances further upon a second oxidation [68].

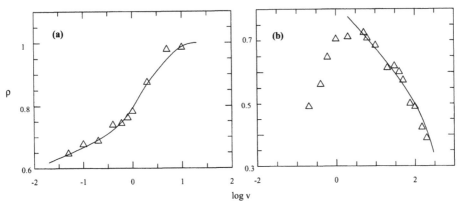

Figure 9. Experimental (symbols) and theoretical (curves) variations of the current ratios ρ as a function of the scan rate (v in V s^{-1}) for the isomers of [Re(NCC$_6$H$_4$F-4)$_2$(dppe)$_2$][BF$_4$]: (a) $\rho = i_p(trans^{2+/+})/ i_p(trans^{+/2+})$ for the *trans* isomer; (b) $\rho = i_p(trans^{2+/+})/ i_p(cis^{+/2+})$ for the *cis* isomer. Adapted from ref. [64].

The above ratios of equilibrium constants correspond to the equilibrium constants of the homogeneous ET cross-reactions 10-12 of the double square scheme 9, and have been estimated by considering $\Delta G^\circ = 0$ for each thermochemical cycle represented in this scheme and the measured E° values of the corresponding redox steps, in the mode indicated in section 2 for a single square process.

$$cis^{2+} + trans^+ \xrightleftharpoons{K_2/K_1} cis^+ + trans^{2+} \tag{10}$$

$$cis^{3+} + trans^{2+} \xrightleftharpoons{K_3/K_2} cis^{2+} + trans^{3+} \tag{11}$$

$$cis^{3+} + trans^+ \xrightleftharpoons{K_3/K_1} cis^+ + trans^{3+} \tag{12}$$

In view of (i) the above linear dependences (Equations 6-9) of E° on the Hammett's σ_p constant for the phenyl substituent in the aromatic nitrile complexes and of (ii) the general linear relationship (Equation 5) between the ratio of the isomeric equilibrium constants and ΔE° for each square process, the above ratios of the isomeric equilibrium constants at the various redox levels also respond linearly to the electronic effects of that substituent as measured by σ_p (Equations 13-15) [64]. A greater effect is observed for the first oxidation (higher slope of the linear relationship 13 in comparison with that of 14).

$$\ln \frac{K_2}{K_1} = -3.23\,\sigma_p + 4.17 \tag{13}$$

$$\ln \frac{K_3}{K_2} = -1.99\,\sigma_p + 1.05 \tag{14}$$

$$\ln \frac{K_3}{K_1} = -5.22\,\sigma_p + 5.22 \tag{15}$$

The *cis* and *trans* mononitrile chloro-Re(I) complexes [ReCl(NCR)(dppe)$_2$] constitute another system whose electrochemical behaviour follows a double-square scheme (Schemes 8 and 10) with ET-induced *cis* to *trans* isomerization which we have also investigated in detail (complexes with R = C$_6$H$_4$Me-4, onwards simply denoted by *cis* or *trans*) [69].

The *cis* isomer, in thf, undergoes two successive single electron-oxidations at potentials $E^\circ(cis^{\circ/+}) = -0.13$ V and $E^\circ(cis^{+/2+}) = 0.70$ V vs. SCE that are considerably more anodic than the corresponding ones for the *trans* isomer, $E^\circ(trans^{\circ/+}) = -0.31$ V and $E^\circ(trans^{+/2+}) = 0.67$ V vs. SCE [69]. Simplified metal d_π orbital level splittings based on the destabilizing π-electron donor effect of the chloro-ligand and on the stabilizing π-electron acceptor effect of the ligated nitrile (see below the analogous chloro-carbonyl complex, Section 4.3) do not provide an interpretation for this isomeric effect. However, theoretical calculations [68] indicate that the HOMOs in both isomers are of the same type, both being destabilized not only by p(Cl) but also by π(NC) of the NCR ligand (which thus exhibits some π-electron donor charater), being represented as a linear combination of π*(Re-Cl), π(NC) and π*(NC) orbitals (Figure 10). The contributions of the p(N) AOs roughly cancel each other whereas those of the β-carbon orbitals add. The calculations also show [68] that the ionisation potential of the *cis*-acetonitrile complex is higher than that of the *trans* isomer in accord with the experimental results.

Simulation of the cyclic voltammetric behaviour allowed to estimate the following rate and equilibrium constants [69]: $K_1 = 6.9 \times 10^{-4}$, $K_2 = 1.5$ (with rate constants k_2 and k_{-2} smaller than 10^{-4} s^{-1}) and $K_3 = k_3/k_{-3} = 5.6$ s^{-1}/1s^{-1} = 5.6. Therefore the *cis* isomer is thermodynamically more stable than the *trans* one ($K_1 \ll 1$) but a single-electron oxidation reverses the relative isomeric stability ($K_2 > 1$), also in agreement with quantum chemical calculations [68]. However, no isomeric interconversion was detected at either the Re(I) 18-electron or the Re(II) 17-electron level due to the slow kinetics (*e.g.* small k_2). A second oxidation step, to form the Re(III) 16-electron *cis*$^{2+}$ species, is required

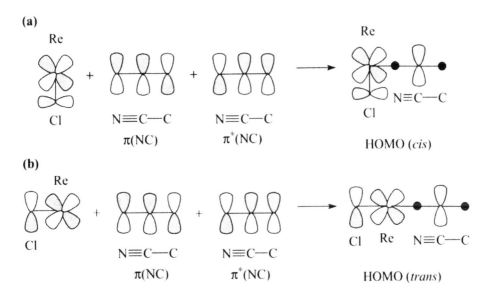

Figure 10. HOMOs of (a) *cis*- and (b) *trans*-[ReCl(NCR)(dppe)$_2$] as linear combinations of π*(Re-Cl), π(NC) and π*(NC) orbitals (only one HOMO of the degenerate pair for *trans* is shown). Adapted from ref. [68].

cis cis^+ cis^{2+}

K_1 k_1 k_{-1} K_2 k_2 k_{-2} K_3 k_3 k_{-3}

$trans$ $trans^+$ $trans^{2+}$

−e   −e

High thermodynamic gain   Low thermodynamic gain
$K_2 / K_1 = 2.2 \times 10^3$ $K_3 / K_2 = 3.7$

High kinetic gain
$k_3 / k_2 > 5 \times 10^4$

Scheme 10   (R = C_6H_4Me-4)

to effectively induce the isomerization due to the high kinetics acceleration ($k_3/k_2 > 5\times10^4$) although with only a concomitant modest thermodynamic gain ($K_3/K_2 = 3.7$).

These and other double-square systems can be of significance towards developing a "molecular hysteresis" behaviour [18c, 18d, 64, 68], in which the more stable non-oxidized 18-electron isomer and the more stable oxidized 16-electron one (*cis* and *trans*$^{2+}$, respectively, in the above example, Scheme 10) would represent, in a bistable system, the stable states whose interconversion could be triggered by ET.

The opposite type of anodically promoted geometrical isomerization, *i.e.* *trans*-to-*cis*, is also known in different systems with reversed relative isomeric stabilities, as reported [70] for the redox series [Os(N$_2$O$_2$)L$_2$]^n [N$_2$O$_2$ = basic form (4−) of 1,2-bis(3,5-dichloro-2-hydroxybenzamido)-4,5-dichlorobenzene; L = 4-substituted pyridine; n = 2− to 1+] which follow a triple square scheme whose kinetics however were not investigated.

4.3 Electron-Transfer-Chain (ETC)-Catalyzed Isomerizations

In the systems discussed above, the ET-generated isomer (*trans*$^+$ in Scheme 6, or, more generally, species B$^+$ in Scheme 1) belongs to a redox couple with a lower redox potential than that of the parent species, *i.e.* E$^\circ$ (*trans*$^{\circ/+}$) < E$^\circ$ (*cis*$^{\circ/+}$) or, in a more general way, E$^\circ$(B$^{\circ/+}$) < E$^\circ$ (A$^{\circ/+}$), and therefore ΔE$^\circ$ > 0 and K$_2$/K$_1$ > 1 (see Equations 3 or 1, respectively).

However, if the opposite happens, *i.e.* E$^\circ$(*trans*$^{\circ/+}$) > E$^\circ$(*cis*$^{\circ/+}$) or E$^\circ$(B$^{\circ/+}$) > E$^\circ$(A$^{\circ/+}$), then ΔE$^\circ$ < 0 and K$_2$/K$_1$ < 1, *i.e.* the *cis*-to-*trans* isomerization (or the A-to-B) is *not* promoted by oxidation (in fact, it should be hampered). Nevertheless, an interesting sequence of further ET reactions occurs and an ET-triggered isomerization can proceed without an overall redox change. This results from the fact that any *trans*$^+$ generated upon oxidation of *cis* at E$^\circ$(*cis*$^{\circ/+}$) (even if formed in a relatively small amount in view of the unfavourable relationship K$_2$ < K$_1$) should be reduced to *trans*, either by the electrode (heterogeneously) which is at a cathodic potential relatively to *trans*$^+$ [note that E$^\circ$(*cis*$^{\circ/+}$) < E$^\circ$(*trans*$^{\circ/+}$)] or by *cis* (homogeneous cross-redox reaction) with catalytic regeneration of *cis*$^+$ (Scheme 11). Hence, the process is catalysed by the electron which simply triggers the isomerization of *cis* to *trans* and corresponds to an ET-chain (ETC) catalyzed process that can be denoted by $\overrightarrow{E}C\overleftarrow{E}$ where $\overrightarrow{E}$ is the initial ET step, C is the chemical propagation (isomerization) step and $\overleftarrow{E}$ is the "back ET", the second propagation step that occurs, as mentioned above, either at the electrode or in solution (chain propagation cross-redox reaction) [8]. ETC-catalysis was first found in Inorganic Chemistry by Taube [71] and also developed by Basolo and Pearson [72]. The electrochemical $\overrightarrow{E}C\overleftarrow{E}$ electrocatalytic cycle was rationalized by Feldberg [73], and pioneering applications in Electrochemistry were also reported by Savéant and Amatore [74].

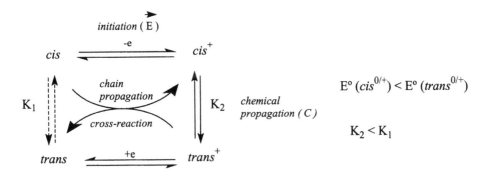

Scheme 11

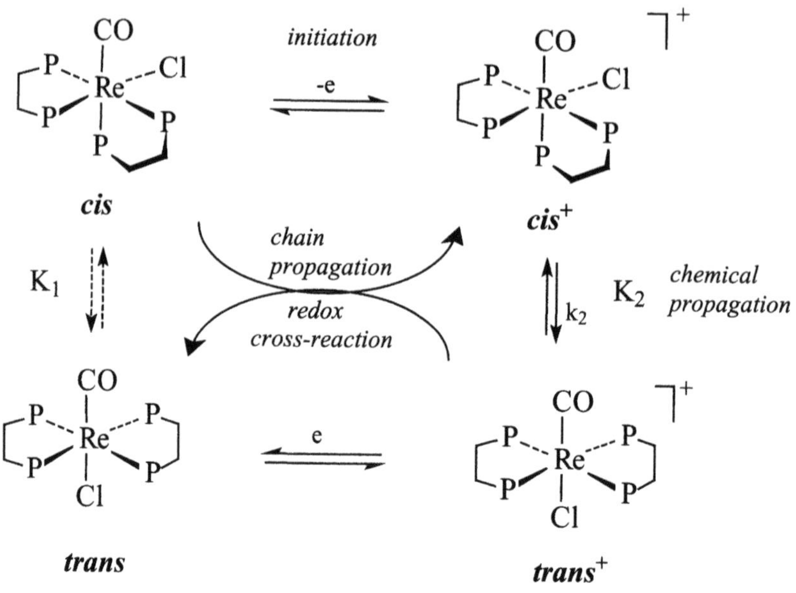

Scheme 12

An ETC catalysed isomerization we have studied in detail [75] involves the chloro-carbonyl *cis-/trans-*[ReCl(CO)(dppe)$_2$] isomeric couple (Scheme 12). In fact and in contrast with the cases mentioned in the previous section, the *cis* isomer is oxidized at a lower potential than the *trans* one, *i.e* E°(*cis*$^{\circ/+}$) = 0.41 V < E°(*trans*$^{\circ/+}$) = 0.68 V *vs.* SCE, in CH$_2$Cl$_2$, what can be interpreted by a simplified d$_\pi$ orbital level splitting (Figure 11) for complexes of the general type [MXLL'$_4$] (X = strong π-electron donor, L = strong π-electron acceptor, L' = 2e-donor ligand that is neither a π-electron donor nor an effective π-acceptor, *i.e.* in our case, X = Cl, L = CO, L' = ½dppe) [75]. Destabilization of a metal d$_\pi$ orbital by the chloride π-donor ligand without a compensating stabilization by CO (x = 1) occurs only for the d$_{xy}$ orbital of the *cis* isomer which thus lies at a higher energy than that of the HOMO of the *trans* isomer.

Theoretical calculations corroborate these expectations indicating [68] that the HOMOs of the *cis* and *trans* isomers (Figure 12) are π*[d$_{xy}$(Re)–p(Cl)] and the linear combination of π*[d$_{xz}$ or d$_{yz}$(Re)–p(Cl)] with π*(CO) and, to a smaller extent, with π(CO), respectively. As expected, CO is a stronger π-acceptor than a nitrile and the MO of *cis-*[ReCl(CO)(dppe)$_2$] that corresponds to the HOMO in *cis-*[ReCl(NCR)(dppe)$_2$] (Figure 10a) has a stronger π*(CO) contribution than that of π*(CN) in the latter MO resulting in its stabilization to an energy level

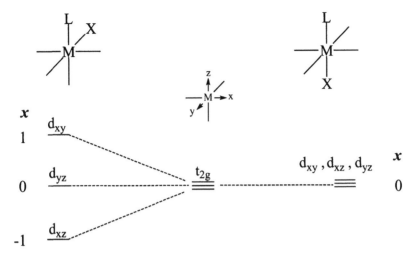

Figure 11. Metal d_π orbital level splitting for the *cis* and *trans* isomers of complexes of the general formula [MXLL'$_4$] (X = strong π-electron donor, L = strong π-electron acceptor, L' = neither a π-donor nor an effective π-acceptor, *x* = difference between the number of destabilizing interactions with filled X orbitals and the number of stabilizing interactions with empty L orbitals).

below that of $\pi^*[d_{xy}(Re)-p(Cl)]$ (Figure 12a) where such a stabilization is not possible and that MO then becomes the HOMO.

The higher stabilizing contribution of the $\pi^*(CO)$ orbital in the HOMO of the *trans* carbonyl complex relatively to $\pi^*(NC)$ in the *trans* nitrile complexes is also in accord with the higher oxidation potential of the carbonyl compared with the nitrile complexes.

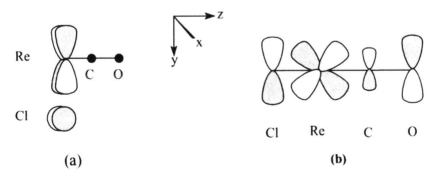

(a)   (b)

Figure 12. HOMOs of (a) *cis*- and (b) *trans*-[ReCl(CO)(dppe)$_2$] (only one of the pair of degenerate HOMOs is shown for *trans*). Adapted from ref. [68].

The quantum chemical studies also show [68] that the ionization potential of the *trans* carbonyl isomer is higher than that for the *cis*, consistent with the experimental E° data. Moreover, and in contrast to the nitrile complexes, the *trans* isomer is more stable than the *cis* one [68], in accord with the observed [75] conversion of the latter into the former, at room temperature, whereas the reverse reaction was not detected. However, the spontaneous conversion of *cis* into *trans* is not a clean reaction and other decomposition products are also detected in relative amounts that are temperature dependent. To avoid such a decomposition of *cis*, the electrochemical study was undertaken [75] at low temperature, typically –40 °C, in CH_2Cl_2.

The clean ET-induced conversion of *cis* to *trans* is shown by multiple scans cyclic voltammetry which reveals the occurrence of isopotential points (Figure 13) [75]. Oxidation of *cis* to *cis*⁺ is followed by its isomerization to *trans*⁺ (Scheme 12) which is reduced either heterogeneously by the electrode, at the potential it is generated, E°(*cis*°/⁺), or in an homogeneous way by *cis* (redox cross reaction) to yield *trans* and *cis*⁺ (Equation 16).

$$trans^+ + cis \underset{k_-}{\overset{k}{\rightleftharpoons}} trans + cis^+ \qquad\qquad (16)$$

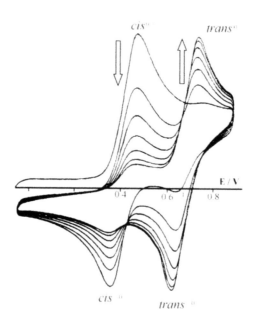

Figure 13. Multiple scans cyclic voltammetry for a solution of *cis*-[ReCl(CO)(dppe)₂] in CH_2Cl_2 with 0.2 M [NBu₄][BF₄], at –40 °C (scan rate of 0.2 V s⁻¹). Adapted from ref. [75].

Hence, the overall reaction consists on the oxidative conversion of *cis* to the non-oxidized *trans* isomer.

The redox cross reaction is thermodynamically favoured, since its equilibrium constant is the inverse of K_2/K_1 given by Equation 5, *i.e.* $K_1/K_2 = 6.7 \times 10^5$ at $-40\ ^\circ$C. Therefore, in sharp contrast with the systems mentioned in the previous section, the oxidation of *cis* does *not* result into a shift of the isomerization equilibrium   towards the oxidized *trans* geometry (formation of *trans*$^+$), *i.e.* $K_2 \ll K_1$. Theoretical calculations [68] confirm the decrease of the relative stability of the *trans* carbonyl isomer upon oxidation. However, the *cis*$^+ \rightarrow$ *trans*$^+$ isomerization (chemical propagation step, Scheme 12) occurs on account of the effects of other reactions (consumption of *trans*$^+$ in the two ways indicated above) and of the expected faster kinetics at the 17-electron complexes than at the starting 18-electron species.

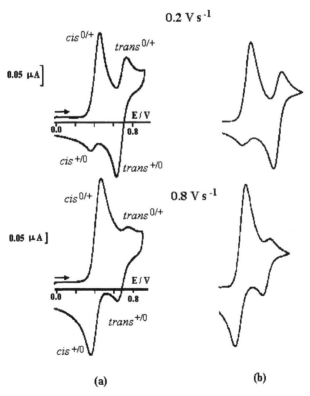

Figure 14. Experimental (a) and simulated (b) cyclic voltammograms of *cis*-[ReCl(CO)(dppe)$_2$] in CH$_2$Cl$_2$ with 0.2 M [NBu$_4$][BF$_4$], at $-40\ ^\circ$C and at a Pt disc electrode (scan rate of 0.2 and 0.8 V s^{-1}). Adapted from ref. [75].

The catalytic and quantitative isomeric *cis* → *trans* conversion was achieved by controlled potential electrolysis at the oxidation wave of *cis*: it was fast and involved only a very small current and charge, being completed after the consumption of less than 0.018 F mol^{-1} (coulombic efficiency greater than 56) [75]. A small charge was sufficient to trigger all the conversion.

Digital simulation of the cyclic voltammograms in a range of scan rates (illustrated in Figure 14 for 0.2 and 0.8 Vs^{-1}) has allowed to estimate (i) the forward rate constant for the cross redox reaction 17, $k = (9\pm2) \times 10^{-3}$ M^{-1} s^{-1} [the backward rate constant, $k_- = (1.3\pm0.9) \times 10^{-2}$ M^{-1} s^{-1}, could then be estimated from the knowledge of the equilibrium constant of this reaction, see above] and (ii) the rate of the *cis*$^+$ → *trans*$^+$ isomerization at various temperatures, *e.g.* $k_2 = 0.25\pm0.10$, 0.85 ± 0.15, 12.5 ± 1.5 and 80 (by extrapolation) s^{-1} at −50, −40, −20 and 0 °C, respectively [75]. The Eyring and Arrhenius activation parameters were estimated (log A = 13.4±1.3, $E_a = 60\pm6$ kJmol^{-1}, $\Delta H^{\neq} = 58\pm6$ kJ mol^{-1} and $\Delta S^{\neq} = -4\pm20$ JK^{-1}mol^{-1}) [75] and the low $\Delta S^{\neq}$ value (in spite of the pronounced uncertainty) may be indicative of the occurrence of the isomerization *via* an intramolecular twist rather than a dissociative mechanism.

A comparison of the isomerization rate constant of this carbonyl complex with those of related compounds will be mentioned below.

Further examples of electrocatalyzed structural rearrangements are barely known, namely a cobalt/iridium carbonyl cluster [17a] or various di- or poly-carbonyl complexes [76]. The field is still underdeveloped, even in comparison with other reaction types (mainly ligand exchange) that have been electrocatalyzed [8], and deserves further exploration.

4.4   Isomerization Rate Factors

Among the above *cis* carbonyl and nitrile Re complexes, *cis*-[ReCl(CO)(dppe)$_2$] is the one that exhibits the highest propensity for isomerization. In fact, the *cis*-to-*trans* isomerization, which is already detected even without oxidation at temperatures above *ca.* −20 °C, occurs, upon a single-electron oxidation (*cis*$^+$ → *trans*$^+$), at the highest rate constant, *i.e* $k_i = k_2 = 80$ [75] or 7.3×10^2 s^{-1} (values extrapolated for 0 or 25 °C, respectively), whereas for the singly oxidized nitrile complexes *cis*-[ReCl(NCR)(dppe)$_2$]$^+$ [69] or *cis*-[Re(NCR)$_2$(dppe)$_2$]$^{2+}$ [64] such a rate is much lower than 10^{-4} s^{-1} and the isomerization is not yet detected. In order to undergo ready isomerization, these nitrile complexes require a further oxidation to the 16-electron level, *i.e.* to form *cis*-[ReCl(NCR)(dppe)$_2$]$^{2+}$ ($k_i = k_3 = 5.6$ s^{-1} at 0 °C for R = C$_6$H$_4$Me-4 [69]) or *cis*-[Re(NCR)$_2$(dppe)$_2$]$^{3+}$ ($k_i = k_3$ in the range 0.4×10^2 − 25×10^2 s^{-1} *e.g.* 1.5×10^3 s^{-1} for R = C$_6$H$_4$Me-4, at 25 °C [64]), respectively.

Moreover, within the aromatic dinitrile complexes, the isomerization rate increases with the electron-withdrawing ability of the phenyl substituent as measured by the Hammett's σ_p constant [64].

Those observations suggest that the geometrical rearrangement, which conceivably occurs via an intramolecular twist mechanism (see above), is promoted by an increase of the electron-acceptor character of the ligand (CO > NCR when comparing the carbonyl with the nitrile complexes, and enhancement of the σ_p value for the phenyl substituent within the aromatic nitrile series). This electron-acceptance conceivably results in the labilization of the metal-P (diphosphine) bonds with a promoting effect on the isomerization.

Similarly, the isomerization rate of the dioxidized alkyl dinitrile complexes increases from CNBu^t to CNPr^i [64].

Interestingly, the kinetics of the ET-induced metal-hydride bond cleavage can respond similarly to electronic factors as discussed above (section 3.2) for the hydride-isocyanide complexes *trans*-[FeH(CNR)(dppe)$_2$]$^+$ which, upon oxidation, undergo H$^+$ loss (by Fe-H bond cleavage) whose rate constant correlates [33] with CNR electronic constants (Hammett's σ_p or Taft σ^* constants, and P$_L$ ligand parameter).

However, the *alkyl* dinitrile rhenium complexes (dioxidized level) isomerize at rates ($k_i = k_3$ in the 1.0×10^3 to 1.3×10^3 range [64]) that are not slower than those of the aromatic complexes, in disagreement with the behaviour expected on the basis of the above electronic arguments (alkyl nitrile ligands should be weaker π-electron acceptors than the aromatic ones). The anomaly conceivably can be accounted for by steric effects which can play a dominant role in the alkyl dinitrile complexes, mainly at the higher metal oxidation state, Re(III), when π-electronic effects for the ligated aliphatic nitriles should be minimized. The steric effects are expected to provide a further driving force to the *cis*-to-*trans* isomerization resulting into higher rates than those expected by taking into consideration only electronic arguments.

5   FINAL REMARKS

The situations discussed above show that the metal-hydrogen bond and the isomeric form of a coordination compound are particularly sensitive to its electronic configuration, structure and composition. They can be readily activated by oxidation of the complex and hydride compounds often can undergo *e.g.* oxidative heterocyclic metal-H bond cleavage with H$^+$ loss that can be followed by further ET and/or chemical steps, whereas geometrical isomerizations can also readily be induced by ET.

A particularly interesting conclusion of some of our studies is that those apparently quite distinct processes (metal-H bond cleavage and geometrical isomerization) can respond similarly to electronic factors and follow related types of mechanisms, thus in fact constituting comparable processes.

The differences in the redox potentials of the involved hydride/conjugate base or *cis/trans* isomeric couples are a proeminent factor of the relative stability of the species involved, thus playing a major role on the control of the effect of ET on the equilibrium constants concerning the (inter)conversions of the members of those couples. Therefore such reactions are dependent on all the factors that influence the redox potential and common rationalizations are expected.

Some have already been proposed for certain types of isomers, namely based on simplified metal orbital level splittings, but other types have not yet been adequately interpreted and/or neither experimental studies nor theoretical calculations have been applied in a systematic way.

Electrochemical methods can provide convenient tools for the study of the above processes, not only by inducing the ET (electrode as the electron-acceptor) but also by detecting intermediates and products and establishing the involved mechanisms provided the redox processes of the hydride complex and its conjugate base, or of the involved isomers, occur at distinct potentials.

Digital simulation of cyclic voltammetry, in a range of scan rates, constitutes a particularly promising approach as demonstrated above for some systems which we have studied in detail.

The mechanisms can be of a considerable complexity, following *e.g.* EC sequences or other E_xC_y combinations. In some cases a single ET is not sufficient to induce the metal-H bond cleavage or the isomerization and a further ET is required. Square-type mechanisms have been established and fully analysed in some systems, with the estimate of acid or isomerization equilibrium and rate constants, allowing the quantification of the thermodynamic and kinetic gains associated to the chemical steps induced by ET. Thermodynamics and kinetics appear to respond to both electronic and steric factors, in a delicate and not always predictable manner. Nevertheless, we have shown that, in related complexes, the rate constants of the anodically-induced metal-hydrogen bond cleavage and of the geometrical isomerization increase with the net electron acceptor character of the ligands which thus favours such reactions.

In accord with this, complexes with carbonyl ligands (CO is a strong π-acid) have been the most studied ones. However, the use of ligands with different electronic properties (*e.g.* weak π-acceptors like nitriles or even π-donors like halides) can lead to quite distinct behaviours, in particular involving more than one ET *e.g.* in double-square schemes or related ones (see our diiron hydride complexes with bridging dinitriles and rhenium nitrile complexes).

Other types of studies that deserve to be further explored include poly-ET processes (*e.g.* extended square or ladder schemes), ETC-catalyzed processes (still very little investigated in the above reactions in spite of their high synthetic significance with a minimum energy consumption) and complementary theoretical investigations (only scarcely applied but rather promising towards the understanging and even prevision of the various mechanistic steps as we have shown for a dinuclear dihydride system and for some isomeric couples).

ACKNOWLEDGEMENTS

The authors are indebted to all the co-authors cited in the references, as well as to Prof. C. Amatore (École Normale Supérieure, Paris) for training on digital simulation methods (the ET-induced isomerizations of the rhenium nitrile complexes were performed jointly). Thanks are also due to Prof. Fraústo da Silva for his general support. The work has been supported by the Foundation for Science and Technology (FCT), the PRAXIS XXI and POCTI (FEDER funded) programmes.

REFERENCES

1. Vlcek, A., Jr. *Chemtracts-Inorg. Chem.*, *5*, 1 (1993).
2. Geiger, W.E. *Acc. Chem. Res.*, *28*, 351 (1995).
3. Bond, A.M.; Colton, R. *Coord. Chem. Rev.*, *166*, 161 (1997).
4. Battaglini, F.; Calvo, E.J.; Doctorovich, F. *J. Organometal. Chem., 547*, 1 (1997)
5. Pombeiro, A.J.L.; A.J.L.; Guedes da Silva, M.F.C.; Lemos, M.A.N.D.A. *Coord. Chem. Rev., 219-221*, 53 (2001).
6. Pombeiro, A.J.L. *Portugaliae Electrochim. Acta, 1*, 19 (1983).
7. Pombeiro, A.J.L.; McCleverty, J. (Eds.) *Molecular Electrochemistry of Inorganic, Bioinorganic and Organometallic Compounds*, NATO ASI Series, Kluwer Academic, Dordrecht, 1993.
8. Astruc, D. *Electron Transfer and Radical Processes in Transition-Metal Chemistry*, VCH, New York, 1995.
9. Savéant, J.-M. *Advances in Physical Organic Chemistry*, *35*, Academic Press, London, 2000, pp. 117-192.
10. Crabtree, R.H. *The Organometallic Chemistry of the Transition Elements*, 3rd ed., John Wiley, Toronto, 2001.
11. Astruc, D. *Chimie Organométallique*, EDP Sciences, Les Ulis, 2000.
12. Mathey, F.; Sevin, A. *Molecular Chemistry of the Transition Elements*, John Wiley, Chichester, 1996.
13. Fraústo da Silva, J.J.R.; Williams, R.J.P. *The Biological Chemistry of the Elements*, Oxford Univ. Press, 2nd ed., 2001.
14. Kaim, W.; Schwederski, B. *Bioinorganic Chemistry: Inorganic Elements in the Chemistry of Life*, John Wily, Chichester, 1994.

15. Peruzzini, M.; Poli, R. (Eds.) *Recent Advances in Hydride Chemistry*, Elsevier, Amsterdam, 2001.

16. (a) Ruiz, J; Ogliaro, F.; Saillard, J.-Y; Halet, J.-F; Varret, F.; Astruc, D. *J. Am. Chem. Soc.*, *120*, 11693 (1998). (b) Roth, T.; Kaim, W. *Inorg. Chem.*, *31*, 1930 (1992).

17. (a)Geiger, W.E.; Shaw, M.J.; Wünsch, M; Barnes, C.E.; Foersterling, F.H. *J. Am. Chem. Soc. 119*, 2804 (1997). (b) Delville-Desbois, M.-H.; Mross, S.; Astruc, D.; Linares, J.; Varret, F.; Rabaâ, H.; Le Beuze, A.;. Saillard, J.-Y; Culp, R.D.; Atwood, D.A.; Cowley, A.H. *J. Am. Chem. Soc. 118*, 4133 (1996). (c) Delville-Desbois, M.-H.; Mross, S.; Astruc, D. *Organometallics, 15*, 5598 (1996). (d) Powell, D.W.; Lay, P.A. *Inorg. Chem.*, *31*, 3542 (1992).

18. (a) Szczepura, L.F.; Kubow, S.A.; Leising, R.A.; Perez, W.J.; Huynh, M.H.V.; Lake, C.H.; Churchill, D.G.; Churchill, M.R.; Takunchi, K.J., *J. Chem. Soc. Dalton Trans.*, 1463 (1996). (b) Tomita, A.; Sano, M., *Inorg. Chem.*, *33,* 5825 (1994). (c) Sano, M.; Taube, H. *Inorg. Chem.*, *33,* 705 (1994). (d) Sano, M.; Taube, H. *J. Am. Chem. Soc.*, *113*, 2327 (1991). (e) Roth, T.; Kaim, W. *Inorg. Chem.*, *31*, 1930 (1992).

19. (a) Pierce, D.T.; Hatfield, T.L.; Billo, E.J.; Ping, Y. *Inorg. Chem.*, *36*, 2950 (1997). (b) Villeneuve, N.M.; Schroeder, R.R.; Ochrymowycz, L.A.; Rorabacher, D.B. *Inorg. Chem. 36*, 4475 (1997). (c) Robandt, P.V.; Schroeder, R.R.; Rorabacher, D.B. *Inorg. Chem.*, *32,* 3957 (1993). (d) Cárdenas, D.J.; Livoreil, A.; Sauvage, J.-P. *J. Am. Chem. Soc., 118,* 11980 (1996). (e) Livoreil, A.; Dietrich-Buchecker, C.O.; Sauvage, J.-P. *J. Am. Chem. Soc.*, *116*, 9393 (1994).

20. Wytko, J.A.; Boudon, C.; Weiss, J.; Gross, M. *Inorg. Chem.*, *35*, 4469 (1996).

21. Jacq, J. *J. Electroanal. Chem. Interfacial Electrochem.*, *29*, 149 (1971).

22. (a) Evans, D.E. *Chem. Rev.*, *90*, 739 (1990). (b) Astruc, D. *Angew Chem. Int. Ed. Engl.*, *27*, 643 (1988). (c) Geiger, W.E. *Progress Inorg. Chem.*, *33*, 275 (1985). (d) Vallat, A.; Person, M.; Roullier, L.O.; Laviron, E. *Inorg. Chem.*, *26*, 332 (1987).

23. Pombeiro, A.J.L. *Inorg. Chem. Commun.*, *4*, 585 (2001).

24. Pombeiro, A.J.L. ; Guedes da Silva, M.F.C. *J. Organometal. Chem.*, *617-618*, 65 (2001)

25. Almeida, S.S.P.R.; Guedes da Silva, M.F.C.; Fraústo da Silva, J.J.R.; Pombeiro, A.J.L. *J. Chem. Soc., Dalton Trans.*, 467 (1999).

26. Guedes da Silva, M.F.C.; Lemos, M.A.N.D.A.; Fraústo da Silva, J.J.R.; Pombeiro, A.J.L.; Pellinghelli, M.A.; Tiripicchio, A. *J. Chem. Soc., Dalton Trans., 373* (2000).

27. Lemos, M.A.N.D.A.; Guedes da Silva, M.F.C.; Pombeiro, A.J.L. *Inorg. Chim. Acta*, *226*, 9 (1994).

28. (a) Ryan, O.B.; Tilset, M.; Parker, V.D. *J. Am. Chem. Soc.*, *112*, 2618 (1990). (b) Tilset, M. *J. Am. Chem. Soc.*, *114*, 2740 (1992). (c) Skagestad, V.; Tilset, M. *J. Am. Chem. Soc.*, *115*, 5077 (1993). (d) Tilset, M.; Hamon, J.-R.; Hamon, P. *Chem. Commun.*, 765 (1998). (e) Tilset, M.; Fjeldahl, I.; Hamon, J.-R.; Hamon, P.; Toupet, L.; Saillard, J.-Y.; Costuas, K.; Haynes, A. *J. Am. Chem. Soc.*, *123*, 9984 (2001).

29. Marken, F.; Bond, A.M.; Colton, R. *Inorg. Chem.*, *34*, 1705 (1995).

30. Wang, D.; Angelici, R.J. *J. Am. Chem. Soc.*, *118*, 935 (1996).

31. Berning, D.E.; Noll, B.C.; DuBois, D.L., *J. Am. Chem. Soc. 121*, 11432 (1999).

32. Lemos, M.A.N.D.A.; Pombeiro, A.J.L. *J. Organometal. Chem.*, *438*, 159 (1992).

33. Lemos, M.A.N.D.A.; Pombeiro, A.J.L. *J. Organometal. Chem.*, *332*, C17 (1987)

34. Venâncio, A.I.F.; Kuznetsov, M.L.; Guedes da Silva, M.F.C.; Martins, L.M.D.R.S.; Fraústo da Silva, J.J.R.; Pombeiro, A.J.L. *Inorg. Chem.*, *41*, 6456 (2002).

35. Klinger, R.J.; Huffman, J.C.; Kochi, J.K. *J. Am. Chem. Soc.*, *102*, 208 (1980).

36. Pedersen, A.; Tilset, M. *Organometallics*, *13*, 4887 (1994).

37. Ryan, O.B.; Tilset, M. *J.Am. Chem. Soc.*, *113*, 9554 (1991).

38. Zlota, A.A.; Tilset, M.; Caulton, K.G. *Inorg. Chem.*, *32*, 3816 (1993).

39. Smith, K.-T.; Tilset, M.; Kuhlman, R.; Caulton, K.G. *J. Am. Chem. Soc.*, *117*, 9473 (1995).

40. Westerberg, D.E.; Rhodes, L.F.; Edwin, J.; Geiger, W.E.; Caulton, K.G. *Inorg. Chem.*, *30*, 1107 (1991).

41. Smith, K.-T.; Romming, C.; Tilset, M. *J. Am. Chem. Soc.*, *115*, 8681 (1993).

42. Detty, M.R.; Jones, W.D. *J. Am. Chem. Soc.*, *109*, 5666 (1987).

43. Amatore, C.; Fraústo da Silva, J.J.R.; Guedes da Silva, M.F.C.; Pombeiro, A.J.L. ; Verpeaux, J.-N. *J. Chem. Soc., Chem. Commun.*, 1289 (1992).

44. Guedes da Silva, M.F.C.; Fraústo da Silva, J.J.R.; Pombeiro, A.J.L., *Inorg. Chem.*, *41*, 219 (2002).

45. Baptista A.A.; Cordeiro, L.A.C.; Oliva, G. *Inorg. Chim. Acta*, *203*, 185 (1993).

46. Menglet, D.; Bond, A.M.; Coutinho, K.; Dickson, R.S.; Lazarev, G.G.; Olsen, S.A.; Pilbrow, J.P. *J.Am. Chem. Soc.*, *120*, 2086 (1998).

47. Bianchini, C.; Peruzzini, M.; Ceccanti, A.; Laschi, F.; Zanello, P. *Inorg. Chim. Acta*, *61*, 259 (1997).

48. Guedes da Silva, M.F.C.; Hitchcock, P.B.; Hughes, D.L.; Marjani, K.; Pombeiro, A.J.L.; Richards, R.L. *J. Chem. Soc., Dalton Trans.*, 3725 (1997).

49. Bond, A.M.; Colton, R.; Cooper, J.B.; Traeger, J.C.; Walter, J.N.; Way, D.M. *Organometallics*, *13*, 3434 (1994).

50. Bond, A.M.; Grabaric, B.S.; Jackowski, J.J. *Inorg. Chem.*, *17*, 2153 (1978).

51. Cummings, D.A.; McMaster, J.; Rieger, A.L.; Rieger, P.H. *Organometallics*, *16*, 4362 (1997).

52. Wimmer, F.L.; Snow, M.R.; Bond, A.M. *Inorg. Chem.* 13, 1617 (1974).

53. Abdel-Hamid, R.; El-Samahy, A.A.; Rabia, M.K.M.; Taylor, N.; Shaw, B.L. *Bull. Chem. Soc. Jpn.*, *67*, 321 (1994).

54. Bond, A.M.; Colton, R.; Humphrey, D.G.; Mahon, P.J.; Snook, G.A.; Tedesco, V.; Walter, J.N. *Organometallics 17*, 2977 (1998).

55. Bond, A.M.; Colton, R.; Gable, R.W.; Mackay, M.F.; Walter, J.N. *Inorg. Chem. 36*, 1181 (1997).

56. Conner, K.A.; Walton, R.A., *Inorg. Chem.*, *25*, 4422 (1986).

57. Carriedo, G.A.; Connelly, N.G.; Crespo, M.C.; Quarmby, I.C.; Riera, V.; Worth, G.H. *J. Chem. Soc., Dalton Trans.*, 315 (1991).

58. Carriedo, G.A.; Connelly, N.G.; Crespo, M.C.; Quarmby, I.C.; Riera, V. *J. Chem. Soc., Chem. Commun.* 1806 (1987).

59. Barrado, G.; Carriedo, G.A.; Diaz-Valenzuela, C.; Riera, V. *Inorg. Chem.*, *30*, 4416 (1991).

60. Bursten, B.E. *J. Am. Chem. Soc.*, *104*, 1299 (1982).

61. Bond, A.M.; Colton, R.; Kevekordes, J.E. *Inorg., Chem.*, *25*, 749 (1986).

62. (a) Menon, M.; Pramanik, A.; Bag, N.; Chakravorty, A. *J. Chem. Soc., Dalton Trans.*, 1543, (1995). (b) Bag, N.; Lahiri, G.K.; Chakravorty, A. *J. Chem. Soc., Dalton Trans.*, 1557 (1990).

63. Pramanik, A.; Bag, N.; Chakravorty, A. *J. Chem. Soc., Dalton Trans.*, 237 (1993).

64. Guedes da Silva, M.F.C. ; Fraústo da Silva, J.J.R.; Pombeiro, A.J.L.; Amatore, C.; Verpeaux, J.-N. *Inorg. Chem.*, *37*, 2344 (1998).

65. Guedes da Silva, M.F.C.; Fraústo da Silva, J.J.R.; Pombeiro, A.J.L., *J. Organometal. Chem.*, *526*, 237 (1996).

66. Guedes da Silva, M.F.C. ; Duarte, M.T.; Galvão, A.M.; Fraústo da Silva, J.J.R.; Pombeiro, A.J.L. *J. Organometal. Chem.*, *433*, C14 (1992).

67. Guedes da Silva, M.F.C. ; Pombeiro, A.J.L.; Hills, A.; Hughes, D.L.; Richards, R.L. *J. Organometal. Chem.*, *403*, C1 (1991).

68. Kuznetsov M.L.; Pombeiro, A.J.L. *Dalton Trans.*, 738 (2003).

69. Guedes da Silva, M.F.C.; Fraústo da Silva, J.J.R.; Pombeiro, A.J.L.; Amatore, C.; Verpeaux, J. -N. *Organometallics*, *13*, 3943 (1994).

70. Anson, F.C.; Collins, T.J.; Gipson, S.L.; Keech, J.T.; Krafft, T.E.; Prake, G.T. *J. Am. Chem. Soc.*, *108*, 6593 (1986).

71. Rich, R.L.; Taube, H. *J. Am. Chem. Soc.*, *76*, 2608 (1954).

72. Basolo, F.; Pearson, R.G. *Mechanisms of Inorganic Reactions*, John Wiley, New York, 1967 and refs. therein.

73. Feldberg, S.W.; Jeftic, L. *J. Phys. Chem.*, *76*, 2439 (1972), and refs. therein.

74. (a) Savéant, J.M. *Acc. Chem. Res.*, *13*, 323 (1980). (b) Amatore, C.; Pinson, J.; Savéant, J.M.; Thiébault, A. *J. Electroanal. Chem., Interfacial Electrochem.*, *107*, 59 (1980). (c) Amatore, C.; Pinson, J.; Savéant, J.M.; Thiébault, A. *J. Am. Chem. Soc.*, *104*, 817 (1982), and refs. therein.

75. Guedes da Silva, M.F.C.; Ferreira, C. M.P.; Fraústo da Silva, J.J.R.; Pombeiro, A.J.L. *J. Chem. Soc., Dalton Trans.*, 4139 (1998).

76. (a) Geiger, W.E., in: Zuckermann, J.J. (Ed.) *Inorganic Reactions and Methods*, Vol. 15, VCH, New York, 1986, p. 88. (b) Connelly, N.G.; Raven, S.J.; Carriedo, G.A.; Riera, V. *J. Chem. Soc. Chem. Commun.*, 992 (1986).(c) Acewgoda, G.M.; Robinson, B.H.; Simpson, J. *J. Chem. Soc., Chem. Commun.*, 284 (1982). (d) Rieke, R.D.; Kojima, H.; Öfele, K. *Angew. Chem. Int. Ed. Engl.*, *19*, 538 (1980).

Part 3
Bioelectrochemistry

Chapter 6

Towards Exploiting the Electrochemistry of Cytochrome P450

Dejana Djuricic, Barry D. Fleming, H. Allen O. Hill and Yanni Tian

Department of Chemistry, University of Oxford,
South Parks Road, Oxford UK OX1 3QH

1   INTRODUCTION

One cannot but fail to be impressed and intrigued by the range of substrates successfully oxidised by the cytochromes P450 [1]. These enzymes, present in *all* aerobic organisms, oxidise intractable substrates, often of limited solubility in aqueous media. In humans, their oxidative talents are used to transform, for example, aromatic hydrocarbons but therein lies the paradox: in doing so, it is thought that they convert a harmful material into a potent carcinogen as in the transformation of benzo[*a*]pyrene to benzo[*a*]pyrene 7,8 diol-9,10-epoxide [2]. This marked ability to catalyse difficult oxidations renders cytochromes P450 invaluable for all organisms in taking care of either noxious materials or accessing recalcitrant sources of energy. There is another paradox in this story: to generate a sufficiently powerful oxidant, dioxygen or, effectively, its complex with the cytochrome, must be *reduced*. That is, the peroxide formed generates a species that is a sufficiently powerful oxidant and, combined with the ability of the enzyme to arrange the substrate in a kinetically advantageous position, leads to the transformation of essentially all materials. One can state, with some confidence, that the advent of genetically engineered forms of the enzyme renders few compounds safe from attack.

The required presence of a source of reducing agents opens up the possibility of supplying electrons directly to the enzyme and substrate. In Nature, they derive electrons from NADH or NAD(P)H and are transferred *via* partner protein(s). There has been some success in providing them artificially, even electrochemically (*vide infra*): whether the latter give rise to a form of electroenzymology exploiting cytochrome P450 is the subject of this chapter.

1.1   Cytochrome P450s

The cytochrome P450 group of enzymes comprises a variety of hæm-containing monooxygenases present in the majority of prokaryotic and eukaryotic organisms [3]. As shown in Equation (1), the fundamental reaction catalysed by these enzymes is *monooxygenation*. These enzymes also catalyze sulphur oxidation, epoxidation, cleavage of carbon-carbon bonds, alkyl group migration and aromatisation, as well as a few other reactions (aryl and alkenic epoxidation, heteroatom dealkylation). The cytochrome P450 system is also important in detoxification of foreign substances (xenobiotic compounds). The action of P450 enzymes is however not always beneficial, as referred to above; some of the most dangerous carcinogens are converted *in vivo* into chemically reactive forms. In Equation (2) below, cytochrome P450s primary catalysis function is substrate–specific and stereo-specific hydroxylation of unreactive C-H bonds. It is the selectivity of substrate and product binding which controls the final product, while the orientation and strength of the substrate binding direct the remarkable stereospecificity.

$$R\text{-}CH_3 + O_2 + NAD(P)H + H^+ \xrightarrow{\quad P450 \quad} R\text{-}CH_2\text{-}OH + NAD(P)^+ + H_2O \quad (1)$$

$$R\text{-}H + O_2 \xrightarrow{\quad P450,\, 2e^- + 2H^+ \quad} R\text{-}OH + H_2O \quad (2)$$

Hydroxylation requires activation of dioxygen. The two-electron reducing agent in biological systems is almost exclusively NADPH or NADH, and this reductive activation of oxygen by P450 occurs in two separate one-electron additions. NADPH is the primary reducing agent; it transfers a proton and two electrons to an electron transfer (ET) flavoprotein. The ET proteins can then release electrons to cytochrome P450 one at a time as required. In mammalian P450s only one ET protein is required (Figure 1(a)), while in mitochondrial and bacterial P450 enzymes a second ET protein containing an $[Fe_2S_2]$ cluster is required (which receives one electron from the flavoprotein and carries it to the P450) (Figure 1(b)).

The cytochrome P450s are membrane-anchored proteins that contain a hæm prosthetic group and function as the terminal component of an electron transport chain in adrenal mitochondria and liver microsomes (Figure 2). In this assembly, it is hydroxylation, and not oxidative phosphorylation, that is the major role.

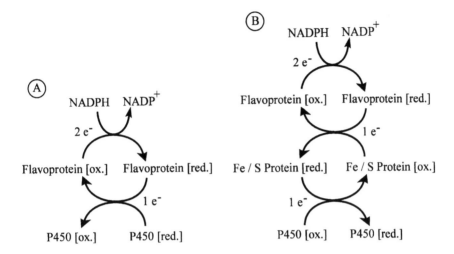

Figure 1. Cytochrome P450 electron transfer routes.

NADPH transfers its high potential electrons to a flavoprotein, which are then transferred to an adrenodoxin (a non-hæm iron protein). An electron is then transferred by an adrenodoxin to reduce the ferric form of P450 (1) (substrate free enzyme containing a ferric hæm iron octahedrally coordinated with a porphyrin, proximal cysteine and H_2O (or OH^-)) to the ferrous form (3).

The Fe^{II} intermediate immediately binds O_2 to form the formally ferrous dioxygen adduct (4), which then accepts a second electron from adrenodoxin to give a hæm Fe^{III} peroxo species (5). The electron transfer from adrenodoxin to the hæm is the rate-limiting step. Compound (5) then accepts two protons, and one of the bound oxygen atoms is reduced to water. If the oxidation equivalents are localized on the hæm, the resultant 'activated oxygen' species (6) can be denoted as oxoiron(V), $[(Por)Fe^V=O]$, or as an oxoiron(IV) porphyrin radical, $[(Por^{+\bullet})Fe^{IV}=O]$. The exact nature of (5) and (6) are still unknown, and the other bound electron is required to form ROH, the hydroxylated product.

An important observation relating to this catalytic cycle is that species (3), (4) and (5) can be bypassed by adding exogenous oxygen sources such as hydrogen peroxide to the substrate bound enzyme (the so-called "Peroxide Shunt" reaction). A characteristic of all cytochrome P450 enzymes is that, after the first electron reduction, species (3) will bind very strongly to carbon monoxide giving an octahedrally coordinated ferrous hæm-CO complex. This complex, when analyzed by UV/Vis spectroscopy, exhibits a strong Sorét band at ~ 450 nm. This observation, while also responsible for the name of this enzyme group, serves as a useful means of characterisation.

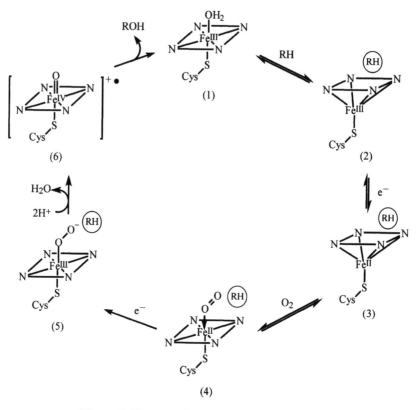

Figure 2. The cytochrome P450 catalytic cycle.

The mechanism by which the enzyme catalyses the transformation of a vast range of substrates has been much studied. The provision of the two electrons, with the aid of putidaredoxin, has been shown to be the slowest steps in the catalytic cycle. For example, the rate of the first electron transfer from putidaredoxin to cytochrome P450$_{cam}$ is considerably slower in the absence than when the substrate is present. There is thought to be a concomitant change in the formal redox potential on substrate binding and this, less negative $E^{o'}$ is thought to be associated with the change in rate for the reduction process. However, it has been pointed out that [4], in the presence of dioxygen, the reduction potential of *both* the substrate-bound *and* substrate-free will increase by approximately 60 mV per decade increase in dioxygen activity. Why then, is the *rate* of electron transfer to the substrate-free slower that that to the substrate-bound? That there is change in the geometry of the iron upon reduction in the substrate-free state has been shown on a number of occasions: this contrasts with the substrate-bound state in which the iron is five-coordinate in both instances. The change in the rate has been associated, therefore, with the reorganisation energy barrier in the semi-classical Marcus equation, being larger for the substrate-free form.

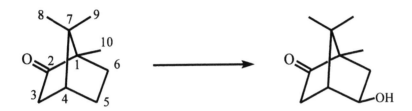

Figure 3. Regio- and stereo-specific hydroxylation reaction of camphor.

The most completely characterized cytochrome P450 is the soluble bacterial P450$_{cam}$, which is responsible for the regio- and stereo-specific hydroxylation of camphor to 5-*exo*-hydroxycamphor (Figure 3). This hæm protein, found in the soil bacterium *Pseudomonas putida*, has a characteristic red colour, consists of a 414 amino acid polypeptide chain and is of molecular weight ~ 45 kDa. The high-resolution structure of cytochrome P450$_{cam}$ has been determined by X-ray crystallography [5, 6]. Cytochrome P450$_{cam}$ is an asymmetrically shaped protein resembling a triangular prism [7], which is about 30 Å thick with about 60 Å sides, as shown in Figure 4.

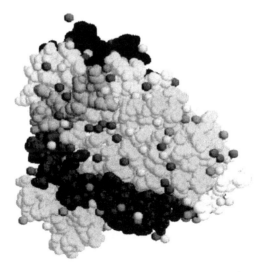

Figure 4. Space-filling representation of the crystal structure of cytochrome P450$_{cam}$.

The main features of cytochrome P450$_{cam}$ are:

- Hæm positioned in the centre of the protein.
- Hæm proximal ligand, a cysteine thiolate group, stationed at position 357 in the amino acid sequence.
- 5 surface cysteine residues (8 in total): C58, C85, C136, C148, and C334.
- The tyrosine residue at position 96 has a phenolic side chain in the hæm pocket which forms a hydrogen bond with the camphor carbonyl group and contributes to the selective camphor binding ($K_{Diss.} = 2.13$ µM).

1.2   Scanning Probe Microscopy

Scanning Probe Microscopes (SPMs) is a generic term that encompasses a wide range of high-resolution surface mapping instruments. Scanning probe microscopes are the newest entry into the surface characterising tools. Initially, SPMs were primarily used to measure surface topography, and this is still their primary application. In last 10 years, SPMs have developed to become one of the most powerful tools for surface metrology of our time. In contrast to the optical microscopes and SEMs, they do measure surface features in all three dimensions on the nanometer-to-micron scale.

One of the most valuable aspects of these techniques is that they are carried out in a variety of environments, including ambient air, vacuum and in liquids. In view of the rapidly expanding scale of applications and techniques, it is clear that SPM will continue to grow in power and application, providing information about molecular environment not directly observed before, and steadily renewing the concept of microscopy as it does so.

Recently, the application of SPM to the characterisation of electrode surfaces and the chemical processes that occur on them has provided a priceless means of improving our knowledge and understanding of these interactions. The combination of electrochemistry and environmental control gives the SPM instruments an extreme advantage in the study of solid-liquid interfaces, either at atomic/molecular resolution or at larger scales.

1.2.1   Scanning Tunnelling Microscopy

The STM was the first instrument to produce real space images of surfaces with an atomic resolution [8-11]. A sharpened, conductive tip is scanned approximately one nanometer or less above a conductive surface. At such separations, electrons from the surface begin to "tunnel" through the 10 Å gap into the tip or vice versa, depending upon the sign of the applied bias voltage. The resulting tunnelling current varies with tip-sample separation, and is the signal used to generate an STM image. For tunnelling to occur, both the tip and sample must be semiconductors or conductors. The current is extremely sensitive to the composition of the tunnel gap, so its variation can be used to generate an image of material adsorbed on the conductive surface below the tip.

The major factors influencing the resolution of STM are the geometry of the tip, the sample being imaged, and on the respective electronic properties of the tip and sample. The maximum resolution in the x and y axis plane is established by the geometry and size of the probe itself, the probe object distance and its control. When imaging extremely flat surfaces this is determined by the diameter of an atom (or atoms) at the probe's tip.

One of the most valuable characteristics of STM is that high-resolution imaging can be carried out in fluid media, and under full electrochemical control through use of a four-electrode configuration (EC-STM). One of the main advantages of being able to operate under a selected fluid is to eliminate, or rather minimize, many of the problems associated with contamination of the sample under ambient operation. In addition, water plays a crucial role in the stabilization of bio-molecular structure and surface tension-induced structural change during drying of the specimen.

1.2.2 Biological Applications of Scanning Tunnelling Microscopy

The atomic resolution attainable by STM has generated considerable excitement and interest through its potential use in the imaging of bio-molecular structure. STM and its offspring techniques like AFM were proving to be really attractive to chemists and biochemists because of their ability to resolve the molecular structure in three dimensions as the molecule is absorbed at an interface. In 1986, the discovery that STM can operate under fluid gave rise to an opportunity to study biological substances in a near-physiological conditions [8, 9]. In 1989, STMs capability to image under fluid was experimentally confirmed when DNA was imaged on a gold surface in an aqueous environment [10]. Since then, a large selection of biological samples, ranging from small biomolecules like DNA [10, 11], DNA bases and their derivatives, amino acids and peptides to entire cells have been imaged by STM [12-16]. In spite of the promise of high resolution and the ability to image under solution, there remains much to be understood about interactions of the scanning probe and biological material. As with any form of microscopy, there are significant problems associated with biological applications that must be addressed. These problems can essentially be categorised into three groups: the identification problem, fixation of the bio-molecule to the substrate and bio-molecule electrical conductivity.

There are a number of distinct aspects to the imaging process that contribute to interpretation difficulties [17,18]. Firstly it is essential to ensure one has the ability to distinguish between topographic features of the substrate and the geometry of the adsorbed molecules themselves. Protein molecules, for example, under imaging conditions, are rarely "structurally recognisable" or possess well-defined features. Most of the time they look like "blobs" whose size may or may not closely agree with crystallographically determined size (the molecules may, for example, appear larger than they are due to tip-induced flattening and convolution effects). It is therefore possible to confuse molecules imaged with small particles of contamination. In order to minimise the chance of making such errors, it is necessary (and common in practice) to carry out extensive and time-consuming studies of bare substrates or substrate surfaces treated with "control" samples.

While experimentally simple, physical adsorption methods may not be the ideal solution to the problem outlined above since they can lead to a loss of native protein structure, and are also not specific enough. Chemical modification methods, whereby the proteins are covalently bound to the solid phase, have been developed as possible ways to controlling protein binding. Anchoring bio-molecules to an underlying substrate surface has become a highly used approach for high-resolution imaging that would otherwise be difficult due to either inherent or tip-induced mobility of the molecule [19]. There are two main ways this can be accomplished, either by modification of the electrode surface or of the surface of the protein. In a cytochrome P450 study, proteins were modified by site directed mutagenesis and immobilised on gold *via* thiol-Au (111) covalent bonding [20]. This is a novel approach to the immobilisation of proteins.

It has been argued repeatedly, that, since most biological material is electrically insulating, the successful application of tunnelling methods should be questioned. However, even though proteins are generally classed as electronic insulators, they are able to display electronic [21], ionic [22], and protonic conduction [21, 23]. It has subsequently been demonstrated that successful imaging by STM is possible, at least for deposits of a few nanometers thick. A number of models based on an electron–tunnelling mechanism have been proposed to explain the contrast in the images [24-26], though there is no currently widely accepted mechanism that supports electrons tunnelling through molecule several nm thick.

2   ELECTROCHEMICAL INVESTIGATIONS

In 1977 the pioneering work of Eddowes & Hill [27] and Yeh & Kuwana [28], showed that the problem of slow electron transfer between an electrode and metalloprotein could be overcome. Since then much effort has been directed into developing suitably mediated or modified electrode systems, which facilitate biological electrochemistry. This has been motivated by a fundamental need to understand the mechanism of biological redox reactions, and the potentially fruitful practical outcomes in the area of bioreactors and biosensing. Given the vast array of compounds already known to be catalytically reactive with the cytochrome P450s, it is no wonder that they have been, and will continue to be, the target of electrochemical exploitation.

2.1 Evidence for Indirect Electrochemistry

Several reports have been made of replacing NAD(P)H, the natural electron donor to P450, with more readily available redox molecules or mediators. The mediators that have so far shown to be effective have included methylviologen [29], ferrocene [30], Co(III) sepulchrate [31] and the natural redox partner of P450$_{cam}$, putaredoxin (Pdx) [32]. In each case the mediator is electrochemically reduced at an appropriate electrode, with the reduced form of the mediator then becoming the electron donor, as would NADPH. A prime example of the effectiveness of this approach was shown with the Co(III) sepulchrate mediator by the Estabrook group [31]. Using a recombinant fusion protein system containing P450$_{4A1}$ and NADPH-P450 reductase, they showed that electrons supplied by Co(sep)$^{2+}$ could be effectively coupled to the ω-hydroxylation of lauric acid to 12-hydroxydodecanoic acid. Production formation rates using the Co(sep)$^{2+}$ mediator were comparable to those with NADPH. This work emphasised the virtues of using P450s in electrocatalytically driven bioreactors. The challenge was then to see whether similarly rapid electron transfer rates could be obtained without the use of mediators, that is, the direct electrochemistry of P450 enzymes.

2.2 Evidence for Direct Electrochemistry

The first evidence for direct electrochemistry of a cytochrome P450 enzyme was reported in 1996 [33]. For cytochrome P450$_{cam}$ it was shown that reversible electron transfer from an edge plane graphite electrode to the hæm Fe^{III}-Fe^{II} couple of the enzyme was possible. The potential associated with the redox couple, both in the camphor-free and camphor-bound forms of the enzyme, was in good agreement with previous potentiometric measurements. It was proposed that the same series of positively charged residues on the surface of P450$_{cam}$ that were likely to be the binding site to its natural redox partner, PdX, also facilitated the interaction with the negatively charged electrode.

The same group has also had some success obtaining direct electrochemistry on unmodified surfaces (bare EPG and gold) using a series of mutated and modified P450$_{cam}$ molecules [34-36]. Site–directed mutagenesis on the surface of the enzyme was performed to facilitate its immobilisation on a gold electrode surface and enhance electron transfer to P450$_{cam}$ [35, 36]. The direct electrochemistry of P450$_{cam}$ having redox active labels coupled into its active site has shown promise for understanding the mechanism of substrate entry. The significant positive shift in the hæm redox potential observed for the modified P450$_{cam}$ compared with camphor-bound wild-type enzyme was linked to the position of the iron in relation to the plane of the porphyrin ring [34].

Obtaining the direct electrochemistry of P450 enzymes on unmodified electrodes is fraught with many difficulties (as for most proteins). The hæm group is often deep within the protein structure, and favourable orientation at the electrode surface must occur to ensure electron transfer. Only freshly purified enzyme has reportedly given reliable and reproducible results [33]. This is likely the consequence of proteinaceous impurities and/or denatured enzyme rapidly adsorbing to and minimising the available electrode surface area for protein interaction and electron transfer. Much of these concerns have been alleviated for a wide-range of hæm containing proteins including P450s through the exploitation of a series of modified electrode surfaces. These have included biomembrane-like films [37], alternate layer-by-layer films [38] and modification with clay minerals [39]. These approaches provide favourable environments for the protein, in terms of being conducive to protein structure integrity and facilitating rapid electron transfer. Also, these modified surfaces would be expected to be less favoured by solution contaminants that commonly plague electrochemical experiments involving 'bare' electrodes.

Rusling's group have had success with P450$_{cam}$ electrochemistry using cast biomembrane-like films on electrode surfaces [40, 41]. This approach has been further utilised by other groups interested in P450 electrochemistry [42, 43] and biosensor applications [44]. These easily prepared films, usually consisting of multi-layers of the synthetic surfactants DDAB or DMPC, have been cast on to pyrolytic graphite, gold, platinum and screen-printed rhodium-graphite surfaces from either organic solutions or aqueous dispersions to form stable lamellar crystal films. In some cases additional components such as glutaraldehyde, bovine serum albumin or a sol-gel have been used to aid the immobilisation process, and riboflavin to act as an electron tunnelling relay [45, 46]. When incorporated into the surfactant layer, the direct, rapid and reversible electron transfer between the P450 hæm and electrode has been observed. The enzyme is concentrated close to the electrode surface within the biomembrane, and this has been confirmed by the thin film electrochemical behaviour. Despite the difficulties in accurately measuring electron transfer rate constants in these thin film systems, a relative measure of the rate (k_s' ~ 26 s^{-1} for P450$_{cam}$ in a DDAB film on PG electrode) highlights the greater efficiency of electron transfer in films than in solution [40].

Recently the direct electrochemistry of $P450_{cam}$ assembled in multilayered films of oppositely charged polyions was reported [38, 47]. In these studies a MPS-modified Au electrode was first coated with either a bilayer of PEI/PSS or PDDA. This provided the 'electrostatic glue' to hold the $P450_{cam}$ monolayer close to the electrode surface. Successive alternate layers were then added, with the amount of electroactive enzyme still continuing to increase even after seven layers had been constructed (~ 90 nm) [48]. Within these layers the enzyme has shown improved stability (response unchanged after 2 weeks storage in buffered solution) and the actual amount of enzyme required is small (10^{-10} mol cm^{-2}).

Another approach that has been highly successful in obtaining the direct electrochemistry of P450 is with clay modified electrodes [39]. Substrate free $P450_{cam}$ was observed to readily adsorb onto a sodium montmorillonite-modified glassy carbon electrode. Within this environment the rapid and reversible exchange of electrons to the hæm was observed. Some similarities were drawn between the negatively charged clay and putaredoxin, its natural redox partner.

The variety of approaches that has lead to the direct electrochemistry of $P450_{cam}$ has shown significant differences in the formal potential of the hæm. Table 1 shows a comparison of the redox potential (E_m) for substrate-free $P450_{cam}$ measured by cyclic voltammetry for a series of modified electrodes. The types of electrode, surfactant, polyion and clay all have some influence over the hæm potential. When incorporated in a positively charged DDAB layer, the E_m values for $P450_{cam}$ were 180-320 mV more positive than that measured by

Table 1. Electrochemically measured hæm redox potential (E_m) of $P450_{cam}$ in substrate-free buffered (pH 7) solution using modified and unmodified electrodes (in order of ascending potential).

electrode	E_m / mV .v. SCE	reference
$P450_{cam}$ solution*	-544	[49]
EPG#	-526	[33]
Clay-GC	-405	[39]
Sol-Gel/DDAB-GC^	-400	[44]
DMPC-PG	-357	[40]
($P450_{cam}$/PEI)$_6$-PG	-320	[48]
DDAB-GC^	-260	[46]
PSS/PEI/MPS-Au	-250	[38]
DDAB-PG	-238	[40]

*Measured in solution by redox titration, # pH 7.4, ^ pH 7.5

redox titration [40]. This can be explained in terms of the electrical double-layer effect on the potential experienced by the enzyme, as well as some contribution from protein-surfactant interaction. A positive shift in the formal redox potential of the clay adsorbed P450$_{cam}$ (cf. substrate-free solution value) was suggested to be the result of partial dehydration caused by the clay.

Cyclic voltammetry of P450 enzymes is typically performed under strict anaerobic conditions (< 1.5 ppm). Those familiar to this area will know the incredible sensitivity that P450s have to molecular oxygen. Given the chance O$_2$ will rapidly bind to the reduced hæm, with the concomitant rapid electrochemical catalysis to H$_2$O$_2$ (Equation 3). Hydrogen peroxide can have a detrimental effect on the enzyme and in most cases is an undesired product (those seeking peroxide driven catalysis being the exception) [48].

$$P450Fe^{III} + e^- \longrightarrow P450Fe^{II}$$

$$P450Fe^{II} + O_2 \longrightarrow P450Fe^{II}\text{–}O_2 \tag{3}$$

$$P450Fe^{II}\text{–}O_2 + 2e^- + 2H^+ \longrightarrow P450Fe^{II} + H_2O_2$$

Figure 5 highlights the potential effects of trace oxygen on cyclic voltammetry data for P450$_{BM3}$. At low cyclic voltammetry scan rates (< 500 mVs^{-1}) the electrochemistry can be dominated by reduction of even trace oxygen. An increased reductive current is observed (in this case at a slightly more positive potential) at the expense of the oxidative current. This has been the generally

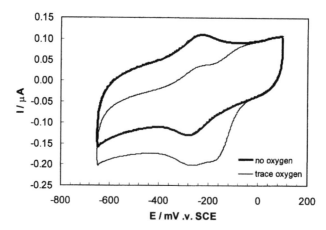

Figure 5. Cyclic voltammetry data for P450$_{BM3}$ monooxygenase domain immobilised in a DDAB layer on an EPG electrode. Voltammogram recorded at 10 mV/sec in 40 mM phosphate buffer at pH 7.4.

reported experience despite all the best efforts to minimise and maintain a highly anaerobic atmosphere (solutions purged of O_2 with inert gas e.g. nitrogen or argon, experiment performed in a glove box). At faster scan rates the oxygen present at the electrode surface is rapidly consumed, and any oxygen present in solution is unable to diffuse rapidly enough back to the surface before the potential cycle is repeated.

3   SCANNING PROBE INVESTIGATIONS OF ELECTRODE SURFACES

The local investigation of electrode surfaces and processes by SPM techniques has become an important aspect in electrochemistry. Special emphasis has in the last few years been given to the correlation of structural electrode properties, investigated by SPMs, with the electrochemical system studied by classical electrochemical methods. It must be pointed out that one of the most important features of SPM, especially STM, as structure analysis tools in electrochemistry is that the high-resolution images of electrode surfaces can be obtained in electrolyte solutions.

3.1   Modified electrode surfaces

SPM has proven to be an extremely useful tool for determining the order and structure of self-assembled monolayers and multi-layers deposited on several substrates, most importantly gold. Patterned self-assembled monolayers (SAMs) which can be fabricated by various techniques, such as photolithography, micro-machining, micro-writing, electrochemical stripping as well as micro-contact printing, are very convenient molecular templates for protein patterning on gold. As mentioned in § 2.0, the direct electrochemistry of proteins is often hindered by unfavourable interactions with the electrode and that the proteins absorbed onto the bare electrode surface act as self-blocking (insulating) material. With SAMs it is possible to achieve the fast electron transfer, provided that the surface is functionalised in a way which allows specific, favourable and reversible binding of the redox protein.

Over the years a succession of compounds called surface facilitators or promoters have been synthesised and used. The surface modifiers appear to prevent direct adsorption of a protein onto the bare electrode surface. The choice of the facilitator(s) directly governs the surface properties of the electrode surface, enabling the "tailoring" of the electrode surface to suit the individual requirements of each protein, and resulting in fast electron transfer between the protein and the electrode surface. Traditional spectroscopic techniques have been used to characterise SAMs, but it is SPM (mainly STM) that has provided the most striking detail of the structural form of the facilitators on the electrode surfaces. SPM has a significant role to play in the characterisation of surfaces modified by facilitators with SPM. For example, many electrode processes may take place at surfaces with less than monolayer coverage of a facilitator, in which case the nature of the electrochemical response will be inevitably complicated by the fact that the uncovered parts of the electrode surface are apparently electro-inactive. SPM allows the visualisation of the SAM, identifying the extent of the facilitator coverage.

3.2   Adsorbed Cytochrome P450

It is known from crystallographic studies that cytochrome $P450_{cam}$ have a triangular prismatic shape, approximately 6.5 nm on each side and 3.5 nm thick (crystallographically reported values). We previously reported the results of a tunnelling study carried out on both the engineered surface cysteine mutant (K344C) and the wild type cytochrome $P450_{cam}$ [36]. We have developed a procedure whereby proteins and enzymes are anchored to gold electrode surfaces through site-specifically introduced surface cysteine residues. The application of genetically engineered methods permits the introduction of single immobilisation sites on the surface of an enzyme. The biomolecule can then be immobilised on electrode surfaces in a reproducible and ordered manner, with a desired orientation and in many cases in active form. It was concluded in this study, that more reproducibly ordered adsorption, at high coverage, occurs with K344C mutant than the wild-type enzyme. At low magnification, the surface features appeared to be of regular quasi-spherical shape, as shown in Figure 6(a). At higher magnification, however, a significant number of adsorbed molecules appear to have a recognisable "triangular" shape (Figure 6(b)). On reference to a computer modelling representation of the structure (Brookhaven Protein Database) (with the introduced cysteine assumed to be "face down" on the gold) one indeed expects such shape. Tunnelling images obtained under fluid conditions at a gold/fluid interface gave molecular "triangular" dimensions for K344C $P450_{cam}$ of, on average, 4-5 nm (width) x 5-6 nm (length) i.e. close to the reported crystallographic values.

Electrochemical studies indicated that electronic coupling of the mutant enzyme (in which the hæm is somewhat buried in the protein matrix and as a result difficult to address) to the underlying gold electrode surface is enhanced in comparison to the wild type structure (for which there is reproducible coupling). This is most probably due to the position of the anchoring residue; the introduced cysteine lies on a proximal side of the enzyme where the enzyme is nearest to the structure surface (the cysteine residues of the native protein do not sit in this region).

(a)

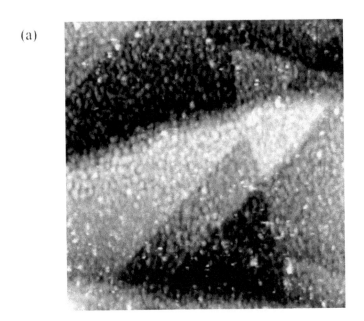

(b)

Figure 6. Cytochrome K344C P450$_{cam}$ molecules adsorbed to a Au(111) electrode surface imaged under potassium phosphate buffer by STM. (a) Scan size 345 nm × 345 nm, bias 1.126 V, tunnel current 187 pA, scan rate 6.1 Hz, z-scale 0-5.8 nm. (b) Scan size 14.0 nm × 17.2 nm, bias 0.7 V, tunnel current 245 pA, scan rate 3.5 Hz, z-scale 0-4.3 nm.

Figure 7 shows the appearance of a cytochrome P450$_{BM3}$ coated gold electrode under fluid conditions (deionised water, unadjusted pH). The coverage of enzyme over the surface is quite extensive, with close interaction between enzyme molecules evident. In some instances, it is possible to see 'piling up' of molecules

Figure 7. Cytochrome P450$_{BM3}$ molecules adsorbed to a Au(111) electrode surface imaged under deionised water by STM. Scan size 60.9 nm × 36.5 nm, bias 1.1 V, tunnel current 10 pA, scan rate 4.9 Hz, z-scale 0-2.3 nm.

on top of each other. There is a reasonable degree of variation in the size, shape and orientation of the adsorbed molecules. This is related to the structural mobility of the protein in solution. An analysis of 132 individual molecules showed that the average shape was 'oval', having dimensions of 4.4 ± 1.1 nm (length) by 3.4 ± 0.9 nm (width). Comparisons with crystallographic data suggest these figures to be quite reasonable. The STM imaging of wild-type cytochrome P450$_{BM3}$ shows that the enzyme does bind to a gold electrode surface. However, attempts to obtain the direct electrochemistry of P450$_{BM3}$ on a bare gold electrode have yet to show any success. This suggests that the bound protein is not in a favourable orientation to facilitate electron transfer. On the other hand, the incorporation of P450$_{BM3}$ in a DDAB modified pyrolytic graphite electrode, has proven to be a highly successful means to obtain the direct electrochemistry of the enzyme. This will be the subject of a forthcoming article.

4   BULK ELECTROCHEMISTRY

4.1   *Exploiting the Electrochemistry of Cytochrome P450 for Reductive Transformations*

It is obvious that the impediment in exploiting the cytochromes P450 in electroenzymology is associated with having dioxygen as substrate: it tends to 'hijack' the electrons intended for the enzyme, forms hydrogen peroxide and although some reaction can take place with the latter as substrate, it tends to lead to the formation of species that destroy the enzyme. That being the case, why not exploit the inherent properties of the enzyme, substrate-binding and manipulation, together with the provision of reducing equivalents, and in that manner engage in some valuable pursuit? Nearly twenty years ago, Castro *et al.*[50], reported the reactions of halomethanes with *reduced* cytochrome P450: later, Li and Wackett [51] reported that halogenated methane and ethane substrates could be dehalogenated using both natural and artificial reductants. These reactions could be effected by the electrochemical reduction of cytochrome P450: the reduction of a pentachlorophenol was, however, efficiently catalysed by the direct electrochemistry of the flavin-containing monooxygenase, pentachlorophenol hydroxylase [52].

4.2 Exploiting the Electrochemistry of Cytochrome P450 for Oxidative Transformations

There is now no difficulty effecting the one-electron reduction of P450 and there have been several reports that appear to indicate that two-electron reduction can take place in the presence of dioxygen. However, many who have tried to achieve this goal, know that it represents a tantalising yet frustrating target: if the reaction is indeed observed, there is always difficulty in ascertaining whether it proceeds in a manner analogous to the scheme outlined in Figure 2 or *via* the peroxide shunt mechanism. The work that employed the cobalt(III) sepulchrate is a fine example of the effectiveness of this method: whether it really can be transformed into electrochemically-driven bioreactors, thus exploiting electroenzymology, remains to be seen.

5   FINAL REMARKS

Such a versatile enzyme seems a wonderful aid to the transformation of a wide variety of recalcitrant substrates: why isn't it exploited more often? In particular, since it requires electrons to be effective, why cannot these be supplied electrochemically? Apart from the few exceptions reported above, there have been few successful electrosyntheses reported. The problems appear to be connected with the difficulty of one-electron transformations to the enzyme in the obligatory presence of dioxygen. The latter reacts slowly with two-electron donors but much more rapidly with one-electron donors: the presumed mechanism of the P450 reactions *requires* the one-electron thus dioxygen is always likely to be a successful competitor for the electrons giving rise, ultimately, to hydrogen peroxide. So, why not accept this and relinquish the desire to use the electroenzymological method and, instead, use hydrogen peroxide directly? One can do so but, till now, the enzyme didn't survive long in the presence of hydrogen peroxide, not surprisingly since the enzyme has, for example, a number of reactive cysteines and other residues. Why not genetically engineer a really rugged protein, resistant to hydrogen peroxide, (and the reactivity of H_2O_2 is often enhanced by the presence of trace amounts of iron or copper)? Indeed, since it is possible that H_2O_2 reacts with P450 *via* the peroxide anion, why not alter the protein such that it has a higher rate at increased pH values? The combination of these approaches might realise the long sought objective: a really tough catalyst of difficult transformations that makes use on hydrogen peroxide, perhaps supplied electrochemically, finally giving the electroenzymological ideal.

ACKNOWLEDGEMENTS

We would like to thank the European Union for the collaborative grant, ECEnzymes, BASF and Dr. Habicher, Oxford Biosensors Ltd., and Dr. L-L. Wong for support and helpful discussions.

REFERENCES

1. Guengerich, F. P., *Nat. Rev. Drug Discov.*, *1*, 359, (2002).
2. Gelboin, H. V., *Physiol. Rev.*, *60*, 1107, (1980).
3. Ortiz de Montellano, P. R., *Cytochrome P-450: Structure, mechanism and biochemistry*, Plenum Press, New York, (1995).
4. Honeychurch, M. J., Hill, H. A. O. & Wong, L.-L., *FEBS Lett.*, *451*, 351, (1999).
5. Poulos, T. L., in *Cytochrome P-450*, Eds. Ortiz de Montellano, P. R., Plenum Publishing Corporation, New York, (1986).
6. Poulos, T. L., Finzel, B. C. & Howard, A. J., *Biochemistry*, *25*, 5314, (1986).
7. Hasemann, C. A., Kurumbail, R. G., Boddupali, S. S., Peterson, J. A. & Diesenhofer, J., *Structure*, *3*, 41, (1995).
8. Sonnenfeld, R. & Hansma, P. K., *Science*, *232*, 211, (1986).
9. Liu, H. Y., Fan, F. R. F., Lin, C. W. & Bard, A. J., *J. Am. Chem. Soc.*, *108*, 3838, (1986).
10. Lindsay, M. S., Thundat, T., Nagahara, L., Knipping, U. & Rill, R. L., *Science*, *244*, 1063, (1989).
11. Lindsay, M. S., Tao, N. J., DeRose, J. A., Oden, P. I., Lyubchenko, Y. L., Harrington, R. E. & Shlyakhtenko, L., *Biophys. J.*, *61*, 1570, (1992).
12. Wandlowski, T., Lampner, D. & Lindsay, S. M., *J. Electroanal. Chem.*, *404*, 215, (1996).
13. Srinivasan, R., Murphy, J. C., Fainchtein, R. & Pattabiraman, N., *J. Electroanal. Chem.*, *312*, 293, (1991).
14. Tao, N. J. & Shi, Z., *J. Phys. Chem.*, *98*, 1464, (1994).

15. Katsumata, S. & Ide, A., *Jpn. J. Appl. Phys.*, *33*, 3723, (1994).
16. Katsumata, S. & Ide, A., *Jpn. J. Appl. Phys.*, *34*, 3360, (1995).
17. Salmeron, M., Beebe, T., Odriozola, J., Wilson, T., Ogletree, D. F. & Siekhaus, W., *J. Vac. Technol. A*, *8*, 635, (1990).
18. Navaz, Z., Cataldi, T. R. I., Smekh, J. K. R. & Pethica, J. B., *Surf. Sci.*, *265*, 139, (1992).
19. Rao, S. V., Anderson, K. W. & Bachas, L. G., *Microchim. Acta*, *128*, 127, (1998).
20. Davis, J. J., Halliwell, C. M., Hill, H. A. O., Canters, G. W., van Amsterdam, M. C. & Verbeet, M. P., *New J. Chem.*, *10*, 1119, (1998).
21. Pethig, R., *Dielectric and Electronic Properties of Biological Materials*, J. Wiley and Sons, New York, (1979).
22. Morgan, H. & Pethig, R., *J. Chem. Soc., Faraday Trans.*, *82*, 143, (1986).
23. Bruni, F., Careri, G. & Leopold, A. C., *Phys. Rev. A*, *40*, 2803, (1989).
24. Yuan, J., Shao, Z. & Gao, C., *Phys. Rev. Lett.*, *67*, 7863, (1991).
25. Lindsay, S. M., Jing, T. W., Pan, J., Vaught, A. & Rekesh, D., *Nanobiology*, 3, 17, (1994).
26. Garcia, R. & Garcia, N., *Chem. Phys. Lett.*, *173*, 44, (1990).
27. Eddowes, M. J. & Hill, H. A. O., *J. Chem. Soc., Chem Commun.*, *771*, (1977).
28. Yeh, P. & Kuwana, T., *Chem. Lett.*, 1145, (1977).
29. Scheller, F., Renneberg, R., Strnad, G., Pommerening, K. & Mohr, P., *Bioelectrochem. Bioenerg.*, *4*, 500, (1977).
30. Vilker, V. L., Khan, F., Shen, D., Baizer, M. M. & Nobe, K., in *Redox Chemistry and Interfacial Behaviour of Biological Molecules*, Eds. Dryhurst, G. and Niki, K., Plenum, New York, (1988).
31. Faulkner, K. M., Shet, M. S., Fisher, C. W. & Estabrook, R. W., *Proc. Natl. Acad. Sci. USA*, *92*, 7705, (1995).
32. Reipa, V., Mayhew, M. P. & Vilker, V. L., *Proc. Natl. Acad. Sci. USA*, *94*, 13554, (1997).
33. Kazlauskaite, J., Westlake, A. C. G., Wong, L.-L. & Hill, H. A. O., *Chem. Commun.*, *18*, 2189, (1996).
34. Di Gleria, K., Nickerson, D. P., Hill, H. A. O., Wong, L.-L. & Fulop, V., *J. Am. Chem. Soc.*, *120*, 46, (1998).
35. Lo, K. K.-W., Wong, L.-L. & Hill, H. A. O., *FEBS Lett.*, *451*, 342, (1999).
36. Davis, J. J., Djuricic, D., Lo, K. K. W., Wallace, E. N. K., Wong, L.-L. & Hill, H. A. O., *Faraday Discuss.*, *116*, 15, (2000).
37. Rusling, J. F. & Nassar, A.-E. F., *J. Am. Chem. Soc.*, *115*, 11891, (1993).
38. Lvov, Y. M., Lu, Z., Schenkman, J. B., Zu, X. & Rusling, J. F., *J. Am. Chem. Soc.*, *120*, 4073, (1998).
39. Lei, C., Wollenberger, U., Jung, C. & Scheller, F. W., *Biochem. Biophys. Res. Commun.*, *268*, 740, (2000).
40. Zhang, Z., Nassar, A.-E. F., Lu, Z., Schenkman, J. B. & Rusling, J. F., *J. Chem. Soc., Faraday Trans.*, *93*, 1769, (1997).
41. Rusling, J. F., *Acc. Chem. Res.*, *31*, 363, (1998).
42. Koo, L. S., Immos, C. E., Cohen, M. S., Farmer, P. J. & Ortiz de Montellano, P. R., *J. Am. Chem. Soc.*, *124*, 5684, (2002).

43. Aguey-Zinsou, K.-F., Bernhardt, P. V., De Voss, J. J. & Slessor, K. E., *Chem. Commun.*, 418, (2003).
44. Iwuoha, E., Kane, S., Ania, C. O., Smyth, M. R., Ortiz de Montellano, P. R. & Fuhr, U., *Electroanalysis*, *12*, 980, (2000).
45. Shumyantseva, V. V., Bulko, T. V., Usanov, S. A., Schmid, R. D., Nicolini, C. & Archakov, A. I., *J. Inorg. Biochem.*, *87*, 185, (2001).
46. Iwuoha, E. I., Joseph, S., Zhang, Z., Smyth, M. R., Fuhr, U. & Ortiz de Montellano, P. R., *J. Pharm. Biomed. Anal.*, *17*, 1101, (1998).
47. Zu, X., Lu, Z., Zhang, Z., Schenkman, J. B. & Rusling, J. F., *Langmuir*, *15*, 7372, (1999).
48. Munge, B., Estavillo, C., Schenkman, J. B. & Rusling, J. F., *ChemBioChem*, *4*, 82, (2003).
49. Sligar, S. G., Cinti, D. L., Gibson, G. G. & Schenkman, J. B., *Biochem. Biophys. Res. Commun.*, *90*, 925, (1979).
50. Castro, C. E., Wade, R. S. & Belser, N. O., *Biochemistry*, *24*, 204, (1985).
51. Li, S. & Wackett, L. P., *Biochemistry*, *32*, 9355, (1993).
52. Xie, W., Jones, J. P., Wong, L.-L. & Hill, H. A. O., *Chem. Commun.*, 2370, (2001).

Chapter 7

Chemistry, Electrochemistry and All-Iron Hydrogenase

Christopher J. Pickett

Department of Biological Chemistry, John Innes Centre,
Norwich Research Park, Colney, Norwich, NR4 7UH.

1   INTRODUCTION

The reversible reduction of protons to dihydrogen :

$$2H^+ + 2e \rightleftharpoons H_2$$

is deceptively the simplest of reactions but one which requires multi-step catalysis to proceed at practical rates. How the metal-sulfur clusters of the hydrogenases catalyse this interconversion is currently the subject of extensive structural, spectroscopic and mechanistic studies of the enzymes, of synthetic assemblies and of *in silico* models. This is driven both by curiosity and by the view that an understanding of the underlying chemistry may inform the design of new electrocatalytic systems for hydrogen production or uptake, pertinent to energy transduction technology in an 'Hydrogen Economy'. Can new materials be designed to replace the (unsustainable) platinum metal catalysts of fuel cells and which are based on Ni, Fe and S cluster assemblies as found in the natural systems? Here some first steps towards artificial systems related to the di-iron catalytic sub-site of all-iron hydrogenase are described.

2   THE ACTIVE SITE OF ALL-IRON HYDROGENASE

X-ray crystallographic structures of Fe-only hydrogenases from *D. desulfuricans* and *Clostridium pasteurianum* [1,2], together with spectroscopic data on Fe-only hydrogenase from *D. vulgaris* [3], show that the H-cluster, the active site at which protons are reduced to dihydrogen, is a conventional $\{Fe_4S_4\}$-

cluster linked by a bridging cysteinyl sulfur to an 'organometallic' $\{Fe_2S_3\}$ sub-site, Figure 1. At the sub-site a terminal carbon monoxide, a bridging carbon monoxide and a cyanide ligand are bound at each iron atom which also share two bridging sulfur ligands of a 1,3-propanedithiolate or possibly the related di(thiomethyl)amine unit. The Fe-atom distal to the $\{Fe_4S_4\}$-cluster has a coordinated water molecule (or vacancy) in the resting paramagnetic oxidised state of the enzyme, $\{H_{ox}\}$. This site is occupied by carbon monoxide in the CO inhibited form of the enzyme $\{H_{ox}(CO)\}$ and is therefore thought to be where hydride/dihydrogen is bound during turnover.

The oxidation states of the two Fe centres in the sub-site of the enzyme has been the subject of some debate [4,5]. In the epr active $\{H_{ox}\}$-state, the proximal iron-atom is almost certainly diamagnetic Fe^{II} because only weak magnetic coupling of the paramagnetic $\{4Fe4S\}$ cluster to the sub-site is observed [6]. Thus the spin-density in the sub-site resides on the distal Fe-atom. The question is whether this metal atom has a formal oxidation state of Fe^{III} or is in a biologically unprecedented Fe^I state. It has been argued on the basis of Mössbauer data that the distal iron is in a conventional Fe^{III} state, however Fe^I could not be excluded. On the other hand, FTIR ^{13}CO labelling studies of the enzyme by DeLacey and coworkers strongly support an Fe^I oxidation state for the distal iron centre in $\{H_{ox}(CO)\}$ because the uncoupled v(CO) stretch at this centre is *lower* in energy than that for the CO bound at the proximal Fe^{II} site [7]. Clear support for an $\{Fe^I_{distal}\text{-}Fe^{II}\}$ arrangement comes from studies of synthetic $\{2Fe3S\}$-systems [8] and from DFT calculations [9] as discussed below.

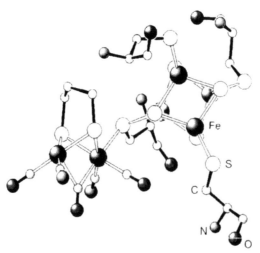

Figure 1. Representation of the active site of all-iron hydrogenase from the composite crystallographic data of Peters and Fontecilla-Camps.

Whether the bridging dithiolate at the sub-site is 1,3-propanedithiolate or a di(thiomethyl)amine unit is still an open question. Re-analysis of the first structures of Fe-only hydrogenases marginally favoured the latter ligand and led to the speculation that the amine group could function in proton delivery to the active site [10]. However, a recent higher resolution structure of Fe-only hydrogenase from *Clostridium pasteurianum* has not allowed discrimination between NH and CH_2 at the '2-position' of the bridging dithiolate [11].

3   SYNTHETIC SUB-SITES

Studies of synthetic analogues of the active sites of the hydrogenases in which for example proton, hydride or dihydrogen bonding interactions are stabilised or 'frozen-in' by tuning redox state and/or ligand environment should provide crucial insights into the enzymic hydrogen evolution/uptake chemistry. Other key questions which model studies are beginning to address are: why are two metal centres necessary for the biocatalyses; why are the biologically unusual ligands CO and CN necessary structural elements at the active sites; what is the mechanistic role of bridging/terminal CO interconversion during turn-over of Fe-only hydrogenase; what are the likely metal atom redox states involved in biocatalysis; and what are the minimum structural requirements necessary for synthetic assemblies to display hydrogenase chemistry? How do the energetics and kinetics of catalysis of the natural system compare with artificial clusters and with electrocatalysis by Pt and other metals. Beyond providing chemical precedence for putative structures, intermediates and mechanisms in the biocatalyses, synthetic analogues of the active sites provide a means of validating interpretations of the spectroscopy of the enzyme systems and to 'ground' *in silico* DFT estimates of structural and spectroscopic properties of the active sites.

The crystallographic characterisations of Fe-only hydrogenase [1,2] revealed the close resemblance of the sub-site of the H-cluster to known $[Fe_2(\mu\text{-}SR)_2(CO)_6]$ (R = organic group) complexes. This type of {2Fe2S}-assembly, first discovered by Reihlen and co-workers [12] more than 70 years ago, together with the chemistry developed by Seyferth and co-workers in the 1980's [13], opened the way for the synthesis of sub-site models. Within a very short time our group and two US groups [14-16] independently reported the synthesis of the dianion $[Fe_2(SCH_2CH_2CH_2S)(CO)_4(CN)_2]^{2-}$ **1** . Elegant work by Rauchfuss and coworkers [17] subsequently showed that the related $[Fe_2(SCH_2NRCH_2S)(CO)_4(CN)_2]^{2-}$ dianions **2** (R = H, Me) are also accessible. The 'butterfly' arrangement of the dithiolate ligands in both **1** and **2** are closely similar to that in the sub-site and the {Fe(CO)$_2$(CN)} motifs in the complexes reasonably model the distal iron of the sub-site in the CO inhibited form of the enzyme $\{H_{ox}(CO)\}$. Two studies have addressed the mechanism of formation of **1** from the hexacarbonyl precursor. Darensbourg and coworkers have shown that the monocyanide $[Fe_2(SCH_2CH_2CH_2S)(CO)_5(CN)]^{1-}$ can be indirectly synthesised [18] by nucleophilic attack of $[(Me_3Si)_2NLi]$ on $[Fe_2(SCH_2CH_2CH_2S)(CO)_6]$ and that it reacts with CN^- to give **1** suggesting the monocyanide is a plausible intermediate in the overall di-cyanation reaction. However, Rauchfuss and coworkers [19] have proposed a di-cyanation pathway which invokes fast attack on an undetected bridging CO monocyanide intermediate because the rate of formation of **1** from the parent hexacarbonyl and CN^- is apparently *faster* than from the monocyanide. The mechanism of cyanation of $\{2Fe3S\}$-carbonyls is rather better understood as discussed below [20].

Complexes **1** and **2** have essentially undifferentiated ligation at the two iron sites and can be viewed as possessing a $\{2Fe2S\}$- rather than the $\{2Fe3S\}$-unit as is observed within the protein. Pickett and coworkers [21] have described the synthesis of propanedithiolate ligands functionalised with an appended thioether group and showed that these allow the assembly of $\{2Fe3S\}$-carbonyls $[Fe_2-\{RSCH_2C(Me)(CH_2S)_2\}(CO)_5]$ (R = Me or $PhCH_2$). X-ray crystallography of **3** revealed the close similarity of the $\{2Fe3S\}$-unit in the synthetic model and that in the enzyme, Figure 2 [21,22].

The pentacarbonyl **3** is readily converted to a monocyanide derivative **4** in which cyanation takes place regioselectively at the Fe atom distal to the thioether ligand, Scheme 1 and Figure 3. The reaction of **4** with further cyanide results

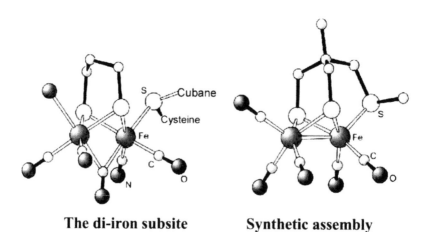

The di-iron subsite   **Synthetic assembly**

Figure 2. Comparison of the structure of the active sub-site of all-iron hydrogenase with that of the first synthetic 2Fe3S - assembly

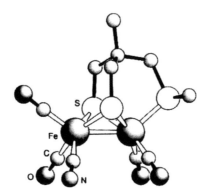

Figure 3. X-ray crystalographic structure of monocyanide 2Fe3S assembly

in the formation of a moderately stable intermediate $[Fe_2\{RSCH_2C(Me)(CH_2\text{-}S)_2\}(\mu\text{-}CO)(CO)_3(CN)_2]^{2-}$ **5** possessing a bridging CO, a cyanide ligand at each Fe atom, and differential ligation at the Fe atoms. These are the key structural elements of the CO inhibited sub-site of the H-cluster. **5** slowly rearranges to **6** a close {2Fe2S}- analogue of **1**, Scheme 1.

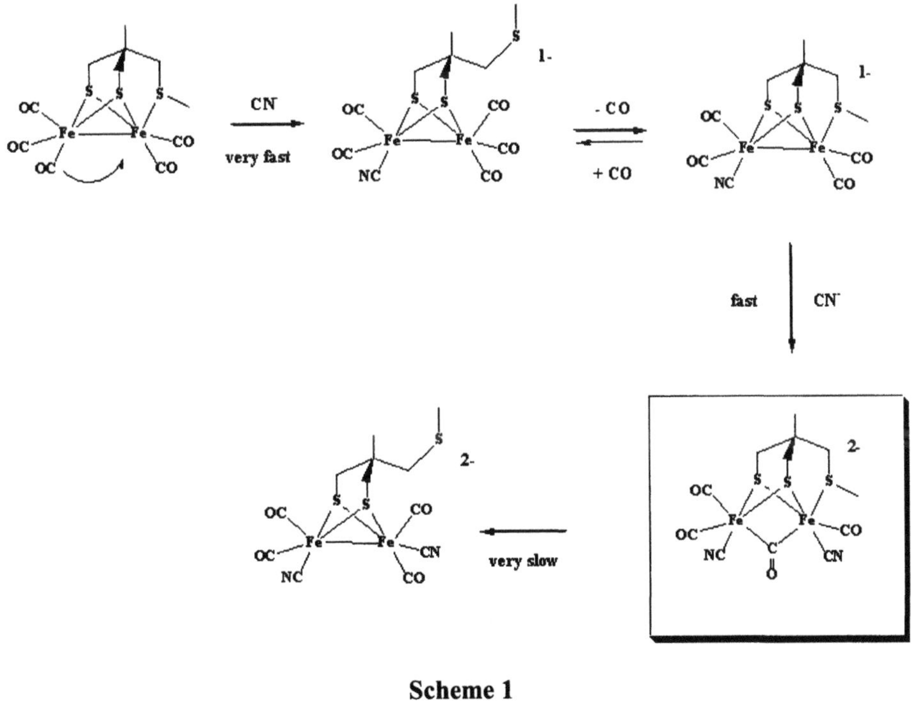

Scheme 1

The detailed mechanism of cyanation of the carbonyl **3** has been studied by stopped-flow FTIR and UV visible spectroscopy [20]. This has: shown 'on-off' thioether ligation plays a key role in the primary substitution step; unequivocally defined the monocyanide **4** as an intermediate; and shown the bridging carbonyl species **5** is on the pathway to **6**, Scheme 1.

4   ELECTRON-TRANSFER CHEMISTRY AND ELECTROCATALYSIS

Whereas spectroscopy shows that the intermediate **5** possesses structural elements of the sub-site of $\{H_{ox}(CO)\}$, electronically it differs in being a diamagnetic Fe^I-Fe^I rather than a paramagnetic Fe^I-Fe^{II} or Fe^{III}-Fe^{II} system. Generating paramagnetic species by oxidation of precursors such as **1** has been problematic, no stable mixed valence species have been isolated. Best, Pickett and coworkers [8] have shown that a transient Fe^I-Fe^{II} species can be spectroscopically characterised using FTIR spectroelectrochemical and stopped-flow methods. Thus one-electron oxidation of **6** generates a short-lived paramagnetic species with a bridging CO group and $v(CO)$ bands at wavenumbers close to those of $\{H_{ox}(CO)\}$, Scheme 2. This strongly supports the argument of DeLacey and coworkers [7] that the $\{H_{ox}\}$ and $\{H_{ox}(CO)\}$ states of the enzyme

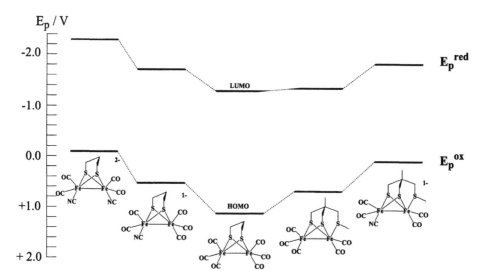

Figure 4. Evolution of HOMO and LUMO of {2Fe3S}- assemblies as qualitatively measured by solution oxidation and reduction potentials E_p^{ox} and E_p^{red} versus SCE respectively.

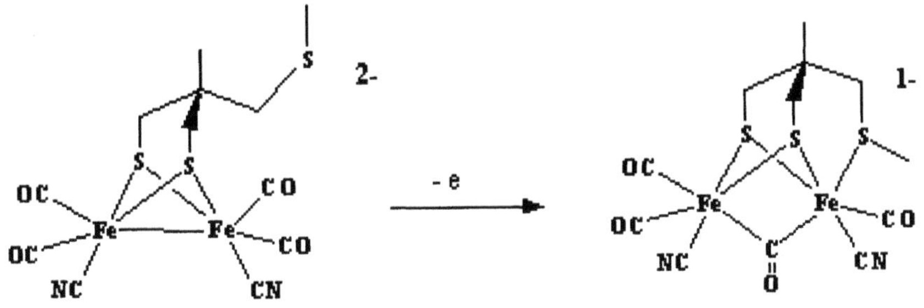

Scheme 2

have a mixed valence Fe^I-Fe^{II} unit with the spin density located on the distal Fe centre, as deduced from magnetic studies by Munck *et al* [5], rather than an Fe^{III}-Fe^{II} arrangement [4,5]. *In silico* DFT calculations on model assemblies by Cao and Hall [9], and subsequently by Liu and Hu [23], also conclude that {H_{ox}} and {$H_{ox}(CO)$} most likely consist of an Fe^I-Fe^{II} pair. Taken together the enzymic, model complex and DFT calculations all point to an unexpected rôle for Fe^I-Fe^I/Fe^I-Fe^{II} systems in biology. It is the π-acid ligands, CO and CN, which undoubtedly allow access to these low oxidation/spin-state levels.

Redox potential data for a range of di-iron dithiolate complexes are collected in Table 1; how the redox orbitals of such systems evolve on cyanation is illustrated by Figure 4. The general effect of successive substitution of CO for CN^- is that E_p^{ox} and E_p^{red} are shifted to more negative values by around 400 - 500 mV for both $2Fe_2S$ and $2Fe_3S$ assemblies as illustrated by Figure 4 and evident from the Table 1 (compare entries 2,8; 3,6; and 4,7). This is most likely a predominately electrostatic effect which shifts the HOMO – LUMO redox manifold in tandem as the negative charge density on the complex is increased, Figure 4. Substitution of CO by CN^- at closed-shell mononuclear centres usually invokes a shift of around 1000mV, as quantified by the ligand parameter P_L, which is 0.0 and 1.00 V for CO and CN^- respectively [24]. This smaller shift in the binuclear systems is consistent with delocalisation of charge over the two-iron centres and this is supported by the general shift in all the FTIR terminal carbonyl frequencies to lower values upon substitution of CO by CN^- [25].

Formally replacing neutral CO by a neutral pendant thioether ligand leads to a relatively small negative shift in E_p^{red} of less than 100 mV but a considerable negative shift in E_p^{ox} of nearly 500 mV, Table 1 and Figure 4. Thus the LUMO, which is almost certainly an Fe-Fe anti-bonding (σ*) orbital [26], can have little CO or thioether ligand character. In contrast, replacing the strong π -acid CO by

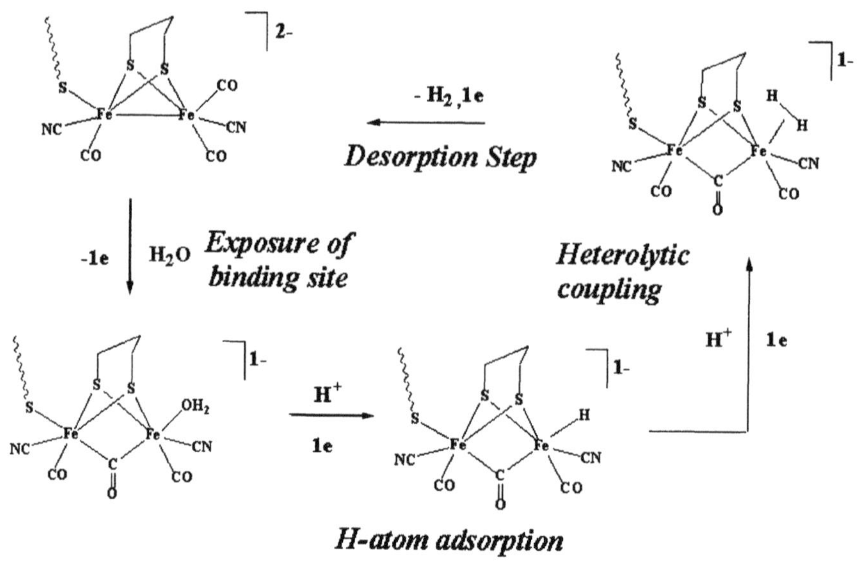

Scheme 3

the thioether ligand significantly perturbs the HOMO which is evidently very sensitive to the Fe atom environment.

It is becoming clear from FTIR studies on the enzyme system that the bridging CO of $\{H_{ox}\}$ becomes terminally bound on reduction [7]. The mechanistic implications of this may be related to the exposure of a proton binding site at the distal iron when CO adopts the bridging mode, with subsequent elimination of dihydrogen driven by rearrangement to a terminal CO bonding mode with concomitant metal-metal bond formation, Scheme 3, with adsorption – desorption steps perhaps providing some analogy with Pt electrocatalysis.

This Scheme invokes formation of terminal hydride/dihydrogen intermediates. Such intermediates remain spectroscopically undetected in the enzymic system although indirect evidence for the involvement of hydrides in the biological system come from studies of H^+/D_2 exchange reactions [27,28]. To date no synthetic di-iron thiolate models with terminally bound hydrides have been isolated. However, tertiarylphosphine di-iron dithiolate assemblies possessing a bridging hydride have been characterised crystallographically and are shown to be capable of catalysing the H^+/D_2 exchange process [29]. Related synthetic tertiarylphosphine di-iron complexes are beginning to show some evidence for catalysis. For example, Rauchfuss and Gloaguen [30] have provided preliminary evidence that a monocyanide derivative is capable of electrocatalytic proton

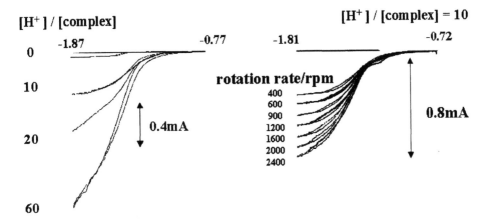

Figure 5. Current-potential curves at an Hg(Au) rotating disc electrode scanned at 10mV/s for electrocatalytic reduction of proton in the presence of 0.067 mM $[Fe_2(SCH_2CH_2CH_2S)(CO)_6]$ in MeCN – 0.1M $[NBu_4][BF_4]$. The first set of curves shows the response in the absence and with increasing concentration of tosylic acid, the second set shows the dependence of the response on the disc rotation rate, ω. From the latter experimental data it is found that the plateau current varies linearly with $\omega^{1/2}$ confirming that the reduction is limited by the rate of diffusion of proton to the electrode interface.

reduction to dihydrogen and that the catalysis may involve both protonation at the metal-metal bond forming a bridging hydride and also protonation at the cyanide ligand.

That the simplest of di-iron dithiolates show capacity for electrocatalytic reduction of protons is illustrated by some of our recent work. Figure 5 shows the rotating disc response for the reduction of $[Fe_2(SCH_2CH_2CH_2S)(CO)_6]$ **7** in the presence of tosylic acid at increasing concentrations of the acid. The linear dependence of the limiting current on the square root of the rotation rate shows that the process is controlled by diffusion of protons to the electrode at least up to a [H⁺]/[catalyst] ratio of 10. Detailed electrochemical, spectroelectrochemical and synthetic studies of this system with Stephen Best's group at Melbourne are revealing a fascinating electron-transfer chemistry involving the formation of bridging carbonyl intermediates in proton coupled reduction and a role for hydridic intermediates [31].

5   FINAL REMARKS

Preparative chemistry is now affording a wide range of di-iron materials related to the catalytic machinery of the active site of the all-iron hydrogenase.

Electrochemical studies of these are beginning to reveal chemistry both pertinent to understanding the enzymic catalysis and the identification of electrocatalytic systems. The challenges are to design assemblies which reversible catalyse hydrogen production/ uptake at high rates and at low overpotentials paralleling the high exchange current densities observed for platinum electrocatalysis [31].

Table 1. Primary redox potentials of {2Fe₃S}-carbonyl / cyanide assemblies referenced against calomel, at 25°C, in MeCN (0.1 M [NBu₄][BF₄]).

Entry	Compound	E_p^{red} (V)	E_p^{ox} (V)
1	$[Fe_2(SCH_2C(CH_2OH)S)(CO)_6]$	-1.28	1.3
2	$[Fe_2(SC_3H_6S)(CO)_6]$	-1.32	1.15
3	$[Fe_2\{MeSCH_2C(Me)(CH_2S)_2(CO)_5]$	-1.38	0.67
4	$[Fe_2\{PhCH_2SCH_2C(Me)(CH_2S)_2(CO)_5]$	-1.36	0.77
5	$[Fe_2(SC_3H_6S)(CN)(CO)_4(PMe_3)]^-$	-1.97	
6	$[Fe_2\{MeSCH_2C(Me)(CH_2S)_2\}(CN)(CO)_4]^-$	-1.83	0.17
7	$[Fe_2\{PhCH_2SCH_2C(Me)(CH_2S)_2\}(CN)(CO)_4]^-$	-1.83	0.12
8	$[Fe_2(SC_3H_6S)(CN)(CO)_5]^-$	-1.84	0.46
9	$[Fe_2(SC_3H_6S)(CN)(CO)_4(PMe_3)]^-$	-2.25	-0.06
10	$Fe_2(\mu\text{-}CO)\{MeSCH_2C(Me)(CH_2S)_2\}(CN)_2(CO)_3]^{2-}$	≤ -2.40	-0.10b
11	$[Fe_2\{MeSCH_2C(Me)(CH_2S)_2\}(CN)_2(CO)_4]^{2-}$	≤ -2.40	-0.25
12	$[Fe_2\{PhCH_2SCH_2C(Me)(CH_2S)_2\}(CN)_2(CO)_4]^{2-}$	≤ -2.40	-0.26
13	$[Fe_2(SCH_2C(CH_2OH)S)(CN)_2(CO)_4]^{2-}$	-2.36	-0.08
14	$[Fe_2(SC_3H_6S)(CN)_2(CO)_4]^{2-}$	-2.33	-0.11

ACKNOWLEDGEMENTS

I gratefully acknowledge the major contributions that Stephen Best, Stacey Borg, Saad Ibrahim, Mathieu Razavet, Xiaoming Liu, Cedric Tard, David Evans, David Hughes and Shirley Fairhurst have made towards this work and the BBSRC, JIF and the ARC for funding.

REFERENCES

1. Nicolet, Y.; Piras, C.; LeGrand, P.; Hatchikian, C.E.; Fontecilla-Camps, J.C. *Structure*, 7, 13, (1997).
2. Peters, J.W.; Lanzilotta, W.N.; Lemon, B.J.; Seefeldt, L.C. *Science*, 282, 1853, (1998).
3. Pierik, A.J.; Hulstein, M.; Hagen, W.R.; Albracht, S.P.J. *Eur. J. Biochem.*, 258, 572, (1998); De Lacey, A.L.; Stadler, C.; Cavazza, C.; Hatchikian, E.C.; Fernandez, V.M. *J.Am.Chem.Soc.*, 122, 11232, (2000); Nicolet, Y.; De Lacey, A.L.; Vernede, X.; Fernandez, V.M.; Hatchikian, E.C.; Fontecilla-Camps, J.C. *J. Am. Chem. Soc.*, 123, 1596, (2001).
4. Pereira, A.S.; Tavares, P.; Moura, I.; Moura, J.J.G.; Huynh, B.H., *J. Am. Chem. Soc.*, 123, 2771, (2001).
5. Popescu, C.V.; Munck, E. *J. Am. Chem. Soc.*, 121, 7877, (1999).
6. Surerus, K.K.; Chen, M.; Vanderzwaan, J.W.; Rusnak, F.M.; Kolk, M.; Duin, E.C.; Albracht, S.P.J.; Munck, E. *Biochemistry*, 33, 4980, (1994).
7. DeLacey, A.L.; Stadler, C.; Cavazza, C.; Hatchikian, E.C.; Fernandez, V.M. *J. Am. Chem. Soc.*, 122, 11232, (2000).
8. Razavet, M.; Borg, S.J.; George, S.J.; Best, S.P.; Fairhurst, S.A.; Pickett, C.J. *Chem. Commun.*, 700, (2002).
9. Cao, Z.X.; Hall, M.B. *J. Am. Chem. Soc.*, 123, 3734, (2001).
10. Nicolet, Y.; Lemon, B.J.; Fontecilla-Camps, J.C.; Peters, J.W. *Trends Biochem. Sci.*, 25, 138, (2000).
11. Peters, J.W. personal communication.
12. Reihlen, H.; Gruhl, A.; Hessling, G. *Liebigs Annalen Der Chemie*, 472, 268, (1929).
13. Seyferth, D.; Womack, G.B.; Archer, C.M.; Dewan, J.C. *Organometallics*, 8, 430, (1989).
14. Lyon, E.J.; Georgakaki, I.P.; Reibenspies, J.H.; Darensbourg, M.Y. *Angew. Chem., In.Ed.*, 38, 3178, (1999).
15. Schmidt, M.; Contakes, S.M.; Rauchfuss, T.B. *J. Am. Chem. Soc.*, 121, 9736, (1999).
16. Le Cloirec, A.; Best, S.P.; Borg, S.; Davies, S.C.; Evans, D.J.; Hughes, D.L.; Pickett, C.J. *Chem. Commun.*, 2285, (1999).
17. Lawrence, D.; Li, H.X.; Rauchfuss, T.B.; Benard, M.; Rohmer, M.M. *Angew.Chem., Int. Ed.*, 40, 1768, (2001).
18. Lyon, E.J.; Georgakaki, I.P.; Reibenspies, J.H.; Darensbourg, M.Y. *J. Am. Chem. Soc.*, 123, 3268, (2001).
19. Gloaguen, F.; Lawrence, J.D.; Schmidt, M.; Wilson, S.R.; Rauchfuss, T.B. *J.Am.Chem.Soc.*, 123, 12518, (2001).
20. George, S.J.; Cui, Z.; Razavet, M.; Pickett, C.J. *Chemistry: A European Journal*, 8, 4037, (2002).
21. Razavet, M.; Davies, S.C.; Hughes, D.L.; Pickett, C.J. *Chem. Commun.*, 847, (2001).
22. Razavet, M.; Davies, S.C.; Hughes, D.L.; Barclay, J.E.; Evans, D.J.; Fairhurst, S.A.; Liu, X.; Pickett, C.J. *J.Chem.Soc., Dalton Trans.*, 586, (2003).

23. Liu, Z.P.; Hu, P. *J. Am. Chem. Soc.*, *124*, 5175, (2002).
24. Chatt, J.; Kan, C.T.; Leigh, G.J.; Pickett, C.J.; Stanley, D.R. *J.C.S Dalton*, 2032, (1980).
25. Teo, B.K.; Hall, M.B.; Fenske, R.F.; Dahl, L.F. *Inorg. Chem.*, *14*, 3103-3117, (1975).
26. Zhao, X.; Georgakaki, I.P.; Miller, M.L.; Yarbrough, J.C.; Darensbourg, M.Y. *J. Am. Chem. Soc.*, *123*, 9710, (2001).
27. Collman, J.P.; Wagenknect, P.S.; Hembre, R.T.; Lewis, N.S.; *J. Am. Chem. Soc.*, *112*, 1294, (1990).
28. Zhao, X.; Georgakaki, I.P.; Miller, M.L.; Mejia-Rodriguez, R.; Chiang, C.Y.; Darensbourg, M.Y. *Inorg.Chem.*, *41*, 3917, (2002).
29. Gloaguen, F.; Lawrence, J.D.; Schmidt, M.; Wilson, S.R.; Rauchfuss, T.B. *J. Am. Chem. Soc.*, *123*, 12518, (2001).
30. Best, S.P.; Borg, S.; Razavet, M.; Liu, X.; Tard, C.; Pickett, C.J. unpublished results
31. Evans, D.J.; Pickett, C.J. *Chem.Soc.Rev.*, *in press*, 2003.

Part 4
Supramolecular Electrochemistry

Chapter 8

Supramolecular Electrochemistry of Coordination Compounds and Molecular Devices

Massimo Marcaccio, Francesco Paolucci and Sergio Roffia

Dipartimento di Chimica "G. Ciamician", University of Bologna,
via Selmi, 2, 40126 Bologna, Italy.

Dedicated to the memory of Maria Gabriela Teixeira.

1   INTRODUCTION

1.1   From Molecular Electrochemistry to Supramolecular Electrochemistry

Supramolecular electrochemistry concerns the description of the electrochemical properties of supramolecular systems. The study of supramolecular systems has developed enormously over the last two decades and, for its extensive interdisciplinarity, has lead to a very large integration of modern chemistry with materials science [1a]. The classical definition of supramolecular chemistry, given by J.-M. Lehn, as "the chemistry beyond the molecule" [1b-e] underlines the concept that "the association of two or more chemical species held together by intermolecular forces" may give rise to a new chemical entity whose properties are totally new and perhaps even unexpected with respect to those of the individual components. Orthodoxy would distinguish between supramolecular systems and multicomponent systems, the former being assemblies in which the individual components are kept together by relatively weak, non-covalent, bonds. One should however consider that in most cases a supramolecular system intrigues more for the chemical or physical functions it is able to perform rather than for the way the single components are brought together. From a functional point of view, then, a different definition has been proposed [2], and is now widely accepted [3], which underlines the fact that, regardless the nature of the bonds that link the molecular units, a supramolecular system, at variance with a *large molecule*, is a chemical species in which the electronic interactions between the units are weak enough to leave most of their properties almost unchanged (with respect to the separated units), and, at the same time, strong enough to promote the novel and unique properties of the supramolecular assembly.

Supramolecular electrochemistry deals with such systems by applying methods and concepts that are typical of molecular electrochemistry. In their search for the subtle and sometimes elusive effects that the supramolecular environment exerts on the redox properties of the molecular components,[1] supramolecular electrochemists can profitably use the wealth of experimental techniques, methodologies and theoretical knowledge, accumulated over a period of many decades, that made molecular electrochemistry such a highly refined science [4], as largely exemplified throughout this book. At the same time, experimental and theoretical methodologies proper to supramolecular electrochemistry have been developing over the years [3d,e,5]. Because of the high complexity typical of most of the supramolecular systems, extensive use of molecular models is usually made in their electrochemical, as well as spectroscopic and photophysical, characterization ("bottom-up" approach). Comparison with models is fundamental in order to evidence small changes in the electronic properties of the single units due to the supramolecular framework. An appropriate experimental and theoretical approach will comprise the use of several electrochemical techniques, cyclic voltammetry (CV), steady-state voltammetry, chronoamperometry and coulometry and electrochemical impedance spectroscopy. The electrochemical study will also be supported by spectroscopic and photochemical investigations, also by use of coupled spectroelectrochemical or photoelectrochemical techniques. High vacuum procedures, with opportunely realized equipment, and chemical treatment of the solvents to minimize their protic and nucleophilic content, allow high reproducibility of the experimental conditions and permit very wide potential windows in both the negative and positive regions. Reproducibility is a very important prerequisite for the "bottom-up" approach, where the supramolecular species and its molecular components are often investigated in separated experiments. Low temperatures

---

[1] To the best of our knowledge, the expression "supramolecular electrochemistry" was used for the first time by J.-M. Savéant who entitled "Elements of Supramolecular Electrochemistry" his lecture given at the "Giornate della Elettrochimica Italiana 1986" (Atti, Bologna, 1986, 15-16). The idea, exposed in his lecture, that the reactivity of a redox center (namely, an iron porphyrin) may be modulated by the introduction of "local superstructures" was later on developed by Savéant in [5a].

and high scan rates (using ultramicroelectrodes) allow the investigation of electro-generated intermediates with micro and sub-microseconds lifetimes. Moreover, the experimental approach is often complemented by the use of digital simulation of CV experiments for the description of the mechanism of electrode processes, by employing either commercial softwares (such as DigiSim, by Bioanalytical Systems) or home-made programs (e.g., Antigona, by Dr. L. Mottier, [5f]).

The electrochemical study, supported by spectroscopic, photochemical and photoelectrochemical techniques, may provide fundamental information (i) on the spatial organisation of the redox sites within the molecular or supramolecular structure, along with the energy location of the corresponding redox orbitals; (ii) on the entity of the interactions between units in the supramolecular framework and therefore on the thermodynamical feasibility of such processes as intramolecular electron/energy transfers and electrochemically-induced isomerization and conformational changes, (iii) on the kinetic stability of reduced/oxidised and charge-separated species, (iv) on the interaction of such species with conducting and semiconducting materials in view of the realization of electrochemical, photoelectrochemical and photovoltaic devices. Finally, the modeling of the supramolecular and molecular species by the methods of quantum-chemistry and of molecular mechanics also plays a fundamental role in the understanding of the electronic properties and of the dynamics of supramolecular systems.

As an example that anticipates most of the concepts that will be developed in the following Sections, the anodic behavior of the series of coordination compounds $[Ru(bpy)_2(CN)_2]$, $[NC-Ru(bpy)_2-CN-Ru(bpy)_2-CN]^+$, $[NC-Ru(bpy)_2-CN-Ru(bpy)_2-NC-Ru(bpy)_2-CN]^{2+}$ and $[NC-Ru(bpy)_2-CN-Ru(DCE-bpy)_2-NC-Ru(bpy)_2-CN]^{2+}$ (Figure 1) will be described [5b,c, 6].

In cyclic voltammetry (CV), the above species undergo Ru(II)-centered oxidation processes, as schematically summarized in the diagram of Figure 1. The identification of the *redox sites* (see Section 1.4) was made by taking the electronic asymmetry of the CN bridge into account. In fact, since the N-terminal of the bridge has amine-type properties, a lower E° is expected for the Ru(II) coordinating the higher number of Ns, i.e. the CN-Ru-CN center in the dinuclear complex, and the CN-Ru-NC center in the trinuclear ones. Such a localization of the first oxidation was confirmed by comparing the two trinuclear complexes. As expected because of the electronic perturbation exerted by the four electron-withdrawing carboxyethyl groups on the bpy's of the central unit, the first oxidation is shifted, in the DCE-complex, towards more positive potentials. The following oxidations involve obviously the remaining Ru(II) centers. Such processes are significantly shifted towards more positive potentials as a consequence of the electron-withdrawing effect exerted by the Ru(III) moiety mediated by the CN bridge. The same effect also explains the considerable separation between second and third oxidation in trinuclear complexes, that, although involving two identical moieties, greatly exceeds the expected statistical factor (36 mV at 25 °C) [7]. The energy distribution of the MLCT states within the supramolecular structure, as deduced from the electrochemical properties, explains the efficient energy funneling, observed in the trinuclear species, from the peripheral chromophores to the central unit (*antenna effect*). This prompted the use of the related trinuclear complex [NC-Ru(bpy)$_2$-CN-Ru(DCH$_2$-bpy)$_2$-NC-Ru(bpy)$_2$-CN]$^{2+}$ (DCH$_2$-bpy = 2,2'-bipyridine-4,4'-dicarboxylic acid) for the sensitization of mesoporous TiO$_2$ thin films in photoelectrochemical solar cells that gave an overall conversion efficiency of ca. 7% with turnover numbers of at least five million without decomposition [8].

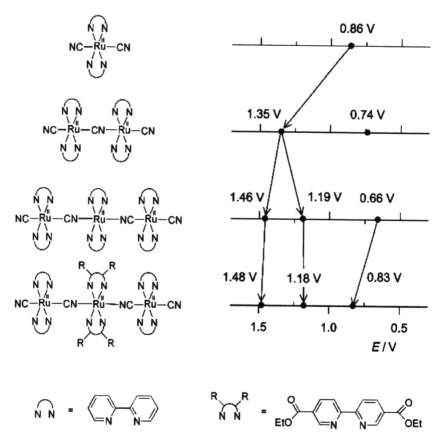

Figure 1. Genetic diagram relative to the metal-based oxidation processes of a series of polynuclear cyanide-bridged Ru(II) complexes. E$_{1/2}$ values (*vs.* SCE) obtained in acetonitrile, tetraethylammonium tetrafluoroborate solutions; T = 25 °C.

A comprehensive coverage of all the aspects of supramolecular electrochemistry is beyond the scopes of the authors. The selected examples are taken from the authors' contribution to this research field and reflect otherwise their personal preferences. Comprehensive reviews and books [3d, 9-12], also covering many other aspects of this fascinating area, are available to which the reader is directed for further information.

1.2   Self-Organization and Supramolecular Chemistry

The systems described in this Chapter are to be considered supramolecular systems according to the definition given above since (i) they are well-defined oligomolecular assemblies that result from the intermolecular association of a few components, (ii) the molecular components keep, in the supramolecular assembly, most of their physical and chemical properties but new and unique properties arise, with respect to the separate subunits, triggered by the (weak) interactions between them.

The concept of molecular recognition is central to supramolecular chemistry. By exploiting ligand coordination, π donor-π acceptor (charge-transfer, CT) interactions, hydrophobic or solvophobic forces and hydrogen bonding, molecular recognition-directed self-organization has opened new synthetic routes for the construction of a large variety of supramolecular systems with increasing architectural complexity [1d, 11]. The synthetic procedures for the construction of the supramolecular entities herein described greatly rely on self-organization of their molecular precursors and on their non-covalent 3-D positioning to which covalent bond formation follows.

1.3   Molecular Devices and Molecular Machines

A *molecular device* [2c, 12], i.e. a device at the molecular level, is an assembly of a finite number of molecular components organized on the spatial, temporal and energetic scales so that it may perform a useful and specific function. While each molecular component can perform a single act, the supramolecular assembly is able to perform complex functions through the concerted cooperation between the various components.

A *molecular machine* [2c, 12] is a particular type of molecular device. Analogously to its macroscopic equivalent, a molecular machine is a combination of parts so connected as to have constrained motion and to be capable of transmitting or transforming energy. A molecular machine works, i.e. performs a mechanical function, by dissipating the energy supplied to it in the form of chemical energy (as most biomolecular motors do [13]), light or electrical/electrochemical stimuli [12]; in other words, molecular machines cannot be heat engines [12, 13]. They may perform linear movements, rotary motions, changes in molecular structure, assembly/disassembly, translocation of parts, contraction/extension [12], and, in principle, various functions can be performed exploiting their movements, from transport of given molecules, for instance across a membrane, to information storage and processing [12].

1.4   Localization of Redox Sites in Multicomponent Systems

A *redox site* is a domain of atoms, within the molecular structure, over which the redox orbital is delocalised. The redox site concept is instrumental in the definition of *redox series*, i.e. a set of compounds with identical chemical composition that only differ for the overall number of electrons. The redox series concept originated in the 1970s mainly in connection with the electrochemistry of coordination compounds, and was formalized by A. A. Vlček at the beginning of 1980s in a series of fundamental papers where the rules that govern, in particular, the pattern of ligand-based redox series were laid [14]. A widespread diffusion of this concept in both the inorganic and organic electrochemical literature was then observed, mainly due to the fact that it has permitted a unified description and rationalization of the redox behavior of families of complex systems in terms of localized redox processes, and, at the same time, the precise evaluation of the mutual electronic interactions occurring between different redox sites. A molecule, and, *a fortiori*, a supermolecule, may contain several redox orbitals that belong to different part of space spanned by the molecule. In cyclic voltammetry, a redox series is manifested by a succession of reversible voltammetric peaks, and the separation between them is related to the interactions between the various sites. Coordination chemistry offers several examples that illustrate the different types of redox series. In transition metal coordination compounds, it is usually possible to discriminate between metal-centred and ligand-centred redox series respectively. Ligand-ligand interactions occurring either through space or mediated by the metal core are responsible for the separation usually observed between the subsequent reduction peaks relative to identical ligands in mononuclear complexes, and give rise to the typical multiplets observed in the CV patterns of such species[2]. $[Ru(bpy)_3]^{2+}$ (Figure 2) is a typical example of a coordination com-

---

[2] Although it is normal practice to use the separation between subsequent redox processes as a guide to the extent of the electronic interaction between redox sites, the effects of solvent and counterions on such a separation may be large and should always be taken into account especially when comparing redox series investigated in different media [Barrière, F.; Camire, N.; Geiger, W.E.; Mueller-Westerhoff, U.T.; Sanders, R. *J. Am. Chem. Soc.*, *124*, 7262 (2002)].

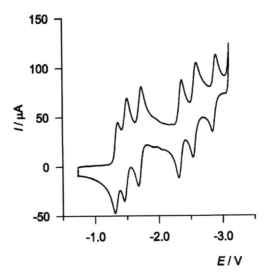

Figure 2. Cyclic voltammetric curve of $[Ru(bpy)_3]^{2+}$ (1×10^{-3} M) in dimethylformamide (0.05 M tetraethylammonium tetrafluoroborate); T = -54 °C, scan rate: 0.2 V/s; working electrode: Pt wire.

pound in which metal-ligand and ligand-ligand interactions are weak enough that its reduction behavior is easily related to that of the free ligands [5b].

The interaction with the metal stabilizes the bpy-centred redox orbital and consequently bpy is reduced at less negative potentials in the coordinated state than in the free one. Due to their chemical equivalence, the three ligands should be reduced at the same potential but ligand-ligand interactions intervene provoking their splitting in a triplet of peaks, one for each bpy reduction. The intramultiplet potential gaps reflect the electronic repulsion energy between electrons that are located on different ligands, while the intermultiplet potential gap relates to the electronic coupling energy of two electrons entering the same redox orbital, and, to a first approximation, is equal to that measured for the free ligand. It should be noted that polypyridine ligands are strong bases when doubly-reduced. This means that a symmetric, reversible pattern such as that shown in Figure 2, can only be obtained when very dry conditions are used in the CV experiments, otherwise follow-up protonation reactions bring about the early termination of the redox series, i.e., a smaller number of reduction peaks is observed, displaying a much lower reversibility. Extracting information of thermodynamical significance from the CV experiments is straightforward when a reversible behavior is observed, while it may be done in the presence of kinetic complications only by digital simulation procedures. The use of strictly aprotic conditions represents therefore a convenient choice, although sometimes highly demanding from the experimental point of view[3].

More complex redox series than that shown in Figure 2 are usually observed in the case of heteroleptic coordination compounds, in polynuclear complexes and supramolecular systems, where the various subunits may have different and also partly superimposing redox patterns. In the case of polynuclear metal complexes, for instance, additionally to the above types of interactions, metal-metal interactions (direct or mediated by bridging ligands, that can be either electroactive or not) and remote ligand-ligand interactions have also to be considered.

In most cases, the electronic interaction between subunits in supramolecular systems is weak; therefore their voltammetric behavior is described with the assumption that their ground state redox properties can be represented with a sufficient degree of approximation within the localized molecular orbitals model. In general, the analysis of the voltammetric curves, also by the use of digital simulation techniques, allows to obtain the redox standard potentials for overlapping processes in multielectron waves. The localization of all the redox process, i.e. their attribution to the various electroactive subunits present in the supramolecular structure, is based on the comparison with the redox properties of suitable model compounds. In the case of polynuclear metal complexes, the effects of either increasing nuclearity or replacement of a ligand in the coordination sphere on the CV pattern are also considered, with the construction of *genetic diagrams*, i.e. diagrams in which each redox potential for the various processes observed in the supramolecular system is correlated to that of an equivalent process observed in an appropriate model. Additionally, the absorption characteristics of the reduced/oxidised species generated during spectroelectrochemical experiments and the results of quantum chemical calculations may support the attribution of the redox processes, as is described in another Chapter of this book.

In Figure 3, an example of genetic diagram, with the typical generation of correlated multiplets, is illustrated in the case of the family of heteroleptic mononuclear complexes of general formula $[Ru(2,5\text{-}dpp)_n(bpy)_{3-n}]^{2+}$ (n=1-3) [5h].

---

[3] "The presence of an uncontrolled amount of water is obviously the main reason for large discrepancies in the literature both on the reversibility of the processes and on the values of the corresponding half-wave potentials" [Krejčik, M. Vlček, A. A. *J. Electroanal. Chem.*, *313*, 243 (1991)]. The procedures adopted in our laboratory, based on the chemical pretreatment of the solvents and on the use of high-vacuum techniques for the preparation of the electrochemical experiments, have allowed us to observe the largest redox series so far reported in the fields of coordination compounds and of fullerenes electrochemistry [5e, 25b,c].

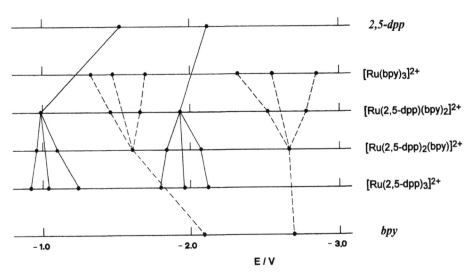

Figure 3. Genetic diagram relative to ligand-based reductions of a series of heteroleptic Ru(II)-polypyridine complexes; $E_{1/2}$ values (*vs.* SCE) obtained in dimethylformamide, tetraethylammonium tetrafluoroborate solutions; $T = -54\,°C$.

The identification of the ligand-based redox sites for each species belonging to the $[Ru(2,5\text{-dpp})_n(bpy)_{3-n}]^{2+}$ family is obtained by correlating the $E_{1/2}$ values of the various reduction steps along the whole family. The diagram also includes $[Ru(bpy)_3]^{2+}$ and the two free ligands bpy and 2,5-dpp. For both the homoleptic species, i.e., $[Ru(2,5\text{-dpp})_3]^{2+}$ and $[Ru(bpy)_3]^{2+}$, the six reductions arrange into two triplets. The first three processes represent the first reduction of each ligand and the following three reductions, in the second triplet, correspond to the electron coupling into the same redox orbital. As a matter of fact, the two triplets are separated by 560 mV, which is very close to the electron pairing energy found for the free ligands. The larger electron pairing energy (by 0.06 eV) observed in $[Ru(bpy)_3]^{2+}$ with respect to $[Ru(2,5\text{-dpp})_3]^{2+}$ may be attributed to the better ability of the latter to delocalize the injected charge [5h].

2   COORDINATION COMPOUNDS IN SUPRAMOLECULAR SYSTEMS

Polypyridine Ruthenium(II) and Osmium(II) complexes have been extensively investigated in the past two decades because of their outstanding luminescence and electrochemical properties [15]. They have high absorption coefficient in the visible spectral region, a well known and highly stable photophysics and are able to give energy and electron transfer processes when coupled to suitable acceptor units. In addition they show great chemical stability and relative ease of derivatization. For all of the above, these species are ideal building blocks for the development of molecular assemblies that may play an important role for many fundamental and practical purposes such as obtaining a better understanding of natural photosynthesis and the creation of molecular devices for the information storage and processing or the conversion of light into chemical/electric energy.

2.1   *Metal-based Dendrimers*

Dendrimers are macromolecules characterized by a highly branched three-dimensional globular structure, exhibiting a well-defined surface [3d,16]. Dendrimers are usually built with repetitive synthesis, and from the topological

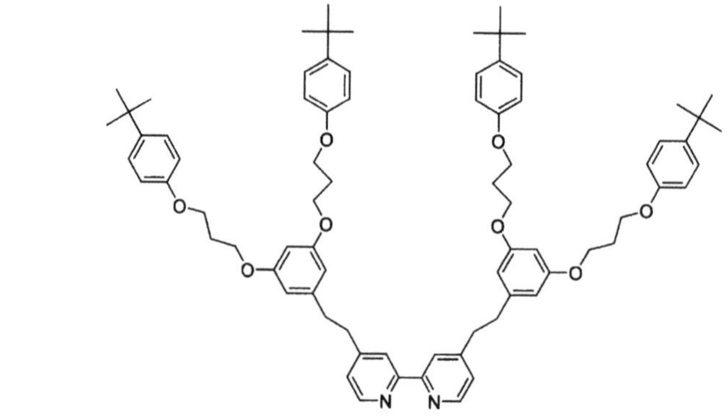

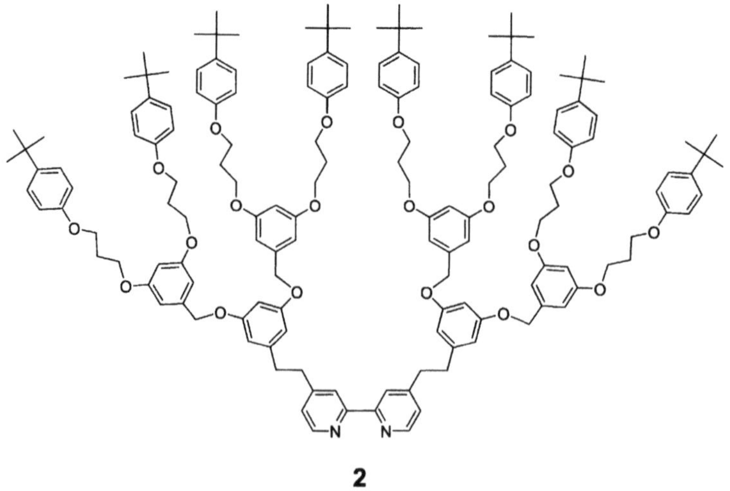

Figure 4. Dendrimeric ligands.

point of view they are characterized by a core unit and a number of branches radiating from it and terminating with end groups. Specific chemical, redox and photophysical properties may be attached at each or all of the above constitutive parts. A great variety of fully-organic and hybrid dendrimeric species, i.e. containing organometallic or coordination compounds, have been developed, as reported in many recent reviews [16]. Here only selected examples of redox- and photoactive dendrimers containing metal complexes will be presented.

2.1.1 Mononuclear systems

Dendrimers that contain a single metal core, typically $[Ru(bpy)_3]^{2+}$, have been designed principally to achieve two effects: (i) to protect the luminescence of the metal complex from external quenchers by the dendrimer branches or (ii) to sensitize its luminescence by chromophoric groups located at the periphery of the dendrimer (*antenna effect*) [16].

The protection of the dendrimer branches on the luminescent ^{3}MLCT level of the Ru(II)-bpy core from quenching reactions with various redox quenchers, 1,1'-dimethyl-4,4'-bypiridinium dication (MV^{2+}), tetrathiafulvalene (TTF) and anthraquinone-2,6-disulfonate (AQ^{2-}) was systematically investigated in the case of $[Ru(bpy)_2(1)]^{2+}$, $[Ru(bpy)_2(2)]^{2+}$ and $[Ru(2)_3]^{2+}$ (Figure 4) [17].

Electron-transfer quenching reactions may involve either the oxidation or the reduction of the excited state

$$D^* + A \rightarrow D^+ + A^- \qquad \text{oxidative quenching}$$
$$D + A^• \rightarrow D^+ + A^- \qquad \text{reductive quenching}$$

The thermodynamic driving force of such reactions ($\Delta G°$) is estimated (in eV) from the relationship

$$\Delta G° = E_{00} + e(E°_{D+/D} - E°_{A/A-})$$

where E_{00} is the spectroscopic energy of the excited state and $E°$s are the standard potentials of the two redox couples involved in the process, as obtained, for instance, by CV. In the present case, the $E°$ values relative to the metal-based oxidation and bpy-based reduction of dendrimers were found to be almost coincident with the respective values in $[Ru(bpy)_3]^{2+}$, i.e., 1.33 and -1.33 V respectively, and they have also the same excited-state energy (E_{00}= 2.12 eV). The values of $E°$(A/A-) are -0.44 and -0.57 V for MV^{2+} and AQ^{2-} respectively, while for TTF, $E°$(D+/D) = 0.33 V. By substituting these values in the above equation it was found that, in all the cases considered, the redox quenching process is thermodynamically allowed and has identical driving force. Quenching constants were however found to decrease significantly with increasing number and size of the dendrimer branches. The effect was not the same for all the quenchers and a charge effect

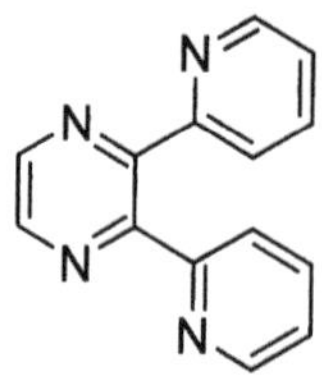

2,3-dpp

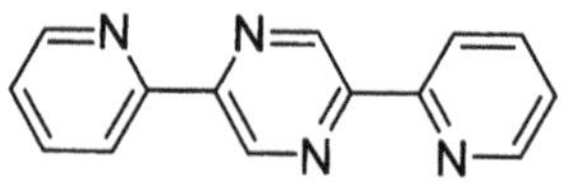

2,5-dpp

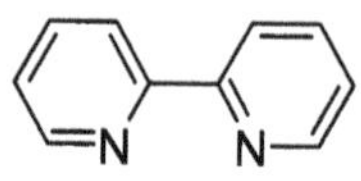

bpy

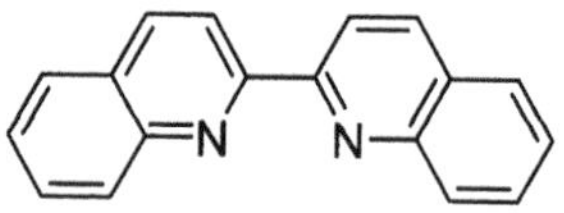

biq

Figure 5. Structural formulas of polypyridine ligands used as building blocks for the construction of dendrimeric polynuclear transition metal complexes.

was also evidenced: the rate constant decreased more in the case of the positively-charged MV^{2+} than for the negatively-charged AQ^{2-}. Among the various factors that may explain the experimental observations are (i) the dependence of diffusion constant of the dendrimer on its size, (ii) effect of dendrimer size and structure on diffusion of quenchers within the dendrimer, (iii) competition between solvent and dendrimer branches for the solvation of the metal complex, (iv) coulombic interaction between metal core and charged quencher.

2.1.2 Polynuclear systems

Several examples of dendrimeric polynuclear metal complexes have been reported [11e, 16a,b, 18]. Such systems contain a metal at each branching point of the dendrimer structure and have been investigated especially for their unique

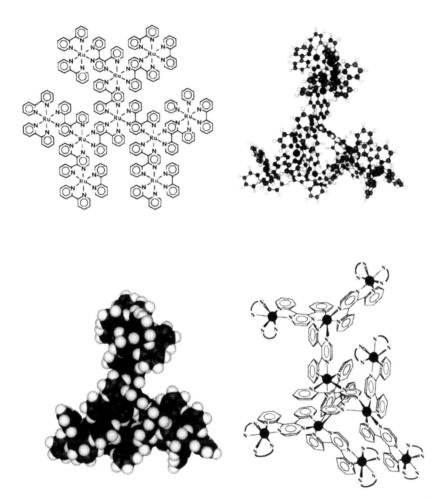

Figure 6. Different representations of the decanuclear dendrimeric polypyridine Ru(II) complex.

light harvesting properties. Polypyridine ligands have extensively been used to design and prepare polynuclear complexes with dendritic structures [11e, 18]. By using the building blocks depicted in Figure 5, probably the largest class of light-harvesting dendrimers investigated so far was obtained.

Dendrimers containing up to 22 chromophores, i.e. 22 metal atoms, 24 terminal ligands and 21 bridging ligands, have been synthesized by using the approach of using complexes as ligands and complexes as metals (various representations of the decanuclear species are shown in Figure 6) [11e]. In these species, because of the small but still not negligible electronic interactions between nearby units, very fast, exoergonic, and vectorial energy transfer processes occur.

Such processes quench the luminescence of the units with higher ^{3}MLCT levels and sensitize that of the units with lower ^{3}MLCT levels: the direction of the energy transfer processes (from the periphery of the system to the center or vice versa) can be predicted on the basis of the electrochemical and photophysical properties of the dendrimer and can be modulated by changing ligands and metal centers. Knowledge of the electrochemical properties of metal-based dendrimers is therefore of paramount importance to understand fully their properties and to foresee possible applications. For example, the study of successive ligand-centered reductions and ligand- or metal-centered oxidations is important (i) to understand the degree of metal-metal interaction within the dendritic array, (ii) to know which metal has the highest occupied d orbital and

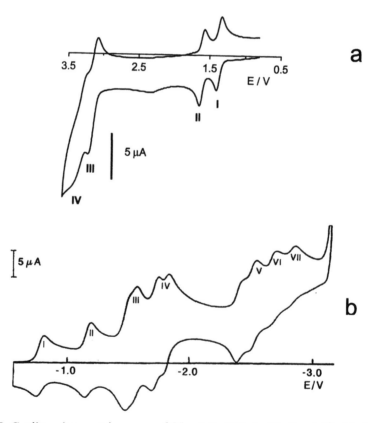

Figure 7. Cyclic voltammetric curve of [(bpy)$_2$Ru(2,3-dpp)Ru(bpy)$_2$]$^{4+}$ (a) 1 mM in liquid SO$_2$ (0.1 M tetraethylammonium hexafluoroarseniate), T = –70 °C; (b) 0.5 mM in dimethylformamide (0.05 M tetraethylammonium tetrafluoroborate); T = –54 °C. Scan rate: 0.5 V/s; working electrode: Pt wire.

which ligand has the lowest π^* orbital, and (iii) to predict the direction and the rate of a possible intramolecular energy or electron transfer. Electrochemical data, in terms of (i) localization of redox sites within the supramolecular structure and (ii) determination of the corresponding standard potentials, may therefore help in the assignment of ligand-centered and MLCT transitions of the above species, finally allowing the mapping of MLCT excited states within the dendrimeric structure. For the presence of many metal centers and polypyridine ligands, each capable of several subsequent reduction processes, very complex cyclic voltammetric patterns are found for such systems. By taking advantage of the wide anodic and cathodic potential windows respectively made available in liquid SO_2 and DMF at low temperature (up to ca. 4.3 V and –3.1 V vs SCE respectively), up to 14 reversible metal- or ligand-centered reversible oxidations and 26 reversible ligand-centered reductions, for a total of 40 redox steps were observed (in the case of the hexanuclear complex $\{[(bpy)_2Ru(2,3-dpp)]_2Ru(2,3-dpp)Ru[(2,3-dpp)Ru(bpy)_2]_2\}^{12+}$), i.e. the largest redox series so far known for well-defined molecular systems [5d,e].

In Figure 7, the CV curves showing the oxidation and reduction processes observed in the case of the dinuclear species $[(bpy)_2Ru(2,3-dpp)Ru(bpy)_2]^{4+}$ are reported.

Peak I and II in Figure 7a correspond to the first and second metal-centered oxidation processes. The splitting between the two metal-centered oxidations and the positive displacement of the first metal oxidation with respect to the mononuclear analogs [19] are common features in dinuclear complexes and reflect the extent of metal-metal interaction through the bridge [20]. This can be described within the superexchange theory [20, 21], where the overlap between metal orbitals is mediated by overlapping with those of the bridging ligand. Two different modes of metal-metal communication through a bridge

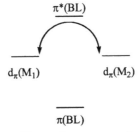

Figure 8. Orbital energy diagram illustrating superexchange interaction between two metal ions (M_1 and M_2) through a bridging ligand (BL). $d\pi$ (M_1), $d\pi$ (M_2), π (BL), π^* (BL) represent orbitals of the two metal centers and the LUMO and HOMO of the bridging ligand, respectively.

are possible: (i) an electron transfer mode across low-lying π^*_{BL} orbitals and (ii) a hole transfer mode across occupied π_{BL} orbitals. For complexes having low-lying π^*_{BL} orbitals, such as 2,3- and 2,5-dpp, mode (i) is the dominant one [20]. In this case, metal-metal interaction depends on the energy gap between the dπ-metal orbitals and the LUMO bridging ligand orbitals (see Figure 8). The double coordination of BL lowers the π^*_{BL} orbital (in fact, a large positive shift for the first reduction of the bridging ligand between mono- and dinuclear species is observed, vide infra), so it can couple more efficiently with d$_\pi$-metal orbitals as compared to the mononuclear analogs. Therefore, stabilization of d$_\pi$-metal orbitals via backbonding can explain the observed positive shift of the first metal-centered oxidation in dinuclear complexes with respect to mononuclear analogs. Electrostatic interaction between charged metal centers may in part play a role. The relative importance of through-bond and coulombic interactions is however difficult to estimate in the presence of counterions, which can effectively diminish the extent of through-space interactions.

At more positive potentials, the ligand-centered oxidations are observed (peaks III and IV). The first ligand-based processes probably involve bpy ligands rather than the bridging ligand on the basis of the analogous behavior of the previously reported mononuclear compounds [19]. Furthermore, the bridging ligand is linked to two positively charged metal subunits, so the dpp π orbitals are more stabilized than the corresponding bpy orbitals.

In Figure 7b, the ligand-centered reduction processes are shown. It should be noted that on the basis of electrochemical studies carried out on the mononuclear building blocks of the present polynuclear systems [22], redox series made up of at least twelve one-electron reduction processes would be expected. 2,3-dpp can undergo up to four reductions [22], when coordinated to a Ru(II) center. Coordination to two metal centers should expectedly anticipate such processes, and, as a possible additional consequence of bis-coordination, an increase of the number of the bridge-centered redox processes (because of the stabilization of a higher redox orbital) rather than its decrease should be observed, at variance with the experimental findings. Quantum chemical calculations [5e] have shown that the second dpp-centered LUMO, responsible in the mononuclear species for the third and fourth reductions, shifts significantly to higher energies because of the geometric constraints imposed by the coordination to a second metal center, thus making the ligand able to accept only two electrons within the available potential window. The electron density distributions (ZINDO-INDO/1) in one-electron reduced $[(bpy)_2Ru(2,3\text{-}dpp)Ru(bpy)_2]^{4+}$ and $[(bpy)_2Ru(2,5\text{-}dpp)Ru(bpy)_2]^{4+}$ and three-electron reduced $[(bpy)_2Ru(2,3\text{-}dpp)Ru(bpy)_2]^{4+}$ are shown in Figure 9 that evidences that, while the first two electrons are localized on the bridging ligand, the third electron involves a terminal bpy ligand.

Interestingly in this respect, the comparison of the energy diagrams of MO's involved in the bpy-centered redox processes for the two dinuclear species, $[(bpy)_2Ru(2,3\text{-}dpp)Ru(bpy)_2]^{4+}$ and $[(bpy)_2Ru(2,5\text{-}dpp)Ru(bpy)_2]^{4+}$, shows several important features which are in good agreement with the electrochemical results. This is shown in particular for the six-electron reduced species in Figure 10. In agreement with the experimental findings, (i) the bpy-centered MO's of the dinuclear species with 2,3-dpp are at lower energy (i.e., easier to reduce) than the corresponding ones in the species containing 2,5-dpp, and (ii) a lower spin pairing energy onto dpp-centered MO results from the calculations for the 2,3-dpp species with respect to the 2,5-dpp one.

The energy difference between two MO's localized onto bpy ligands bound to different metal centers (i.e. corresponding to processes comprised in peaks III and IV of CV curves) is larger in the case of 2,3-dpp than 2,5-dpp. Vice-versa, the

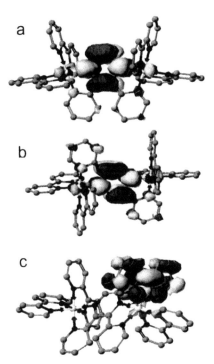

Figure 9. Molecular orbital surfaces showing the localization of relevant orbitals involved in the redox processes. (a) Singly occupied MO of $[(bpy)_2Ru(2,3\text{-}dpp)Ru(bpy)_2]^{3+}$ (one-electron reduced species) centered on the 2,3-dpp bridging ligand; (b) singly occupied MO of $[(bpy)_2Ru(2,5\text{-}dpp)Ru(bpy)_2]^{3+}$ centered on the 2,5-dpp bridging ligand; (c) singly occupied MO of $[(bpy)_2Ru(2,3\text{-}dpp)Ru(bpy)_2]^{+}$ (triply one-electron reduced species) centered on a terminal bpy ligand.

separation between the couples of two closely spaced MO's, centered onto bpy coordinated to the same metal atom, is larger for the species with 2,5-dpp than in that with 2,3-dpp. Within the superexchange model for the interaction between remote (bpy-centered) redox sites in the doubly-reduced complexes, calculations evidence that 2,3-dpp is a better mediator, possessing a local unoccupied MO at lower energies than the 2,5-isomer.

As nuclearity, and therefore the number of redox steps increases, the CV behavior becomes more and more complex. However, the localization of each redox process and the mutual interactions of the redox centers in the tri-, tetra- and hexanuclear species were analogously elucidated through the comparison of the redox series of the various compounds of increasing nuclearity.

In Figure 11a, the electrochemical behavior of the hexanuclear compound, $\{[(bpy)_2Ru(2,3-dpp)]_2Ru(2,3-dpp)Ru[(2,3-dpp)Ru(bpy)_2]_2\}^{12+}$ is shown.

Oxidation peak I involves the four external Ru ions while peak II is due to the

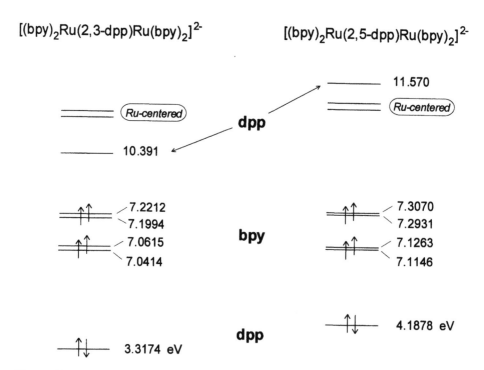

Figure 10. Energy diagrams of MO's obtained for the $[(bpy)_2Ru(2,3-dpp)Ru(bpy)_2]^{2-}$ and $[(bpy)_2Ru(2,5-dpp)Ru(bpy)_2]^{2-}$ compounds considering the first two one-electron reductions as coupling into the same orbital and the successive four processes as unpaired electrons.

two internal ones [5d]. Importantly, the two internal Ru oxidations occur at the same potential. This was unexpected on the basis of the behavior of the dinuclear complex, $[(bpy)_2Ru(2,3\text{-}dpp)Ru(bpy)_2]^{4+}$, in which the two Ru^{II} oxidations occur at distinct potentials ($\Delta E_{1/2}$ = 230 mV). A possible explanation is that, in the four-electron oxidized hexanuclear compound, the four strong electron-withdrawing peripheral groups $[Ru^{III}(2,3\text{-}dpp)(bpy)_2]^{3+}$ cause a lowering of the $d\pi$ orbitals of the two central Ru^{II} ions and therefore a lower overlap between metal-$d\pi$ and π^*_{BL} orbitals (Figure 8) than in the dinuclear analogs. This results in a smaller interaction between the two central Ru ions and in virtually coincident oxidation potentials.

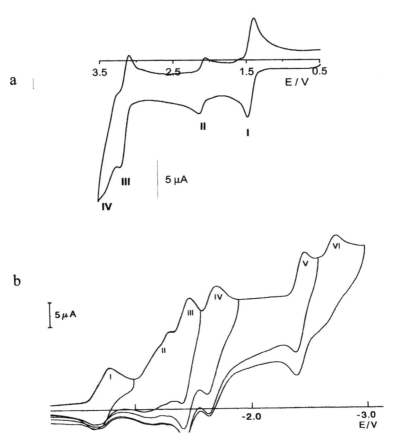

Figure 11. Cyclic voltammetric curve of $\{[(bpy)_2Ru(2,3\text{-}dpp)]_2Ru(2,3\text{-}dpp)Ru[(2,3\text{-}dpp)Ru(bpy)_2]_2\}^{12+}$ (a) 0.6 mM in liquid SO_2 (0.1 M tetraethylammonium hexafluoroarseniate), T = −70 °C; (b) 0.5 mM in dimethylformamide (0.05 M tetraethylammonium tetrafluoroborate); T = −54 °C. Scan rate: 0.5 V/s; working electrode: Pt wire.

As to the cathodic processes [5e], comparison of the CV curve of the hexanuclear complex (Figure 11 b) with those of the trinuclear species, suggested that the four processes in peak I are localized on the four "external" 2,3-dpp ligands, since the interactions between redox centers destabilize the ligands bound to the same metal center. The first out of the five processes comprised in peak II is localized on the "central" 2,3-dpp, because the separation between peak I and II is unusually small. The other four processes in peak II represent the electron pairing into the same redox orbital of "external" 2,3-dpp. The electron pairing for the "central" bridging ligand is represented by the first process contained in peak III, i.e. the tenth electron exchanged. The remaining four processes of peak III and those in peak IV are the first reduction localized on each of eight bpy. Finally, the last two voltammetric peaks (V and VI) represent the second reduction of the terminal ligands. The separation of about 0.6 V between peaks IV and V confirms this assignment.

In conclusion, the electrochemical characterization evidenced that, in the family of dendrimeric metal complexes, metal-centered and ligand-centered oxidations and, among these latter processes, dpp-centered and bpy-centered ones, are energetically well separated, and the same holds for ligand- (bpy or dpp) centered reductions. Furthermore, the electronic properties of ligands and metal centers also depend strongly on their location within the molecular structure, i.e. whether in the periphery or in an inner position. In principle the layering of the injected charge, either positive or negative, over concentric shells within the dendrimeric structure is therefore possible, making such dendrimeric species a kind of molecular-sized capacitors.

2.2. Photoactive Dyads

Photoactive dyads are molecular devices designed to perform, through the generation of charge separated states, the conversion of light energy to chemical energy, in analogy to photosynthesis. An efficient artificial photosynthetic system, mimicking the natural ones, needs to include, possibly within a supramolecular structure, various complementary functions: light absorption, energy transfer, charge separation and charge stabilisation need to be accomplished in a proper way in order to ensure the maximum conversion of photonic energy into chemical potential [23]. The primary event on which artificial photosynthetic systems are based, is the photoinduced electron transfer process involving an excited state of the chromophore and a nearby electron donor (D) or acceptor (A). In most cases, the chromophore, also acting as either electron donor or acceptor, is covalently linked to the corresponding A or D unit through a suitable spacer (S). A great variety of organic and inorganic chromophores, D/A units and spacers have been investigated, and examples will be found in many recent reviews [10a, 24].

2.2.1 Fullerene-porphyrin systems

Fullerenes have widely been used as redox and photoactive units in D-S-A photoactive dyads, triads and polyads [24a,d-g,i,j]. In contrast with their poor ability to form stable cations, (up to C_{60}^{3+} can howevr be observed under very low nucleophilic conditions [25c]) fullerenes and their derivatives exhibit good electron acceptor properties (Figure 12): [60]-fullerene (C_{60}) [25] and its derivatives [26a] can accept reversibly up to 6 electrons in aprotic media, the first reduction potential being similar to that of quinones, the biological redox couples involved in natural photosynthetic processes. Fullerene derivatives have been coupled to electron donor groups in dyads where the fullerene also acts as the primary photosensitizer [24a]. However, the poor absorption properties of fullerenes in the visible region has prompted to the development of D-S-C_{60} dyads (and triads) where the main chromophoric function was attached to the D unit rather than to the fullerene.

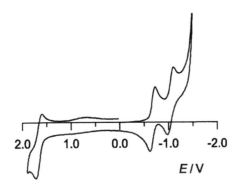

Figure 12. Cyclic voltammetric curve of C_{60}, 0.5 mM in tetrachloroethane (0.05 M tetrabutylammonium hexafluorophosphate). Scan rate: 0.5 V/s. T = 25 °C; working electrode: Pt disc.

Porphyrins and metalloporphyrins have been preferentially used as donor/chromophores in such systems, for the unique combination of favorable photophysical and redox properties of these compounds. Furthermore, these compounds, particularly appealing for their similarity with natural chromophores, are stable and allow high synthetic availability to many structural variations and therefore great flexibility in the design of D-S-A systems [24a, e-g, i, j, 27]. Since 1994, a great number of different porphyrin-C_{60} dyads, triads and more complex systems, also including supramolecular assemblies, have been investigated (see Figure 13 for some examples) and many examples are collected in a very recent special issue of the *Journal of Materials Chemistry* dedicated to *Functionalised Fullerene Materials* [24i].

Investigation of photoinduced intramolecular ET processes in such systems has in general shown that, compared to analogous porphyrin/quinone systems, (i) charge separation occurs with higher efficiency and (ii) charge-separated states have longer lifetimes, Figure 14. Such behavior is the result of the combination of two favorable characteristics of fullerenes compared to quinones and other normal donor units, namely (i) the curvature of fullerene surface that allows better electronic coupling between the acceptor unit and the hydrocarbon bridge and, ultimately with the D unit [23, 24a], (ii) the smaller reorganization energy associated to fullerenes, with respect to most acceptors, as a consequence of their large size and rigid framework in either the ground, excited or reduced states.

Most photoinduced charge separation processes are only slightly exoergonic, i.e. the process occurs generally in the normal Marcus region; a smaller reorganization energy therefore implies, for a given driving force, a faster process. Vice-versa, charge recombination is often a highly exoergonic process; this places the process in the Marcus inverted region and a smaller reorganization energy makes, for a given driving force, the kinetic constant of the charge recombination process smaller (Figure 14) [28].

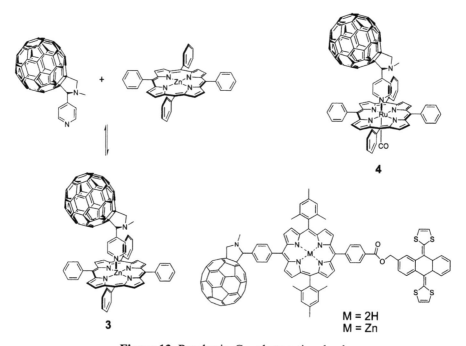

Figure 13. Porphyrin-C_{60} photoactive dyads.

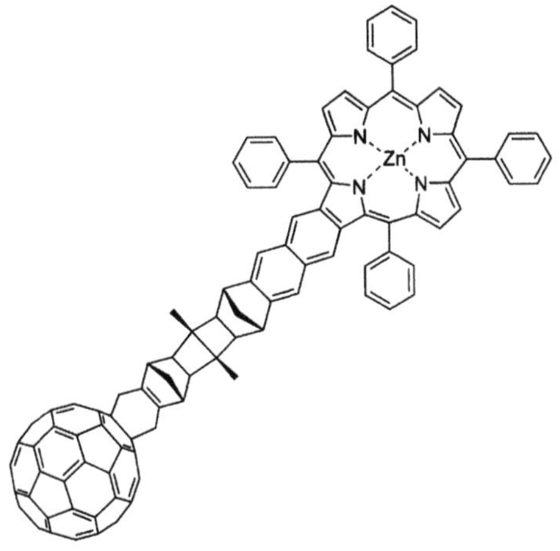

$$k_{cs} / s^{-1} = 1.0 \times 10^{10}$$

$$k_{cr} / s^{-1} = 2.4 \times 10^{6}$$

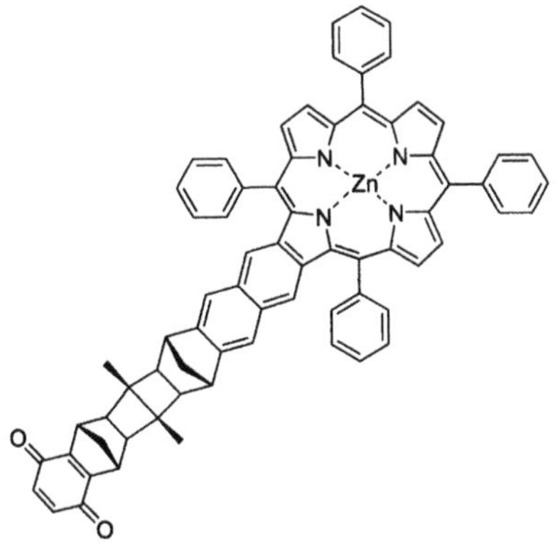

$$k_{cs} / s^{-1} = 2.7 \times 10^{9}$$

$$k_{cr} / s^{-1} = 2.0 \times 10^{8}$$

Figure 14. Rates of photoinduced charge separation and subsequent charge recombination in metalloporphyrin-based dyads (benzonitrile).

A metalloporphyrin-fullerene dyad has recently been reported where, in order to slow down the charge recombination process, a supramolecular approach has been adopted. Dyads **3** and **4** (Figure 13) have different relative stabilities: complexation of Zn-porphyrin by the fullerene unit is reversible and both the complexed and uncomplexed forms coexist in solution [29]. By contrast, the Ru-based dyad is chemically stable as also shown by the CV pattern, Figure 15 [30].

Upon excitation in apolar solvents, highly efficient triplet-triplet energy transfer occurs in the Ru-based dyad, while electron transfer from the porphyrin to fullerene prevails in polar solvents [30]. In the Zn-based dyad electron transfer occurs both intramolecularly in the complex and intermolecularly in the uncomplexed form. In the latter case, radical pair Zn-porphyrin$^{\cdot+}$/C$_{60}$$^{\cdot-}$ have lifetimes of 10 μs in THF and about 50 μs in benzonitrile [29].

The search for fullerene-based materials has prompted the investigation of fullerene-containing thermotropic photoactive liquid crystals (Figure 16) [31]. The electrochemical and photophysical studies on the mixed methanofullerene-ferrocene material **5** were reported.

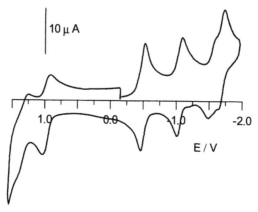

Figure 15. Cyclic voltammetric curve of dyad **4**, 0.4 mM in tetrahydrofuran (0.05 M tetrabutylammonium tetrafluoroborate). Scan rate: 0.5 V/s. T = –60 °C. Working electrode: Pt wire.

The dyad gives rise to a smectic A phase and to fast photoinduced electron transfer from ferrocene to fullerene with the formation of a relatively long-lived charge-separated state [32]. Porphyrin- and ferrocene-containing fulleropyrrolidines **6** have also been synthesized (Figure 16) that are promising candidates for the elaboration of molecular switches [31].

2.2.2 *Fullerene-[Ru(bpy)$_3$]$^{2+}$ systems*

Ru tris-bipyridine complexes have been widely used in photoactive dyads to study photoinduced electron- and energy transfer processes for their unique photophysical and redox properties [15]. Furthermore, the chemistry of bipyridine is well developed and allows a wide range of possible functionalized derivatives [33]. By combining the standard potential for the Ru(II)-based oxidation process (E° ~ +1.3 V) and its ^{3}MLCT excited state energy (E$_{00}$ ~ 2.0 eV), the standard potential of excited state E°[Ru(III)/*Ru(II)] is ~ –0.7 V. This latter may therefore ignite the photoinduced electron transfer to fullerene derivatives (E° ~ –0.4 V).

The nature and strength of the electronic interaction between a fulleropyrrolidine and the [Ru(bpy)$_3$]$^{2+}$ moiety, and the influence of the bridging spacer on such an

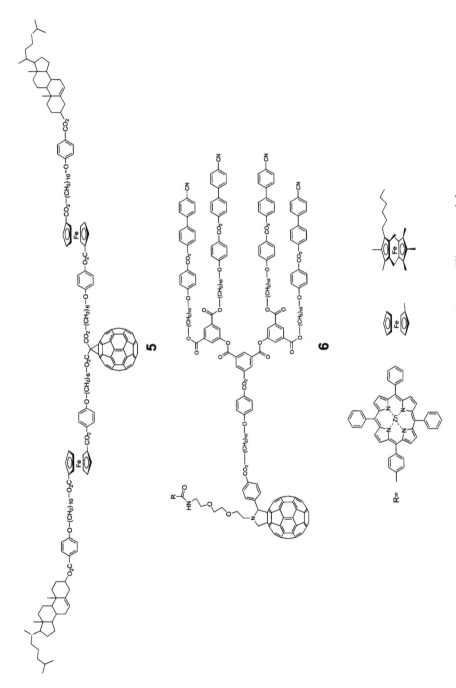

Figure 16. Fullerene-based liquid crystalline materials.

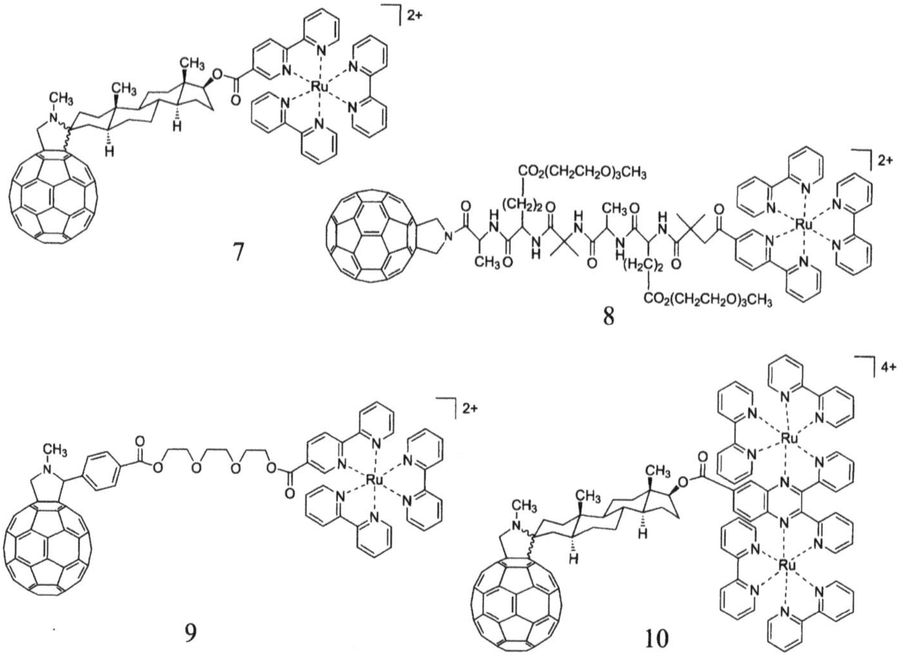

Figure 17. Fullerene-[Ru(bpy)$_2$)]$^{2+}$-based photoactive dyads.

interaction, were investigated by electrochemistry, steady state and time-resolved UV-Vis-NIR absorption spectroscopy and emission spectroscopy in a series of dyads where the units were covalently linked to each other by spacers showing different flexibility (Figure 17): a rigid androstane bridge that suppresses conformational freedom in the dyad, (**7**) [34], a hexapeptide (**8**) [35] that allows variable dimensional freedom upon temperature and/or solvent changes, and finally a highly flexible triethylene glycol spacer (**9**) [36].

Steady-state absorption and CV evidenced no significant interaction between the C$_{60}$ and [Ru(bpy)$_3$]$^{2+}$ units in the ground state. For example, Figure 18 shows the CV curves for dyad **7** and its separate components (**7A-7C** in Figure 19).

A reversible one-electron transfer at + 1.34 V is observed, assigned to the metal-centered Ru^{II}/Ru^{III} oxidation. The complex cathodic pattern comprises ten successive reversible one-electron transfer, whose $E_{1/2}$ values were obtained also by digital simulation of the CV curves. To identify the redox sites and to obtain information about the interactions between the two subunits, the electro-chemical behavior of **7** was compared to that of the models **7A-7C**. All these compounds show reversible processes (Figure 18). The cathodic processes were assigned as shown in the genetic diagrams of Figure 20: the electrochemical

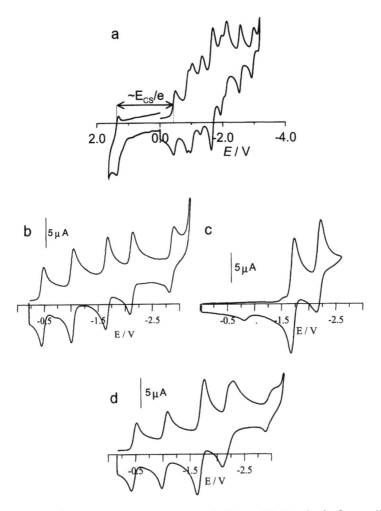

Figure 18. Cyclic voltammetric curves of 0.5 mM tetrahydrofuran (0.05 M tetrabutylammonium hexafluorophosphate) solutions of (a) **7**; (b) **7C**; (c) **7A** and (d) **7B**. T = −60 °C, scan rate = 0.2 V/s, working electrode: Pt.

behavior of this dyad is the sum of that of its two subunits, the C_{60} core and the Ru^{II} tris-bipyridyl complex, thus excluding any significant electronic interaction between the two moieties in the ground state. The first reduction processes (at −0.47 V) involves the fullerene moiety. The driving force for the photoinduced ET process, from Ru ^{3}MLCT to fulleropyrrolidine, was then calculated in two media (CH_2Cl_2 and ACN), according to the formula $-\Delta G°_{ET} = E_{00} − E_{CS}$. E_{00} is the energy of the ^{3}MLCT excited state of **7** (1.97 eV). $E_{CS} = e[E_{1/2}(Ru^{II/III})$ − $E_{1/2}(C_{60}/C_{60}^-)] − \Delta G_s$ is the free energy of the charge-separated state corrected by the ion-ion and ion-solvent electrostatic interaction term (ΔG_s). $E_{1/2}(Ru^{II/III})$

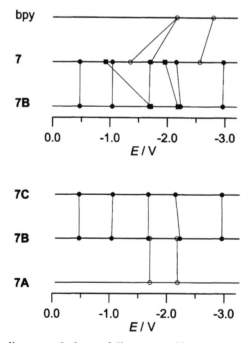

Figure 19. Building blocks of dyad **7**.

Figure 20. Genetic diagram relative to fullerene- and bpy-centered reductions in dyad **7** and its building blocks. Data obtained under the conditions described in Figure 18.

and $E_{1/2}(C_{60}/C_{60}^-)$ are the standard potentials for Ru-centred oxidation (1.34 V, in ACN) and fulleropyrrolidine-centred reduction (−0.41 V, in ACN).

The electrostatic term accounts for the different behavior observed in media of different polarity and is calculated within the dielectric continuum model which assumes two spherical ions separated by a distance and immersed in a dielectric continuum of proper dielectric constant. In the present case, ΔG°_{ET} was −0.02 eV in CH_2Cl_2 and −0.24 eV in ACN. As expected on the basis of the favorable driving force, electron transfer from the ruthenium complex to the fulleropyrrolidine was found to quench the photoexcited $[Ru(bpy)_3]^{2+}$ MLCT state. Also expectedly, the excited state dynamics was strongly affected by the solvent polarity. Charge-separated state was formed from the Ru ^{3}MLCT excited state, via a through-bond mechanism, with intramolecular rate constants of 0.69×10^9 s^{-1} and 5.1×10^9 s^{-1} in CH_2Cl_2-toluene (1:1, v/v) and ACN respectively. Time-resolved absorption spectra showed the superimposed signatures of both the Ru^{3+} radical cation and fulleropyrrolidine radical anion. Interestingly, the CS state decayed to the ground state with different rates and along different routes depending on the solvent polarity, as displayed in Figure 21.

In CH_2Cl_2 (solvent of low polarity) back electron transfer occurred, quite unusually [37], to the pristine Ru(II) ^{3}MLCT excited state, with $t_{1/2}$ = 210 ns. In polar ACN, charge recombination produced instead the triplet-excited state of fulleropyrrolidine, with $t_{1/2}$ = 100 ns. The fact that charge recombination produces excited states instead of leading directly to the ground state is a manifestation of the Marcus inverted character of such processes. As shown in Figure 20, the driving force values for charge recombination to ground state are 1.95 and 1.73 eV in CH_2Cl_2 and ACN respectively, and therefore such are highly inverted processes. Population of excited states in highly exoergic electron transfer reactions often prevents the observation of the inverted behavior and can also lead to chemiluminescence [38].

A different mechanism for electron transfer than in **7** was suggested in the case of **9**. Emission spectroscopy and time-resolved absorption spectroscopy showed the effective quenching of Ru MLCT excited state via fast electron transfer to the fullerene. The rate constant of such a process depends only weakly on the solvent polarity, increasing by less than a factor of 2 from CH_2Cl_2/toluene (2.3×10^9 s^{-1}) to ACN (4.1×10^9 s^{-1}), at variance with the case of **7**, where, over the same change in polarity, the rate constant increased by an order of magnitude (from 0.69×10^9 s^{-1} to 5.1×10^9 s^{-1}). The longer distance between $[Ru(bpy)_3]^{2+}$ and fulleropyrrolidine in **9** (20 σ-bonds) than in **7** (17 σ-bonds) would suggest intramolecular electron transfer be much slower in the former than in the latter, at variance with the experimental findings. An intramolecular quenching mechanism was proposed that involved a conformational rearrangement of

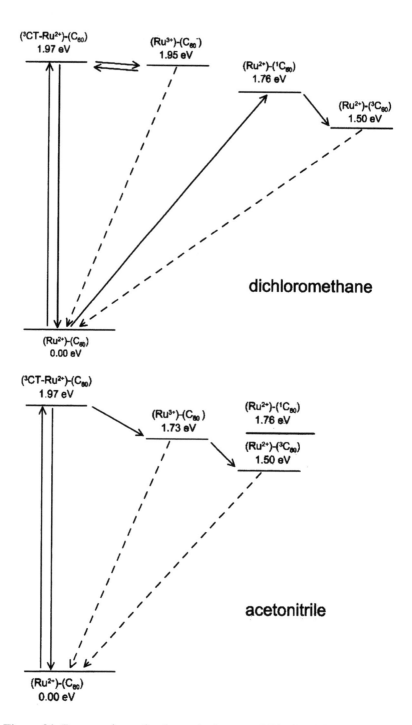

Figure 21. Energy scheme for the excited states of **7** in CH$_2$Cl$_2$ and acetonitrile.

the dyad and likely the formation of a transient exciplex. Accordingly, the two moieties, *Ru(II) and C_{60}, are brought into close contact, after the initial excitation process, thus facilitating the exoergonic electron transfer.

In **8**, the peptide spacer acts as a molecular switch: In non-protic solvents, the spacer adopts a fixed conformational geometry which locks the two moieties in well-defined positions, with an edge-to-edge distance of ca. 12 Å, and electron transfer occurs from the ruthenium MLCT excited state to the fullerene. In contrast, after addition of a strongly protic solvent, the helical structure of the spacers is disrupted leading to an increase in the distance between $[Ru(bpy)_3]^{2+}$ and fullerene. This prevents either the through bond or exciplex mechanism to support the electron transfer, that is therefore suppressed.

While intramolecular electron transfer is the prevailing processes occurring in **7**, energy transfer was only observed in the case of **10**, where the dinuclear complex $-[(bpy)_2Ru(dpq)Ru(bpy)_2]^{4+}$ (dpq = 2,3-bipyridin-2-yl-quinoxaline) replaces the $[Ru(bpy)_3]^{2+}$ moiety in dyad **7** [39a]. The CV study evidenced that, in contrast to the mononuclear dyad **7**, the first reduction is a ligand (dpq)-centered rather than a fullerene-centered process, the first reduction involving the fullerene moiety being located 260 mV towards more negative potentials than the dpq-centered one. In line with the lower energy of the redox LUMO associated to the bridging ligand, the MLCT transition is red-shifted in **10** with respect to dyad **7** from 460 nm to 625 nm. This makes the electron transfer from

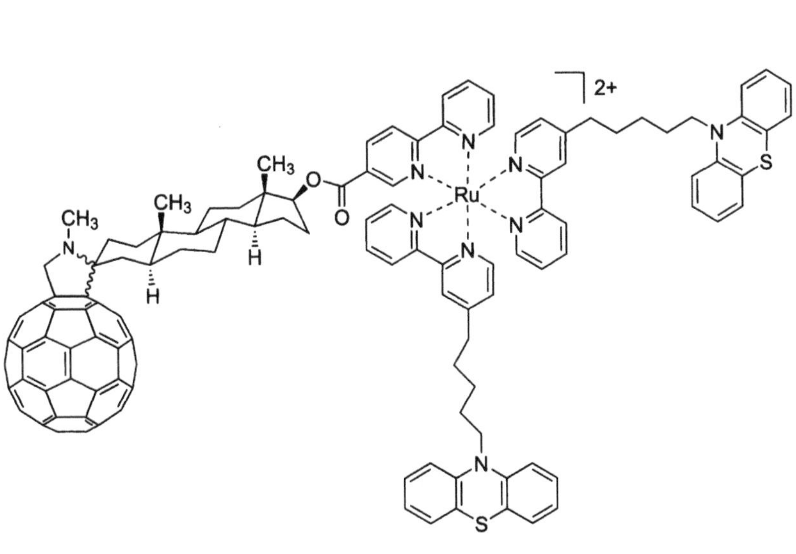

11

Figure 22. A C_{60}-bpy-PTZ photoactive triad.

Ru(II) excited state to fullerene thermodynamically prohibited in all solvents. Time-resolved absorption spectroscopy revealed that MLCT excited state transforms slowly (4.5×10^7 s^{-1}) into the fullerene triplet excited state.

As a conceptual development of dyad **7**, the triad **11** was obtained where the strong electron donor phenothiazine (PTZ) was covalently linked to the [Ru(bpy)$_3$]$^{2+}$ moiety to act as secondary electron donor capable to stabilize the charge-separated state [39b].

Upon photoexcitation of the [Ru(bpy)$_3$]$^{2+}$ chromophore, the PTZ-Ru(III)-C$_{60}^{\cdot-}$ pair develops, with similar intramolecular kinetics (2.8×10^9 s^{-1} in ACN) as that observed in **7**, followed by formation of the the PTZ$^{\cdot+}$-Ru(II)-C$_{60}^{\cdot-}$ species. Importantly in view of the practical exploitation of photogenerated chemical potential, lifetimes of charge-separated state are significantly longer than those observed for **7**, 1.29 µs and 304 ns in CH$_2$Cl$_2$ and ACN respectively.

2.2.3 Enhancing the electron acceptor properties of fullerenes

As a consequence of the saturation of its double bonds, most of C$_{60}$ mono and polyadducts have decreased electronegativity with respect to pristine fullerene [25]. It is clear that the improvement of the electron-accepting properties of C$_{60}$, coupled with the versatility of the organic chemistry of fullerenes, may lead to new and, hopefully, more efficient behavior in charge-transfer processes. Recently, a series of pyrrolidinium salts, mono and bis-adducts of C$_{60}$, have been shown to possess enhanced electron-accepting properties with respect to both the parent pyrrolidine derivatives and C$_{60}$ (Figure 23) [26a]. The first reduction of **13** is in fact located more positive than the corresponding process in **12** and C$_{60}$ by 180 and 60 mV respectively. Furthermore, CV measurements performed at low temperatures and fast scan rates, using ultramicroelectrodes, have allowed the observation, for the first time in fullerene derivatives, of six C$_{60}$-centered reductions. For their improved electron-accepting properties, fulleropyrrolidinium salts may found various applications. Also taking advantage of their amphiphilic properties, due to the presence of the positive charges and of the triethylene glycol chains, an amperometric biosensor was, for instance, developed based on the immobilization of the fullerene bisadduct **14**, together with glutathione reductase, within a polymer matrix onto the electrode surface [26b].

Fluorination of C$_{60}$ was also found to greatly enhance its electron affinity. C$_{60}$F$_{48}$, e.g., is reduced in CH$_2$Cl$_2$ at +0.79 V, i.e. ca. 1.29 V more positive than C$_{60}$ [40], thus prompting the investigation of fluorinated fullerenes as potential electron-acceptor units in dyads. C$_{60}$F$_{18}$ is a unique member of the fluorofullerene family because it possesses a flattened hemisphere comprising an aromatic face surrounded by a fluorinated crown; and a curved "normal" hemisphere akin to its all-carbon parent [41].

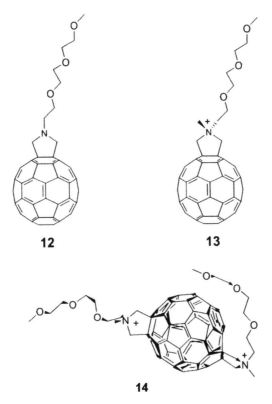

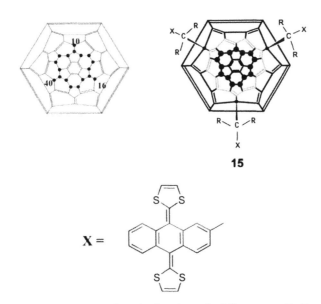

Figure 23. Fulleropyrrolidines: mono and bisadducts

X =

Figure 24. Schlegel diagrams of poly-fluorinated fullerenes: $C_{60}F_{18}$ (left) and [18]trannulene-AnthTTF dyad (right). ● = F. R = CO_2Et.

Substitution chemistry to $C_{60}F_{18}$ allows the attachment of a plethora of functionalities in three precise locations on the fullerene surface leading to derivatives of general formula $C_{60}F_{15}X_3$. Such species are characterized by the presence of an all-*trans* 18π annulenic chromophore (they are named *trannulenes*), and, at odds with most fullerenes, show strong visible light-absorption (e.g., ε_{608} = 13,265 $M^{-1}cm^{-1}$). Recently, the electrochemical and photophysical properties of the dyad (tetrad) **15**, carrying the extended analogue of TTF as electron donor was reported (Figure 24) [42]. Light photoexcitation of the fullerene moiety generates a low-lying (0.54 eV) and long-lived (870 ns) charge-separated state via a rapid intramolecular electron transfer from the TTF moiety to the fullerene moiety.

2.2.4 Other D-S-A Systems

Recently, the electrochemical and photophysical properties of a series of transition-metal β-ketoenolato derivatives, in which at least one of the ligands is linked to a fluorophore through a spacer, have been reported [43]. The fluorescence emission spectra of the complexes **16-19** (Figure 25) are quite similar to that of 9-methylanthracene, which suggest that in all cases, the fluorescence emission is located on the anthrylic group. Although the spectra are similar, the emission intensity is drastically quenched (less than 1%) in the dyads, with only

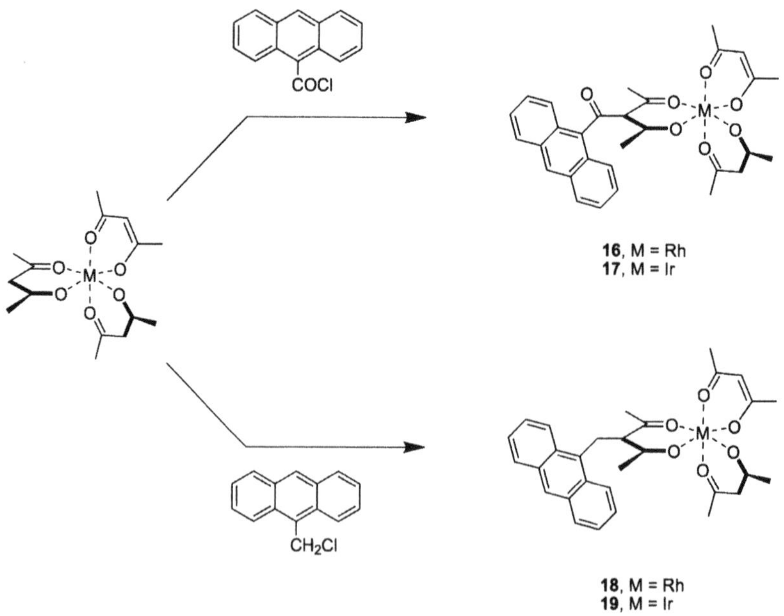

Figure 25. Rh(III) and Ir(III) β-ketoenolato-anthracene dyads.

a small intensity increment passing from complexes **16** and **17**, which have a carbonyl group as spacer, to **18** and **19**, in which the two dyad components are connected by a methylene group.

The introduction of the anthrylic moiety also causes significant changes in the CV behavior of the Rh-β-ketoenolato moieties (Figure 26): a large anodic shift of metal-centered reductions (e.g., E_p = –1.88 V in **18** vs. –2.21 V in [Rh(acac)$_3$]) is observed, and the half-width of the two-electron reduction peak is modified, the much narrower reduction peaks in **16** and **18** (Figure 26 a and b) being fully compatible with a Nernstian (fast) process in contrast with the broad

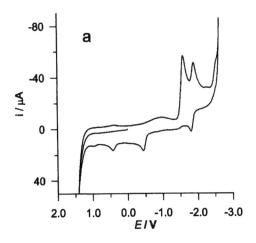

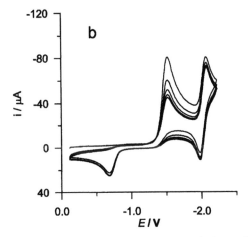

Figure 26. Cyclic voltammetric curve of a 1.0 mM solution of (a) **16** or (b) **18** in tetrahydrofuran (0.05 M tetrabutylammonium hexafluorophosphate). T = 25 °C, scan rate : 0.5 V/s; working electrode: Pt.

peak associated to the slow heterogeneous electron transfer in [Rh(acac)$_3$]. The effects observed on both the location and the half-width of the metal-centered reduction in **16** and **18**, when compared with those observed for [Rh(acac)$_3$], stem from a significant interaction between the anthryl moiety and the metal containing component. The reversible and fast anthrylic unit-based redox couple seems to play an important role in promoting the reduction of the kinetically-impeded rhodium(III) center, by shuttling the incoming electrons to the metal core [44]. Such a process, that would occur intramolecularly, as indicated by the absence of any concentration effect on the CV pattern, would be driven by the very large overpotential for the rhodium(III) reduction [$E_{1/2}$ (Rh^{III}/Rh^{II}) = –0.68 V]. The anticipation of the reduction peak is larger in the case of the dyad **16**, containing the carbonyl group as the spacer between the fluorophore and the metal-carrying component, than in the case of the dyad **18** suggesting that carbonyl is more effective than methylene in mediating the electronic interaction between subunits.

Very recently, a combinatorial approach has been adopted for the construction of a series of photoactive D-S-A dyads, based on the [Ru(bpy)$_3$]$^{2+}$ unit [45]. Such a unit may play the role of either an energy donor when it is coupled to an anthracene unit acting as energy acceptor or as electron acceptor if connected to an amino derivative. D-S-A systems self-assemble in the solution containing Sc(III)(acac)$_3$ (acac = acetyl acetonato) and the D or A-acac free ligands shown in Figure 27. By ligand exchange, various dyads containing different combinations of D and A units are formed, i.e., a statistical library of donor-acceptor systems is dynamically formed. Notably, the Sc(III) complex plays only a structural role, since it does not show any absorption band below 33500 cm^{-1} and does not undergo redox processes within the potential window where D/A units are electroactive. Despite the complexity of such systems, due to the coexistence of many different dyads, the occurrence of fast photoinduced intramolecular energy or electron transfer processes was evidenced [45].

A very interesting system has been recently reported that is the artificial replica of the electron donor triad in photosystem II (PSII), the key enzyme in the light-driven reactions in photosynthesis of plants, algae and cyanobacteria [46] (Figure 28). Both the structure of the components involved in the processes and the energetics of the artificial photosynthetic system are strikingly similar to those involved in PSII. When the aqueous solution containing the [Ru(bpy)$_3$]$^{2+}$ complex covalently linked to L-tyrosine was irradiated, in the presence of Co(III) pentamine chloride as sacrificial electron acceptor, the Mn(III,III) dimer was photooxidized to Mn(III,IV). The oxidation occurred via the tyrosine radical cation generated by intramolecular ET to Ru(III), in turn formed by the intermolecular photoreduction of Co(III).

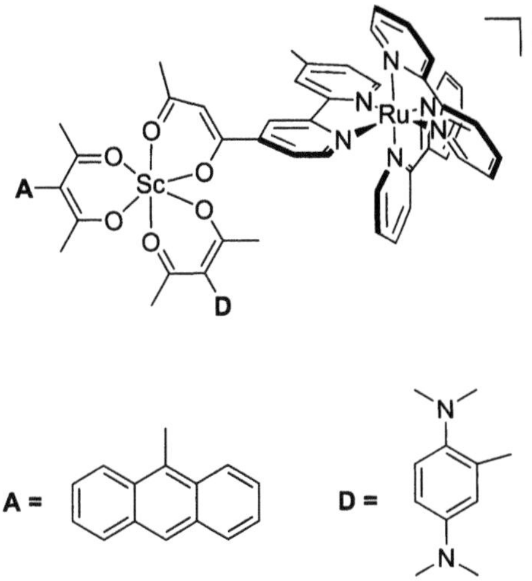

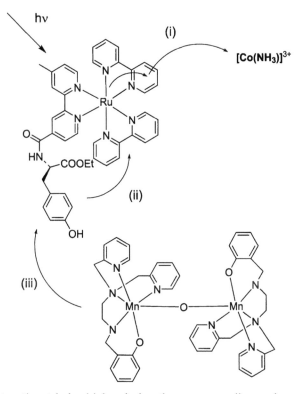

A =   **D =**

Figure 27. Formulas of the components and schematic representation of self-assembled dyads via the coordination to Sc(III).

Figure 28. Photoactive triad which mimics the corresponding redox components in natural PSII.

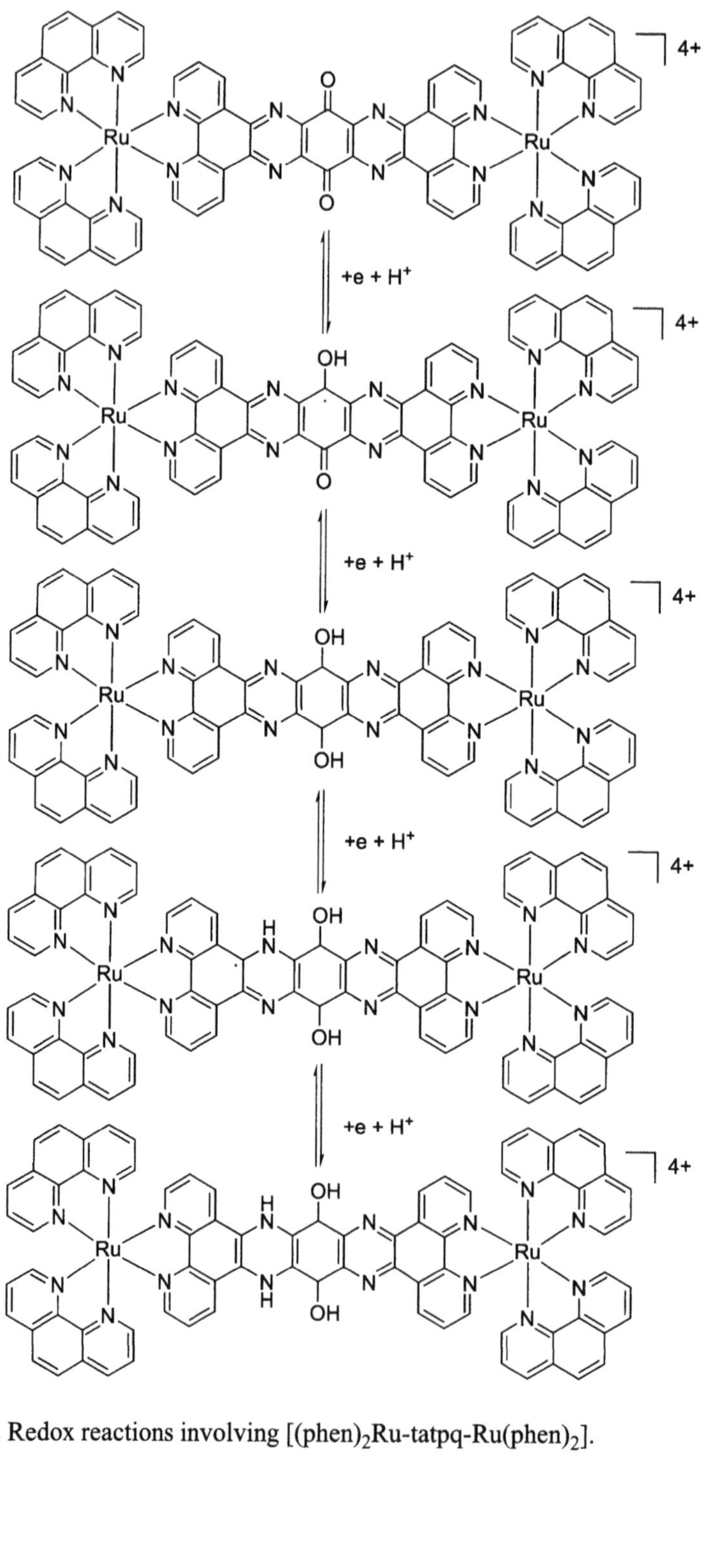

Figure 29. Redox reactions involving [(phen)₂Ru-tatpq-Ru(phen)₂].

2.3   Multielectron catalysis

While research has principally focussed on the development of artificial systems mimicking natural antennas and reaction centers, only few reports on another important aspect of natural photosynthesis, i.e. light-driven multielectron catalysis, are so far known. Very recently, the photoactive Ru(II) dimer, [(phen)$_2$Ru-tatpq-Ru(phen)$_2$] shown in Figure 29 (phen = 1,10-phenanthroline), has been described which is capable to reversibly store up to four electrons in the central bridging ligand, upon visible irradiation in the presence of trietylamine as both electron and proton donor [47].

The four electrons are stored in two overlapping molecular orbitals largely located on the tatpq ligand and this suggests that, similarly to natural photosynthetic systems, they may be delivered to a suitable substrate in a concerted way.

3   ELECTROCHEMICALLY-DRIVEN MOLECULAR-LEVEL MACHINES

A molecular-level machine is an assembly of a discrete number of molecular components, organized in either a supramolecular or mechanically-linked molecular architecture, designed to perform a specific mechanical function [9-12]. In a mechanically-linked system the single components are covalently interlocked to one another. Conversely, noncovalent "weak" intercomponent interactions, such as hydrogen-bonding, charge-transfer interactions, solvo-phobic interactions, prevent instead dissociation of single components in supramolecular systems.

3.1   Pseudorotaxanes

A pseudorotaxane is the supramolecular product of the self-assembly of two components (i.e,. a [2]pseudorotaxane), a molecular ring (macrocycle) and a molecular wire (thread) threaded into it. The occurrence of specific interactions between the two components lowers the free energy of the assembled system with respect to the separate components and drives the self-assembly of the pseudorotaxane (threading process). Threading occurs generally in solution and, upon changes of solvent polarity or other external stimuli, dethreading of the pseudorotaxane may be induced: dethreading/rethreading of the wire and ring components is reminiscent of the way a piston moves within a cylinder.

A large number of pseudorotaxanes, catenanes and rotaxanes have been prepared based on cyclophanes, i.e., macrocycle made of aromatic ring subunits that show considerable affinity for π-donor systems [12a, 48]. Furthermore, if the aromatic groups in the cyclophane are electroactive, then electrochemical control of its binding affinity may be achieved. The tetracationic cyclophane

20 (Figure 30) shows two bielectronic reduction processes, corresponding to the first and second reductions respectively of each bipyridinium unit. Most of the work with this cyclophane has focussed on its inclusion complexes with biphenyl derivatives; it forms with biphenol and benzidine stable host-guest pairs. Interestingly, the first reduction processes involving the cyclophane are sensibly shifted to more negative potentials in the presence of a π-donor guest, as a consequence of the charge-transfer (CT) stabilization of the tetracationic form of the host. The effect of guests on both the first and second reduction of cyclophane has been used to monitor threading/dethreading processes [48].

An interesting example of dual-mode electrochemical control of threading/dethreading dynamics in a pseudorotaxane system was based on cyclophane **20** and a tetrathiafulvalene derivative containing two polyether chains (TTF), Figure 31. The TTF unit retains the same redox properties as pristine tetrathiafulvalene; however, the association equilibrium constant with cyclophane is about 50 times higher than in the case of pristine tetrathiafulvalene [48], likely due to hydrogen bonding of the oxygen atoms in the polyether chains with the α-hydrogen atoms of bipyridinium units. In this system, both the thread and the macrocycle can undergo reversible redox processes. The threading and dethreading equilibria are affected by the strength of charge-transfer interaction between the donor (TTF) and acceptor (bipyridinium) units and may therefore be controlled in a reversible way by either reducing the acceptor unit (path i-ii) or oxidising the donor one (path iii-iv). Furthermore, the rethreading/dethreading processes are accompanied by pronounced changes of the absorption spectra. This latter property makes the present system appealing for applications in electrochromic devices.

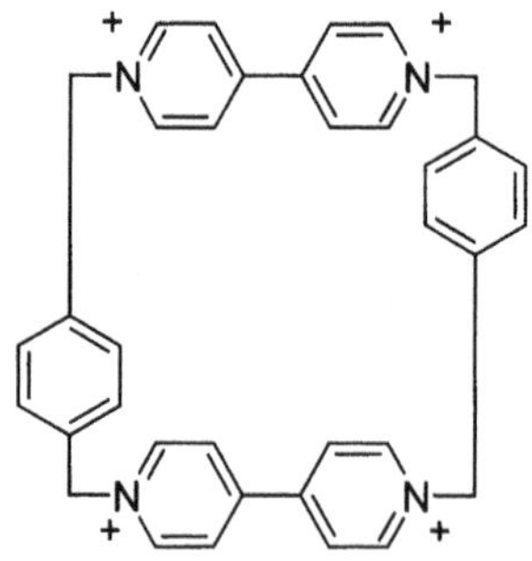

20

Figure 30. Structural formula of a cyclophane.

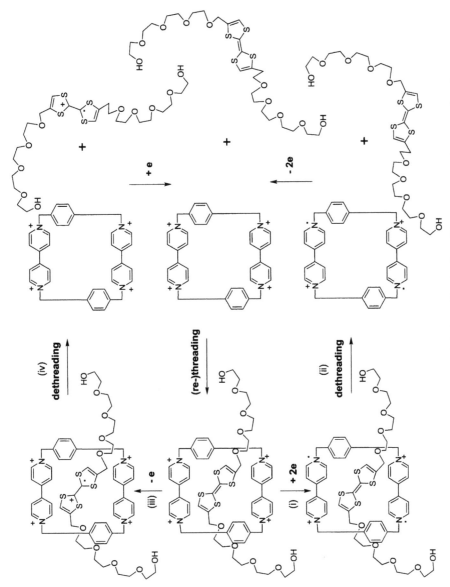

Figure 31. Electrochemically-induced dethreading/rethreading processes in a pseudorotaxane.

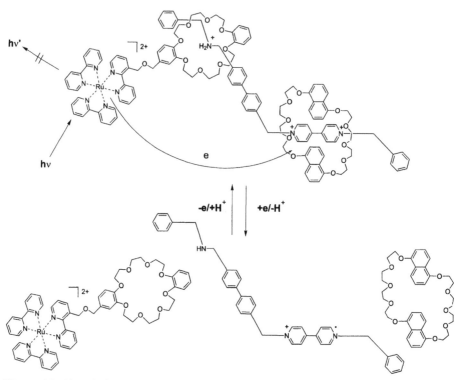

Figure 32. Photoinduced electron transfer in the self-assembled triad and control of the two plug/socket functions by acid/base or redox stimulation.

Along this line of research, very recently, a self-assembling supramolecular triad performing the photoinduced intramolecular electron transfer has been reported [49]. As shown in Figure 32, the system is made up of three components: the linear one contains two different groups, the bipiridinium unit and the secondary dialkylamine group that are suitable (the latter when in its protonated form) for self-assembling with the crown ether moieties present in the other two components, giving a three-component double pseudorotaxane. Oxidation of the bipyridinium moiety, and/or deprotonation of the dialkylammonium group provoke the dethreading processes and therefore assembling/disassembling of the triad can be controlled reversibly and independently by acid/base and redox external inputs.

3.2   Catenanes and Rotaxanes

A catenane consists of two interlocked rings (it is called in this case a [2]catenane) while a rotaxane differs from a pseudorotaxane (see preceding Section) for two bulky stoppers located at the end of the wire that prevent

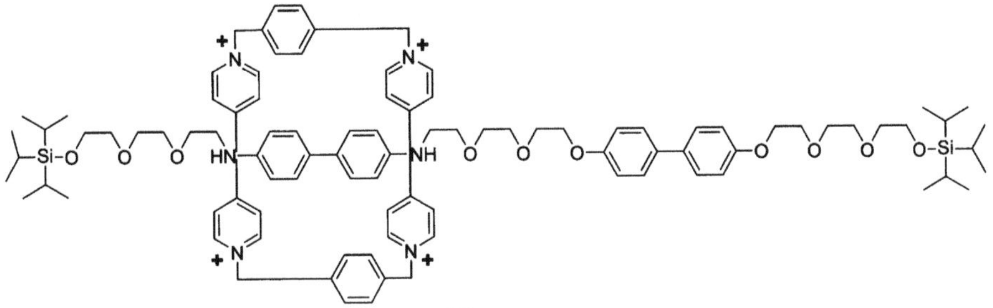

21

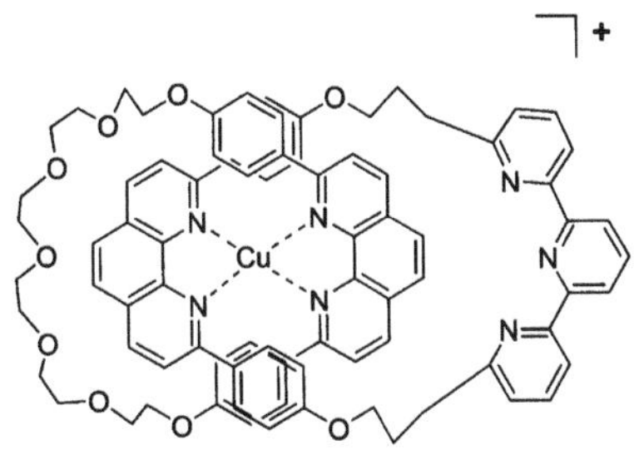

22

Figure 33. Molecular machines: a rotaxane (**21**) and a catenane (**22**).

dethreading. In both systems, separation of the two components may only be obtained by cleavage of at least one covalent bond [11, 12, 50].

The mechanically-interlocked architecture of catenanes and rotaxanes is ideally suited for the construction of molecular machines where large-amplitude movements may be induced by an external stimulus. In a rotaxane, the ring threaded onto the wire can either spin around it (pirouetting) or undergo a translational movement (shuttling). In a catenane, the two interlocked rings can undergo a complex circumrotational movement.

The two structures shown in Figure 33 are examples of a rotaxane [51] and a catenane [52] respectively in which shuttling and circumrotation can be controlled by electrochemical stimuli.

3.2.1 CT-Systems

Rotaxane **21** in Figure 33 was used to illustrate for the first time the concept of an *electrochemically controllable molecular shuttle* [51]. The cyclophane in fact resides preferentially around the benzidine station that is a better π-electron donor than biphenol. Benzidine can however be reversibly oxidised, during for instance a CV experiment, and in its oxidised state is a weaker π-electron donor than biphenol. This drives the displacement of macrocycle along the thread from one station to the other (Figure 34).

In the rotaxane shown in Figure 35, the abacus-like movement of the π-electron donating bis-p-phenylene-34-crown-10 macrocyclic polyether between two π-electron accepting bipyridinium stations may be induced by electron transfer [53]. As shown by combined NMR and CV studies, the macrocycle resides preferentially on the un-methylated bipyridinium station because it is a better π-electron acceptor than the methylated one. For the same reason, the un-methylated station is the easiest to reduce and upon electrochemical reduction of such an unit the macrocycle shuttles to the methylated station. In principle, the shuttling process might be induced in this system by the absorption of light and without the intervention of any other external agent. The $[Ru(bpy)_3]^{2+}$ excited state is in fact quenched by the rapid intramolecular electron transfer to the un-methylated pipyridinium moiety (steps i-ii). The shuttling of macrocycle might than take place (step iii), analogously to the case of electrochemical control. Back electron transfer (step iv) was however found to occur in a much shorter time scale than macrocycle displacement. The photochemically-driven shuttling was instead performed successfully in the presence of a sacrificial reductant that, regenerating Ru(II), suppressed back electron transfer [53].

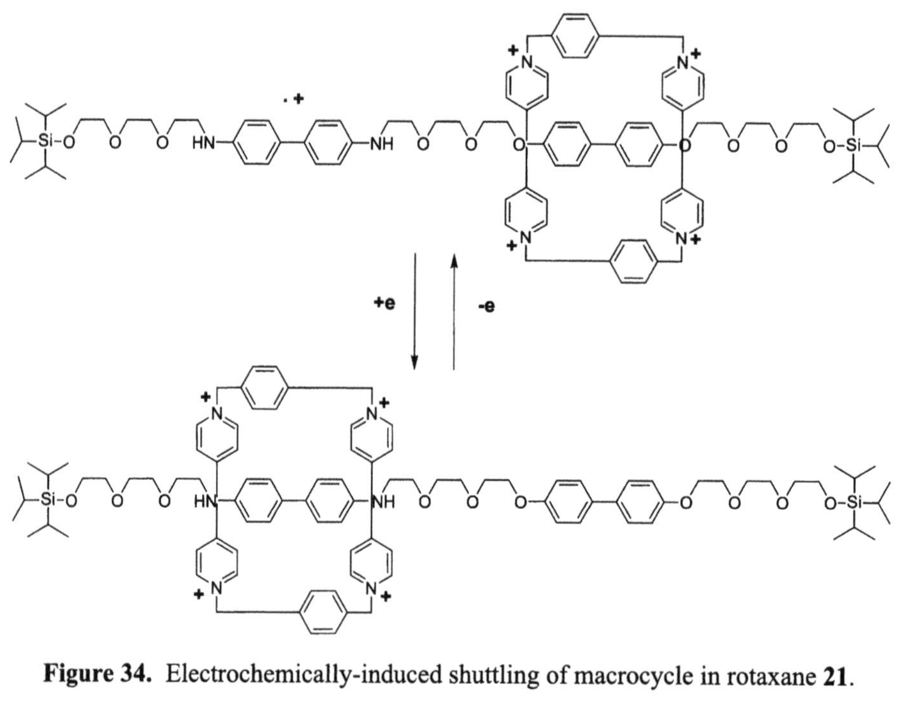

Figure 34. Electrochemically-induced shuttling of macrocycle in rotaxane **21**.

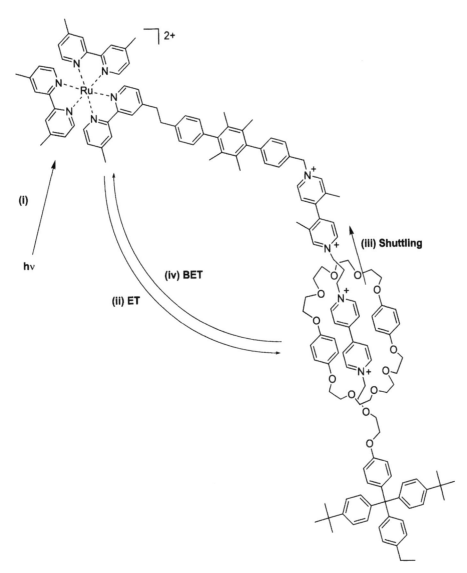

Figure 35. Possible mechanism of reagent-less photoinduced macrocycle shuttling in a CT-based rotaxane.

3.2.2 Transition metal complexes-based molecular machines

In the catenane (or rather catena*te*) **22** shown in Figure 33, circumrotation of the rings may be induced by changing electrochemically the redox state of the metal ion (copper) (Figure 36) [52]. One of its rings contains a phenatroline and a terpyridine unit, while the second ring contains a phenatroline only. Depending on its redox state, copper prefers different coordination geometries: Cu(I) favors

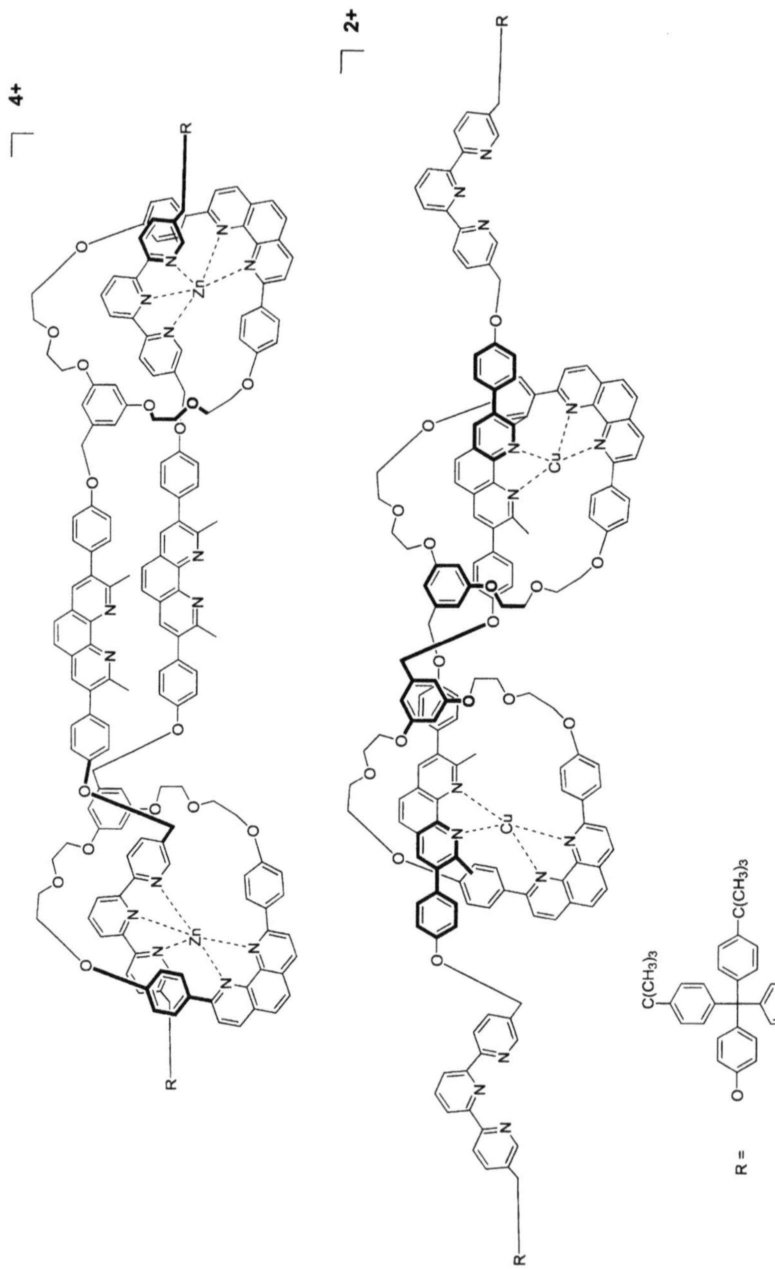

Figure 37. Reversible chemically-induced motions between extended and contracted conformations of a molecular "muscle".

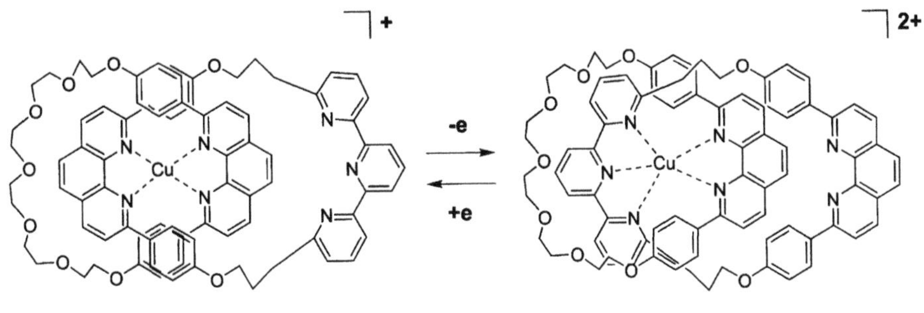

Figure 36. Macrocycles circumrotation induced electrochemically in catenate **22**.

a tetrahedral coordination sphere while Cu(II) adopts preferentially trigonal-bipyramidal geometries. Therefore, when copper is in its +1 oxidation state, the two rings arrange so that the two phenatrolines can coordinate the metal centre. When copper is oxidised to the +2 state, they rearrange so that the terpyridine instead of the phenatroline is coordinated. Reduction of copper back to the +1 state resets the process (Figure 36).

The double-threaded rotaxane system shown in Figure 37 was recently obtained in order to mimic the behavior of a muscle at the molecular level, i.e. it may contract or stretch under an appropriate external stimulus [11d]. The system is made up of two identical components each containing a linear unit (thread) and a macrocycle. The macrocycle is equipped with a phenantroline, while the thread contains both a phenatroline and a terpyridine. Stretching and contracting motions may be reversibly induced in the rotaxane by changing the metal ions and therefore the coordination geometry adopted. So, in the presence of Cu(I) ions, tetrahedral coordination prevails and the structure adopt a stretched conformation, while in the presence of Zn(II) ions, the contracted conformation is preferred. In principle, electrochemical control of stretching/contraction motions could be obtained with the present system, by analogy to the catenate illustrated above, by using the Cu(II)/Cu(I) redox couple.

3.2.3 Hydrogen-bonded systems

Hydrogen bonding is one of the most important non-covalent interaction in supramolecular chemistry. Great advances have occurred over the last decade in the control of hydrogen bond-based host-guest complexation, principally through the systematic modification of the electrostatic properties and geometry of hosts. Supramolecular systems in which intercomponent hydrogen bonding interactions may be modulated by an external stimulus, in particular by electrochemistry, have been reported [5j-m]. Since hydrogen bonds are substantially electrostatic in nature, reduction and oxidation processes that lead to changes in the electron density distribution on one of the partner of host-guest couple have significant effects on the strength of hydrogen bond. At the same time, hydrogen bonding modifies the redox properties of the electroactive species through the stabilization of the reduced species with respect to the oxidation one: reductions become easier, oxidations more difficult.

The effect of hydrogen bonding on the redox properties of guests was investigated using the host-guest systems shown in Figure 38 [5j,k].

Large changes in redox potential of flavin (**23**) were observed upon addition of the diamido-pyridine host, as a consequence of intermolecular hydrogen bonding. The molecular recognition process is described, from the electrochemical point of view, by the *square scheme* shown in Figure 39 that illustrates the interconnection between the host-guest association process and the redox processes. The following equation allows to quantify the strength of intermolecular interaction through the measure of the association constants K_{red} and K_{ox} relative to the two different redox states of flavin.

$$\frac{K_{red}}{K_{ox}} = e^{\frac{nF}{RT}\left(E_{1/2}(bound)-E_{1/2}(unbound)\right)}$$

In the presence of diamido-pyridine, the $E_{1/2}$ relative to flavin reduction is anticipated by 155 mV. This corresponds to a substantial stabilization (3.6 kcal/mol) of flavin radical anion due to H-bonding and to a 500-fold enhancement of binding as a consequence of reduction. Similar changes in redox potential (+ 128 mV) were observed upon addition of the diamido-pyridine host to the naphthalimide guest (**24**, Figure 38).

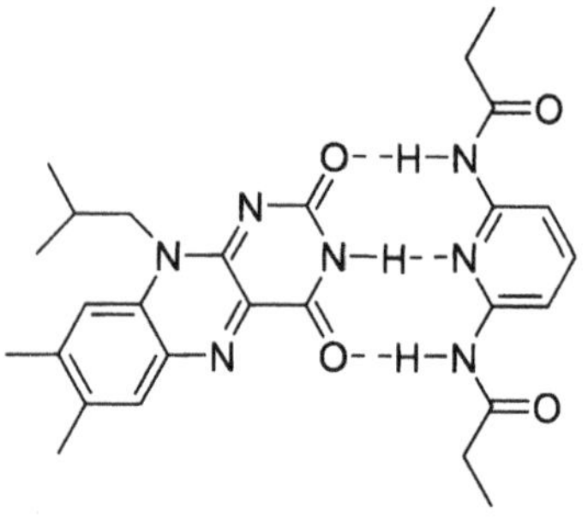

23 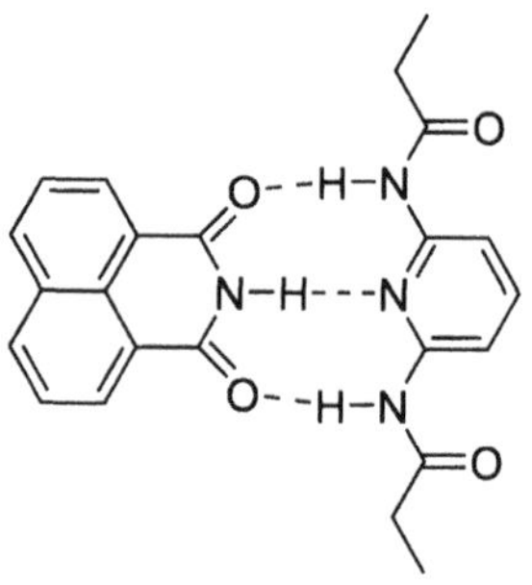   **24**

Figure 38. Molecular recognition based on H-bonding.

A light and redox-driven molecular shuttle closely related to the host-guest systems illustrated above has recently been reported [54]. In the rotaxane **25** shown in Figure 40 a benzylic amide macrocycle is locked onto a molecular thread containing two potential H-bond acceptors, a naphthalimide group and a succinic amide site; the macrocycle resides preferentially onto the latter unit.

Such a submolecular translational process, reminiscent of the way a piston moves within a cylinder, was triggered by the use of a photon and of a suitable electron donor as illustrated in Figure 41 [54].

The reduction of the naphthalimide group brings about the fast shuttling of the macrocycle along the C_{12}-thread down to the naphthalimide unit. The shuttling is then reversed upon the naphthalimide re-oxidation (recombination with oxidised electron donor).

The same intramolecular dynamics was put into action by the reversible transfer of one electron at the electrode, during fast scan rate cyclic voltammetry experiments [55].

The CV curve of thread **26** shows the reversible one-electron reduction of the naphthalimide unit at $E_{1/2} = -1.41$ V. Such a reduction is irreversible in the

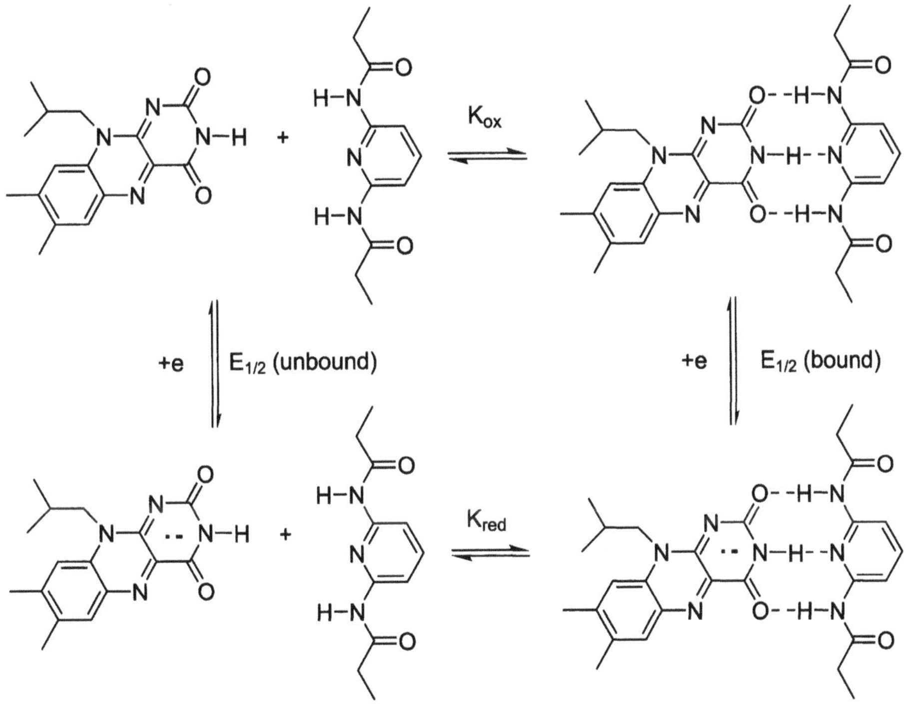

Figure 39. Square scheme illustrating the electrochemical control on molecular recognition processes for **23**.

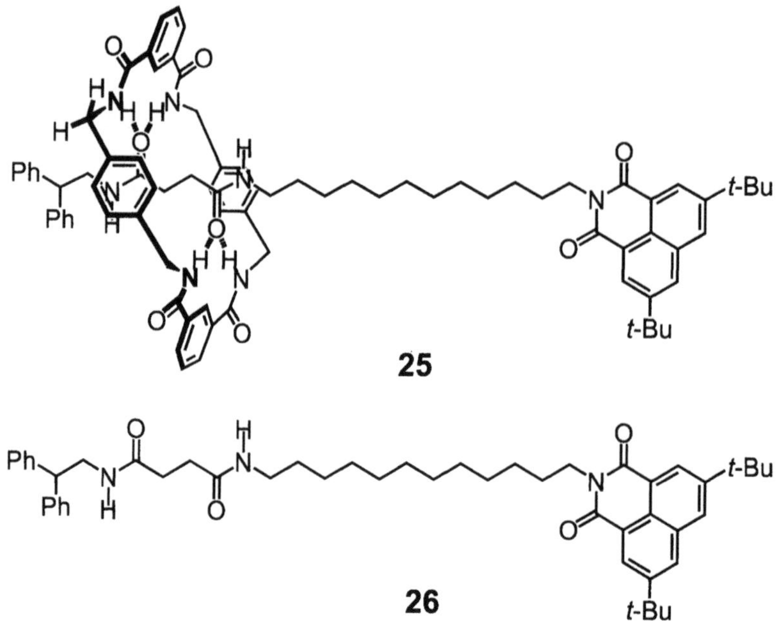

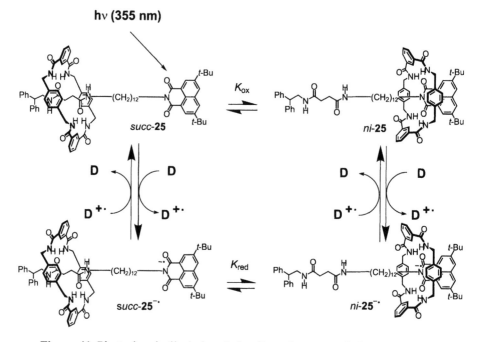

Figure 40. Benzylamide rotaxane and related thread.

Figure 41. Photochemically-induced shuttling of macrocycle in rotaxane **25**.

case of rotaxane: the cathodic peak I at $E = -1.40$ V lacks of its anodic partner and, in addition, a new irreversible anodic peak, A, appears at $E = -0.89$ V. The CV pattern is reproducible upon cycling the potential over the range including both the reduction and the oxidation processes. The complex electrochemical behaviour of **25** is conveniently interpreted, as confirmed by the digital simulation of the cyclic voltammetric curves, by the *square scheme* shown in Figure 43, where the two electron transfer steps are interleaved with two reversible chemical transformations associated to the shuttling processes.

The molecular shuttle **25** adopts at the beginning the co-conformation *succ-***25**, in agreement with the results obtained in the spectroscopic investigations. Subsequently, the reduced *succ-***25**⁻⁺ isomerises to *ni-***25**⁻⁺, before it can be re-oxidised. This conversion is thermodynamically driven by the formation of H-bonds between the macrocycle and the naphthalimide radical anion; as a matter of fact, the oxidation of *ni-***25**⁻⁺ occurs at higher potential with respect to **26**⁻⁺ ($\Delta E° = -510$ mV). Finally, the oxidised *ni-***25** isomerises back to *succ-***25**, again rapidly on the CV time scale. The difference in $E°$ is remarkable and is a measure of the intercomponent binding energy between macrocycle and reduced naphthalimide: 510 mV = 11.76 kcal/mol is the equivalent of four strong, or three extremely strong, hydrogen bonds.

When activated photochemically, pumped by a laser at the frequency of its recovery stroke (10^4 s⁻¹), this molecular machine generates ~ 10^{-15} W of mechanical power per molecule.

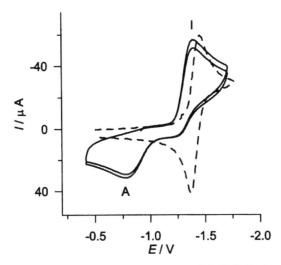

Figure 42. Cyclic voltammetric curves of rotaxane **25** (full line) and corresponding thread **26** (dashed line), 0.5 mM in tetrahydrofuran (0.05 M tetrabutylammonium hexafluorophosphate). Scan rate: 1 V/s, T = 25 °C. Working electrode: Pt disk.

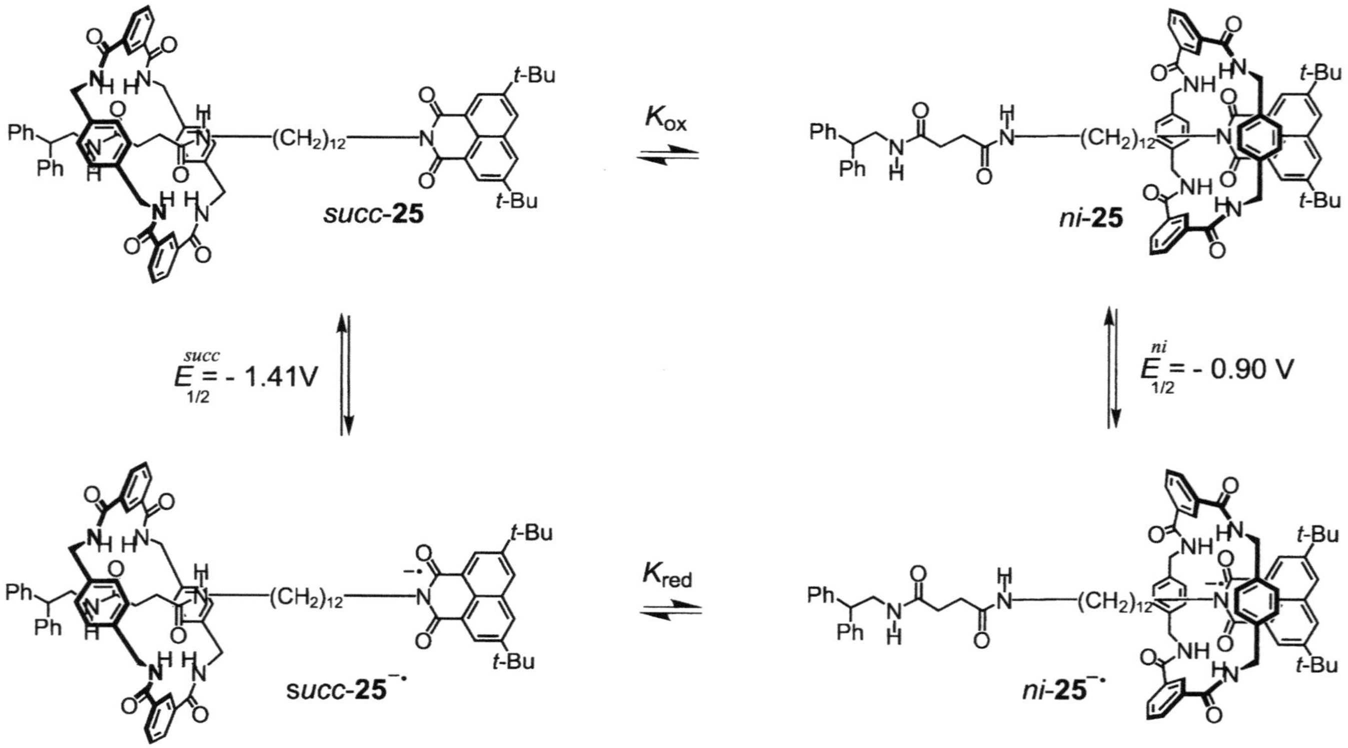

Figure 43. Square scheme illustrating the electrochemically-induced shuttling of macrocycle in rotaxane **25**.

FINAL REMARKS

Central concepts to supramolecular electrochemistry are that control on the supermolecule structure can be exerted by changes in the oxidation states of one or some of its subunits, and that the supramolecular framework affects the kinetics and energetics of electron transfer processes. Such themes have been illustrated by referring to a few selected examples taken from the literature of the last decade. Over the years, electrochemists have learned how to deal with the more and more complex molecular architectures that supramolecular chemists have been manifacturing. Electrochemistry, though, is only part of the game: in most cases, the complexity that is proper to supramolecular systems deserved the contribution from many different disciplines in a joint, collaborative effort.

Organised architectures at a molecular level are attracting increasing attention in materials science for their vital role in emerging technologies related to environmental, energy conversion and catalytic applications, for the development of novel sensors and biosensors and in molecular electronics [1a, 56]. The integration of organized molecular and supramolecular devices into viable and competitive computation and memory architectures, in chip-based arrays of nanostructured sensors and biosensors, and in electrocatalytic and optoelectronic devices still remains challenging. In these regards, self-organization of the supramolecular systems onto technologically relevant substrates, also including state-of-the-art silicon input/output platforms, seems the most promising bottom-up approach towards the realization of an interface between the nanoscopic and macroscopic worlds [57]. The collective fabrication of nanosized dots has been, for instance, very recently obtained by applying a force by the AFM tip onto thin films of amide-based rotaxanes (similar to those investigated by the authors, see Section 3.2.3) [58]. The film reorganization was favored by the intercomponent mobility proper to the rotaxane structure.

The electrochemical interface may play a privileged role in the realization of nanoscale electronic components [56a, d, 57d-g]. In particular, the preparation and characterization of nanostructured thin-film electrode surfaces, in which the supramolecular systems are organized through simple self-assembly methods in ordered structures onto conducting or semiconducting substrates, are growing in importance. By using, for instance, Langmuir-Blodgett techniques or the gold-thiolate approach, also coupled to scanning probe lithographic procedures, several nanometer-scale electronic devices, including nanostructured electrode surfaces, electronic switches, and single-molecule amplifiers and transistors, also employing nanoparticles [59] and single-walled carbon nanotubes [60], have recently been realized [56, 57]. HP and UCLA have recently patented a simple logic gate based on a one-molecule thick layer of electrochemically switchable rotaxanes sandwiched between wires the size of those used in today's computer chips [57h].

Supramolecular chemistry is the area of scientific activity in which, more than in others, people belonging to very disparate fields, chemistry, solid-state physics, biology, informatics and mathematics, materials science, etc., are brought together by the shared goal of transforming a bright idea in a useful, practical device. And, in such an effort, they soon learn to talk a common language.

ACKNOWLEDGEMENT

The authors are indebted to their collaborators whose names appear in the references. This work was supported by MIUR (PRIN 2002035735), FIRB (RBAU017S8R) and the University of Bologna ("Funds for Selected Research Topics").

REFERENCES

1. (a) Jones, W.J.; Rao, C.N.R. (Eds.) *Supramolecular Organization and Materials Design*, Cambridge University Press, Cambridge (2002); (b) Lehn, J.-M. *Supramolecular Chemistry: Concepts and Perspectives*, VCH, Weinheim (1995); (c) Lehn, J.-M. *Angew. Chem.*, *100*, 91 (1988); (d) Lehn, J.-M. *Angew. Chem. Int. Ed. Engl.*, *27*, 89 (1988); (e) Lehn, J.-M. *Science*, *295*, 2400 (2002).

2. (a) Balzani, V; Scandola, F. *Supramolecular Photochemistry*, Horwood, Chichester (1991); (b) Balzani, V.; Scandola, F. in Atwood, J.L., Davies, J.E.D., Macnicol, D.D.; Vögtle, F. (Eds) *Comprehensive Supramolecular Chemistry, Vol. 10*, Pergamon Press, Oxford (1996), p. 687; (c) Balzani, V.; Credi, A.; Venturi, M. *Chem. Eur. J.*, *8*, 5525 (2002).

3. (a) Schneider, H.-J.; Yatsimirsky, A. *Principles and Methods in Supramolecular Chemistry*, Wiley, Chichester (2000); (b) Steed, J.W.; Atwood, J.L., *Supramolecular Chemistry*, Wiley, Chichester (2000); (c) Lehn, J.-M. in Kisakürek, M.V. (Ed.) *Organic Chemistry: Its Language and Its State of the Art*, VCH, Weinheim (1993), p. 77; (d) Kaifer, A.E.; Gómez-Kaifer, M. *Supramolecular Electrochemistry*, Wiley-VCH, Weinheim (1999); (e) Boulas, P. L.; Gomèz-Kaifer, M.; Echegoyen, L. Angew. *Chem. Int. Ed.* 37, 216 (1998).

4. See, for instance: (a) Lund, H.; Hammerich, O. (Eds.) *Organic Electrochemistry*, Marcel Dekker, New York (2001); (b) Bard, A.J.; Stratmann, M. (Eds.) *Encyclopedia of Electrochemistry*, Wiley-VCH, Weinheim (2002); (c) Bard, A.J.; Faulkner, L.R. *Electrochemical Methods, Fundamentals and Applications*, Wiley, New York (2001), and references cited therein.

5. (a) Lexa, D.; Maillard, P.; Momenteau M.; Savéant, J.-M. *J. Phys. Chem. 91*, 1951 (1987) ; (b) Roffia, S.; Casadei, R.; Paolucci, F.; Paradisi, C.; Bignozzi, C.A.; Scandola, F. *J. Electroanal. Chem.*, *302*, 157 (1991); (c) Roffia, S.; Paradisi, C.; Bignozzi, C.A. in Pombeiro, A.J.L.; McCleverty, J.A. (Eds.) *Molecular Electrochemistry of Inorganic, Bioinorganic and Organometallic Compounds*,

Kluwer Academic Publishers, Dordrecht (1993), p. 217; (d) Ceroni, P.; Paolucci, F.; Paradisi, C.; Juris, A.; Roffia, S.; Serroni, S.; Campagna, S.; Bard, A.J. *J. Am. Chem. Soc., 120*, 5480 (1998); (e) Marcaccio, M.; Paolucci, F.; Paradisi, C.; Roffia, S.; Fontanesi, C.; Yellowlees, L.J.; Serroni, S.; Campagna, S.; Denti, G.; Balzani, V. *J. Am. Chem. Soc., 121*, 10081 (1999); (f) Ceroni, P.; Leigh, D.A.; Mottier, L.; Paolucci, F.; Roffia, S.; Tetard. D.; Zerbetto, F. *J. Phys. Chem. B, 103*, 10171 (1999); (g) Carano, M.; Ceroni, P.; Maggini, M.; Marcaccio, M.; Menna, E.; Paolucci, F.; Roffia, S.; Scorrano, G. *Collect. Czech. Chem. Commun., 66*, 276 (2001); (h) Marcaccio, M.; Paolucci, F.; Paradisi, C.; Carano, M.; Roffia, S.; Fontanesi, C.; Yellowlees, L.J.; Serroni, S.; Campagna, S.; Balzani, V. *J. Electroanal. Chem., 532*, 99 (2002); (i) Fabbrizzi, L.; Gatti, F.; Pallavicini, P.; Zambarbieri, E. *Chem. Eur. J., 5*, 682 (1999); (j) Ge, Y.; Lilienthal, R.R.; Smith, D.K. *J. Am. Chem. Soc., 118*, 3976 (1996); (k) Deans, R.; Niemz, A.; Breinlinger, E.C.; Rotello, V.M. *J. Am. Chem. Soc., 119*, 10863 (1997); (l) Niemz, A.; Rotello, V.M. *Acc. Chem. Res., 32*, 44 (1999); (m) Kaifer, A.E. *Acc. Chem. Res., 32*, 62 (1999);.(n) see also: Balzani, V. (Ed.) *Electron Transfer in Chemistry*, Wiley-VCH (2001), volumes: 2, 3, and 4.

6. Teixeira, M.G.; Roffia, S.; Bignozzi, C.A.; Paradisi, C.; Paolucci, F. J. Electroanal. Chem., 345, 243 (1993).

7. Ammar, F.; Savéant, J.-M. *J. Electroanal. Chem.*, 47, 215 (1973).

8. (a) O'Regan, B.; Gratzel, M. *Nature, 353*, 737 (1991); (b) Bignozzi, C.A.; Argazzi, R.; Kleverlaan, C. *J. Chem. Soc. Rev., 29*, 87 (2000); (c) Bard, A.J.; Stratmann, M. (Eds.) *Encyclopedia of Electrochemistry*, Wiley-VCH, Weinheim (2002), volume 6.

9. For recent reviews on molecular devices and machines see the extensive references cited in: Ballardini, R.; Balzani, V.; Clemente-Léon, M.; Credi, A.; Gandolfi, M.T.; Ishow, E.; Perkins, J.; Stoddart, J.F.; Tseng, H.-R.; Wenger, S. *J. Am. Chem. Soc., 124*, 12786 (2002).

10. (a) Balzani, V. (Ed.) *Electron Transfer in Chemistry*, Wiley-VCH (2001); (b) Vögtle, F.; Stoddart, J.F.; Shibasaki, M. (Eds.) *Stimulating Concepts in Chemistry*, Wiley-VCH, Weinheim (2000);

11. (a) Reinhoudt, D.N.; Crego-Calama, M. *Science, 295*, 2403 (2002); (b) Schalley, C.A. *Angew. Chem. Int. Ed., 41*, 1513 (2002); (c) Cooke, G.; Rotello, V.M. *Chem. Soc. Rev., 31*, 275 (2002); (d) Collin, J.-P.; Dietrich-Buchecker, C.; Gaviña, P.; Jimenez-Molero, M.C.; Sauvage, J.-P. *Acc. Chem. Res., 34*, 477 (2001); (e) Serroni, S.; Campagna, S.; Puntoriero, F.; Di Pietro, C.; McClenaghan, N.D.; Loiseau, F. *Chem. Soc. Rev., 30*, 367 (2001); (f) Leigh, D.A.; Murphy, A.; Smart, J.P., Slawin, A.M.Z. *Angew. Chem. Int. Ed. Engl., 36*, 728 (1997); (g) Balzani, V.; Credi, A.; Raymo, F.M.; Stoddart, J.F. *Angew. Chem. Int. Ed., 39*, 3348 (2000).

12. (a) Balzani, V.; Credi, A.; Venturi, M. *Molecular Devices and Machines – Journey in the Nano World*, Wiley-VCH, Weinheim (2003); (b) *Acc. Chem. Res., 34(6)* (2001), Special Issue on Molecular Machines (Ed.: Stoddart, J.F.); (c) *Struct. Bond, 99* (2001), Special volume on Molecular Machines and Motors (Ed.: Sauvage, J.-P.); (d) Kuhn, H.; Försterling, H.-D. *Principles of Physical Chemistry: Understanding Molecules, Molecular Assemblies, Supramolecular Machines*, Wiley, New York (1999).

13. Bustamante, C.; Keller, D.; Oster, G. *Acc. Chem. Res.*, *34*, 412 (2001).

14. (a) Vlček, A. A. *Rev. Chim. Miner.*, *20*, 612 (1983), and references therein cited; (b) Vlček, A. A. *Coord. Chem. Rev.*, *43*, 39 (1982).

15. (a) Juris, A; Balzani, V.; Barigelletti, F.; Campagna, S.; Belser, P.; Zelewsky, A.; *Coord. Chem. Rev.*, *84*, 85 (1988); (b) Sauvage, J.-P.; Collin, J.-P.; Chambron, J.-C.; Gillerez, S.; Coudret, C.; Balzani, V. ; Barigelletti, F. ; De Cola, L ; Flamigni, L. *Chem. Rev.*, *94*, 993 (1994); (c) Balzani; V. Juris, A.; Venturi, M.; Campagna, S.; Serroni, S. *Chem. Rev.*, *96*, 759 (1996).

16. (a) Juris, A. in Balzani, V. (Ed.) *Electron Transfer in Chemistry*, Wiley-VCH (2001), volume 3, p.655; (b) Balzani, V.; Ceroni, P.; Juris, A.; Venturi, M.; Campagna, S.; Puntoriero, F.; Serroni, S. *Coord. Chem. Rev.*, *219-221*, 545 (2001); (c) Newkome, G.R.; Moorefield, C.N.; Vögtle, F. *Dendrimers and Dendrons: Concepts, Syntheses, Applications*, Wiley-VCH, Weinheim (2001); (d) Matthews, O.A.; Shipway, A.N.; Stoddart, J.F. *Prog. Polym. Chem.*, *23*, 1 (1998); (e) Smith, D.K.; Diederich, F. *Chem. Eur. J.*, *4*, 1353 (1998); (f) Bosman, A.W.; Jansen, E.W.; Meijer, E.W. *Chem. Rev.*, *99*, 1665 (1999); (g) Hecht, S.; Fréchet, J.M.J. *Angew. Chem. Int. Ed.*, 40, 74 (2001); (h) an example of supramolecular H-bonded metallo-dendrimer assembly has been recently reported which shows interesting recognition properties towards $H_2PO_4^-$: Daniel, M.-C.; Ruiz, J.; Astruc, D. *J. Am. Chem. Soc.*, *125*, 1150 (2003).

17. Vögtle, F.; Pleovets, M.; Nieger, M.; Azzellini, G.C.; Credi, A.; De Cola, L.; De Marchis, V.; Venturi, M.; Balzani, V. *J. Am. Chem. Soc.*, *121*, 6290 (1999).

18. Campagna, S.; Serroni, S.; Puntoriero, F.; Di Pietro, C.; Ricevuto, V. in Balzani, V. (Ed.) *Electron Transfer in Chemistry*, Wiley-VCH (2001), volume 5, p. 186.

19. Ceroni, P.; Paolucci, F.; Roffia, S.; Serroni, S.; Campagna, S.; Bard, A.J. *Inorg. Chem.*, *37*, 2829 (1998).

20. Giuffrida, G.; Campagna, S. *Coord. Chem. Rev.*, *135-136*, 517 (1994).

21. (a) Halpern, J.; Orgel, L.E. *Discuss. Faraday Soc.*, *29*, 32 (1960). (b) McConnell, H. M. *J. Chem. Phys.*, *35*, 508 (1961). (c) Day, P. *Comments Inorg. Chem.*, *1*, 155 (1981). (d) Miller, J. R.; Beitz, J. V. *J. Chem. Phys.*, *74*, 6746 (1981). (e) Richardson, D. E.; Taube, H. *J. Am. Chem. Soc.*, *105*, 40 (1983). (f) Newton, M. D. *Chem. Rev.*, *91*, 767 (1991). (g) Jordan, K. D.; Paddon-Row, M. N. *Chem. Rev.*, *92*, 395 (1992). (h) Todd, M. A.; Nitzan, A.; Ratner, M. A. *J. Phys. Chem.*, *97*, 29 (1993).

22. Roffia, S.; Marcaccio, M.; Paradisi, C.; Paolucci, F.; Balzani, V.; Denti, G.; Serroni, S. *Inorg. Chem.*, *32*, 3003 (1993).

23. Paddon-Row, M.N. in Balzani, V. (Ed.) *Electron Transfer in Chemistry*, Wiley-VCH (2001), volume 3, p. 179.

24. (a) Paddon-Row, M.N. in Vögtle, F.; Stoddart, J.F.; Shibasaki, M. (Eds.) *Stimulating Concepts in Chemistry*, Wiley-VCH, Weinheim (2000); (b) Wasielewski, M.R. *Chem. Rev. 92*, 435 (1992); (c) Meyer, T.J. *Acc. Chem. Res.*, *22*, 163 (1989); (d) Meijer, M.D.; van Klink, G.P.M.; van Koten, G. *Coord. Chem. Rev.*, *230*, 141 (2002); (e) Martín, N.; Sánchez, L.; Illescas, B.; Pérez, I. *Chem. Rev.*, *98*, 2527 (1998); (f) Diederich, F.; Gómez-López, M. *Chem. Soc. Rev.*, *28*, 263 (1999); (g) Imahori, H.; Sakata, Y. *Adv. Mater.*, *9*, 537 (1997); (h) Lewis, F.D.; Letsinger, R.L.; Wasielewski, M.R. *Acc. Chem. Res.*, *34*, 159 (2001); (i) Prato, M. (Ed.) *J. Mater. Chem.*, *12 (7)*, (2002), *Special Issue on Functionalised Fullerene Materials*; (j)

Fukuzumi, S.; Guldi, D.K. in Balzani, V. (Ed.) *Electron Transfer in Chemistry*, Wiley-VCH (2001), volume 2, p. 270; (k) for systems accomplishing energy transfer processes: Ward, M.D.; Barigelletti, F. *Coord. Chem. Rev., 216-217*, 127 (2001).

25. (a) Echegoyen, L. in Lund, H.; Hammerich, O. (Eds.) *Organic Electrochemistry*, Marcel Dekker, New York (2001), p. 323; (b) Paolucci, F.; Carano, M.; Ceroni, P.; Mottier, L.; Roffia, S. *J. Electrochem. Soc., 146*, 3357 (1999); (c) Bruno, C.; Doubitski, I.; Marcaccio, M.; Paolucci, F.; Paolucci, D.; Zaopo, A. *J. Am. Chem. Soc.* 125, *15738* (2003).

26. (a) Da Ros, T.; Prato, M.; Carano, M.; Ceroni, P.; Paolucci, F.; Roffia. S.; *J. Am. Chem. Soc., 120*, 11645 (1998); (b) Carano, M.; Cosnier, S.; Kordatos, K.; Marcaccio, M.; Margotti, M.; Paolucci, F.; Prato, M.; Roffia, S. *J. Mater. Chem., 12*, 1996 (2002); (c) Carano, M.; Da Ros, T.; Fanti, M.; Kordatos, K.; Marcaccio, M.; Paolucci, F.; Prato, M.; Roffia, S.; Zerbetto, F. *J. Am. Chem. Soc.* 125, *7139* (2003).

27. Imahori, H.; Yamada, H.; Nishimura, Y.; Yamazaki, I.; Sakata, Y. *J. Phys. Chem. B*, *104*, 2099 (2000).

28. Imahori, H.; Hagiwara, K.; Akiyama, T.; Aoki, M.; Taniguchi, S.; Okada, T.; Shirakawa, M.; Sakata, Y. *Chem. Phys. Lett.*, 263, 545 (1996).

29. Da Ros, T.; Prato, M.; Guldi, D.M.; Alessio, E.; Ruzzi, M.; Pasimeni, L. *Chem. Commun., 635* (1999).

30. (a) Da Ros, T.; Prato, M.; Guldi, D.M.; Ruzzi, M.; Pasimeni, L. *Chem. Eur. J., 7*, 816 (2001); (b) Da Ros, T.; Prato, M.; Carano, M.; Ceroni, P.; Paolucci, F.; Roffia, S.; Valli, L.; Guldi, D.M. *J. Organomet. Chem., 599*, 62 (2000).

31. Chuard, T.; Deschenaux, R.; Campidelli, S.; Vázquez, E.; Milic, D.; Prato, M.; Barberá, J.; Guldi, D. M.; Marcaccio, M.; Paolucci, D.; Paolucci, F. *J. Mater. Chem.*, asap (2004).

32. (a) Even, M.; Heinrich, B.; Giollon, D.; Guldi, D.M.; Prato, M.; Deschenaux, R., *Chem. Eur. J., 7*, 2595 (2001); (b) Carano, M.; Chuard, T.; Deschenaux, R.; Even, M.; Marcaccio, M.; Paolucci, F.; Prato, M.; Roffia, S. *J. Mater. Chem., 12*, 829 (2002).

33. (a) Kaes, C.; Katz, A.; Hosseini, M.W. *Chem. Rev., 100*, 3553 (2000); (b) Kröhnke, F. *Synthesis, 1* (1976).

34. Maggini, M.; Guldi, D.M.; Mondini, S.; Scorrano, G.; Paolucci, F.; Ceroni, P.; Roffia, S. *Chem. Eur. J., 4*, 1992 (1998)

35. (a) Polese, A.; Mondini, S.; Bianco, A.; Toniolo, C.; Scorrano, G. Guldi, D.M.; Maggini, M. *J. Am. Chem. Soc., 121*, 3456 (1999).

36. Guldi, D.M.; Maggini, M.; Menna, E.; Scorrano, G.; Ceroni, P.; Marcaccio, M.; Paolucci, F.; Roffia, S., *unpublished results*.

37. Vlček, A. *ChemTracts-Inorg. Chem., 12*, 620 (1999).

38. Balzani, V.; Barigelletti, F.; De Cola, L. *Top. Curr. Chem., 158*, 31 (1990).

39. (a) Guldi, D.M.; Maggini, M.; Menna, E.; Scorrano, G.; Ceroni, P.; Marcaccio, M.; Paolucci, F.; Roffia, S. *Chem. Eur. J., 7*, 1597 (2001); (b) Guldi, D.M.; Maggini, M.; Menna, E.; Scorrano, G.; Marcaccio, M.; Paolucci, F.; Roffia, S., *unpublished results*.

40. Zhou, F.; Van Berkel, G.J.; Donovan, B.T. *J. Am. Chem. Soc.*, *116*, 5485 (1994).

41. Wei, X.-W.; Darwish, A.D.; Boltalina, O.V.; Hitchcock, P.B.; Street, J.M.; Taylor, R. *Angew. Chem. Int. Ed.*, *40*, 2989 (2001); Burley, G. A.; Avent, A. G.; Boltalina, O. V.; Drewello, T.; Goldt, I. V.; Marcaccio, M.; Paolucci, F.; Paolucci, D.; Street, J. M.; Taylor, R. *Org. Biomol. Chem. 1*, 2015 (2003); Burley, G. A.; Avent, A. G.; Gol'dt, I. V.; Hitchcock, P. B.; Al-Matar, H.; Paolucci, D.; Paolucci, F.; Fowler, P. W.; Soncini, A.; Street, J. M.; Taylor, R. *Org. Biomol. Chem. 2*, 319 (2004).

42. Burley, G.A.; Avent, A.G.; Boltalina, O.V.; Gol'dt, I.V.; Guldi, D.M.; Marcaccio, M.; Paolucci, F.; Paolucci, D.; Taylor, R. *Chem. Commun.*, *148* (2003).

43. Carano, M.; Cicogna, F.; Houben, J.L.; Ingrosso, G.; Marchetti, F.; Mottier, L.; Paolucci, F.; Pinzino, C.; Roffia, S. *Inorg. Chem.*, *41*, 3396 (2002).

44. Zi-Rong, Z.; Evans, D.H. *J. Am. Chem. Soc.*, *121*, 2941 (1999).

45. Kercher, M.; König, B.; Zieg, H.; De Cola, L. *J. Am. Chem. Soc.*, *124*, 11541 (2002).

46. Sun, L.; Ammarström, L.; Åkermark, B.; Styring, S. *Chem. Soc. Rev.*, *30*, 36 (2001).

47. Konduri, R.; Ye, H.; MacDonnell, F.M.; Serroni, S.; Campagna, S.; Rajeshwar, K. *Angew. Chem. Int. Ed.*, *41*, 3185 (2002).

48. Venturi, M.; Credi, A.; Balzani, V. in Balzani, V. (Ed.) *Electron Transfer in Chemistry*, Wiley-VCH (2001), volume 3, p. 501.

49. Ballardini, R.; Balzani, V.; Clemente-Léon, M.; Credi, A.; Gandolfi, M.T.; Ishow, E.; Perkins, J.; Stoddart, J.F.; Tseng, H.-R.; Wenger, S. *J. Am. Chem. Soc.*, *124*, 12786 (2002); Badjić, J. D.; Balzani, V.; Credi, A.; Silvi, S.; Stoddart, J. F. *Science 303*, 1845 (2004).

50. (a) Ballardini, R.; Gandolfi, M.T.; Balzani, V. in Balzani, V. (Ed.) *Electron Transfer in Chemistry*, Wiley-VCH (2001), volume 3, p. 539; (b) Armaroli, N.; Chambron, J.-C.; Collin, Dietrich-Buchecker, C.; Flamigni, L.; Kern, J.-M.; Sauvage, J.-P. in Balzani, V. (Ed.) *Electron Transfer in Chemistry*, Wiley-VCH (2001), volume 3, p. 582.

51. Bissell, R.A.; Córdova, E.; Kaifer, A.E.; Stoddart, J.F. *Nature*, *369*, 133 (1994).

52. Livoreil, A.; Dietrich-Buchecker, C.O.; Sauvage, J.-P. *J. Am. Chem. Soc.*, *116*, 9399 (1994).

53. Ashton, P.R.; Ballardini, R.; Balzani, V.; Credi, A.; Ruprecht Dress, K.; Ishow, E.; Kleverlaan, C.J.; Kocian, O.; Preece, J.; Spencer, N.; Stoddart, J.F.; Venturi, M.; Wenger, S. *Chem. Eur. J.*, *6*, 3558 (2000).

54. Brouwer, A.M.; Frochot, C.; Gatti, F.G.; Leigh, D.A.; Mottier, L.; Paolucci, F.; Roffia, S.; Wurpel, G.W.H. *Science*, *291*, 2124 (2001).

55. Altieri, A.; Gatti, F. G.; Kay, E. R.; Leigh, D. A.; Martel, D.; Paolucci, F.; Slawin, A. M. Z.; Wong, J. K. Y. *J. Am. Chem. Soc. 125*, 8644 (2003); for an example of unidirectional rotation in a mechanically interlocked molecular rotor based on a similar system, see: Leigh, D. A.; Wong, J. K. Y.; Dehez, F.; Zerbetto, F. *Nature 424*, 174 (2003).

56. (a) Hodes G. (Ed.) *Electrochemistry of Nanomaterials*, Wiley-VCH, Weinheim (2001); (b) Prasanna de Silva, A. (Ed.) *Molecular Level Electronics* in V. Balzani (Ed.), *Electron Transfer in Chemistry*, Wiley-VCH, Weinheim (2001), vol. 5, Part

1; (c) Nalwa, H.S. *Supramolecular Photosensitive and Electroactive Materials*, Academic Press, San Diego, 2001.

57. (a) Ashkenasy, G.; Cahen, D.; Cohen, R.; Shanzer, A.; Vilan, A. *Acc. Chem. Res.*, *35*, 121 (2002); (b) Tour, J.M. *Acc. Chem. Res.*, *33*, 791 (2000); (c) Fan, F.F.; Yang, J.; Dirk, S.M.; Price, D.W.; Kosynkin, D.; Tour, J.M.; Bard A.J. *J. Am. Chem. Soc.*, *123*, 2454 (2001); (d) Gittins, D.I.; Betthell, D.; Schiffrin, D.J.; Nichols, R.J. *Nature*, *408*, 67 (2000); (e) Collier, C.P.; Mattersteig, G.; Wong, E.W.; Luo, Y.; Beverly, K.; Sampaio, J.; Raymo, F.M.; Stoddart, J.F.; Heath, J.R. *Science*, *289*, 1172 (2000); (f) Gimzewski, J. K.; Joachim, C. *Science*, *283*, 1683 (1999); (g) Kolb, D.M. *Angew. Chem. Int. Ed.*, *40*, 1162 (2001); (g) Welter, S.; Brunner, K.; Hofstraat, J.W.; De Cola, L. *Nature*, *421*, 54 (2003); (h) Pease, A.R.; Jeppesen, J.O.; Stoddart, J.F.; Luo, Y.; Collier, C.P.; Heath, J.R. *Acc. Chem. Res.*, *34*, 433 (2001); (i) *Science*, *283* (1999), Special Issue *on Frontiers in Chemistry: Single Molecules*.

58. Cavallini, M.; Biscarini, F.; León, S.; Zerbetto, F.; Bottari, G.; Leigh, D.A. *Science*, *299*, 531 (2003).

59. (a) Chen, S.; Ingram, R.S.; Hostetler, M.J.; Pietron, J.J.; Murray, R.W.; Schaaf, T.G.; Khoury, J.; Alvarez, M.M.; Wheteen, R.L. *Science*, *280*, 2098 (1998); (b) Templeton, A.C.; Wuelfing, W.P.; Murray, R.W. *Acc. Chem. Res.*, *33*, 27 (2000); (c) Lee, D.; Donkers, R.L.; DeSimone, J.M.; Murray, R.W. *J. Am. Chem. Soc.*, *125*, 1182 (2003).

60. (a) Haddon, R.C. (Ed.) *Acc. Chem. Res.*, *35* (2002), special issue on Carbon Nanotubes; (b) Kong, J.; Franklin, N.R.; Zhou, C.; Chapline, M.G.; Peng, S.; Cho, K.; Dai, H. *Science*, *287*, 622 (2000); (c) Collins, P.G., Bradley, K.; Ishigami, M.; Zetti, A. *Science*, *287*, 1801 (2000); (d) Wilson, N.R.; Cobden, D.H.; MacPherson, J.V. *J. Phys. Chem. B*, *106*, 13102 (2002); (e) Bahr, J.L.; Tour, J.M. *J. Mater. Chem.*, *12*, 1952 (2002); (f) Georgilas, V.; Kordatos, K.; Prato, M.; Guldi, D.M.; Holzinger, M.; Hirsch, A. *J. Am. Chem. Soc.*, *124*, 760 (2002); (g) Hirsch, A., *Angew. Chem. Int. Ed.*, *41*, 1853 (2002); (h) Haughman, R.H.; Zakhidov, A.A.; de Heer, W.A. *Science*, *297*, 787 (2002); (i) Callegari, A.; Marcaccio, M.; Paolucci, D.; Paolucci, F.; Tagmatarchis, N.; Tasis, D.; Vazquez, E.; Prato, M. *Chem. Commun.*, 2576 (2003); (j) Guldi, D. M.; Marcaccio, M.; Paolucci, D.; Paolucci, F.; Tagmatarchis, N.; Tasis, D.; Vazquez, E.; Prato, M. *Angew. Chem. Int. Ed. 42*, 4206 (2003); (k) Callegari, A.; Cosnier, S.; Marcaccio, M.; Paolucci, D.; Paolucci, F.; Georgakilas, N.; Tagmatarchis, N.; Vázquez, Prato, M. *J. Mater. Chem. 14*, (2004) 807; (l) Mellefranco, M.; Marcaccio, M.; Paolucci, D.; Paolucci, F.; Georgakilas, V.; Guldi, D. M.; Prato, M.; Zerbetto, F. *J.Am. Chem. Soc. 126*, 1646 (2004).

Chapter 9

Electrochemistry and Electron-transfer Chemistry of Metallodendrimers

Didier Astruc, Marie-Christine Daniel, Sylvain Nlate, Jaime Ruiz

Group of Molecular Nanosciences and Catalysis, LCOO,
UMR CNRS N°5802, University Bordeaux I
33405 Talence Cedex, France

1   INTRODUCTION

Molecular aspects are crucial in nanoscience for the definition of electric and photonic devices, since the to-down approach using smaller and smaller chips is now reaching its limits. Challenges involving the bottom-up approach concern the design of molecular nanowires, nanoswitches and nanomagnets involving both bulk metals connected by molecular components. Therefore, the study of electron-transfer processes of such nano-systems is of basic importance for applications towards modern nanotechnology. We have engaged a program for the construction of molecular materials, essentially ferrocenyl terminated dendrimers and ferrocenyl-terminated gold colloids. Here, we review aspects dealing with the electrochemistry and electron-transfer chemistry of ferrocenyl dendrimers and other redox-active metallodendrimers.

2   DECORATION OF DENDRIMERS WITH FERROCENYL GROUPS AND REDOX ACTIVITY: A LITERATURE SURVEY

Early examples of redox-active metallodendrimers [1] are Balzani's poly-ruthenium polypyridine systems [2] and stars or dendritic cores containing ferrocenyl or [FeCp(arene)]$^+$ groups [3]. The latter series were made subsequent to the CpFe$^+$-induced polybranching [4-10] using the considerable increase of acidity (about 15 pK_a units in DMSO) of the benzylic protons upon complexation by the cationic group [11]. The Cuadrado group reported dendrimers containing from 4 to 16 equivalent SiMe$_2$RFc groups (R = CH$_2$CH$_2$ or NHCH$_2$CH$_2$) [12,13]. Amperometric biosensors were developed for the titration of glucose in blood. The Madrid group investigated the extension of ferrocene mediators to ferrocene dendrimers in order to circumvent the problem of the instability of ferrocenium in solution [14]. Recently, this group reported recognition of H$_2$PO$_4^-$ and HSO$_4^-$ with dendrimers containing 4, 8 or 16 SiNHCH$_2$CH$_2$Fc groups [15]. The effect of these anions on the ferrocenyl wave was comparable with the previously reported amidoferrocenyl dendrimers described in the preceding section. This group also reported dendrimers terminated with 4 or 8 mixed-valence Si-bridged biferrocene group SiFc$_2^+$ [16]. Finally, this group has reported the branching of carbonyl ferrocene groups onto the commercial DAB polyamine dendrimers, providing polyamidoferrocenyl dendrimers up to the 64-amidoferrocenyl dendrimer.

Cyclic voltammograms showed reversible waves in CH$_2$Cl$_2$, but not in DMF due to the instability of the ferrocenium form in DMF [17]. These metallodendrimers adsorb on electrodes. The thermodynamics and kinetics of adsorption of these dendrimers has been studied using electrochemical quartz crystal microbalance (EQCM) by Abruna's group who also recorded molecularly resolved images on Pt (1,1,1) single crystal electrode using non-contact AFM (AFM Tapping mode) [17a]. Kaifer's group has studied these 4-Fc, 8-Fc and 16-Fc polyamidoferrocenyl dendrimers as guests for inclusion by the hosts β-cyclodextrin and dimethyl-β-cyclodextrin providing aqueous solubility of the hydrophobic dendrimers, increasing with dendritic generation [18]. Only one cyclic voltammetry wave was observed for the 4-Fc and 8-Fc. On the other hand, two waves for complexed and uncomplexed ferrocenyl groups were found for the 16-Fc dendrimer, which indicated that steric effects at the periphery inhibited the complexation of all the ferrocenyl units [18]. The DAB polyamine dendrimers were also sources of polycobaltocenium dendrimers analogues [19a] of the cobaltocenium dendrimers already reported previously by our group [19b]. With these cobaltocenium dendrimers, Kaifer's group studied the β-cyclodextrine-dendrimer assembly leading to solubilization of the neutral cobaltocene form in water which was monitored by the removal of the adsorption peak observed by cyclic voltammetry [19b].The DAB polyamine dendrimers were complexed by the Madrid group with a mixture of ferrocenyl- and cobaltocenyl carbonyl chlorides giving statistical mixtures of ferrocenyl and cobaltocenyl branching [20].

Kaifer's group have synthesized asymmetric dendrimers of different sizes containing a single amidoferrocenyl group [21,22] by reaction of Newkome and Behera's dendritic amine. The heterogeneous electron transfer rate constant, the apparent diffusion coefficient and the standard redox potential $E°$ (from $E_{1/2}$) decrease with increasing dendritic generation. This shows that buried redox centers transfer electron are at lower rates than surface ones, and that the dendritic structure shields the redox center. In addition, the electron transfer rate depends on the orientation of the dendrimer, i.e., how close the redox site is from the electrode surface [22,23]. The voltammetric response of the one-electron oxidation of ferrocene dendrimers depends on the molecular orientation effects that are controlled by the presence of carboxylic vs. cystamine termini. For instance, at high pH, the negatively charge carboxylate dendrimers approach the positively charged monolayer, and the ferrocene-electrode distance is kept maximum, resulting in a decreased electron-transfer rate [23].

Ferrocenes have multiple properties that can be adapted to the dendritic structures [24]. For instance, Togni's group reported dendrimers containing chiral ferrocenyl diphosphines which are efficient for highly enantioselective hydrogenation reactions [25a,b]. The catalytic activity is very similar to that obtained with monomeric catalysts, but dendritic catalysts can be easily removed using commercial nanofiltration membranes due to their nanoscopic size. Ferrocenyldiphosphine ligands at the core of dendrimers having carbosilane tethers were used in palladium catalyzed allylic alkylation [25c].

Shu et al. synthesized Fréchet-type poly(arylether)dendrimers with ferrocenyl groups at the periphery and showed that these groups were independent [26]. Catalano et al. synthesized heterometallic dendrimers containing 4 Pt(IV)-1,2-bipyridine units and 8 ferrocenyl groups [27]. Deschenaux et al. reported the first examples of liquid-crystalline ferrocene dendrimers. These dendrimers were synthesized using 1,3,5-tris(chlorocarbonyl)benzene as a core and cholesterol dendrons containing ferrocenyl units [28]. Phosphorous-containing dendrimers have been synthesized with ferrocenyl units on the core, within the branches and at the periphery. Coulometry showed that up to 1354 ferrocenyl units were found at the ninth generation [29].

Ferrocene dendrimers have been used as mediators with microelectrodes for the determination of analytes in clinical and environmental issues including drug screening. The ferrocene dendrimers are tagged with affinity ligands such as biotin, haptens, carbohydrates, DNA fragments, etc. The system is applied in conjunction with amperometry, chronocoulometry and voltammetry [30]. PAMAM dendrimers have been modified with ferrocenyl groups by reaction of ferrocenecarboxaldehyde in various amounts, and these redox-active dendrimers were applied to the construction of a reagentless enzyme electrode. A multilayer assembly of enzyme was constructed by alternate layer-by-layer deposition of ferrocenyl-tethered dendrimers with periodate-oxidized glucose oxidase. The optimum level of modification of surface amines was found to be 32%. The bioelectrocatalytic signal was shown to be directly correlated to the number of bilayers [31]. An affinity biosensor system based on avidin-biotin interaction on a gold electrode was developed using this strategy [32]. Pulse-field gradient spin-echo (PGSE) measurements on three different ferrocenyl phosphine dendrimers provided a practical alternative to classical methods used in organometallic chemistry for the determination of molecular size [33]. Unsymmetrical dendrimers were analyzed by MALDI TOF mass spectrometry with both conventional and electron-transfer matrixes [34]. Dendrimers containing a single ferrocenyl unit located "off-center" were submitted to interaction with β-cyclodextrin, and the dendritic groups were found to hamper the formation of inclusion complexes [35]. Ferrocenyl dendrimers were synthesized using the formation of quinodimethane intermediates [36]. Cyclic siloxanes were used as cores and frameworks for the construction of ferrocenyl dendrimers containing up to 16 ferrocenyl groups; these dendrimers were mediators for the oxidation of ascorbic acid [37]. Ferrocenyl dendrimers were adsorbed on Au surface, and electrochemical and ac-impedance were measured in order to investigate the porosity towards the redox couple [38].

3   CONSTRUCTION OF LARGE FERROCENYL DENDRIMERS: MOLECULAR BATTERIES

The functionalization of the three allyl chains of a phenol dendron could be achieved in our group by hydrosilylation reaction catalyzed by the Karsted catalyst [39,40]. Indeed, it is very interesting that there is no need to protect the phenol group before performing these reactions. For instance, catalyzed hydrosilylation using ferrocenyldimethylsilane gives a high yield of the triferrocenyl dendron HOp-C$_6$H$_4$C(CH$_2$CH$_2$CH$_2$SiMe$_2$Fc)$_3$ that is easily purified by column chromatography [41-43]. Protection of the phenol dendron using propionyliodide gave the phenolate ester which was hydroborated. Oxidation of the triborane using H$_2$O$_2$/OH$^-$ gave the triol. Then, reaction with SiMe$_3$Cl followed by NaI yielded the tri-iodo compound, and reaction with the tri-ferrocenyl dendron provided the nona-ferrocenyl dendron that was deprotected using K$_2$CO$_3$ in DMF. The nona-ferrocenyl dendron was allowed to react with hexakis(bromomethyl)benzene, which gave the 54-ferrocenyl dendrimer. This convergent synthesis is clean and the 54-ferrocenyl dendrimer gave correct analytical data (Scheme 1) [41].

We have also developed a divergent synthesis of polyallyl dendrimers indicated on Scheme 2 whereby each generation consists in hydroboration, oxidation of the borane to the alcohol, formation of the mesylate, and reaction of

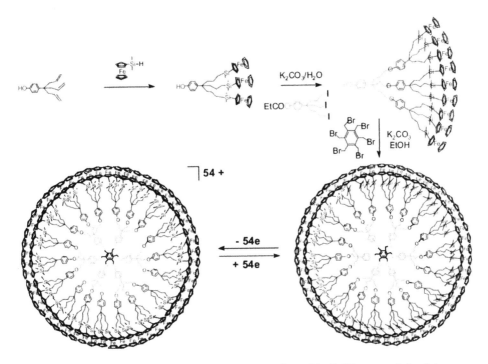

Scheme 1. Convergent synthesis of a redox-robust 54-silylferrocenyl dendrimer.

the phenol dendron with the mesylate. This strategy has allowed us to synthesize dendrimers of generation 0, 1, 2 and 3 with respectively 9 (G$_0$), 27 (G$_1$), 81 (G$_2$) and 243 branches (G$_3$) (Figure 1) [44].

The MALDI TOF mass spectrum of the 27-allyl dendrimer only shows the molecular peak with traces of side product. That of the 81-allyl shows a dominant molecular peak, but also important side products resulting from incomplete branching. That of the 243-allyl could not be obtained, possibly signifying that this dendrimer is polydisperse (correct ^{1}H and ^{13}C NMR spectra were obtained, however, indicating that the ultimate reactions had proceeded to completion). This dendrimer was soluble which indicated that this generation is not the last one, which might be reached. Larger dendrimers have recently been synthesized using a slightly different strategy. The ferrocenylsilylation of all these polyallyl dendrimers was carried out using ferrocenyldimethylsilane in ether or toluene and was catalyzed by the Karsted catalyst [39,40] at 40-45°C. The reactions were complete after two or three days except for the ferrocenylsilylation of 243-allyl that required a reaction time of one week indicating some degree of steric congestion (Scheme 3).

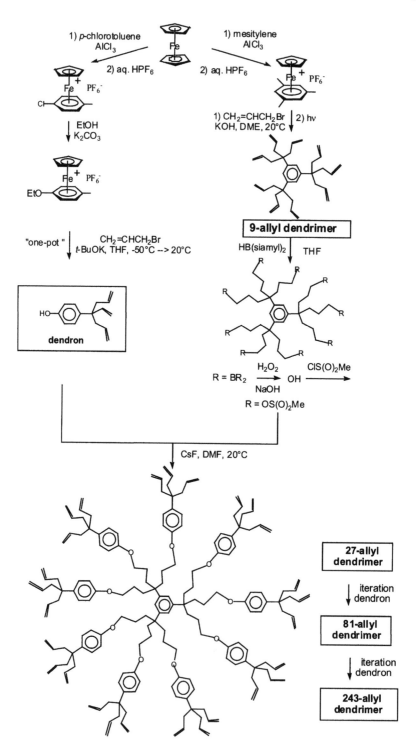

Scheme 2. Strategy for the construction of large dendrimers starting from ferrocene.

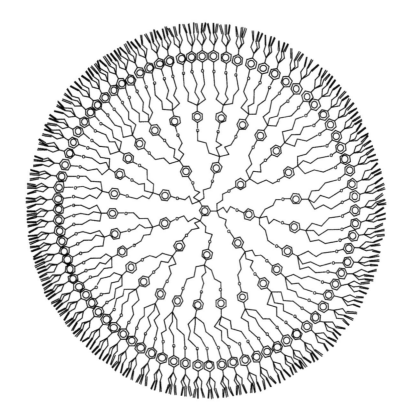

Figure 1. 243-allyl dendrimer (3rd generation, see the construction on Scheme 2).

The ^{1}H and ^{13}C spectra indicated the absence of regioisomer. The solubility in pentane decreased from good for the 9-Fc dendrimer to low for the 27-Fc dendrimer and nil for the superior dendrimers, but the solubility in ether remained good for all the ferrocenyl dendrimers. Likewise, the retention times on plate or column chromatography increased with generation and no migration was observed for the "243-Fc" dendrimer. The silane used here, $HSi(Fc)Me_2$, reported by Pannel and Sharma [45], was already used by Jutzi [46] to synthesize the decaferrocenyl dendrimer $[Fe(CCH_2CH_2SiMe_2Fc)_{10}]$ (with Fc = ferrocenyl) from deca-allylferrocene.

The cyclic voltammetry of all the ferrocenyl dendrimers on Pt anode shows that all the ferrocenyl centers are seemingly equivalent, and only one wave was observed. It was possible to avoid adsorption using even CH_2Cl_2 for the small ferrocenyl dendrimers, but it was required to use MeCN for the medium size ones (27-Fc, 54-Fc and 81-Fc). Finally, adsorption was not avoided even with MeCN for the "243-Fc" dendrimer. From the intensity of the wave, the number

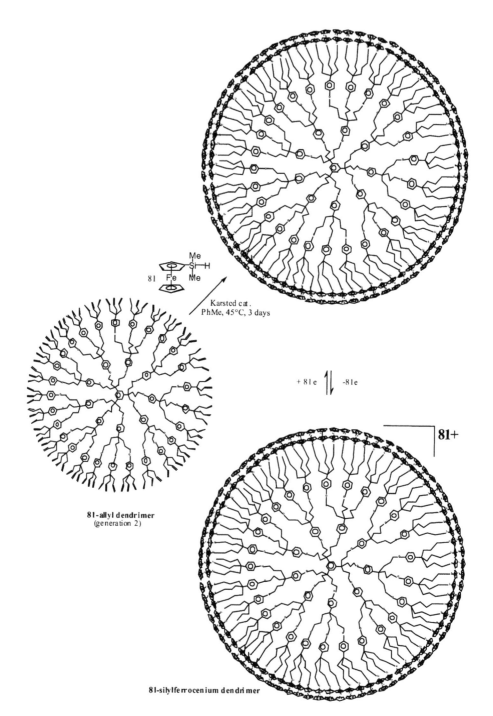

81-allyl dendrimer
(generation 2)

81-silylferrocenium dendrimer

Scheme 3. Ferrocenylsilylation of the polyallyl dendrimers synthesized. Example of the 2nd generation 81-allyl dendrimer.

of ferrocenyl units could be estimated using the Anson-Bard equation [47], and the number found were within 5% of the branch numbers except for the "243-Fc"dendrimer, for which the experimental number was too high (250) because of the adsorption.

We have been seeking to synthesize large ferrocenyldendrimers, which could also withstand oxidation to their ferrocenium analogues. The syntheses of amidoferrocene dendrimers were reported five years ago simultaneously by our group [45,49] and the Madrid group using different cores [50,51]. In our reports, we were able to show the use of these metallodendrimers as redox sensors for the recognition of oxo-anions, with remarkable positive dendritic effects when the generation increased. The amidoferrocenyl dendrimers are not the best candidates for a stable redox activity on the synthetic scale, however, and thus even less so for molecular batteries. Indeed, although they give fully reversible cyclic voltammetry waves, it is known that ferrocenium derivatives bearing an electron-withdrawing substituent are at least fragile, if stable at all. This inconvenient is probably enhanced in the dendritic structures because of the steric effect which forces ferrocenium groups to encounter one another more easily than as monomers. Thus, we have oxidized our silylferrocenyl dendrimers using $[NO][PF_6]$ in CH_2Cl_2 and obtained stable polyferrocenium dendrimers as dark-blue precipitates, as expected from the known characteristic color of ferrocenium itself. These polyferrocenium dendrimers were reduced back to soluble orange polyferrocenyl dendrimers using decamethylferrocene as the reductant [52]. No decomposition was observed either in the oxidation or in the reduction reactions which were very clean, and this redox cycle could be achieved in quantitative yield even with the "243-ferrocenyl" dendrimer. The zero-field Mössbauer spectrum of the 243-ferrocenium dendrimer showed a single line corresponding to the expected spectrum known for ferrocenium itself [53], confirming its electronic structure. Thus, these polyferrocenyl dendrimers are molecular batteries, which could be used, in specific devices. Indeed, as large as they may be, they transfer a very large number of electrons rapidly and "simultaneously" with the electrode. By "simultaneously", we mean that, visually, the cyclic voltammogram looks as if it were that of a monoelectronic wave. One must question the notion of the isopotential for the many ferrocenyl units at the periphery of a dendrimer. In theory, all the standard potentials of the n ferrocenyl units of a single dendrimer are distinct even if all of them are equivalent and independent. This situation arises since the charge of the overall dendrimer molecule increases by one unit of charge every time one of its ferrocenyl units is oxidized to ferrocenium. The next single-electron oxidation is more difficult than the preceding one since, the dendritic molecule having one more unit of positive charge, it is more difficult to oxidize because of the increased electrostatic factor. Thus, the potentials of the n redox units are statistically distributed around an average standard potential centered at the average potential (Gaussian distribution) [47]. In practice, the situation is complicated by the fact that the dendritic molecule, as large as it may be, is rotating much more rapidly than the usual electrochemical time scales [54,55]. Under these conditions, all the potentials are probably averaged. The fast rotation is also responsible for the fact that all the ferrocenyl units come close to the electrode within the electrochemical time scale. Consequently, there is no slowing down of the electron transfer due to long distance from the electrode even in large dendrimers. Indeed, the waves of the ferrocenyl dendrimers always appear fully electrochemically reversible indicating fast electron transfer.

The ferrocenyl dendrimers also adsorb readily on electrodes, a phenomenon already well known with various kinds of polymers [56]. When polymers contain redox centers, the adsorbed polymer have long been shown to disclose a redox wave for which the cathodic and anodic waves are located at exactly the same potential and the intensity of each wave is proportional to scan rate. Continuous cycling shows the stability of the adsorption of the electrode modified in this

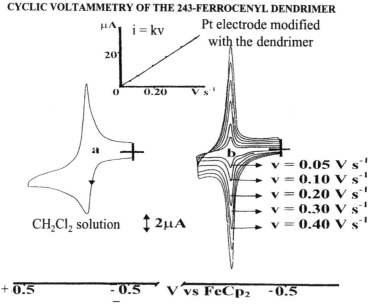

Figure 2. Cyclic voltammogram of the 243-ferrocenyl dendrimer ("243-Fc") in CH_2Cl_2 solution containing 0.1M [n-Bu$_4$N][PF$_6$]: a) in solution (10^{-4} M) at 100 mV. s^{-1} on Pt anode; b) Pt anode modified with "243-Fc" at various scan rates, dendrimer-free clear CH_2Cl_2 solution (inset: intensity as a function of scan rate: the linearity shows the expected behavior of a modified electrode with a fully adsorbed dendrimer).

way. The ferrocenyl dendrimers described show this phenomenon as expected. The stability of the electrode modified by soaking the Pt electrode in a CH_2Cl_2 solution containing the ferrocenyl dendrimer and cyclic scanning between the ferrocenyl and ferrocenium regions is all the better as the ferrocenyl dendrimer is larger. For instance, in the case of the 9-ferrocenyl dendrimer, scanning twenty times is necessary before obtaining a constant intensity, and this intensity is weak. With the 27-, 54-, 81-, and 243-ferrocenyl dendrimers, only approximately ten cyclic scans are necessary before obtaining a constant wave, and the intensity is much larger. When such derivatized electrodes are washed with CH_2Cl_2 and re-used with a fresh, dendrimer-free CH_2Cl_2 solution, the cyclic voltammogram is obtained with $\Delta E_p = 0$. Other characteristic features are the linear relationship between the intensity and scan rate and the constant stability after cycling many times with no sign of diminished intensity (Figure 2).

Under these conditions, one may note that the argument of the fast rotation of the dendritic molecule to bring all the redox centers in turn close to the electrode does not hold for modified electrodes. Some redox centers must be close to the electrode and some must be far. It is probable that a hoping mechanism in the solid state is responsible for fast electron transfer and for averaging all the potentials of the different ferrocenyl groups of a single dendritic molecule around a mean value. The proximity of the ferrocenyl groups at the periphery of the dendrimer is a key factor allowing this hoping to occur since it is known that electron transfer with redox sites which are remote or buried inside a molecular framework is slow, if at all observable [57-63].

Ferrocenes and ferrocenyl dendrimers are poor reductants. On the other hand, the complexes $[Fe(\eta^5\text{-}C_5R_5)(\eta^6\text{-}C_6Me_6)]^{2+/+/0}$ (R = H or Me) have been shown to be efficient for various stoichiometric and catalytic electron-transfer reactions [64,65]. The covalent linkage of this sandwich complex to the Cp ligand by means of a chlorocarbonyl substituent leads, upon reaction with dendritic polyamines, to soluble Fe^{II} metallodendrimers. Moreover, these Fe^{II} metallodendrimers can be reduced to Fe^{I} by $[Fe^{I}Cp(\eta^6\text{-}C_6Me_6)]$. Reduction of the monomeric model $[Fe^{II}(\eta^5\text{-}C_5H_4CONH\text{-}n\text{-}Pr)(\eta^6\text{-}C_6Me_6)][PF_6]$ by Na/Hg in THF (RT) gives the deep-blue-green, thermally stable 19-electron complex $[Fe^{I}(\eta^5\text{-}C_5H_4CONH\text{-}n\text{-}Pr)(\eta^6\text{-}C_6Me_6)]$ that shows the classic rhombic distorsion of the Fe^{I} sandwich family, observable by EPR in frozen THF at 77K (3 g values around 2) [66]. Given this stability, we carried out the same reaction of $[Fe^{II}(\eta^5\text{-}C_5H_4COCl)(\eta^6\text{-}C_6Me_6)][PF_6]$ with the commercial polypropyleneimine dendrimer of generation 5 (64 amino termini) in $MeCN/CH_2Cl_2$: 2/1 in the presence of NEt_3. The polycationic metallodendrimer DAB *dendr*-64-NHCOCpFe^{II}($\eta^6\text{-}C_6Me_6$), was obtained as the PF_6^- salt, soluble in MeCN and DMF (Scheme 4).

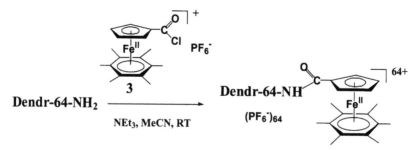

Scheme 4. Covalent linkage of the complex [FeCpCOCl(η^6-C_6Me_6)][PF$_6$] to the DSM polyamine dendrimer of generation 5 (64 branches). Example of the 64-NH$_2$ dendrimer (generation 5).

This dendritic complex was characterized by ^{1}H and ^{13}C NMR and IR spectroscopies and cyclovoltammetry (a single reversible wave in DMF, at $E_{1/2}$ = −1.84 V *vs.* FeCp$_2$$^{0/+}$; ΔEp = 70 mV). Attempts to reduce it with the classic reductants that reduce monomeric complexes [Fe^{II}(η^5-Cp)(η^6-arene)][PF$_6$] such as Na sand, Na/Hg or LiAlH$_4$ in THF or DME failed due to the insolubility of both the metallodendrimer and the reductant in the required solvents. The only successful reductant was the parent 19-electron complex [Fe^{I}Cp(η^6-C_6Me_6)] [67-68] (in pentane or THF) that reduced the metallodendrimer in MeCN at −30°C to the neutral, deep-green-blue 19-electron Fe^I dendrimer in a few minutes (Scheme 5) [69].

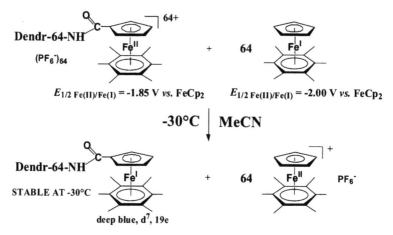

Scheme 5. Exergonic reduction of the cationic Fe^{II} dendritic sandwich groups by the parent 19-electron complex [Fe^{I}Cp(η^6-C_6Me_6)] to the Fe^I dendrimer complex.

The exergonicity of this electron-transfer reaction is 0.16 V, which is due to the electron-withdrawing effect of the juxta-cyclic carbonyl group on the Cp ring that lowers the reduction potential of the metallodendrimer as compared to that $[Fe^I Cp(\eta^6-C_6Me_6)]$. Although the Fe^I dendrimer decomposes a 0°C, it was also characterized by its EPR spectrum at 10 K confirming, as the deep-blue-green color, the Fe^I -sandwich structure analogous to that of the monomeric model (Figure 3).

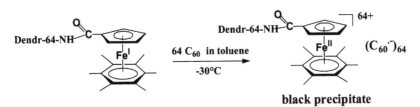

black precipitate

Scheme 6. Exergonic single-electron reduction of C_{60} by the dendritic Fe^I electron-reservoir complex.

Contrary to the case of $[Fe^I Cp(\eta^6-C_6Me_6)]$ [66], however, it was not possible to record the EPR spectrum of the solution of the Fe^I dendrimer above 10 K. This is presumably due to the intramolecular relaxation among the peripheral Fe^I sandwich units. The intermolecular version of this relaxation effect is known to preclude observation of the spectrum of monomeric Fe^I sandwich complexes in the solid state above 4K and in solution above 77K [66]. This acetonitrile solution of the 64-Fe^I dendrimer was used for the reaction with C_{60}, the stoichiometry being Fe^I/C_{60}: 1/1 (64 C_{60} *per* dendrimer). Upon reaction with a toluene solution of C_{60}, the deep-blue-green color of the Fe^I dendrimer disappeared, leaving a yellow solution that contained $[Fe^{II} Cp(\eta^6-C_6Me_6)][PF_6]$ and a black precipitate (Scheme 6). Tentative extraction of this precipitate with toluene yielded a colorless solution, which indicated that no C_{60} was present. The Mössbauer spectra of this black solid at 298K discloses parameters that show the presence of an Fe^{II} sandwich complex of the same family as $[Fe^{II} Cp(\eta^6-C_6Me_6)]^+$ [66-68]. Its EPR spectrum recorded at 77 K shows the same EPR spectrum as that of $[Fe^{II} Cp(\eta^6-C_6Me_6)]^+ C_{60}^-$ [70]. It could thus be concluded that C_{60} had been reduced to its monoanion, as designed for a process that is exergonic by 0.9 V [74]. The [dendr-Fe^{II}]$^+$ C_{60}^- units being very large, they must be located at the dendrimer periphery, presumably with rather tight ion pairs although the number of fullerene layers and overall molecular size are unknown (Figure 3).

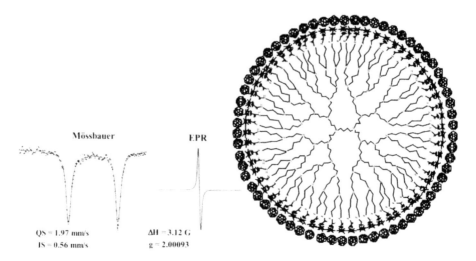

Figure 3: Dendr-64-NHCOCpFe(C$_6$Me$_6$)$^{64+}$, 64 C$_{60}$·$^-$ resulting from the reaction of the 64-Fe^I dendrimer with C$_{60}$ in MeCN/toluene at –30°C yielding the 64-Fe^{II}-C$_{60}^-$ dendrimer with EPR spectrum (bottom, right) in MeCN at 10K and Mössbauer spectrum at 77K (bottom, left) of the latter.

4   WATER-SOLUBLE STAR-SHAPED ORGANOMETALLIC REDOX CATALYSTS: TOWARDS GREEN CHEMISTRY

The design of nanosized molecular catalysts is essential in order to remove then by filtration or ultracentrifugation for green chemistry, i.e., re-use of catalysts known to work according to a mechanism defined for small homogeneous systems. For instance, stars and dendrimers are candidates for such a function. Stars benefit from a reduced steric inhibition as compared to metallodendrimers whose bulk at the periphery may eventually be kinetically counter-productive around the catalytic center, however. Thus stars have been designed in order to synthesize star-shaped hexanuclear redox catalysts for the cathodic reduction of nitrates and nitrites. Polyol stars and dendrimers can be transformed into mesilates and iodo derivatives that are useful for further functionalization. The hexa-iodo star was condensed with *p*-hydroxybenzaldehyde to give an hexa-benzaldehyde star, which could further react with substrates bearing a primary amino group. Indeed, this reaction yielded a water-soluble hexametallic redox catalysts (Scheme 7) which was active in the electroreduction of nitrate and nitrite to ammonia in basic aqueous solution (Scheme 8) [71-73].

The catalytic reduction of nitrate has been known for a long time [74], but the use of the redox catalyst [Fe^{II}(η^5-C$_5$H$_4$CO$_2^-$)(η^6-C$_6$Me$_6$)][PF$_6$] was

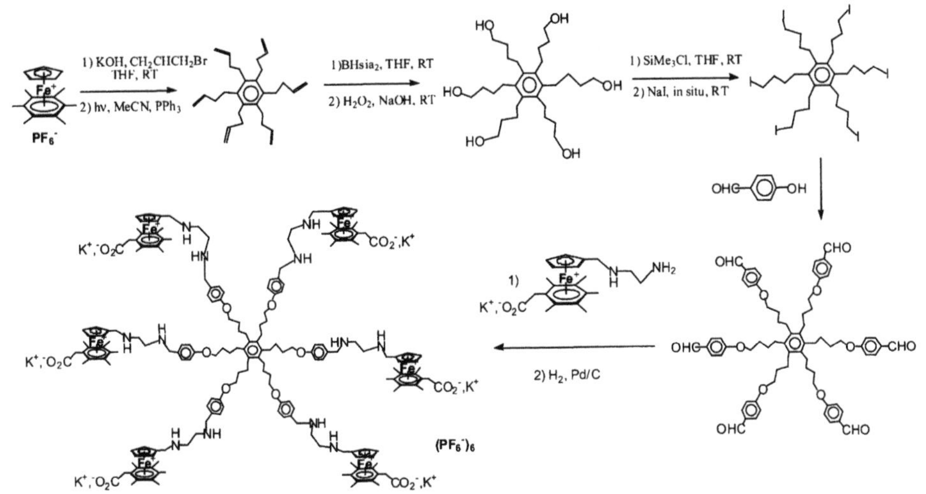

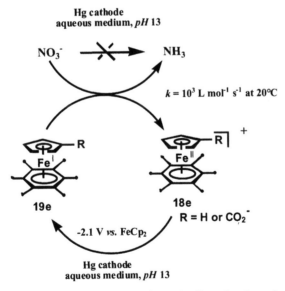

Scheme 7. CpFe⁺-induced hexa-allylation of C_6Me_6 and subsequent hexafunctionalization of the aromatic stars for the synthesis of a water-soluble star-shaped hexanuclear organometallic redox catalyst active for the cathodic reduction of nitrate and nitrite.

Scheme 8. Redox-catalysis mechanism for the cathodic reduction of nitrate (and nitrite) to NH_3 catalyzed by the water-soluble Fe^{II} organometallic sandwich complex.

the first example of an organometallic catalyst for this reaction (Scheme 8). Electrosyntheses catalyzed by $[Fe^I(\eta^5\text{-}C_5H_4CO_2^-)(\eta^6\text{-}C_6Me_6)]$ under these conditions showed that NH_3 was produced in 63% chemical yield and 57% electrical yield with $R = CO_2^-$. Nitrite, hydroxylamine, hydrazine and dinitrogen (minute amounts) were intermediates towards the formation of ammonia. The cathodic reduction stopped at the level of hydroxylamine when the experiment was carried out at *pH* 7 instead of 13. Kinetic studies in homogeneous basic aqueous solution using a polarographic method were carried out with $[Fe^{II}(\eta^5-C_5H_4R)(\eta^6\text{-arene})]$, R = CO_2^- , or when only the cationic Fe^{II} form was soluble in the medium (R = H; arene = benzene, *m*-xylene, hexamethylbenzene). These studies led to the conclusion that the rate of redox catalysis was independent of the nature of the catalyst within this series [75], but have recently been reconsidered, however, with the series of complexes $[Fe^{II}(\eta^5-C_5H_4CO_2^-)(\eta^6\text{-arene})]$ $[PF_6]$, arene = C_6Me_{6-n}, n = 0 to 6. The polarographic, cyclic voltammetry and chronoamperometry techniques were used to investigate the kinetics of the redox catalysis. The three techniques provided similar results. Thus, the rate constant of the redox catalysis can be calculated from the enhancement of the intensity of the cyclovoltammogram wave observed upon addition of the nitrate or nitrite salt into the electrochemical cell (Figure 4). A Marcus-type linear relationship was found between the logarithms of the rate constants and the standard redox potentials of the catalysts, indicating that the electron transfer in solution between the 19-electron Fe^I complex and nitrate or nitrite ion is rate limiting [71-76].

In order to investigate whether the mechanism proceeds by inner-sphere or outer-sphere electron transfer, other catalysts of the type $[Fe^{II}(\eta^5-C_5H_4CO_2^-)(\eta^6\text{-arene})][PF_6]$ were synthesized with bulky arenes such as 1,3,5-tris-*t*-butylbenzene and $C_6(CH_2CH_2p\text{-}C_6H_4OH)_6$. In these catalysts, the redox-active group is at the center of a star- or dendritic core framework. The rate constants were found to be one to two orders of magnitude lower than what would be expected for the same driving force if the steric effect was not interfering, by comparison with the above series of catalysts with polymethylbenzene ligands. This

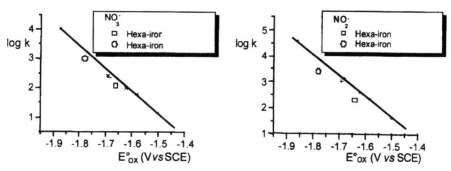

Figure 4. Marcus-type relationship between the kinetics (ln k) and thermodynamics ($E_{1/2}$) for the redox-catalyzed reduction of nitrate indicating that the primary electron-transfer (mainly outer-sphere) is rate-limiting.

showed a significant inner-sphere component to the electron-transfer process, although the kinetic drop would have been even more dramatic with a fully inner-sphere electron-transfer mechanism. It is likely that nitrate and nitrite ions coordinate to the 17-electron form of the Fe^I catalysts and that electron transfer proceeds in such intermediates rather than by outer sphere. However, the bond between an oxygen atom of nitrate or nitrite to such a low oxidation-state center must be very loose and long because π-back bonding is impossible with these ligands [76].

One may now compare the kinetics of a $[FeCp(arene)]^+$-centered star or dendritic core to that of a star bearing $[FeCp(arene)]^+$ catalysts at the periphery. Remarkably, the kinetics of catalysts bearing the $[FeCp(arene)]^+$ moiety at the center of a star or dendritic core is one order of magnitude lower than that of such a star bearing the catalyst at the periphery. We know that, as in cytochromes, electron transfer between an electrode and a redox center located at the center of a dendrimer is slow. On the other hand, the kinetics of the hexanuclear star (Scheme 7) in which the redox moieties are located at the star periphery is about the same as those of mononuclear redox catalysts of the same type and driving force. This category of electron transfer between an electrode and redox centers located at a dendritic periphery is fast due to the fast rotation of the dendrimer as compared to the electrochemical time scale. Note that we have first chosen a star topography rather than a dendritic one in order to avoid the problem of bulk encountered in dendrimers in which the catalytic groups located at the periphery are marred by steric inhibition preventing the substrate to approach the metal coordination sphere. By the way, this redox-catalysis technique could be very useful in order to learn more about this very problem in dendrimers. Work along this line is indeed in progress in our laboratory.

5   DECORATION OF DENDRIMERS WITH RUTHENIUM CLUSTERS: TOWARDS DENDRITIC CATALYSTS

The clean introduction of clusters onto the termini of polyphosphine dendrimers is a real challenge because of the current interest of dendritic clusters in catalysis and of the fact that mixtures usually obtained in thermal reactions of $[Ru_3(CO)_{12}]$ with phosphines. The diphosphine $CH_3(CH_2)_2N(CH_2PPh_2)_2$ (abbreviated P-P below) was used as a simple, model ligand. The reaction between P-P and $[Ru_3(CO)_{12}]$ [77], (molar ratio: 1/1.05) in the presence of 0.1 equiv. $[Fe^ICp(\eta^6\text{-}C_6Me_6)]$ in THF at 20°C led to the complete disappearance of $[Ru_3(CO)_{12}]$ in a few minutes and the appearance of a mixture of chelate [P-P. $Ru_3(CO)_{10}$], monodentate [P-P. $Ru_3(CO)_{11}$] and bis-cluster [P-P. $\{Ru_3(CO)_{11}\}_2$].

These reactions were reported by Bruce with simple diphosphines [78]. On the other hand, the reaction of P-P with $[Ru_3(CO)_{12}]$ in excess (1/4) and only 0.01 equiv. $[Fe^ICp(\eta^6-C_6Me_6)]$ in THF at 20°C led, in 20 minutes, to the formation of

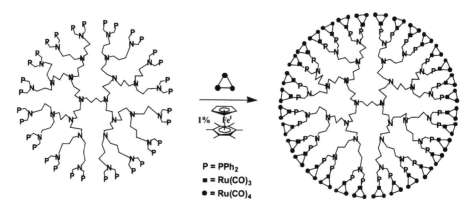

Scheme 9. Electron-Transfer-Chain catalyzed ligand substitution of one Ru-coordinated CO by a dendritic phosphine termini in Reetz's 32-phosphine dendrimer under ambiant conditions leading to the $32\text{-}Ru_3(CO)_{11}$ dendrimer-cluster.

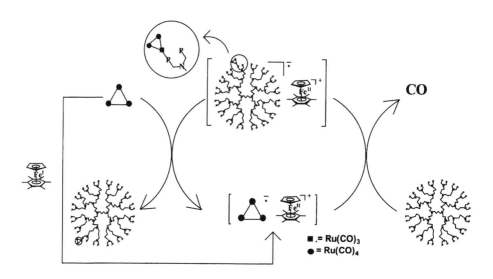

Scheme 10. Electron-Transfer-Chain mechanism for the synthesis of the 96-Ru dendrimer-cluster complex.

the air-stable, light-sensitive bis-cluster [P-P. {Ru$_3$(CO)$_{11}$}$_2$] as the only reaction product. Given the simplicity of the above characterization of the reaction product by ^{31}P NMR and the excellent selectivity of this model reaction when excess [Ru$_3$(CO)$_{12}$] was used, the same reaction between Reetz's dendritic phosphines [79a], derived from DSM's dendritic amines [79b], and [Ru$_3$(CO)$_{12}$] could be more confidently envisaged. This reaction, catalyzed by 1% equiv. [Fe^{I}Cp(η^6-C$_6$Me$_6$)] was carried out in THF at 20°C. The dendrimer-cluster assembly was obtained in 50% yield. This shows the selectivity and completion of the coordination of each of the 32 phosphino ligands of P-P to a Ru$_3$(CO)$_{11}$ cluster fragment (Scheme 9).

ETC mechanism [80-82] proceeds for the introduction of the 32 cluster fragments in the dendrimer for ligation of the first Ru$_3$(CO)$_{11}$ fragment to the dendritic phosphine. Then, this first complex [dendriphosphine.Ru$_3$(CO)$_{11}$] would undergo the same ETC cycle as [Ru$_3$(CO)$_{12}$] initially does to generate the bis-cluster complex [dendriphosphine.{Ru$_3$(CO)$_{11}$}$_2$], and so on (Scheme 10).

Finally, the 64-branch phosphine DAB-*dendr*-G4-[N(CH$_2$PPh$_2$)$_2$]$_{32}$ analogously reacts with [Ru$_3$(CO)$_{12}$] and 1% [Fe^{I}Cp(η^6-C$_6$Me$_6$)] (20°C, THF, 20 min.) to give the dark-red 192-Ru dendrimer. Characterization of the purity of these dendrimer-cluster assemblies is conveniently monitored by ^{31}P NMR. This application should find extension to other metal-carbonyl clusters and other families of phosphine dendrimers.

6   SUPRAMOLECULAR FERROCENYL DENDRIMERS AS EXORECEPTORS FOR THE RECOGNITION OF OXOANIONS

The interaction of amido groups attached to metallocenes with oxoanions allows to influence the redox potential of metallocenes in such a way that these amidometallocenes [83] could be attached to endoreceptors [84]. We have used this property to design amidoferrocenyl dendrimers as exoreceptors that compare with viruses that are natural, albeit toxic, exoreceptors [48]. The comparison of tripodal tris-amidoferrocenyl and 9-amidoferrocenyl and 18-amidoferrocenyl dendrimers allowed us to find dramatic positive dendritic effect in the recognition of the anions. Recognition studies have been carried out by cyclic voltammetry and by ^{1}H NMR. In each case, titration of the ferrocene dendrimers were effected by *n*-Bu$_4$N$^+$ salts of H$_2$PO$_4^-$, HSO$_4^-$, Cl$^-$ and NO$_3^-$. By far, the most informative results were obtained by cyclic voltammetry by scanning the Fe(II/III) wave (Figure 3). Before any titration, the CVs of the 9-Fc and 18-Fc dendrimers show a unique wave at 0.59 V vs. SCE in CH$_2$Cl$_2$ corresponding to the oxidation of the 9 or 18 redox centers, which indicates that, as expected, the 9 or 18 redox centers are approximately electrochemically equivalent, thus independent (when, for instance, two equivalent redox centers are not so far away from each other, two waves are observed at two distinct potentials, even if there is no electronic connection, because of the electrostatic effect). In the present situation, the redox centers are far from one another, thus the electrostatic effect is very weak and not detected. Upon addition of the anion, two situations can arise [85]. In the case of

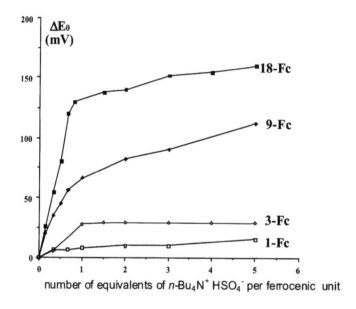

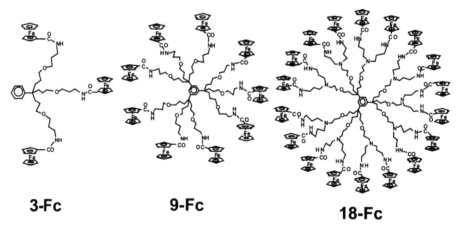

3-Fc   **9-Fc**

18-Fc

Figure 5. Compared shift of ferrocenyl redox potential along the addition of n-[Bu$_4$N][HSO$_4$] to the electrochemical cell containing for a ferrocenyl tripod (3-Fc) or a ferrocenyl dendrimers (9-Fc or 18-Fc). Also compare with the non-dendritic monoamidoferrocenyl compound (1-Fc).

$H_2PO_4^-$, a new wave starts appearing at less positive potentials and correlatively, the intensity of the initial wave starts decreasing. When one equivalent of anion per dendrimer branch has been added, the initial wave has disappeared and, upon addition of the anion, the intensity of the new wave does not increase any longer. In the case of the other anions, no new wave appears, but the initial wave is progressively shifted to less positive potentials upon titration until one equivalent of anion has been added per dendrimer branch. It clearly appears that the shifts $\Delta E°$ of potentials observed after addition of one equiv. anion per dendrimer branch considerably increases in the series : 1-Fc $\longrightarrow$ 3-Fc $\longrightarrow$ 9-Fc $\longrightarrow$ 18 Fc, which shows a dramatic dendritic effect represented in Figure 5 for the titration with the HSO_4^- anion. The magnitude of interaction with the anion increases as follows :

$$H_2PO_4^- > HSO_4^- > Cl > NO_3^-$$

In the amidoferrocene dendrimers, the amide H atom is located on the branch behind the ferrocene unit which provides the surface bulk. Thus the anion must reach the inside of the microcavity formed by the amido-ferrocene units at the surface of the dendrimer. These conditions become optimal for redox sensing and recognition by the close ferrocene units at the 18-Fc generation, since the

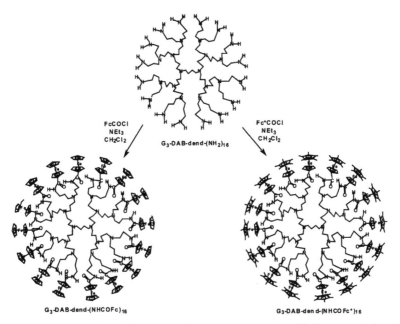

Scheme 11. Synthesis of amidoferrocenyl vs. pentamethylamidoferrocenyl dendrimers (generation 1 to 5) starting from the commercial DSM polyamine dendrimers.

channels allowing the entry of the anions into the surface microcavity to reach the amide H atom are as narrow as possible.

Note that with other metallodendrimers, the recognition of chloride and bromide is selective [86]. Using the same amidoferrocenyl termini, other dendritic cores show different recognition features showing the crucial role of the core topology on the recognition by the amidoferrocenyl groups (Scheme 11, Figure 6). In many instances, this kind of study is marred by adsorption peaks and partial or complete chemical irreversibility of the electrochemical waves. These problems have been circumvented by the synthesis of pentamethylamid oferrocenyl dendrimers. Their use as sensors shows that adsorption peaks are no longer seen and that chemical reversibility is systematically observed. These properties result from the lipophilicity and strong stabilizing effect of the C_5Me_5 ligands at the dendrimer periphery that lead to cleaner voltammograms [87]. With HSO_4^-, however, the stronger interaction with the parent amidoferrocenyl group than with the permethylated one makes the parent dendrimer better choices for recognition using appropriate conditions [88]. A remarkable situation is that found with amidoferrocenyl dendrimers that are assembled by hydrogen bonding between commercial DSM polyamine dendrimers and phenol-containing tris-amidoferrocenyl bricks. These dendrimers also recognize oxo-anions, but the molecular ensemble formed with $H_2PO_4^-$ are so large that the diffusion coefficient dramatically drops at the equivalent point. As expected, such a recognition is very sensitive to dendritic effects, i.e. the phenomenum changes from one dendritic generation to the next due to steric aspects at the dendritic

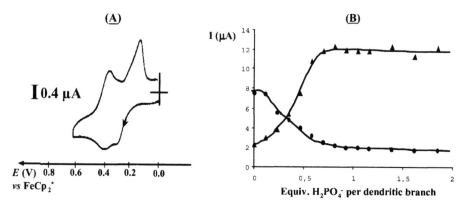

Figure 6. Titration of a 2.7×10^{-4} M solution of the G_3–16Fc* by a 10^{-3} M solution of [n-Bu$_4$N] [H$_2$PO$_4$] in CH$_2$Cl$_2$ in the presence of 0.1 M [n-Bu$_4$N][PF$_6$], Pt anode, 20°C. A: cyclovoltammogram obtained after addition of 0.5 equiv. [n-Bu$_4$N] [H$_2$PO$_4$] per dendritic branch; B: variation of the intensities of the initial (●) and new (▲) waves along the titration.

periphery [89]. Finally, let us just briefly mention here that gold nanoparticle-amidoferrocenylthiols are also good exoreceptors with features distinct from those of the analogous dendrimers [90,91] and that the combination of both nanomaterials, i.e. dendronized gold nanoparticles are excellent sensors because they gather the topological qualities of both types of exo-receptors [89,92,93]

7   CONCLUSION AND PERSPECTIVES

The synthesis of ferrocenyl dendrimers and other redox-active dendrimers brings about nanoscale molecular materials whose electrochemistry and electron-transfer chemistry disclose an impressive body of information in the perspective of their future use in nanodevices. First, large ferrocenyl dendrimers with terminal ferrocenyl groups show a remarkably simple electrochemistry with a single wave for which the number of electrons can be estimated with 5% accuracy as previously shown by the groups of Anson and Bard with ferrocenyl polymers. This property together with the fast heterogeneous electron transfers observed indicate that these nanodendrimers behave as molecular batteries. Indeed, this property can be used to reduce [60] fullerene to its monoanion at each single dendritic branch. It is also useful for redox catalysis towards green chemistry, especially if water is the solvent. The manipulation of electron transfers between an electron-reservoir complex in catalytic amount and a cluster complex in the presence of a dendritic phosphine could allow us to attach the clusters onto the periphery of large dendrimers, a promising property that could find applications in catalysis. Finally, dendrimers containing both a redox group and a supramolecular fonction (for H-bonding) are useful as exoreceptors for the recognition of oxoanions including DNA fragments such as ATP. Further studies in our laboratories upon these lines involving recognition and catalysis, the two functions of enzymes, are in progress.

ACKNOWLEDGMENT

We warmly thank the colleagues, students and post-docs cited in the references who have contributed to the ideas and efforts of this research program. We are especially indebted in this respect to Sylvain Lazarre (LCPT, university Bordeaux I: AFM), to Dr Jean-Claude Blais (University Paris VI: MALDI TOF mass spectrometry) and Pr François Varret (University of Versailles: Mössbauer spectroscopy). Financial support from the Institut Universitaire de France (IUF), the University Bordeaux I, the Centre National de la Recherche Scientifique (CNRS), the Alexander von Humboldt Fundation, Iberdrola, the Picasso and Erasmus-Socrates programs, the European Community and the Region Aquitaine is gratefully acknowledged.

REFERENCES

1. For a recent book, see: Newkome, G. R.; Moorefield, C. N.; Vögtle, F. *"Dendrimers and Dendrons. Concept, Syntheses, Applications"*, Wiley-VCH, Weinheim (2001).
2. Balzani, V.; Campana, S.; Denti, G.; Juris, A.; Serroni, S.; Venturi, M. *Acc. Chem. Res.*, *31*, 26 (1998).
3. For a comprehensive review, see: Ardoin, N.; Astruc, D. *Bull. Soc. Chim. Fr.*, *132*, 875 (1995).
4. Hamon, J.-R.; Saillard, J.-Y.; Le Beuze, A.; McGlinchey, M; Astruc, D. *J. Am. Chem. Soc.*, *104*, 3755 (1982).
5. a) Moulines, F.; Astruc, D. *Angew. Chem. Int. Ed. Engl.*, *27*, 1347 (1988); b) Moulines, F.; Astruc, D. *J. Chem. Soc. Chem. Commun.*, 614 (1989).
6. Alonso, B.; Blais, J.-C.; Astruc, D. *Organometallics*, *21*, 1001 (2000).
7. a) Fillaut, J.-L.; Linares, J.; Astruc, D. *Angew. Chem. Int. Ed. Engl.*, *33*, 2460 (1994); b) Fillaut, J.-L.; Boese, R.; Astruc, D. *Synlett*, 55 (1992); c) Fillaut, J.-L.; Astruc, D. *New J. Chem.*, *20*, 945 (1996).
8. Moulines, F.; Gloaguen, B.; Astruc, D. *Angew. Chem. Int. Ed. Engl.*, *28*, 458 (1992).
9. Marx, H. W.; Moulines, F.; Wagner, T.; Astruc, D. *Angew. Chem. Int. Engl.*, *35*, 1701 (1996).
10. Moulines, F.; Djakovitch, L.; Boese, R.; Gloaguen, B.; Thiel, W.; Fillaut, J.-L.; Delville, M.-H.; Astruc, D. *Angew. Chem. Int. Ed. Engl.*, *105*, 1132 (1993).
11. Trujillo, H.; Casado, C.; Ruiz, J.; Astruc, D. *J. Am. Chem. Soc.*, *121*, 5674 (1999).
12. a) Alonso, B.; Cuadrado, I.; Morán, M.; Losada, J. *J. Chem. Soc., Chem. Commun.*, 2575 (1994); b) Garcia, B.; Casado, C.; Cuadrado, I.; Alonso, B.; Mora, M.; Losada, J. *Organometallics*, *18*, 2349 (1999).
13. Alonso, B.; Morán, M.; Casado, C. M.; Lobete, F.; Losada, J.; Cuadrado, I. *Chem. Mater.*, *7*, 1440 (1995).
14. Losada, J.; Cuadrado, I.; Morán, M.; Casado, C. M.; Alonso, B.; Barranco, M. *Anal. Chim. Acta.*, *251*, 5 (1996); *338*, 191 (1997).
15. Casado, C. M.; Alonso, B.; Morán, M.; Cuadrado, I.; Losada, J. *J. Electroanal. Chem.*, *463*, 87 (1999).
16. Cuadrado, I.; Casado, C. M.; Alonso, B.; Morán, M.; Losada, J.; Belsky, V. *J. Am. Chem. Soc.*, *119*, 7613 (1997).
17. a) Takada, K.; Diaz, D. J.; Abruña, H.; Cuadrado, I.; Casado, C. M.; Alonso, B.; Morán, M.; Losada, J. *J. Am. Chem. Soc.*, *119*, 10763 (1997); b) Cuadrado, I.; Morán, M.; Casado, C. M.; Alonso, B.; Lobete, F.; Garcia, B.; Losada, J. *Organometallics*, *15*, 5278 (1996).
18. a) Castro, R.; Cuadrado, I.; Alonso, B.; Casado, C. M.; Morán, M.; Kaifer, A. *J. Am. Chem. Soc.*, *119*, 5760 (1997); b) Kaifer, A. E.; Gomez-Kaifer, M. *Supramolecular Electrochemistry*, Wiley-VCH, Weinheim, (1999).
19. a) González, B.; Casado, C. M.; Alonso, B.; Cuadrado, I.; Morán, M.; Wang, Y.; Kaifer, A. E. *Chem. Commun.*, 2569 (1998); b) Valério, C.; Ruiz, J.; Fillaut, J.-L.; Astruc, D. *C. R. Acad. Sci.*, Paris, 2, série II c, 79 (1999).

20. a) Casado, C. M.; Gonzales, B.; Cuadrado, I.; Alonso, B.; Morán, M.; Losada, J. *Angew. Chem. Int. Ed. Engl.*, *39*, 2135 (2000); (b) Gonzales, B.; Cuadrado, I.; Casado, C. M.; Alonso, B.; Pastor, C. *Organometallic, 19*, 5518 (2000).

21. Cardona, C. M.; Kaifer, A. E. *J. Am. Chem. Soc.*, *120,* 4023 (1998).

22. Newkome, G. R.; Behera, R. K.; Moorefield, C. N.; Baker, G. R. *J. Org. Chem.*, *56*, 7126 (1991).

23. a) Wang, Y.; Cardona, C.; Kaifer, A. E. *J. Am. Chem. Soc.*, *121*, 9756 (1999); b) Kaifer, A. E.; Gomez-Kaifer, M. *Supramolecular Electrochemistry*; Wiley-VCH: Weinheim (1999).

24. Togni, A.; Hayashi, T., Eds.; *Ferrocenes*, VCH: Weinheim (1995).

25. a) Köllner, C.; Pugin, B.; Togni, A. *J. Am. Chem. Soc.*, *120*, 10274 (1998); b) Schneider, R.; Kollner, C.; Weber, I.; Togni, A. *Chem Commun.*, 2415 (1999); c) Oosterom, E. G.; van Haaren, R. J.; Reek, J. N. H.; Kamer, P. C. J.; van Leewen, P. W. N. M. *Chem. Commun.*, 1119 (1999).

26. Shu, C.-F.; Shen, H.-M. *J. Mater. Chem.*, *7*, 47 (1997).

27. Achar, S.; Immoos, C. E.; Hill, M. G.; Catalano, V. J. *Inorg. Chem.*, *36*, 2314 (1997).

28. a) Deschenaux, R.; Serrano, E.; Levelut, A.-M. *Chem. Commun.*, 1577 (1997); b) Dardel, B.; Descheneaux, R.; Even, M.; Serrano, E. *Macromolecules*, *32*, 5193, (1999).

29. Turrin, C.-O.; Chiffre, J.; de Montauzon, D.; Daran, J.-C.; Caminade, A.-M.; Manoury, E.; Balavoine, G.; Majoral, J.-P. *Macromolecules*, *33*, 7328 (2000).

30. Mosbach, M.; Schuhman, W. *Patent* N° DE 19917052, *CA* 133:293175 (1999).

31. Yoon, H. C.; Hong, M.-Y.; Kim, H.-S. *Anal. Chem.*, *72*, 4420 (2000).

32. Yoon, H. C.; Hong, M.-Y.; Kim, H.-S. *Anal. Biochem.*, *282*, 121 (2000).

33. Valentini, M.; Pregosin, P. S.; Ruegger, H. *Organometallics*, *19*, 2551 (2000).

34. McCarley, T. D.; DuBois, C. J. Jr; McCarley, R. L.; Cardona, C. M.; Kaifer, A. E. *Polym. Prepr. (Am. Chem. Soc., Div. Polym. Chem.)*, *41*, 674 (2000).

35. Cardona, C.; McCarley, T. D.; Kaifer, A. E. *J. Org. Chem.*, *65*, 1857 (2000).

36. Ipatschi, J.; Hosseinzadeh, R.; Schlaf, P. *Angew. Chem. Int. Ed. Engl.*, *38*, 1658 (1999).

37. Casado, C.; Cuadrado, I.; Morán, M.; Alonso, B.; Barranco, M.; Losada, J. *Appl. Organomet. Chem.*, *14*, 245 (1999).

38. Tokuhisa, H.; Zhao, M.; Baker, L. A.; Phan, V. T.; Dermody, D. L.; Garcia, M. E.; Peez, R. F.; Crooks, R. M.; Mayer, T. M. *J. Am. Chem. Soc.*, *120*, 4492 (1998).

39. Marciniec, B. In *"Applied Homogeneous Catalysis with Organometallic Compounds"* (Eds.: B. Cornils, W. A. Herrmann), VCH, Weinheim, Vol. 1, Chap. 2.6 (1996).

40. Lewis, L. N.; Stein, J.; Smith, K. A. In *Progress in Organosilicon Chemistry* (Eds.: Marciniec, B.; Chojnowski, J.) Gordon and Breach, Langhorne, USA, p. 263 (1995).

41. Nlate, S.; Neto, Y.; Blais, J.-C.; Ruiz, J.; Astruc, D. *Chemistry Eur. J.*, *8*, 171 (2002).

42. Nlate, S.; Ruiz, J.; Astruc, D. *Chem. Commun.*, 417 (2000).

43. Nlate, S.; Ruiz, J.; Sartor, V.; Navarro, R.; Blais, J.-C.; Astruc, D. *Chemistry Eur. J.*, *6*, 2544 (2000).

44. a) Sartor, V; Djakovitch, L.; Fillaut, J.-L.; Moulines, F.; Neveu, F.; Marvaud, V.; Guittard, J.; Blais, J.-C.; Astruc, D. *J. Am. Chem. Soc.*, *121*, 2929 (1999); b) Sartor, V.; Nlate, S.; Fillaut, J.-L.; Djakovitch, L.; Moulines, F.; Marvaud, V.; Neveu, F.; Blais, J.-C. *New J. Chem.*, *24*, 351 (2000).

45. Pannel, K. H.; Sharma, H. *Organometallics*, *10*, 954 (1991).

46. Jutzi, P.; Batz, C.; Neumann, B.; Stammler, H. G. *Angew. Chem. Int. Engl.*, *35*, 2118 (1996).

47. Flanagan, J. B.; Margel, S.; Bard, A. J.; Anson, F. C. *J. Am. Chem. Soc.*, *100*, 4248 (1978).

48. Valério, C.; Fillaut, J-L.; Ruiz, J.; Guittard, J.-C.; Blais, J.-C.; Astruc, D. *J. Am. Chem. Soc.*, *117*, 2588 (1997).

49. a) Astruc, D.; Valério, C.; Fillaut, J.-L.; Hamon, J.-R.; Varret, F. In *Magnetism, a Supramolecular Function* (Ed.: O. Kahn), NATO ASAI Series, Kluver, Dordrecht, p. 1107 (1996); b) Valério, C. *PhD Thesis*, Université Bordeaux I, (1996).

50. a) Cuadrado, I.; Morán, M.; Casado, C. M.; Alonso, B.; Lobete, F.; Garcia, B.; Losada, J. *Organometallics*, *15*, 5278 (1996); b) Takada, K.; Diaz, D. J.; Abruña, H.; Cuadrado, I.; Casado, C. M.; Alonso, B.; Morán, M.; Losada, J. *J. Am. Chem. Soc.*, *119*, 10763 (1997).

51. Reviews: a) Casado, C. M.; Cuadrado, I.; Moran, M.; Alonso, B.; Garcia, B.; Gonzales, B.; Losada, J. *Coord. Chem. Rev.*, *185-6*, 53 (1999); b) Cuadrado, I.; Morán, M.; Casado, C. M.; Alonso, B.; Losada, J. *Coord. Chem. Rev.*, *189*, 123 (1999); c) Hershaw, M. A.; Moss, J. R. *Chem. Commun.*, 1 (1999); d) Newkome, G. R.; He, E.; Moorefield, C. N. *Chem. Rev.*, *99*, 1689 (1999).

52. Ruiz, J.; Astruc, D. *C. R. Acad. Sci.* Paris, t. 1, Série II *c*, 21 (1998).

53. Collins, R. L. *J. Chem. Phys.*, *42*, 1072 (1965).

54. Green, S. J.; J. J. Pietron, J. J. Stokes, M. J. Hostetler, H. Vu, W. P. Wuelfing, R. W. Murray, *Langmuir*, *14*, 5612 (1998).

55. Gorman, C. B.; Smith, J. C.; Hager, M. W.; Parhurst, B. L.; Sierzputowska-Gracz, H.; Haney, C. A. *J. Am. Chem. Soc.*, *121*, 9958 (1999).

56. Murray, R. In *Molecular Design of Electrode Surfaces* (Ed.: Murray, R.), Wiley, New York, p. 1 (1992).

57. Dandliker, P. J.; Diederich, F.; Gross, M.; Knobler, B.; Louati, A.; Stanford, E. M. *Angew. Chem. Int. Ed. Engl.*, *33*, 1739 (1994).

58. Newkome, G. R.; Güther, R.; Moorefield, C. N.; Cardullo, F.; Echegoyen, L.; Pérez-Cordero, F.; Luftmann, H. *Angew. Chem. Int. Ed. Engl.*, *34*, 2023 (1995).

59. Chow, H.-F.; Chan, I. Y.-K.; Chan, D. T. W.; Kwok, R. W. M.; *Chem. Eur. J.*, *2*, 1085 (1996).

60. Dandliker, P. J.; Diederich, F.; Chow, H.-F.; Chan, I. Y.-K.; Kwok, R. W. M. ; *Chem. Eur. J.*, *2*, 1085 (1996).

61. Issberner, J.; Vögtle, F.; De Cola, L. ; Balzani, V. *Chem. Eur. J.*, *3*, 706 (1997).

62. Gorman, C. B.; Parkhurst, B. L.; Su, W. Y.; Chen, K. Y. *J. Am. Chem. Soc.*, *119*, 1141 (1997).

63. Smith, D. K.; Diederich, F. *Chem. Eur. J.*, *4*, 2353 (1998).

64. a) For the synthesis of $[Fe^{II}Cp(\eta^6-C_6Me_6)][PF_6]$, see references 18,19 and 35; b) Pauson, P. L.; Watts, W. E. *J. Chem. Soc.*, 2990 (1963); c) Astruc, D.; J.-R. Hamon, J.-R.; Lacoste, M.; Desbois, M.-H.; Román, E. *Organometallic Synthesis* (Ed.: King, R. B.), Vol. IV, p. 172 (1988).

65. Ruiz, J.; Ogliaro, F.; Saillard, J.-Y.; Halet, J.-F.; Varret, F.; Astruc, D. *J. Am. Chem. Soc.*, *120*, 11693 (1998).

66. Rajasekharan, M. V.; Giesynski, S.; Ammeter, J. H.; Oswald, N.; Hamon, J.-R.; Michaud, P.; Astruc, D. *J. Am. Chem. Soc.*, *104*, 129 (1982).

67. Astruc, D.; Hamon, J.-R.; Althoff, G.; Roman, E.; Batail, P.; Michaud, P. ; Mariot, J.-P.; Varret, F.; Cozak, D. *J. Am. Chem. Soc.*, *101*, 5445 (1979). This paper also reports the first CpFe$^+$-induced iterative starburst hexa-alkylation of C_6Me_6.

68. Hamon, J.-R.; Astruc, D.; Michaud, P. *J. Am. Chem. Soc.*, *103*, 758 (1981).

69. Ruiz, J.; Pradet, C.; Varret, F.; Astruc, D. *Chem. Commun.*, 1108 (2002).

70. Bossard, C.; Rigaut, S.; Astruc, D.; Delville, M.-H. ; Félix, G. ; Février-Bouvier, A.; Amiell, J.; Flandrois, S.; Delhaès, P. *J. Chem. Soc., Chem. Commun.*, 333 (1993).

71. Rigaut, S.; Delville, M.-H.; Astruc, D. *J. Am. Chem. Soc.*, *119*, 1132 (1997).

72. Astruc, D. *Acc. Chem. Res.*, *33*, 287 (2000).

73. Astruc, D. In *Electron Transfer in Chemistry* (Ed.: Balzani, V.), Vol II (Matay, J.; Astruc, D. Vol. Eds), Wiley, Weinheim, pp 714-803 (2001).

74. Tokuaka, M. *Collect. Czech. Chem. Commun.*, *4*, 444 (1932); *6*, 339 (1934).

75. Buet, A.; Darchen, A.; Moinet, C. *J. Chem. Soc., Chem. Commun.*, 447 (1979).

76. Rigaut, S.; Delville, M.-H.; Losada, J.; Astruc, D. *Inorg. Chim. Acta*, *334*, 225 (2000), (issue dedicated to Andrew Wojcicki).

77. Alonso, E.; Astruc, D. *J. Am. Chem. Soc.*, *122*, 3222 (2000).

78. a) Bruce, M. I.; Kehoe, D. C.; Matisons, J. G.; Nicholson, B. K.; Rieger, P. H.; Williams, M. L. J. *J. Chem. Soc. Chem. Commun.*, 442 (1982); b) Bruce, M. I.; Mattisons, J. G. ; Nicholson, B. K. *J. Organomet. Chem.*, *247*, 321 (1983).

79. a) Reetz, M. T.; Lohmer, G.; Scwickardi, R. *Angew. Chem. Int. Ed. Engl.*, *36*, 1526 (1997); b) de Brabander-van den Berg, E. M. M.; Meijers, E. W. *Angew. Chem. Int. Ed. Engl.*, **32**, 1308 (1993).

80. First example of recognized ETC catalysis: Rich, R.; Taube, H. *J. Am. Chem. Soc.*, *76*, 2608 (1954).

81. Comprehensive review on ETC catalyzed reactions: Astruc, D. *Angew. Chem. Int. Ed. Engl.*, *27*, 643 (1988). See also ref. 82.

82. a) Astruc, D. *"Electron Transfer and Radical Processes in Transition-Metal Chemistry"*, VCH, New York (1995); b) ref. 80 a), chapter 6: Chain Reactions, pp. 413-478.

83. Beer, P. D.; Gale, P. A. *Angew. Chem. Int. Ed. Engl.*, *40*, 486 (2001).

84. Lehn, J.-M.; *Supramolecular Chemistry: Concepts and Perspectives* VCH, Weinheim (1995).

85. Miller, S. R.; Gustowski, D. A.; Chen, Z.-H.; Gokel, G. W.; Echegoyen, L.; Kaifer, A. E. *Anal. Chem.*, *60*, 2021 (1988).

86. Valério, C.; Alonso, E.; Ruiz, J. ; Blais, J.-C.; Astruc, D. *Angew. Chem. Int. Ed. Engl.*, *38*, 1747 (1999).
87. Ruiz, J.; Ruiz Medel, M.-J.; Daniel, M.-C.; Astruc, D. *Chem. Commun.*, 464 (2003).
88. Daniel, M.-C.; Ruiz, J.; Astruc, D. *Chem. Eur. J.*, *9*, 4371 (2003).
89. Daniel, M.-C.; Ruiz, J.; Astruc, D. *J. Am. Chem. Soc.*, *125*, 1150 (2003).
90. Labande, A.; Astruc, D. *Chem. Commun.*, 1007 (2000).
91. Labande, A.; Ruiz, J.; Astruc, D. *J. Am. Chem. Soc.*, *124*, 1782 (2002)
92. Daniel, M.-C.; Ruiz, J.; Nlate, S.; Blais, J.-C.; Astruc, D. *J. Am. Chem. Soc.*, *125*, 2617 (2003).

Chapter 10

Inorganic Supramolecular Architectures at Surfaces

Masa-aki Haga and Tomona Yutaka

Department of Applied Chemistry, Faculty of Science and Engineering, Chuo University
1-13-27 Kasuga, Bunkyo, Tokyo, 112-8551, Japan.

1   INTRODUCTION

It is well known that the size of integrated circuit chips has followed Moore's law of shrinking transistor dimensions since the 1960s [1]. But transistors cannot be scaled down infinitely; in recent years, the gate size is approaching the ten-nanometer scale, which is close to the size of a large molecule. Therefore, nanometer-sized molecular systems on a surface, as a step towards the realization of molecular electronic devices, are being widely explored in order to overcome the fabrication limit of silicon-based devices [2,3]. The challenge is to prepare new functional materials and to construct well-defined nanometer-sized structures on solid surfaces by self-assembly [4,5]. A two-dimensional (2D) and three-dimensional (3D) topological organization of molecules through metal coordination has led to the development of new metal-organic frameworks in crystal engineering [6,7]. The various frameworks have been constructed by the crystallization of molecular modular units from solutions [8]. On the other hand, molecular assembly at a solid surface makes supramolecular structures at the surface, which have been called self-assembled monolayers (SAM). It is well-known that organic thiols have been self-assembled on a gold surface, and the resulting monolayer films make it possible to control the surface functionality such as wettability, redox activity, photochemical response, etc. On the top of SAM monolayer, another layer can be grown by use of various interactions such as electrostatic interaction, hydrogen bonding interaction, and coordination bonding (Scheme 1).

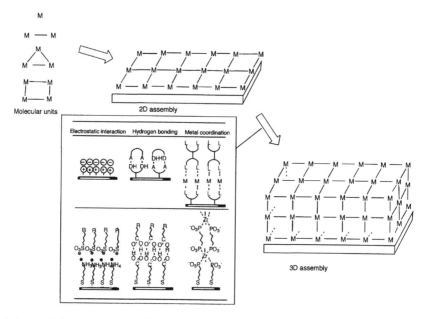

Scheme 1. Modular approach to inorganic supramolecular structure at the surface

In particular, the combination of self-assembly with self-organization by metal complexation has the potential to provide novel well-defined nanostructures and functions on surfaces. In this review, we focus on the fabrication of inorganic supramolecular complexes at the surface and its electrochemistry. Since the chemistry at the interface is an interdisciplinary field, we present here only the perspective of surface inorganic complexes on a molecular scale from the electrochemical viewpoint instead of a comprehensive SAM review. Useful review articles in the related fields have been written elsewhere [9-12].

2 TWO-DIMENSIONAL (2D) AND THREE-DIMENSIONAL (3D) SUPRAMOLECULAR ARCHITECTURES AT SOLID SURFACES

2.1 *Self-assembled Monolayer of Metal Complexes on Solid Surface*

When Allara *et al.* [13-15] found thiol groups attached to gold surfaces, the element of functionality was added to the self-assembled monolayer, and this finding opened potential applications for functional surfaces. For example, many metal complexes such as Ru(bpy)$_3$ [16], ferrocene [17], or porphyrin [18-20] can be self-assembled onto Au or ITO electrodes, and for these systems both the basic electron transfer kinetics and the photocurrent generation or sensing properties have been explored. Synthesis of functionalized thiol compounds allow one to tailor the composition and properties of self-assembled monolayers [21]. In order to synthesize such functionalized thiol compounds, both the auxiliary functional compounds and the connecting organic groups, each of which can be modified independently, have been carefully selected. In addition to the thiol group being self-assembled to Au surface, other organic groups such as carboxylate, phosphonate, and isocyanide [22] are also recognized to act as surface immobilized groups. In particular, each anchoring group shows selectivity for solid surfaces. It has been reported that thiol and disulfide groups bind preferentially to gold substrate [23]; isocyanide and pyridine groups show a preference for the platinum surface; and phosphonate and silanol groups bind preferentially to indium-doped tin oxide and metal oxide surfaces [24]. Thiolate and phosphonate monolayer films are stable in aqueous solution over the pH range 1-10 [25,26]. Table 1 shows typical organic groups immobilized on the solid surface.

Table 1. Surface Selectivity for Various Organic Groups

Organic group	Surface Selectivity	
	Adsorbed	Not Adsorbed
Thiol (R-SH)and Disulfide (R-S-S-R)	Au, Cu, Pt, Hg	Si, Si_3N_4
Isocyanide (R-NC)	Au, Pt	Si, GaAs
Carboxylate (R-CO_2^-)	ITO, Metal oxide,GaAs	
Phosphonate (R-P(O)(OH)$_2$)	ITO, Metal oxide,	Au

Chemical aspects of forming self-assembled structures on surfaces, including molecular orientation and growth mechanisms, is important in order to form uniform films. For alkanethiolate monolayers on Au(111), the self-assembled monolayer structure is commensurate with the underlying gold lattice and is a simple $\sqrt{3} \times \sqrt{3}$ R 30°overlayer. This close-packing and high ordering of alkanethiolate on Au(111) may result from the relatively easy 2D recrystallization process by annealing, as well as from the migration of gold thiolate molecules [10]. Similarly, the growth mechanism of long alkyl chain phoshonic acid on mica or sapphire has been investigated by Schwartz *et al.* [27,28]. They reported that a continuous 2D phase of disordered molecules is initially forms, and that it later evolves into a thicker, more ordered film via the gradual growth of higher structures at room temperature. At lower temperatures, the monolayer forms by nucleation and growth of ligands in which the molecules are close-packed and vertically orientated. The parameters of microscopic film structure and uniformity are important for the control of direct chemical attachments of functional molecules to the surface.

In order to attach and orient molecules between pre-designed dissimilar solid surfaces, selective self-assembly plays an important role in achieving a directional arrangement of molecules. This surface selective modification is referred to as the Wrighton "orthogonal self-assembly method [29]. In order to carry this out, the selection of the immobilized groups plays a key role. The electrochemical method is readily available for the determination of the selectivity of immobilization on dissimilar substrates when an electrochemical active molecular unit is used as an auxiliary component. For example, the oxidation potentials of ferrocene are changed by the substituent on the cyclopentadienyl group; the half-wave potential for alkylferrocenyl is centered at a position about 300 mV negative to that for the acetylferrocenyl center. This potential difference allows easy electrochemical measurements of their surface concentrations even when both species are present on the surface [30].

The shift in potential of redox species, accompanied by proton-coupled electron transfer, is also employed as an indicator for probing the orthogonal self-assembly.

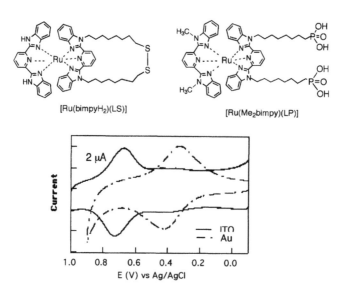

Figure 1. Structures of two Ru complexes with different anchoring groups, and cyclic voltammograms on ITO and Au electrodes after two-hour immersion of the mixed solution of two complexes.

Two Ru complexes, as shown in Figure 1, were prepared with different auxiliary ligands (bimpyH$_2$ and Me$_2$bimpy) and anchoring ligands (LS and LP). Of the two only [Ru(bimpyH$_2$)(LS)] shows a Ru(II/III) oxidation potential with a clear dependence on solution pH. Orthogonal self-assembly can be examined using the simultaneous immersion of Au and ITO electrodes into CH$_3$CN/methanol solution of equimolar amounts of [Ru(bimpyH$_2$)(LS)] and [Ru(Me$_2$bimpy)(LP)] for 2 hours. For the Au electrode, only one oxidation wave was observed at E$_{1/2}$ = +0.41 V, which corresponds to that for [Ru(bimpyH$_2$)(LS)]. On the other hand, for the ITO electrode, a large oxidation wave at +0.71 V and a very small wave at +0.41 V vs Ag/AgCl were observed (Figure 1), which correspond to those of [Ru(Me$_2$bimpy)(LP)] and [Ru(bimpyH$_2$)(LS)], respectively. The relative magnitude of the waves is 0.96 : 0.04. Therefore, [Ru(bimpyH$_2$)(LS)] on the electrode is easily differentiated from [Ru(Me$_2$bimpy)(LP)] by the use of the potential shift created by proton-coupled electron transfer reactions [25].

The extension of this orthogonal self-assembly method will make it possible to select molecular positioning and alignment on pre-existing template terminals [31]. For example, orthogonal self-assembly has been successfully used for the connection of a crossbar between nanorods [32].

Another important factor for the self-assembled monolayer is how to control molecular orientation on the surface. For simple alkanethiol self-assembled monolayers on the Au(111) surface, the closed-packed alkyl chains of thiolates are usually tilted ~26-28° from the surface normal. Therefore, if the molecular packing is low, the molecule lies flat on the surface. Furthermore, simple alkyl chains are flexible, and it is therefore difficult to control their molecular orientation and the distance between the auxiliary metal complex (as the head

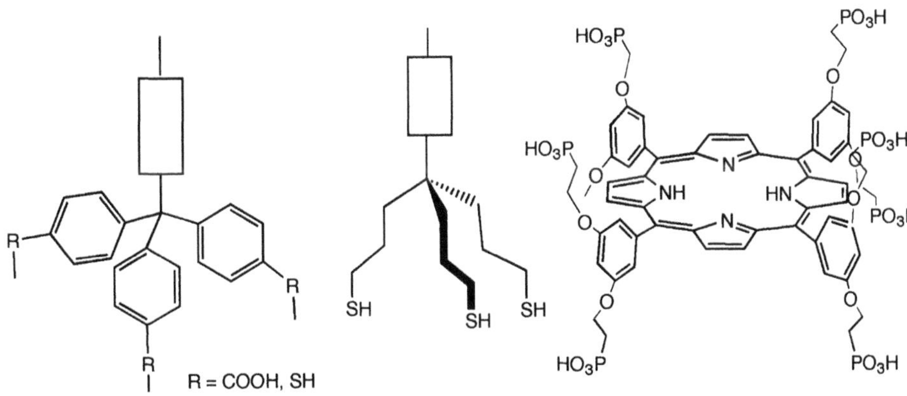

Scheme 2. Examples of tripod and tetrapod free-standing anchor ligands.

group) and the electrode. In order to control molecular orientation, molecular design of both multi-point anchoring compounds and rigid linkers were developed [12,33-36]. Scheme 2 shows some examples of free-standing groups able to anchor to the surface.

The rigid tripod –COOH or –SH surface binding groups having a tetrahedral core such as tetraphenylmethane or 1,3,5,7-tetraphenyladamantane, and as such provide a stable, three-point attachment to the solid surface. This type of free-standing anchoring ligand plays an important role in the interfacial electron transfer reaction at the surface because of the fixed distance between the redox-active end group and electrode. Novel tetrapod anchoring ligands (XP) based on 2,6-bis(benzimidazol-2-yl)pyridine with four phosphonate groups and its metal complexes have been synthesized (see Scheme 5 for the structure of XP ligand). The X-ray structure of a relevant Ru complex revealed that two mesityl groups are perpendicular to the 2,6-bis(benzimidazol-2-yl)pyridine plane and the rotation of methylene group attached to the phosphonate is sterically hindered. A molecular mechanics calculation indicates that the four legs of phosphonate are positioned at the metal-oxide surface, such as tin-doped indium oxide (ITO), and that the auxiliary rigid rod arm is oriented perpendicular to the surface normal [37].

Furthermore, the effect of interconnected molecular linkers between the redox-active molecular unit and electrodes has been investigated from the viewpoint of long-distance electron-transfer reactions involving a single-electron tunneling process. In order to study this effect, ferrocene oligophenylenevinylene methyl thiols have been synthesized and the detailed study for electron transfer rate has been examined [17]. The decay constant, referred to as β, is ~0.06 Å^{-1}, and the ET rate constants are not limited by electronic coupling for bridges up to 2.8 nm long for these oligophenylenevinylene bridges.

2.2 Synthetic Strategy of Surface Coordination Compounds at Interface

Electrochemistry is, of course, the chemistry of the interface between the solution and the electrode, and therefore modification of the electrode surface is now a standard practice. In particular, electrochemical polymerization of redox-active monomers such as $[Ru(phen)_2(vbpy)]^{2+}$ on an electrode have been thoroughly studied by Murray and other researchers [38]. Bilayer and multilayer structures based on the different combination of redox-active Ru/Os complex polymers were also constructed, and rectification and electron hopping within these redox-active polymers have been studied [39]. The film thickness of polymers was roughly submicron in magnitude but remained difficult to control. On the other hand, self-assembled monolayer of functional molecules makes it possible to construct nanometer-scale ultrathin films in a controlled manner. The "bottom-up" approach by use of the self-assembling method allows for the integration of molecular units onto solid surfaces controlled on the nanometer scale. The integration of metal complex units in solution (the "complexes-as-ligands" and/or "complexes-as-metals" strategy) has been developed as a general synthetic method of obtaining polynuclear rod-shaped or dendritic metal complexes [40]. In this method, a building block of metal complex can bind to another metal ion to form a complex of larger size, and the resulting complex is then available to act as the next building block. By continuing this process, nanometer-size polynuclear complexes have been formed. This synthetic strategy can be extended to the solid surface in order to build up the nanometer-scale multilayers or specific multicomponent molecular architectures as shown in Scheme 3.

For this "bottom-up" synthesis on solid surface, the kind of surfaces, anchored groups, functional molecular units and coordinated linker groups for binding the next layer have a fundamental significance. Alkylsilanol, thiol, and

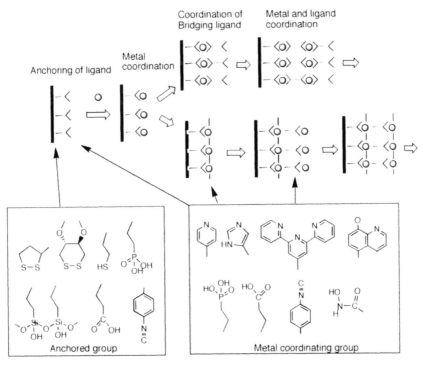

Scheme 3. Layer-by-layer growth by surface coordination chemistry

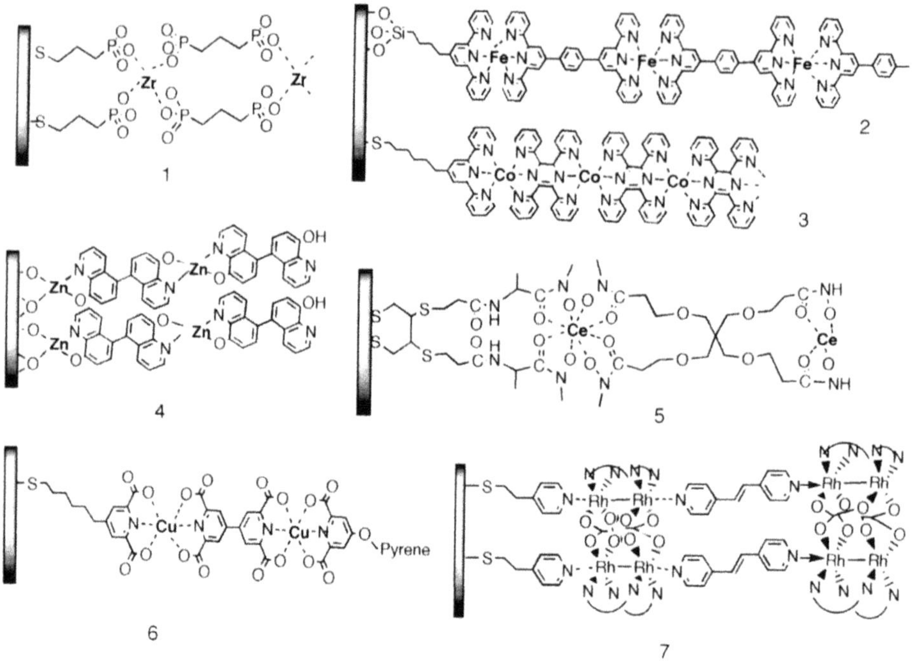

Scheme 4. Examples of layer-by-layer strucrures fabricated by surface coordination

disulfide are generally used as the anchored group to a solid surface such as quartz or Au (see Section 2.1); on the other hand bipyridine, terpyridine, and isocyanide are applied as the metal ion binding agent. Reported examples of layer-by-layer inorganic supramolecular structures are collected in Scheme 4. As an example of this method, Schmehl et al. have synthesized polymetallic complexes on quartz by alternate deposition of Fe(II) complex unit with the dinucleating 1,4-bis(terpyridyl)benzene ligand (btpb) [41]. Similarly, Abruna et al. have demonstrated that a newly synthesized terpyridinethiol ligand allows for the preparation of redox-active mono- and multi-metallic systems capable of sequential self-assembly onto gold surfaces [42]. These studies show the surface coordination ability of anchoring tpy ligand on the surface, which can bind metal ion in the octahedral geometry without any geometrical isomers [43,44].

Furthermore, using two dihydroxamate ligands and 8-coordinating metal ions such as Zr^{4+}, Ce^{4+}, and Ti^{4+}, a new kind of multilayer based on metal-ion coordination was successively built up in a highly controlled step-by-step manner [23]. This methodology is applicable to a variety of C_2-symmetric ligands, possessing various bifunctional coordination groups in combination

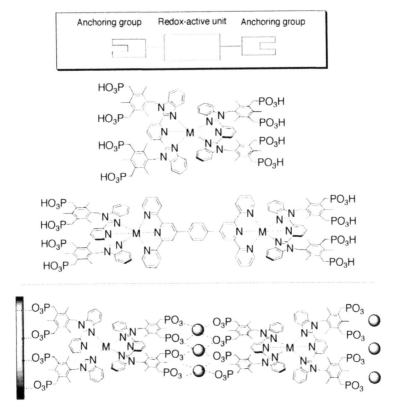

Scheme 5. Novel metal complex units, $[M(XP)_2]$ and $[M_2(XP)_2(btpb)]$ (M = Ru or Os), for the formation and structuring of multilayers incorporating the Zr^{4+} ion.

with a variety of metal ions. The tpy ligands used by Abruna and Schmehl also have C_2 symmetry.

The metal complex itself can also act as a molecular unit for the construction of multilayers. Redox-active Ru/Os complexes with novel tridentate ligands and containing two anchoring groups such as the phosphonate shown in Scheme 5, have been synthesized [45,46]. Ideally these ligands have a C_{2v} symmetry, and the bis-tridentate coordination environment at the metal center forces the anchoring groups to orient themselves in opposing directions. Therefore, the layer-by-layer growing method can be applied using the Ru/Os complexes with phosphonate groups. As expected, the metal complex-based multilayers were constructed by the combination of redox-active $M(L)_2$ and $M_2(L)_2(btpyb)$ (M = Ru(II) or Os(II), L = LP or XP) complexes and the Zr^{4+} or the Cu^{2+} ion. The multilayer formation between phosphonate and metal ions such as Zr^{4+} or

Hf^{4+} was originally described by Mallouk et al. [47,48] on a variety of surfaces, carried out via the sequential adsorption of metal ion and bis(phosphonic acids) from aqueous solution [49].

One of the advantages of layer-by-layer growth in multilayers is that a combinatorial approach is feasible by use of molecular modular units. Since the tuning of the oxidation potential of Ru and Os can be carried out by the change of ligands and central metal ion, modification of the ordering of Ru/Os layers at the surface will lead to different potential distributions or gradients in a controlled manner and on the nanometer-scale. By changing the order of molecular units with different redox potentials on the surface, any potential sequence toward a molecular rectifier or an electron-transfer cascade system can be constructed.

Diisocyanobenzene [50] and 8,8'-dihydroxy-5,5'-bisquinoline [51] have also been used to grow a coordination assembly on the solid surface via a layer-by-layer adsorption (Scheme 4 and 6).

Recently, the multilayers prepared from Cu^{2+} ion with mercaptoalkanoic acid on Au and oxidized Si substrates by metal-organic coordination were used as a molecular ruler for size-controlled resists on predetermined patterns, such as those formed by electron-beam lithography [52].

In the case of the close packed self-assembled monolayer, strong intramolecular van der Waals interactions between the anchored ligands (such as terpyridine alkylthiol) prevent metal coordination because no space is available for accommodating bulky metal complexes. Therefore, the dilution of the anchored ligand with a simple alkyl thiol with a shorter carbon chain is often employed in order to create the free-space required for coordination around the anchored ligand [17,53].

In each of the methods we have discussed so far, the molecular architecture was constructed through metal coordination in the 1D direction from the surface normal. For the formation of 2D lamellar architecture, Mallouk et al. [54] used lamellar metal cyanide networks such as Hofmann clathrates, which has

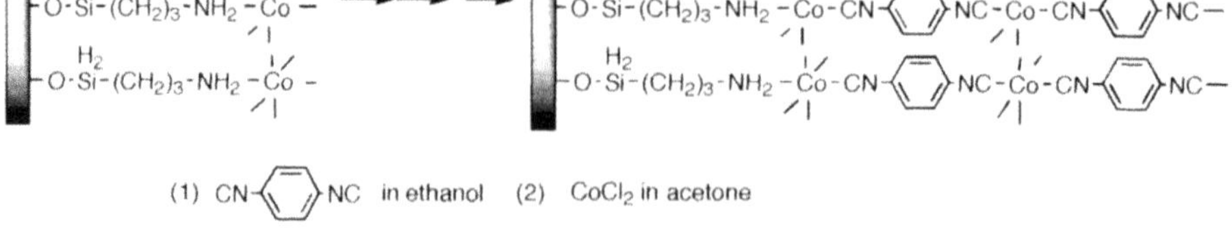

Scheme 6. Layer-by-layer fabrication by complex formation of Co^{2+} and diisocyano-benzene.

a 2D extended metal-ligand layer structure parallel to the surface. A priming monolayer, which was formed by the self-assembly of 4-mercaptopyridine on gold, is used to create a pyridine-rich surface suitable for coordination of metal ions. The planar Pt(CN)$_4$ net of the Hofmann clathrate is then prepared by sequentially adsorbing Ni^{2+} and then Pt(CN)$_4$ ion onto this pyridine SAM surface. The next adsorption step, involving a bridging 4,4'-bipyridine, fills the sixth coordination site of Ni^{2+} and leaves an anchoring ligand for the next planar network. However, the choices of metal ions are relatively limited because of the balance of the ligand substitution kinetics between the surface coordination and removal from the coordination sphere [54].

2.3   Monitoring of Layer Growth and Surface Structures by Physical Measurements.

Many analytical tools can be used to monitor the layer growth on the solid surface [9]. Conventional UV and IR spectra can be used when the molecule under study has a relatively large absorption coefficient. For UV spectra, quartz and indium-tin oxide(ITO) surfaces are most commonly used. ITO glass is made by the chemical vapor deposition of ITO onto the glass surface. The thickness of ITO films affects the quality of UV spectra because of light interference; thicknesses above 300 nm on glass results in a negative absorbance around ~400 nm upon growing of the multilayer films. A typical UV spectral change of Os dinuclear complex, [Os$_2$(XP)$_2$(btpyb)], is given in Scheme 5, accompanied by the layer-by-layer growth as shown in Figure 2. A linear dependence of absorbance at 505 nm vs the number of layers was observed. If the surface sensitive measurements such as ATR or RAS are available, the information of molecular orientation on the surface can be obtained. Ellipsometry and x-ray reflectometry measurements are also used.

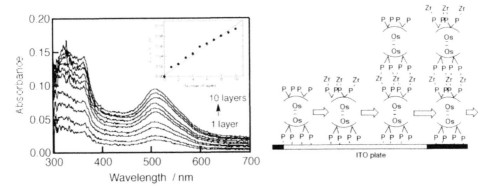

Figure. 2. UV-vis spectral change of Os dinuclear complex, [Os$_2$(XP)$_2$(btpb)], on ITO substrate accompanied by 10 layers growing and its schematic.

As a different approach, piezoelectric quartz crystal microbalance(QCM) measurements can monitor the layer growth as a function of the increase in mass. Electrochemical methods such as cyclic voltammetry are generally used to check for pinholes or defects within self-assembled monolayer; therefore electrochemical studies for multilayers are scarce [55-58]. When redox-active molecular units are incorporated into the multilayer, the redox-activity of the resulting multilayer depends on the electron-hopping or the electron-exchange rate of molecular units and the linker groups. For example, layered films of the Ru complex [Ru(XP)$_2$] reveal a Ru(II/III) oxidation process, even in the multilayer films. The anodic peak current increases linearly with increasing the number of layers within for least 10 layers. However, when the linker group is insulating towards electron transfer, it becomes difficult to monitor the layer growth because the electron-transfer rates decrease exponentially as a function of the distance from the electrode. Conversely, good conducting linkers demonstrate the molecular wire property, which makes it possible to probe the formation of multilayer growth by use of electrochemical techniques.

XPS spectroscopy is another powerful technique for the surface analysis. Since the intensity of XPS signal, I, depends on the collecting angle θ as:

$$I \sim \exp(-\,d\,/\,L\cos\theta)$$

where d is the depth from the sample surface and L the attenuation length of photoelectrons, a segmented multilayer structure can be elucidated from the relative atomic ratio with different takeoff angles. For example, two types of segmented multilayers of Ce/Zr-tetrahydoxamate complex (compound 5 in Scheme 4) were formed from the base layer; i.e., one is a multilayer with three Ce^{4+}-ion-complexed layers followed by three Zr^{4+}-ion-complexed layers (surface$\|$(Zr^{4+}-L)$_3$/(Ce^{4+}-L)$_3$) and the other is a multilayer with three Zr^{4+}-ion-complexed layers followed by three Ce^{4+}-ion-complexed layers (surface$\|$(Ce^{4+}-L)$_3$/(Zr^{4+}-L)$_3$). The construction of these segmented multilayers was proven by the angular dependence of signal intensity ratios of Ce/Zr; i.e., the ratio for the former structure at 0° is 3.13 to 1 while that for the latter is 1 to 3.34 [23]. The surface morphology depends not only on the molecular shape but also on the charge of the molecular component. It has been reported that bisphosphonic acids with cationic organic groups do not follow the typical layered growth motif, but instead form cystallites on the substate surface [59].

These measurements described above give us molecular information about the average surface over a relatively large area. Recent advances in surface analytical techniques such as scanning probe microscopy (scanning tunneling microscope (STM) and atomic force microscope (AFM), etc.) allow us to obtain the detailed images of the surface on an atomic or molecular scale. Even

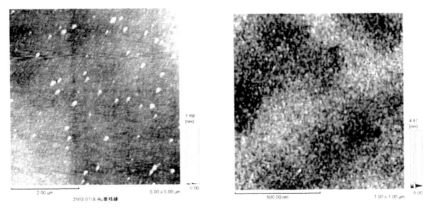

Figure 3. AFM images of [Ru(XP)$_2$] on flat ITO substrate: (a) Immobilized from highly diluted solution (2 × 10^{-7} M) of [Ru(XP)$_2$]. White spots correspond to a domain of complex. (b) Immobilized from 2 × 10^{-6} M solution for 1 hour.

though regular layer growth has been observed from spectral measurements, relatively rough surface images were often obtainable from AFM measurements. Furthermore, domain formation was sometimes observed. One of the typical AFM images for mononuclear Ru complex, [Ru(XP)$_2$], is shown in Figure 3a, in which the immobilized mononuclear Ru complex on a flat ITO surface formed a small domain structure with 50-nm size and ~1.6-nm height, which is consistent with the molecular height obtained from molecular modeling. The number of domains increased with time, and finally a closely packed monolayer was obtained (Figure 3b).

Similar domain formation has been reported for surface coordination on solid surfaces [23,59]. Sometimes the depth or height information of multilayer films cannot be obtained from AFM measurements in the absence of defects. Molecular rulers or artificial crests with defined height can be used for this purpose. While the AFM measurement is a powerful tool for obtaining information about the outermost layer, the depth information for multilayer films is difficult to acquire without scratching the film surface. For surface sensitive physical measurements many fine references are available [9,21,60,61].

3   ELECTROCHEMICAL FUNCTIONS OF INORGANIC SUPRAMOLECULAR ARCHITECTURES ON SURFACES

Inorganic supramolecular multilayered materials based on surface coordination provide applications for a number of different fields such as optics, biotechnology, photo- and electrochemistry; i.e., nonlinear optical materials, electroluminescent devices, enzyme sensors, solar energy storage devices, and molecular rectifiers. Here, we focus on the electrochemical aspects of molecular devices such as molecular switches, photocurrent generation, chemical sensors, and electrochromic devices.

3.1   Molecular Switches

When chemical properties of a molecule are changed by an external stimulus such as electron transfer, proton transfer, or photoirradiation, this molecule becomes a candidate for application as a molecular switch [62,63]. For example, proton-induced molecular switching of Ru/Os complexes containing (2-benzimidazolyl)pyridine derivatives is one system that has been thoroughly studied [64,65]. Scheme 7 shows a typical square scheme of [Ru(bpy)$_2$ (pbimH)]$^{2+}$ (bpy = 2,2'-bipyridien, pbimH = (2-pyridyl)benzimidazole), in which the pKa value of [Ru(bpy)$_2$(pbimH)]$^{2+}$ is shifted from 6.62 to 0.90 by the oxidation of Ru(II) to Ru(III). The oxidation potential of [Ru(bpy)$_2$(pbimH)]$^{2+}$ is changed from +0.86 V to +0.41 V upon deprotonation. Therefore, deprotonation of the benzimidazole N-H groups induces a large energy perturbation in metal complexes. By using this change, the degree of metal-metal interaction in dinuclear Ru/Os complexes can be tuned by protonation/deprotonation from bridging ligand [26,66].

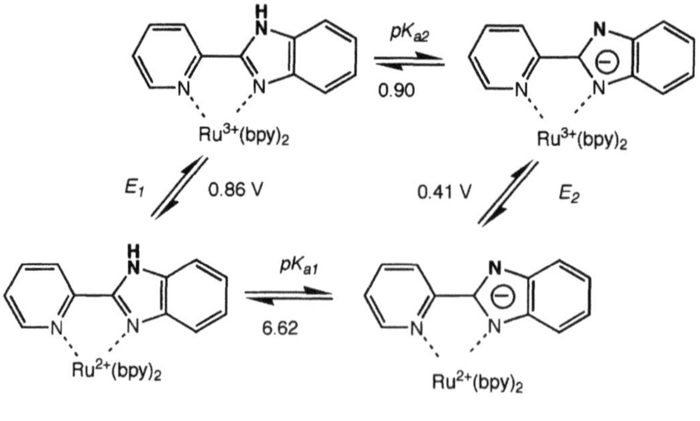

Scheme 7. Electrochemical square scheme of [Ru(bpy)$_2$(pbimH)]$^{2+}$

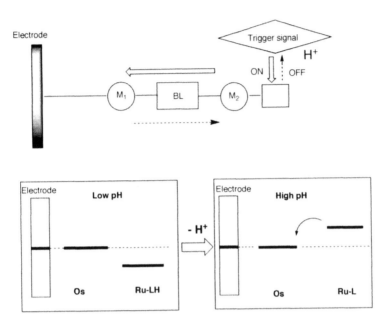

Scheme 8. Schematic of a proton-induced molecular switch and the energetics upon deprotonation.

Furthermore, proton-induced intramolecular electron transfer has been observed in mixed-valence RuOs complexes [67]. This solution chemistry for the proton-induced molecular switch can be extended to that on the surface. When the complex with two redox sites is immobilized on a solid surface and only the outer site can respond to external perturbation such as proton dissociation or photoexcitation (as shown in Scheme 8), this surface can acquire molecular switching properties. When the orbital energy is changed by the external perturbation, the direction of electron transfer can be switched.

Recently, new dinuclear complexes $[Os(bimpy-X)(btpb)Ru(bimpyH_2)]^{n+}$ with an anchoring group ($X = S$ or PO_3H_2) were self-assembled on gold or ITO electrodes. Specifically, it was learned from the study of the solution chemistry of the mixed-valence dinuclear complex, $[Os(II)(bimpy-CH_3)(btpb)Ru(III)(bimpy)]^{2+}$, that the intramolecular electron transfer from Os(II) to Ru(III) site can be induced by protonation of the imino group on the bimpy ligand to form the protonated mixed-valence complex, $[Os(III)(bimpy-CH_3)(btpb)Ru(II)(bimpyH_2)]^{2+}$ (Scheme 9). This intramolecular electron transfer phenomena can be extended to the SAM system on a solid surface, and the molecular-scale proton-gated rectifying RuOs complexes, as shown in Scheme 10, were constructed [26].

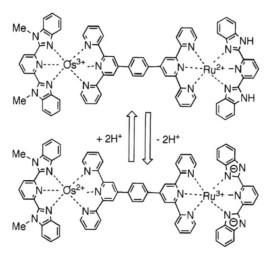

Scheme 9. Proton-induced intramolecular electron transfer in RuOs dinuclear complex containing 2,6-bis(benzimidazol-2yl)pyridine

At pH 2.86, two oxidation waves of OsRu on ITO electrodes are observed at 0.64 V (vs Ag/AgCl) for the Os(III/II) process and 0.85 V for the Ru(III/II) process. At higher pHs, the Ru oxidation potential undergoes a negative shift

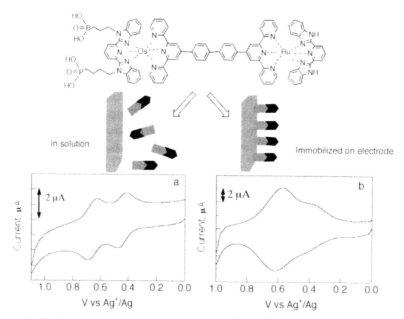

Figure. 4. Cyclic voltammograms of asymmetric RuOs dinuclear complex at pH 9.3: (a) in CH$_3$CN/buffer solution at a glassy carbon electrode, (b) immobilized on ITO electrode.

while the potential for Os site remains almost constant. Above pH 4, the two waves are coalesced, and only one wave is observed. Thus, the first oxidation site is switched from the Os to the Ru center above pH 5.5. Figure 4 shows a clear difference of cyclic voltammograms between the solution and immobilized systems. The immobolization of asymmetric RuOs dinuclear complexes produces the vectorial molecular arrangement, which leads to the rectification effect. Some pioneering works for proton-gated devices based on metal complex polymer and organic compounds such as the quinone-methyl viologen polymeric system have been reported [68,69], however the real molecular scale devices are still rare.

3.2 *Photovoltaic Devices.*

The molecular design of artificial photosynthetic molecular systems for solar energy conversion has attracted our attention for many years.   The aim is to mimic the elaborate molecular system for light harvesting, photoinduced charge separation, and catalytic multielectron oxidation or reduction of substrates such as water or CO_2. In particular, many supramolecular systems that mimic the stepwise nature of photosysnthetic charge separation have been reported on solid supports [70-74]. For example, the dye-sensitized photoelectrochemical cell reported by Gratzel et al. [75] employs Ru complexes anchored to mesoporous nanocrystalline TiO_2 surface together with suitable redox electrolytes or amorphous organic hole conductors. The conversion efficiency is governed

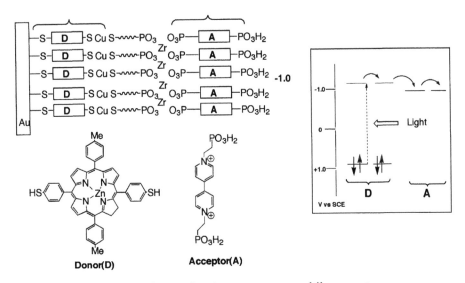

Scheme 10. Photoactive donor-acceptor multilayer systems

by the product of the efficiency for each process such as light harvesting, photoinduced charge separation, and charge injection.

Organization of the redox components on solid supports has resulted in high quantum yields and longer lifetimes of the photoinduced charge-separated states relative to those in solution [76]. Using self-assembly methods, it was possible to organize molecular redox components in a stable lamellar structure at various solid supports [24,77]. Electron and hole can move through a gradient of chemical potential in a molecular array; therefore a judicious molecular arrangement is selected by considering the redox potentials and optical energy gaps of each molecular units. It has been reported that gold electrodes modified with Zr^{4+} bisphosphonate multilayers of viologen on top of copper dithiolate multilayers of porphyrin derivatives were photoelectroactive and produced efficient and stable photocurrents using visible light. In this system, the porphyrins were used as the electron donor layers and viologen as the electron acceptor layers, and the whole was organized in an energy-cascade arrangement with the aid of Zr phosphonate (Scheme 10) [59,78].

Taking yet another example, novel multilayer films containing Cu(II), 4,4'-bipyridyl-2,2',6,6'-tetracarboxylic acid, and capped with the pyrene-containing ligand (see Scheme 4, compound 6) were reported. Photoexcitation of this film causes the generation of a cathodic photocurrent [79].

3.3   Chemical Sensors.

The assembly of molecular sensing components into supramolecular architectures on solid surfaces promises to yield nanometer-size sensor devices. The principle of detection for molecular sensors is based on the requirement that structure or electronic states be affected by the presence of specific substrates. Some of the sensors can use a field-effect scheme, in which numerous adsorbed molecules on the gate induce a potential change in the channel region [80]. For example, a novel NO biosensor based on the surface derivatization of GaAs by hinged iron porphyrins has been reported [81]. The GaAs surface is derivatized by two-component iron(III) tetraphenylporphyrin chloride (FeTPPCl) moieties "hinged" remotely from the surface by bifunctional ligands (Scheme 11). These ligands bind axially, through imidazolyl groups, to the fifth and sixth coordination sites of the Fe(III) porphyrin, and simultaneously to the GaAs surface of the molecular controlled semiconductor resistor (MOCSER) through anchoring groups-disulfides or carboxylic acids. Changes in the MOCSER's current as NO binds to the outer Fe-porphyrin component are monitored to follow the binding events *in situ*.

For immobilized redox-active complexes on the solid surface, ion movement is accompanied by redox reactions. By using these phenomena, new ion-gated

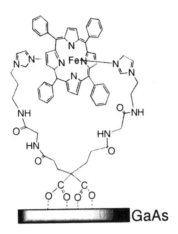

Scheme 11. NO biosensor based on the surface derivatization of GaAs

electron transfer or ion penetration into self-assembled layers can be controlled and subsequently used as a ion sensor [82,83].

Immobilization of single-stranded DNA on an electrode covered with an aluminum alkanebisphosphonate film, obtained by immersion into a solution of single-stranded DNA, was demonstrated for single-stranded DNA labeled with $[Ru(bpy)_3]^{2+}$. This was detected by observing the electrogenerated chemiluminescence (ECL) produced upon oxidation in a solution containing tri-n-propylamine. Hybridization of unlabeled DNA could be detected by treatment with $[Ru(phen)_3]^{2+}$ and by then measuring the ECL of this species associated with the double-stranded DNA. In this manner, the extent of DNA hybridization or the amount of target single stranded DNA that can be immobilized on the electrode is detectable by observing the ECL intensity [84].

3.4 Electrochromic Devices.

An electroactive molecule often exhibits new optical absorption bands during a reaction in which it either gains or loses an electron; this phenomenon is termed 'electrochromism'. In terms of applications of this effect, a wide range of inorganic, organic, and polymeric materials have been investigated for potential use as shutters and displays. Some metal complexes show strong metal-to-ligand charge transfer (MLCT) or ligand-to-metal charge transfer (LMCT) bands in the visible region. These bands are often shifted or bleached upon redox reaction; therefore these complexes are promising candidates for electrochromic devices. Novel dinuclear cyano-bridged mixed valence compounds such as $[Ru^{II}(py)_4(CN)]\text{-}CN\text{-}Ru^{III}(NH_3)_4pyCOOH$ chemisorbed onto transparent

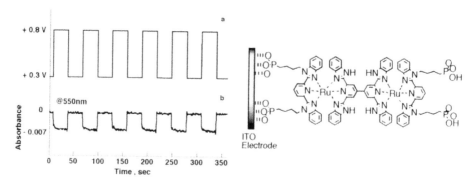

Figure 5. Electrochromic behavior of Ru dinuclear complex immobilized on ITO electrode; (a) the sequence for applied potential and (b) the corresponding absorbance change at 550 nm.

nanocrystalline SnO_2 or TiO_2 films on conductive glass exhibit electrochromic behavior [85]. Two colored states of these complexes can be interconverted in a narrow potential range (-0.5 to +0.5 V vs SCE) with a high stability. The switching times between the two limiting colors is on the order of milliseconds.

Similarly, the self-assembled monolayer or multilayers of dinuclear Ru complex on ITO electrode shows a electrochromic response at 550 nm in sandwich type cells. The performance of this cell is shown in Figure 5. [46]

4   FUTURE PERSPECTIVES FOR SURFACE COORDINATION CHEMISTRY AS APPLIED TO NANOMOLECULAR SCIENCE

The construction of well-defined nanometer-scale supramolecular structures on surfaces is of great importance to the fabrication of molecular devices. One of the potential fabrication methods is the use of surface coordination chemistry as described in the previous Sections. The combination of self-assembly at the surface with subsequent metal coordination (Scheme 12) provides the requisite flexibility to electronically "hook up"active molecules between nanometer-scale contacts [2,31,86,87].

At first, molecules can be hooked up at the designated positions on the surface of the substrate by choosing the appropriate anchoring groups, and then other functional molecular components with ligating sites can be connected through metal coordination. For example, the layer-by-layer growth of Rh-Rh bonded dinuclear complexes onto electrodes was applied for the connection of two gold terminals separated by 60 ~ 80 nm [57]. This type of new molecular device will offer an intriguing potential alternative for electronic applications.

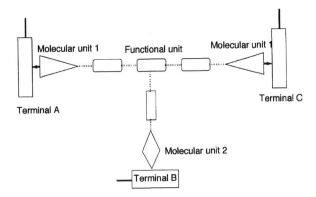

Scheme 12. Schematic for three terminal connection for molecular devices

During testing as a potential molecular capacitor for next-generation information storage media, it was demonstrated that the porphyrinic molecules attached to the electroactive surfaces in Scheme 13 can store charge for significant periods of time upon disconnection from the source of applied potential. This solid-surface-based molecular system will now be adapted to satisfy the prerequisite conditions for use as a dynamic random access memories [88].

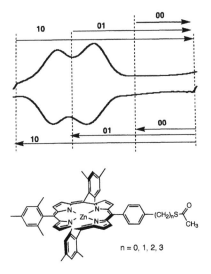

Scheme 13. Schematic of molecular information storage device.

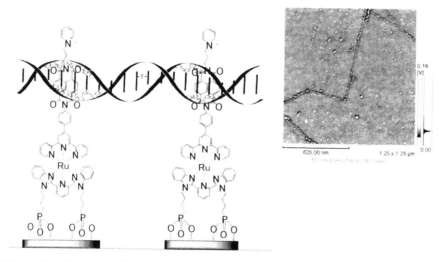

Figure 6. Schematic of DNA nanowiring between two rod-shaped Ru complexes and AFM image for DNA trapping on mica surface (The central bending line is DNA hooked up the immobilized Ru complex).

The host/guest chemistry of inorganic porous network materials on surfaces is another field that will benefit from the application of surface coordination chemistry [89]. For example, electrochemical size-selective guest transport has been observed on glassy carbon or gold electrodes covered by thin films comprised of neutral molecular squares of the form $[Re(CO)_3(Cl)(\mu\text{-}L)]_4$ (L = difunctional imine or azine ligand) [90]. This is one example of how porous or network-type materials on solid surfaces might open up new applications such as chemical sensing devices and catalysis [90].

Biomolecules such as DNA are attractive materials for use as a wire scaffold with predetermined length. For example, a double-stranded DNA was captured by an immobilized Ru complex with DNA intercalating abilities in a point-to-point manner on a mica surface (Figure 6) [37]. This captured DNA acts as a metallization template for DNA-based nanowires.

Controlled assembly by surface coordination chemistry will open new avenues in the fabrication of inorganic supramolecular systems on surfaces and will contribute to future materials science with wide-ranging applications such as sensors, photochemical energy conversion, and molecular devices.

5   ACKNOWLEDGMENTS

We sincerely thank the co-workers who have contributed to this research and whose names are listed in the references. M.H. gratefully acknowledges financial support from the Institute of Science and Engineering at Chuo University and the Ministry of Education, Science, Sports and Culture for a Grant-in-Aid for Scientific Research (No. 15310076).

6   REFERENCES

1. Lundstrom, M. *Science*, *299*, 210 (2003).
2. Wada, Y.; Tsukada, M.; Fujihira, M.; Matsushige, K.; Ogawa, T.; Haga, M.-a.; Tanaka, S. *Jpn. J. Appl. Phys.*, *39*, 3835 (2000).
3. Carroll, L.; Gorman, C. B. *Angew. Chem. Int. Ed.*, *41*, 4378 (2002).
4. Joachim, C.; Gimzewski, J. K.; Aviram, A. *Nature*, *408*, 541 (2000).
5. Jortner, J.; Ratner, M. *Chemistry for the 21st Century "Molecular Electronics"*; Blackwell Science: Oxford (1997).
6. Moulton, B.; Zaworotko, M. J. *Chem. Rev.*, *101*, 1629 (2001).
7. Fujita, M.; Umemoto, K.; Yoshizawa, M.; Fujita, N.; Kurukawa, T.; Biradha, K. *Chem. Commun.*, 508 (2001).
8. Michl, J., Ed. *Modular Chemistry*; Kluwer Academic Publishers: Dordrecht, Vol. 499 (1997).
9. Ulman, A. *An Introduction to Ultrathin Organic Films from Langmuir-Blodgett to Self-Assembly*; Academic Press, Inc.: San Diego (1991).
10. Ulman, A. *Chem. Rev.*, *96*, 1533 (1996).
11. Kuzmenko, I.; Rapaport, H.; Kjaer, K.; Als-Nielsen, J.; Weissbuch, I.; Lahav, M.; Leiserowitz, L. *Chem. Rev.*, *101*, 1659 (2001).
12. Li, G.; Fudickar, W.; Skupin, M.; Klyszcz, A.; Draeger, C.; Lauer, M.; Fuhrhop, J.-H. *Angew. Chem. Int. Ed.*, *2002*, 1828 (2002).
13. Nuzzo, R. G.; Allara, D. L. *J. Am. Chem. Soc.*, *105*, 4481 (1983).
14. Nuzzo, R. G.; Fusco, F. A.; Allara, D. L. *J. Am. Chem. Soc.*, *109*, 2358 (1987).
15. Bain, C. D.; Troughton, E. B.; Tao, Y.-T.; Evall, J.; Whitesides, G. M.; Nuzzo, R. G. *J. Am. Chem. Soc.*, *111*, 321 (1989).
16. Sato, Y.; Uosaki, K. *J. Electroanal. Chem.*, *384*, 57 (1995).
17. Sikes, H. D.; Smalley, J. F.; Dudek, S. P.; Cook, A. R.; Newton, M. D.; Chidsey, C. E. D.; Feldberg, S. W. *Science*, *291*, 1519 (2001).
18. Offord, D. A.; Sachs, S. B.; Ennis, M. S.; Eberspacher, T. A.; Griffin, J. H.; Chidsey, C. E. D.; Collman, J. P. *J. Am. Chem. Soc.*, *120*, 4478 (1998).
19. Li, D.; Moore, L. W.; Swanson, B. I. *Langmuir*, *10*, 1177(1994).
20. Finklea, H. O.; Hanshew, D. D. *J. Am. Chem. Soc.*, *114*, 3173 (1992).
21. Finklea, H. O. In *Electroanalytical Chemistry*; Bard, A. J., Rubinstein, I., Eds., Marcel Dekker: New York; Vol. vol 19(1996).
22. Hickman, J. J.; Zou, C.; Ofer, D.; Harvey, P. D.; Wrighton, M. S.; Laibinis, P. E.; Bain, C. D.; Whitesides, G. M. *J. Am. Chem. Soc.*, *111*, 7271 (1989).

23. Hatzor, A.; Moav, T.; Cohen, H.; Matlis, S.; Libman, J.; Vaskevich, A.; Shanzer, A.; Rubinstein, I. *J. Am. Chem. Soc.*, *120*, 13469 (1998).
24. Cao, G.; Hong, H.-G.; Mallouk, T. E. *Acc. Chem. Res*, *25*, 420(1992).
25. Haga, M.; Hong, H.; Shiozawa, Y.; Kawata, Y.; Monjushiro, H.; Fukuo, T.; Arakawa, R. *Inorg. Chem.*, *39*, 4566 (2000).
26. Haga, M.; Shiozawa, Y.; Suzuki, S.; Inoue, M. *Mat. Res. Soc. Symp. Proc*, *679E*, B531(2001).
27. Woodward, J. T.; Ulman, A.; Schwartz, D. K. *Langmuir*, *12*, 3626 (1996).
28. Messerschmidt, C.; Schwartz, D. K. *Langmuir*, *17*, 462 (2001).
29. Hickman, J. J.; Ofer, D.; Zou, C.; Wrighton, M. S.; Laibinis, P. E.; Whitesides, G. M. *J. Am. Chem. Soc.*, *113*, 1128 (1991).
30. Gardner, T. J.; Frisbie, C. D.; Wrighton, M. S. *J. Am. Chem. Soc.*, *117*, 6927 (1995).
31. Allara, D. L.; Dunbar, T. D.; Weiss, P. S.; Bumn, L. A.; Cygan, M. T.; Tour, J. M.; Reinerth, W. A.; Yao, Y.; Kozaki, M.; Jones II, A. L. *Ann. New York Acad. Sci*, 349 (1999).
32. Kovtyukhova, N. I.; Mallouk, T. E. *Chem. Eur. J.*, *8*, 4355 (2002).
33. Galoppini, E.; Guo, W.; Zhang, W.; Hoertz, P. G.; Qu, P.; Meyer, G. J. *J. Am. Chem. Soc.*, *124*, 7801 (2002).
34. Long, B.; Nikitin, K.; Fitzmaurice, D. *J. Am. Chem. Soc.*, *125*, 5152 (2003).
35. Hirayama, D.; Takimiya, K.; Aso, Y.; Otsubo, T.; Hasobe, T.; Yamada, H.; Imahori, H.; Fukuzumi, S.; Sakata, Y. *J. Am. Chem. Soc.*, *124*, 532 (2002).
36. Hu, J.; Mattern, D. L. *J. Org. Chem.*, *65*, 2277 (2000).
37. Haga, M.; Sakiyama, D.; Sakai, K.; Yutaka, T.; Hashimoto, Y.; Hanakura, K. *to be published*, (2004).
38. Pickup, P. G.; Murray, R. W. *J. Am. Chem. Soc.*, *105*, 4510 (1983).
39. Denisevich, P.; Willman, K. W.; Murray, R. W. *J. Am. Chem. Soc.*, *103*, 4727(1981).
40. Balzani, V.; Juris, A.; Venturi, M.; Campagna, S.; Serroni, S. *Chem. Rev.*, *96*, 759 (1996).
41. Liang, Y.; Schmehl, R. H. *J. Chem. Soc., Chem. Commun.*, 1007 (1995).
42. Maskus, M.; Abruna, H. D. *Langmuir*, *12*, 4455 (1996).
43. Brewer, K. J. *Commnets Inorg. Chem.*, *21*, 201 (1999).
44. Constable, E. C.; Thompson, A. M. W. C. *J. Chem. Soc. Dalton Trans.*, 3467 (1992).
45. Hashimoto, Y.; Inoue, M.; Shindo, H.; Haga, M. *J. Inst. Sci. Eng. Chuo Univ*, *7*, 29(2001).
46. Haga, M.; Takasugi, T.; Tomie, A.; Ishizuya, M.; Yamada, T.; Hossain, M. D.; Inoue, M. *Dalton Trans.*, 2069 –2079(2003).
47. Yang, J. C.; Aoki, K.; Hong, H.-J.; Sackett, D. D.; Arendt, M. F.; Yau, S.-L.; Bell, C. M.; Mallouk, T. E. *J. Am. Chem. Soc.*, *115*, 11855 (1993).
48. Hong, H.-G.; Mallouk, T. E. *Langmuir*, *7*, 2362(1991).
49. Katz, H. E.; Schilling, M. L.; Ungashe, S.; Putvinski, T. M.; Chidsey, C. E. *ACS Symposium Series*, *499*, 24(1992).

50. Ansell, M. A.; Zeppenfeld, A. C.; Yoshimoto, K.; Cogan, E. B.; Page, C. J. *Chem. Mater.*, *8*, 591 (1996).

51. Thomsen III, D. L.; Phely-Bobin, T.; Papadimitrkopoulos, F. *J. Am. Chem. Soc.*, *120*, 6177(1998).

52. Hatzor, A.; Weiss, P. S. *Science*, *291*, 1019(2001).

53. Abe, M.; Michi, T.; Sato, A.; Kondo, T.; Zhou, W.; Ye, S.; Uosaki, K.; Sasaki, Y. *Angew. Chem. Int. Ed.*, *42*, 2912 (2003).

54. Bell, C. M.; Arendt, M. F.; Gomez, L.; Schmehl, R. H.; Mallouk, T. E. *J. Am. Chem. Soc.*, *116*, 8374(1994).

55. Diaz, D. J.; Bernhard, S.; Storrier, G. D.; Abruna, H. D. *J. Phys. Chem. B*, *105*, 8746 (2001).

56. Song, W.; Okamura, M.; Kondo, T.; Uosaki, K. *J. Electroanal. Chem.*, *554-555*, 385 (2003).

57. Lin, C.; Kagan, C. R. *J. Am. Chem. Soc.*, *125*, 336 (2003).

58. Calvo, E. J.; Wolosiuk, A. *J. Am. Chem. Soc.*, *124*, 8490 (2002).

59. Snover, J. L.; Byrd, H.; Suponeva, E. P.; Vicenzi, E.; Thompson, M. E. *Chem. Mater.*, *8*, 1490(1996).

60. Mallouk, T. E.; Kim, H.-N.; Ollivier, P. J.; Keller, S. W. In *Comprehensive Supramolecular Chemistry*, Pergamon; Vol. 7; pp 189 (1996).

61. Rubinstein, I., Ed. *Physical Electrochemistry*; Marcel Dekker, Inc.: New York (1995).

62. Feringa, B. L., Ed. *Molecular Switches*; Wiley-VCH: Weinheim (2001).

63. Kaifer, A. E.; Gomez-Kaifer, M. *Supramolecular Electrochemistry*; Wiley-VCH: Weinheim (1999).

64. Haga, M.; Ano, T.; Kano, K.; Yamabe, S. *Inorg. Chem.*, *30*, 3843 (1991).

65. Haga, M.; Ali, M. M.; Maegawa, H.; Nozaki, K.; Yoshimura, A.; Ohno, T. *Coord. Chem. Rev.*, *132*, 99 (1994).

66. Haga, M.; Ali, M. M.; Koseki, S.; Fujimoto, K.; Yoshimura, A.; Nozaki, K.; Ohno, T.; Nakajima, K.; Stufkens, D. J. *Inorg. Chem.*, *35*, 3335(1996).

67. Haga, M.; Ali, M. M.; Arakawa, R. *Angew. Chem. Int. Ed.*, *35*, 76 (1996).

68. Vining, W. J.; Surridge, N. A.; Meyer, T. J. *J. Phys. Chem.*, *90*, 2281 (1986).

69. Palmore, G. T. R.; Smith, D. K.; Wrighton, M. S. *J. Phys. Chem. B*, *101*, 2437 (1997).

70. Hagfeldt, A.; Gratzel, M. *Acc. Chem. Res.*, *33*, 269(2000).

71. Sakomura, M.; Fujihira, M. *Thin Solid Films*, *273*, 181(1996).

72. Byrd, H.; Suponeva, E. P.; Bocarsly, A.; Thompson, M. E. *Nature*, *380*, 610(1996).

73. Uosaki, K.; Kondo, T.; Zhang, X.-Q.; Yanagida, M. *J. Am. Chem. Soc.*, *119*, 8367 (1997).

74. Imahori, H.; Norieda, H.; Yamada, H.; Nishimura, Y.; Yamazaki, I.; Sakata, Y.; Fukuzumi, S. *J. Am. Chem. Soc.*, *123*, 100(2001).

75. Nazeeruddin, M. K.; Kay, A.; Rodici, I.; Humphry, B.; Mueller, E.; Liska, P.; Valchopoulos, N.; Graetzel, M. *J. Am. Chem. Soc.*, *115*, 6382(1993).

76. Bignozzi, C. A.; Schoonover, J. R.; Scandola, F. *Prog. Inorg. Chem*, *44*, 1 (1997).

77. Kaschak, D. M.; Lean, J. T.; Waraksa, C. C.; Saupe, G. B.; Usami, H.; Mallouk, T. E. *J. Am. Chem. Soc.*, *121*, 3435 (1999).

78. Abdelrazzaq, F. B.; Kwong, R. C.; Thompson, M. E. *J. Am. Chem. Soc.*, *124*, 4796 (2002).
79. Soto, E.; MacDonald, J. C.; Cooper, C. G. F.; McGimpsey, W. G. *J. Am. Chem. Soc.*, *125*, 2838 (2003).
80. Ashkenasy, G.; Cahen, D.; Cohen, R.; Shanzer, A.; Vilan, A. *Acc. Chem. Res*, *35*, 121 (2002).
81. Wu, D. G.; Ashkenasy, G.; Shvarts, D.; Ussyshkin, R. V.; Naaman, R.; Shanzer, A.; Cahen, D. *Angew. Chem. Int. Ed.*, *39*, 4496 (2000).
82. Rowe, G. K.; Creager, S. E. *Langmuir*, *7*, 2307(1991).
83. Campbell, D. J.; Herr, B. R.; Hulteen, J. C.; Van Duyne, R. P.; Mirkin, C. A. *J. Am. Chem. Soc.*, *118*, 10211 (1996).
84. Xu, X.-H.; Bard, A. J. *J. Am. Chem. Soc.*, *117*, 2627 (1995).
85. Biancardo, M.; Schwab, P. F. H.; Argazzi, R.; Bignozzi, C. A. *Inorg. Chem.*, *42*, 3966 (2003).
86. Chen, J.; Reed, M. A.; Rawlett, A. M.; Tour, J. M. *Science*, *286*, 1550(1999).
87. Park, J.; Pasupathy, A. N.; Goldsmith, J. I.; Chang, C.; Yalsh, Y.; Petta, J. R.; Rinkoski, M.; Sethna, J. P.; Abruna, H. D.; McEuen, P. L.; Ralph, D. C. *Nature*, *417*, 722 (2002).
88. Roth, K. M.; Dontha, N. R.; Dabke, B.; Gryko, D. T.; Clausen, C.; Lindsey, J. S.; Bocian, D. F.; Kuhr, W. G. *J. Vac. Sci. Technol. B*, *18*, 2359 (2000).
89. Hagrman, P. J.; Hagrman, D.; Zubieta, J. *Angew. Chem. Int. Ed.*, *38*, 2638 (1999).
90. Belanger, S.; Hupp, J. T.; Stern, C. L.; Slone, R. V.; Watson, D. F.; Carrell, T. G. *J. Am. Chem. Soc.*, *121*, 557 (1999).

Part 5
Spectroelectrochemistry

Chapter 11

Spectroelectrochemical Techniques

Enzo Alessio[1], Simon Daff[2], Marie Elliot[2], Elisabetta Iengo[1], Lorna A. Jack[2], Kenneth G. Macnamara[2], John M. Pratt[3] and Lesley J. Yellowlees*[2]

[1] Dipartimento di Scienze Chimiche, Università di Trieste, 34127 Trieste, Italy

[2] School of Chemistry, University of Edinburgh, Edinburgh EH9 3JJ, Scotland

[3] Department of Chemistry, Imperial College of Science, Technology and Medicine London SW7 2AY

1   INTRODUCTION

The powerful combination of *in situ* spectroscopic and electrochemical techniques are referred to as spectroelectrochemical methods. Such experiments have long been important tools in the investigation of organic, inorganic and biological systems. Spectroelectrochemical results are particularly helpful in the characterisation of short-lived intermediates and when redox products are unstable. To date, the most used solution spectroscopic technique combined with the electrochemical experiment is uv/vis detection, due to the simplicity of the experimental set-up. However numerous spectroscopic techniques have been used successfully, for example, IR [1-13], luminescence [14], epr, nmr [15], CD [16-18], Raman [19,20], X-ray absorption spectroscopy [21], XANES [22]. An IUPAC review of *in situ* spectroelectrochemical techniques was published in 1998 [23,24]. This report will concentrate on uv/vis/nir and epr *in situ* spectroelectrochemistry with detailed studies taken from the authors' laboratories.

The importance of temperature on the results will be detailed since lowering the temperature considerably increases the stability of many reactive redox products. However, the design of variable temperature *in situ* cells adds to the complexity of the apparatus. Different practitioners have built different cells for the same spectroscopic technique. Common to all cells is ease of assembly and a robust constitution. They usually have as small a pathlength as possible for quick electrogeneration and have the electrodes close together to minimise iR drop. Rarely are such cells available commercially with most custom designed to allow study in the spectrometer of choice of the redox active species of interest.

Spectroelectrochemical data may be used to identify the site of redox activity, to determine the formal electrode potential, E' and the number of electrons transferred in the process, n and to characterise the frontier orbital(s) within the redox active species under study. Examples of each of these are detailed below. Although one such study may yield considerable information, comparative studies of related redox active species are more likely to result in significant insights. Furthermore, the combination of two or more spectroelectrochemical techniques can prove more powerful still.

2   EXPERIMENTAL

All reactions were carried out under an atmosphere of dry, oxygen free nitrogen. All cells were loaded either in a glove box or under a positive pressure of N_2. The Ru-dmso compounds were prepared by Alessio *et al* [25]. The diaquo-Co(II)-cobinamide and aquohydroxo-cobinamide were prepared from cyanocobalamine by Pratt *et al* [26-28]. The neuronal NO synthase, nNOS, was purified from E. coli expressing the nNOS FMN, FMN = flavin mononucleotide, domain and Calmodulin protein as a one-to-one protein: protein complex [29]. The syntheses of [Pt(4,4'- ^{t}Bu$_2$-bpy)(C$_2$(C$_6$H$_4$-p-NO$_2$))$_2$], bpy = 2,2'-bipyridine [30], 5,10,15, 20-[N-benzyl-N'''-(4-benzyl-4,4'-bipyridinium-4-pyridyl]-triphenylporphyrin tris-(hexafluorophosphate) [31], [Pt(4,4'-X$_2$-bpy)Cl$_2$], X = NO$_2$[32], H [33], are described in the literature.

Dichloromethane was pretreated with KOH pellets and distilled over P_2O_5 immediately prior to use. All other solvents were used as obtained without further purification. [^{n}Bu$_4$N]BF$_4$ was recrystallised from water/ethanol and dried at 350 K under vacuum for three days before electrochemical use.

All electrochemical studies were performed using a Dell GX110 personal computer with General Purpose Electrochemical System (GPES) version 4.8 software connected to an Autolab system containing either a PSTAT 20 or PSTAT 30 potentiostat. Cyclic voltammetric experiments employed a Pt micro-working electrode (we, 0.5 mm diameter), a Pt counter electrode (ce) and either a Ag/AgCl reference electrode (re) in a 0.45 M [^{n}Bu$_4$N][BF$_4$]/0.05 M [^{n}Bu$_4$N]Cl/dichloromethane solution or a saturated calomel electrode (sce). The ferrocinium/ferrocene couple was measured at +0.55 V *vs* the Ag/AgCl reference electrode. Coulometric experiments utilised an H-type cell with a Pt basket working electrode fritted from the Pt counter electrode. The exact composition of the solution under study is detailed as appropriate in the text. Dissolved oxygen is removed from the solvent by N_2 saturation prior to study. Unless stated otherwise the scan rate used for all cyclic voltammetric studies is 100 mV s⁻¹.

The *in situ* spectroelectrochemical uv/vis/nir optically transparent electrode (OTE) cell used in the detailed studies reported herein is as described previously [34], and is shown in Fig. 1.

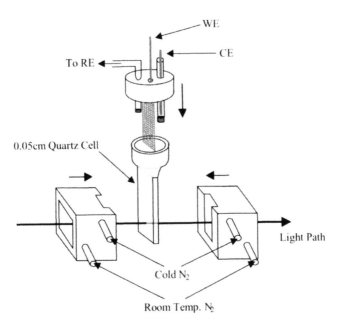

Figure 1. Schematic of in situ uv/vis/nir OTE cell and holder for variable temperature experiments.

The working electrode is composed of a Pt/Rh gauze (Englehard, transparency ~ 40%) placed in a 0.5 mm pathlength quartz cuvette (Heraeus). The quartz cup fused to the cuvette holds the Pt counter electrode and Ag/AgCl reference electrode both of which are separated from the bulk solution by low porosity frits. The temperature of the cell may be altered by passing prechilled nitrogen gas over the outside of the OTE cell. The total cell volume is 1 ml. Spectra are recorded every five minutes during the electrogeneration experiment until there is no change to the absorption. The generating potential is then reversed such that the starting material is recovered and the spectrum returns to that of the initial spectrum so as to ensure no chemical reaction has taken place during the spectroelectrochemical experiment and hence the spectral changes have resulted only from the electron transfer reaction. Uv/vis/nir spectra were recorded on a Perkin-Elmer λ9 spectrometer using UVwinlab software.

In situ epr spectroelectrochemical experiments were performed in a flat epr cell (Wilmad). The Pt/Rh gauze working electrode is placed in the flat region of the cell with the Ag/AgCl reference electrode and Pt counter electrode placed as near to the working electrode as the cell geometry will allow, see Fig. 2. The total cell volume is 2 ml. The temperature of the flat cell may be varied by placing it inside the variable temperature epr cavity and flowing prechilled N_2 gas over its surface.

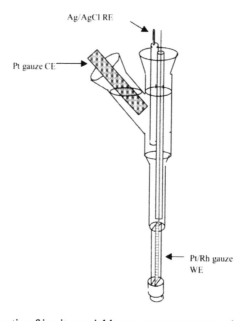

Figure 2. Schematic of in situ variable temperature spectroelectrochemical epr cell.

Variable temperature work was carried out using a Bruker ER4111VT unit. Epr spectra were recorded on an X-band Bruker ER200D spectrometer using epr acquisition system version 2.42 software [35]. All epr spectra were corrected for dpph, $g = 2.0036 \pm 0.0003$.

Molecular orbital calculations were performed using the modelling program CAChe, version 3.2 [36]. Epr solution spectra were simulated using WIN-EPR SimFonia software [37].

3 UV/VIS/NIR SPECTROELECTROCHEMISTRY

Thin layer electrochemistry with an optically transparent electrode (OTE) or an optically transparent thin layer electrode (OTTLE) enables the simultaneous monitoring of both the electrochemical and optical responses of the system [38-40]. There are two main types of transparent electrodes; namely thin films of conductive materials such as platinum, gold, tin oxide, indium tin oxide (ITO) deposited on a substrate, for example glass or quartz and the minigrid electrode made from gold, silver, platinum in which the transparency is due to holes in the minigrid structure.

The first OTTLE cell was described by Murray, Heineman and O'Dom in 1967 and was constructed from a Au minigrid sandwiched between teflon spacers enclosed in microscope slides [41]. The bottom of the cell was placed in a sample reservoir containing the redox active solution, the counter and reference electrodes. Many other OTE and OTTLE cell designs have been published since 1967 with most sharing common features with the system described above. The incorporation of variable temperature capabilities in several designs has permitted the study of unstable, reactive redox species. Some examples from the recent literature detailing studies on inorganic, organic and biological systems are discussed below.

3.1 Recent Developments in uv/vis/nir Spectroelectrochemistry

3.1.1 Gold Working Electrode

Heineman *et al* have been long time exponents of the gold OTTLE cell [42]. A recent publication describes the electrochemistry and spectroelectrochemistry of $[Re(dmpe)_3]^+$, where dmpe = 1,2-bis(dimethylphosphino)ethane [43].

Toma makes extensive use of the Au mingrid room temperature OTE, pathlength 0.25 mm [44,45]. For example, the electronic character of $[Ru_3-(O)(CH_3CO_2)_6(py)_2(NO)]PF_6$, py = pyridine, has been investigated [46]. Nitric oxide is a non-innocent ligand and may bind to the metal centre as NO^+, $NO^•$ or NO^- [47-50]. The trinuclear NO complex may be regarded as having the configuration $Ru(III)Ru(III)Ru(III)NO^•$ or $Ru(III)Ru(III)Ru(II)NO^+$ and the one-electron process at -0.44 V *vs.* Ag/Ag^+ is either NO or Ru based. Uv/vis and ir spectroelectrochemistry define the starting trinuclear complex as $Ru(III)Ru(III)Ru(III)NO^•$ and the redox process at -0.44 V as $Ru(III) \rightarrow Ru(II)$ based. Toma *et al* also report on the electrochemistry and spectroelectrochemistry of $[Fe(imox)(Nmin)_2]^+$, where imox = 2,3,9,10-tetramethyl-1,4,8,11-tetraazaundecane-3,8,10-tetraen-11-ol-1-olate and Nmin = N-methylimidazole [51]. Oxidation of $[Fe(imox)(Nmin)_2]^+$ ultimately yields the redox series $[\{(imox)-(Nmin)Fe\}_2(^•-OH)]^{n+}$, $n = 1, 2, 3$.

A gold sputtered OTE has been used in the construction of a bidimensional uv/vis/nir cell which allows simultaneous measurements in both normal and parallel directions to the working electrode surface [52].

3.1.2 Carbon Working Electrode

A reticulated pyrolytic graphite electrode (minigrid area ~ 0.8 × 2.8 cm^2) has been employed to study Cu_2Co_2SOD, SOD = superoxide dismutase, and confirmed the redox process at +0.311 V was due to reduction of Cu(II) [53].

3.1.3 ITO Working Electrode

ITO glass electrodes covered with electrochemically deposited films were used to obtain *in situ* spectra of poly-CuL$_2$ where HL is the Schiff base shown below [54].

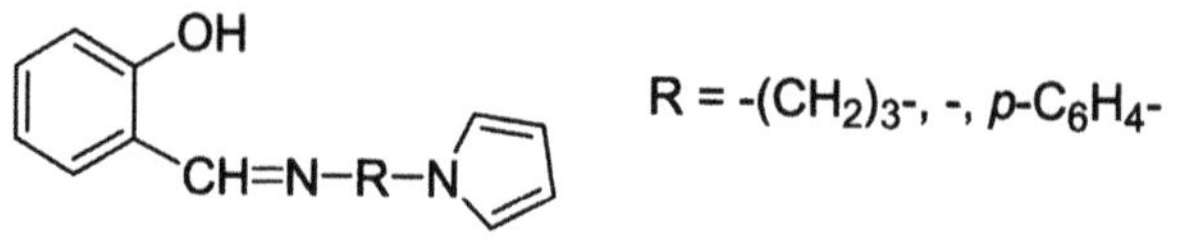

$$R = -(CH_2)_3-, -, p\text{-}C_6H_4-$$

3.1.4 Platinum Working Electrode

The spectroelectrochemical studies of $(S_2COEt)^-$, $Ni(S_2COEt)_2$ and $[Ni(S_2COEt)_2]^-$ using a Pt wire working electrode show that oxidation is followed by dimerisation of free (S_2COEt) to $(S_2COEt)_2$ [55]. Hill and Mann describe an OTTLE cell consisting of vapour deposited Pt working and pseudo-reference electrodes and a Pt wire counter electrode [56]. Hill *et* al have probed the dimerisation reaction which follows reduction of $[Pt(terpyCl]^-$, where terpy = 2,2':6',2"-terpyridine, in this cell [57]. The cell was also used by Barton *et al* to deduce that the site of reduction in $[Ir(bpy)(phen)(phi)]^{3+}$, phen = 1,10-phenanthroline, phi = 9,10-phenanthroline quinone, is the phi ligand [58].

A room temperature OTE cell using a Pt gauze working electrode has been described by Crayston and used to assign the site of oxidation in the Cu species below to Cu(II)/Cu(III) [59].

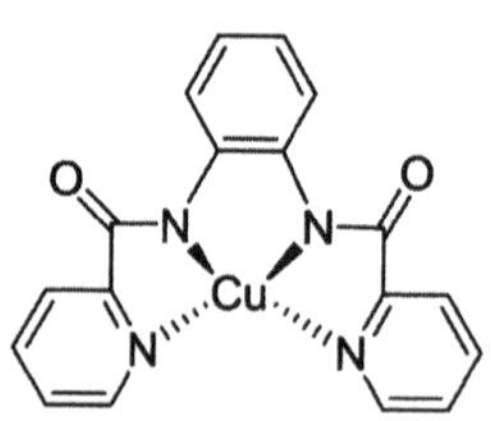

A variable temperature OTTLE cell for use in uv/vis and IR spectroelectrochemical studies has been described by Hartl *et al* [60]. An early room temperature uv/vis and IR cell has also been detailed [3]. The same basic design is used for both spectroscopies by changing the CaF_2 windows (IR) for quartz ones (uv/vis). The electrodes comprise Pt minigrid working and counter electrodes and a Ag pseudo-reference electrode. *In situ* IR and uv/vis studies of the oxidation products of $[Mn(CO)_3(DBCat)]^-$, DBCat = 3,5-di-tert-butyl catecholate in CH_2Cl_2/H_2O and the reduction products of $[Ru(I)(Me)(CO)_2(^iPr-DAB)]$, ^iPr-DAB = N,N'-diisopropyl-1,4-di-aza-1,3-butadiene in THF at 293 and 183 K are discussed [60].

The Hartl designed variable temperature [60] and room temperature [3] cells have been extensively used by Kaim [61-65]. Some examples of uv/vis/nir studies taken from his work include the characterisation of the tetranuclear complexes $\{(\mu_4\text{-}tcnx)[Cu(Me_3TACN)]_4\}(BF_4)_4$, where tcnx = tcne, tcnq, tcnb (1,2,4,5-tetracyanobenzene), Me_3TACN = 1,4,7-trimethyl-1,4,7-triazacyclononane. The redox active site is determined to be primarily tcnx based [66]. The uv/vis and epr spectroelectrochemical study of the mixed metal species shown below shows that the one electron oxidation is based on the Fe(II) site and not on the M(II) centre [67, 68].

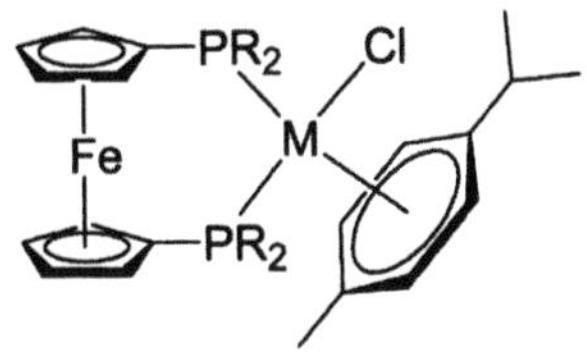

M=Os(II), Ru(II)
R = Ph, Et, ^iPr

A combined IR, uv/vis and epr spectroelectrochemical study of the one-electron oxidation of $[Cr(CO)_4(tmp)]$, tmp = 3,4,7,8-tetramethyl-1,10-phenanthroline, at 193 K in ^nPrCN shows that $[Cr(CO)_4(tmp)]^+$ is best considered as containing a low spin Cr(I), t_{2g}^5 metal centre [69]. The Pt_4 squares shown below have a two-electron reversible reduction process and a two-electron reversible oxidation step. Spectroelectrochemical experiments suggest the site of the oxidations are the dianionic anthracene ligands and the reductions are based on the π-accepting N-donor ligands [70].

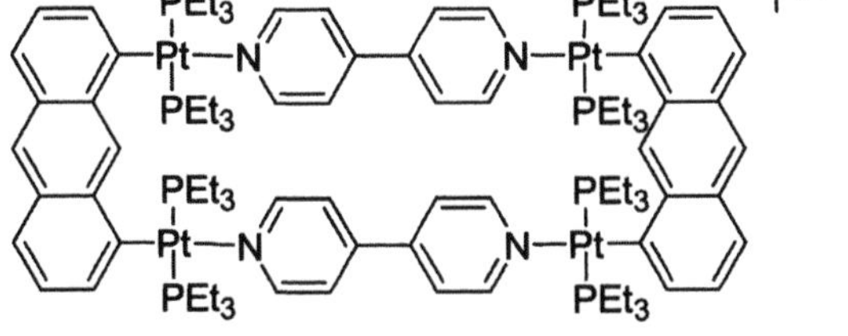

Klein *et al* also employ Hartl's cell to probe, for example, doubly cyclometallated binuclear Pt(II) and Pt(IV) compounds [71].

Kadish and Lin describe a vacuum-tight spectroelectrochemical cell witth a double Pt gauze working electrode, a Pt gauze counter electrode and a sce reference electrode with a 0.2 mm path length [72]. Such a cell has been employed to study the reduction of ferric porphyrins (P) in the presence of nitrite ions which result in formation of Fe(P)(NO) and [Fe(P)]$_2$O [73]. Reduction of hypervalent phosphorus(V) octaethylporphyrins show spectral changes which are consistent with a ring-centred electron transfer leading to formation of a porphyrin π-anion radical [74].

Wieghardt *et al* have characterised the square planar complexes [M(L^{ISQ})$_2$]^n, M = Cu, Ni, Pd, Pt, n = 2–, 1–, 0, 1+, 2+ by variable temperature uv/vis spectroelectrochemistry and concluded that all the redox processes are ligand-based [75, 76]

[L^{ISQ}]$^{1-}$ =

[L^{AP}-H]$^{2-}$ =

A related study (uv/vis, epr spectroelectrochemistry) of [V(V)(L^{ISQ})(L^{AP}-H)$_2$] concluded that the one-electron oxidation and first one-electron reduction were both ligand-based. The second one-electron reduction to form the dianion is metal centred producing [V(IV)(L^{AP}-H)$_3$]$^{2-}$ [77].

Ward *et al* use a uv/vis/nir cell based on the one described herein [34]. They have undertaken a large number of spectroelectrochemical studies on a wide range of redox-active coordination compounds [78,79]. For example, the dinuclear ferrocenyl-molybdenum complexes [Mo(Tp^{Me,Me})(O)Cl(=NFc)] and [Mo(Tp^{Me,Me})(O)Cl(=N-C$_6$H$_4$N=NC$_6$H$_4$-Fc)], where Tp^{Me,Me} = hydrotris(3,5-dimethylpyrazol-1-yl)borate undergo a reversible ferrocenyl Fe(II)/Fe(III) based oxidation [80]. A study of the related complexes [Mo(Tp^{Me,Me})(O)Cl(O-C$_6$H$_4$N=NC$_6$H$_4$-Fc)] and [Mo(Tp^{Me,Me})(O)Cl(O-C$_6$H$_4$C≡CC$_6$H$_4$-Fc)] shows the first one-electron oxidation is based on the Fc fragment with the second oxidation being Mo based (Mo(V) → Mo(VI)) [81]. The di-oxidised species absorbs strongly in the near infra-red, λ_{max} ~ 900 nm (ε = 30 000 M^{-1} cm^{-1}). Other studies on the binuclear complexes [(bpy)$_2$Ru(II)BLRu(II)(bpy)$_2$]$^{2+}$ and [(bpy)$_2$Ru(II)BL'Ru(II)(bpy)$_2$]$^{4+}$ confirm, using uv/vis spectroelectrochemical techniques, that the two reversible oxidations at +0.12 and +0.35 V, in the complex with bridging ligand BL, are based on BL whereas when the bridging ligand is BL' the oxidations at +1.25 and +1.70 V are Ru(II) → Ru(III) based [82,83].

BL =

BL' =

Uv/vis/nir spectroelectrochemical studies by Heath *et al* also employ a cell similar to that described in Fig. 1 [84]. This group has investigated the spectral changes accompanying stepwise redox processes for a number of different systems. For example, in $[RuCl_3(CH_3CN)_3]^{1+/0}$ the Cl^- ⟶ Ru charge transfer moves from 24800 cm^{-1} in Ru(III) to 20200 in Ru(IV) [84]. The degree of metal-metal interactions in buta-1,3-diyne bridged octaethylporphyrin dimers (Co_2, Ni_2, Cu_2, Zn_2, Pd_2, Pt_2, Co/Ni and Ni/Zn) have been determined and typically exhibit an intense intervalence band around 4500 cm^{-1} [85-87]. The metal complexes $[\{Cp(PPh_3)_2Ru\}_2(\mu\text{-}C\equiv C\text{-}C\equiv C)]$ and $[\{Cp(PPh_3)(PMe_3)Ru\}_2(\mu\text{-}C\equiv C\text{-}C\equiv C)]$ can be oxidised to give a series of mono-, di-, tri- and tetracations which have been characterised by ir and uv/vir/nir spectroelectrochemical techniques [88]. It has been concluded that the redox centres are based over the $Ru\text{-}C_4\text{-}Ru$ fragment. Electrochemical, epr and uv/vis/nir spectroscopic studies at 233 K confirmed that the one-electron oxidation of the ethyl and cyclohexyl substituted tris(dithiocarbamato)cobalt(III) complexes produces the Co(IV) species $[Co(IV)(S_2CNR_2)_3]^+$ [89].

3.2   Results and Discussion

The uv/vis/nir spectroelectrochemical cell used in the studies presented below and in representative recent reports is shown in Fig. 1 [90-94]. An important feature of the design is the ability to alter the temperature. Decreasing the temperature of the cell can result in an increase in the stability of the redox product. For example, $[ReCl_6]^{2-}$ undergoes a reversible one-electron oxidation and an irreversible one-electron reduction at room temperature in $CH_2Cl_2/0.1$ M $[^nBu_4N]BF_4$ [95]. If the temperature is decreased then the electron transfer couple $[ReCl_6]^{2-/3-}$ is reversible and the uv/vis spectrum of $[ReCl_6]^{3-}$ can be recorded at 182 K using the spectroelectrochemical cell (note the temperature required to ensure chemical reversibility).

Figure 3a and b show the spectral changes accompanying the oxidation $[ReCl_6]^{2-/1-}$ and reduction $[ReCl_6]^{2-/3-}$ processes, respectively. Note the clear isosbestic points in both series of spectra which indicate a clean conversion between the redox couples. Reversal of the electrogeneration potential to 0 V after complete oxidation or reduction results in complete regeneration of the $[ReCl_6]^{2-}$ spectrum. Comparison of the spectra of $[ReCl_6]^{1-}$, $[ReCl_6]^{2-}$ and $[ReCl_6]^{3-}$ aids in the assignment of the absorption bands to electronic transitions. In all three spectra the major absorption feature can be assigned to ligand-to-metal charge transfers (LMCT), Cl^- (π) $\rightarrow$ Re dπ [96]. As the oxidation state of the Re metal decreases from Re(V), d^2 to Re(IV), d^3 to Re(III), d^4 the energy of the LMCT bands shift to higher energy. The following case studies, taken from our laboratories, are further examples of the uses of uv/vis/nir spectroelectrochemistry.

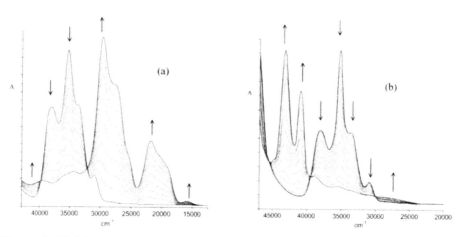

Figure 3. Uv/vis spectroelectrochemical results for (a) oxidation of $[ReCl_6]^{2-} \rightarrow$ $[ReCl_6]^{1-}$ (E_{gen} = +1.6 V) and (b) reduction of $[ReCl_6]^{2-} \rightarrow [ReCl_6]^{3-}$ (E_{gen} = -1.45 V) in CH_2Cl_2/0.5 M [$^nBu_4N$]BF_4 at 182 K.

3.2.1   Mixed Valence Compounds

UV/Vis spectroelectrochemistry is frequently used to study mixed valence compounds that contain two (or more) metal centres linked together either directly or *via* a bridging ligand. Such complexes contain two metal centres; most commonly, the metal centres are of the same metal but with differing numbers of electrons and usually exhibit intervalence charge transfer (IVCT) bands. The IVCT bands arise from excitation of an electron from one metal centre to the other as shown in Scheme 1. Detailed analysis of the IVCT band gives information regarding the strength of interaction between the metal sites.

$$M^{n+} \sim M^{(n-1)+} \xrightarrow[\text{IVCT}]{hv} M^{(n-1)+} \sim M^{n+}$$

Scheme1

Most mixed valence compounds are generated from the parent compound, which contains two equivalent metal centres, by one-electron oxidation or reduction.

The Creutz-Taube (CT) ion $[(NH_3)_5Ru(pyz)Ru(NH_3)_5]^{5+}$, pyz = pyrazine, was one of the first and most studied mixed valence compounds [97] and is formed from the one-electron oxidation of its 4+ parent ion. There has been considerable debate over whether the CT ion is a Class II or Class III compound where the Class II/III nomenclature is as described by Robin and Day [98]. Class II compounds have a weak interaction between the metal centres whereas Class III species are strongly interacting, delocalised systems. Current opinion assigns the CT ion to the Class III delocalised valence description [99].

There have been many studies of mixed valence compounds related to the CT ion containing $M(NH_3)_5$ fragments bridged by organic ligands [100,101]. Other extensively studied fragments include $M(CN)_5$ and M(polypyridyl) moieties which contain π-accepting cyanide and polypyridyl ligands [102-107]. This work will detail the spectroelectrochemical study of mixed valence compounds containing the $RuCl_4$(dmso-S) fragment and their monomeric analogues (Figure 4) to assess the effect of strong σ-donor, π-donor ligands on metal-metal communication.

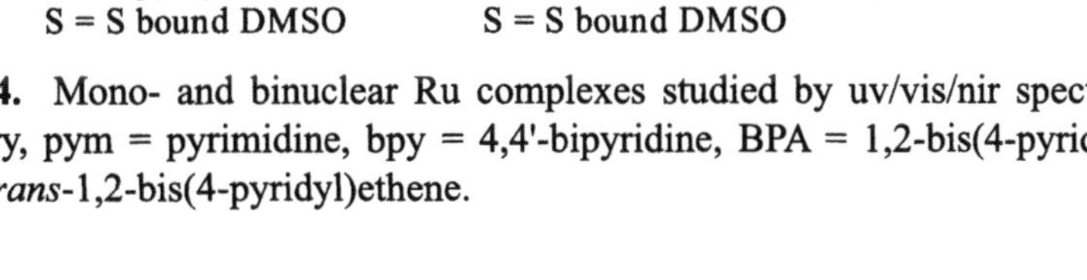

L-L =pyz, pym, bpy
S = S bound DMSO

L-L =pyz, pym, bipy, BPA, BPE
S = S bound DMSO

Figure 4. Mono- and binuclear Ru complexes studied by uv/vis/nir spectroelectro–chemistry, pym = pyrimidine, bpy = 4,4'-bipyridine, BPA = 1,2-bis(4-pyridyl)ethane, BPE = *trans*-1,2-bis(4-pyridyl)ethene.

The complexes were synthesised by Alessio *et al* and X-ray diffraction studies confirm the coordination sphere around the Ru centre as four equitorial chlorides, one S-bonded dimethyl sulfoxide and one heterocyclic N-ligand [25]. The cyclic voltammogram of $Na[RuCl_4(dmso-S)pyz]$ in DMF/0.1 M $[^nBu_4N][BF_4]$ is shown in Figure 5.

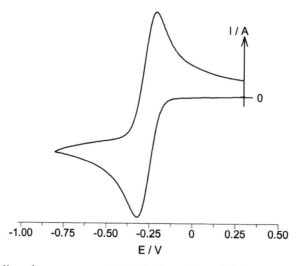

Figure 5. Cyclic voltammogram of $[RuCl_4(dmso-S)pyz]^-$ in DMF/0.1 M $[^nBu_4N][BF_4]$ at 298 K.

Coulometry confirms the number of electrons transferred as one. The half wave potential for the one-electron reduction is measured at –0.22 V *vs.* Ag/AgCl. The related compound *trans*-$[RuCl_4(dmso-S)_2]^-$ undergoes a one-electron reduction at –0.18 V in dmso [108]. Note that replacing one dmso ligand for a pyrazine ligand results in a negative shift of the half wave potential in agreement with the stronger π-accepting properties of the pyz ligand compared to dmso-S. The electrochemical behaviour of $[RuCl_4(dmso-S)pym]^-$ is similar to that of $[RuCl_4(dmso-S)pyz]^-$ and the half wave potential is given in Table 1.

Complexes $[RuCl_4(dmso-S)_2]^-$ and $[RuCl_4(dmso-S)py]^-$ are reported to undergo slow hydrolysis reactions in water at room temperature [109]. However there was no evidence of ligand substitution in $[RuCl_4(dmso-S)L]^-$, L = pyz, pym, in DMF at room temperature. Spectroelectrochemical experiments were performed at reduced temperatures to ensure the chemical integrity of the complexes remained constant. In every case once the spectrum of the reduced species was recorded the compound was re-oxidised back to the starting material and the original spectrum was fully recovered. Fig. 6 shows the spectral changes accompanying the reduction of $[RuCl_4(dmso-S)L]^-$, L = pyz, pym.

Table 1. Half wave potential and K_c data for Ru(III/II) couples for $C_n[\{RuCl_4(dmso-S)\}_n L]^{n-}$

C	L	n	$E_{1/2}$[a]	Solvent	K_c[b]
Na	pyz	1	-0.22	DMF, DMSO	
Na	pym	1	-0.255	DMF	
Na	pym	1	-0.215	DMSO	
$^nBu_4N$	bpy	1	-0.41	DMF	
$^nBu_4N$	bpy	1	-0.28	CH_2Cl_2	
Na	dmso-S [108]	1	-0.18	DMSO	
Na	Py [109]	1	+0.065	H_2O	
Na	μ-pyz	2	-0.16, -0.24	DMF	20
$^nBu_4N$	μ-pyz	2	-0.28, -0.43	DMF	340
$^nBu_4N$	μ-pyz	2	-0.15, -0.32	CH_2Cl_2	750
Na	μ-pym	2	-0.28	DMF	
Na	μ-bpy	2	-0.30	DMF	
Na	μ-bpe	2	-0.28	DMF	
Na	μ-bpa	2	-0.32	DMF	

[a] $E_{1/2}$ values measured against ferrocenium/ferrocene and quoted *vs.* sce,
[b] K_c = comproportionation constant = $\exp[\Delta E_{1/2}nF/(RT)]$.

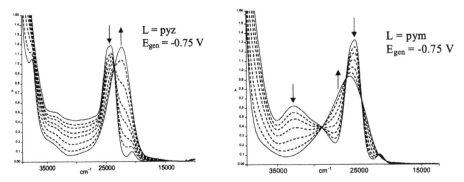

L = pyz
E_{gen} = -0.75 V

L = pym
E_{gen} = -0.75 V

Figure 6. Spectral changes accompanying the one-electron reduction process $[RuCl_4(dmso-S)L]^{1-} \rightarrow [RuCl_4(dmso-S)L]^{2-}$, L = pyz or pym in DMF/0.1 M $[^nBu_4N][BF_4]$ at 224 K.

The spectra of the monoanions (formally containing a Ru(III) centre) are dominated by an intense band around 24 000 cm^{-1} ($\varepsilon \sim 7\ 000$ M^{-1}cm^{-1}) with a weaker band at 20 700 cm^{-1} ($\varepsilon \sim 500$ M^{-1}cm^{-1}) and the absorptions are assigned to chloride to Ru(III) ligand-to-metal charge transfer (LMCT) transitions [96].

On reduction, the LMCT band collapses and a new band grows in at 22300 (L = pyz) and 24 900 cm^{-1} (L = pym). Since reduction of Ru(III) to Ru(II) results in full occupation of the Ru t_{2g} orbitals the transitions in the visible region of the spectrum of the Ru(II) complexes are assigned to metal-to-ligand charge transfers (MLCT) ie Ru(II) $t_{2g}^6 \longrightarrow$ pyz(π^*) and Ru(II) $t_{2g}^6 \longrightarrow$ pym(π^*) respectively. Both sets of spectra in Figure 6 exhibit clean isosbestic points indicating the formation of only one product species, in each case, on reduction. Band maxima and assignments are given in Table 2.

Table 2. Absorption band characteristics and electronic assignment for C[RuCl$_4$(dmso-S)L]$^{z-}$ complexes

C	L	z	Cl$^-$ → Ru(III)	Ru(II) → L
Na	pyz	1	24050 (7200)^a	
Na	pyz	2 (-0.75)^b		22260 (7100)
Na	pym	1	24200 (6550)	
Na	pym	2 (-0.75)		24900 (4600)
^{n}Bu$_4$N	bpy	1	24750 (6550)	
^{n}Bu$_4$N	bpy	2 (-0.50)		20550 (6550)

^{a} band maximum in cm^{-1} (ε in M^{-1} cm^{-1}), ^{b} E$_{gen}$ potential (V)

The electrochemical behaviour of [^{n}Bu$_4$N][RuCl$_4$(dmso-S)bpy] in DMF is less reversible at room temperature at 0.1 V s^{-1} than that of [RuCl$_4$(dmso-S)L]$^-$, L = pyz, pym but becomes reversible at 0.8 V s^{-1} at 292 K or 0.1 V s^{-1} at 258 K. In CH$_2$Cl$_2$ the reversibility of the one-electron reduction is improved at 292 K. Electrochemical and spectroscopic data for [^{n}Bu$_4$N][RuCl$_4$(dmso-S)bpy] are reported in Tables 1 and 2 respectively. Note that the half wave potential is solvent dependent and correlates with Gutmann's acceptor number [110]. [^{n}Bu$_4$N][RuCl$_4$(dmso-S)bpy] has a Cl$^-$ → Ru(III) charge transfer band at 24750 cm^{-1} and a Ru(II) → bpy π^* transition at 20 550 cm^{-1} in its reduced state.

The electrochemical behaviour of [{Ru(dmso-S)Cl$_4$}$_2$(μ-pyz)]$^{2-}$ is shown in Fig. 7 and tabulated in Table 1.

The half wave potentials are both solvent and counter ion dependent. Bulk electrolysis at –0.60 V confirms that the number of electrons transferred are two. The shift of the redox couples to more positive potentials, in DMF, on replacing the ^{n}Bu$_4$N$^+$ counter ion with Na$^+$ can be understood by considering the level of ion pairing between the counter ion and the binuclear compound. If the ^{n}Bu$_4$N$^+$ counter ion dissociates from the binuclear complex in DMF and the sodium salt is not fully dissociated on dissolution then the intimate presence of the positive charge in the sodium species will result in the reduction potentials shifting to lower energy. Upon reduction the overall negative charge of the binuclear complex increases thereby increasing the association of the complex and the sodium ion. Thus the second reduction potential will shift to more positive potentials (compared to the $^{n}Bu_4N^+$ salt in the same solvent) and hence the $E_1 - E_2$ separation is smaller for the sodium salt as noted.

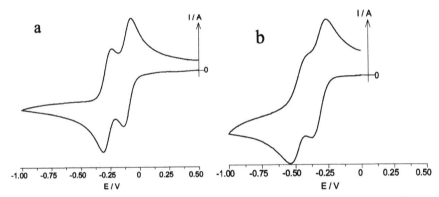

Figure 7. Cyclic voltammograms of $[^{n}Bu_4N]_2[\{Ru(dmso\text{-}S)Cl_4\}_2(\mu\text{-pyz})]^{2-}$ (a) in $CH_2Cl_2/0.4$ M $[^{n}Bu_4N]BF_4$ at 298 K and (b) in DMF/0.1 M $[^{n}Bu_4N]BF_4$ at 298 K.

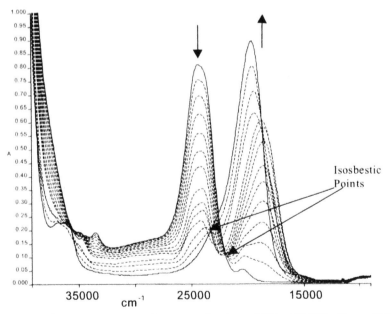

Figure 8. Spectral response to two electron reduction of $[^{n}Bu_4N]_2[\{Ru(dmso\text{-}S)Cl_4\}_2$-$(\mu\text{-pyz})]$ in DMF/0.1 M $[^{n}Bu_4N]BF_4$ at 217 K, $E_{gen} = -0.60$ V

The observation of two redox processes and the value of the comproportionation constant, K_c, (Table 1) indicate that the mixed valence compound [{Ru(dmso-S)Cl$_4$}$_2$(μ-pyz)]$^{3-}$ is a Class II species ie that there is a weak metal-metal interaction which will be mediated through the pyz bridge. Thus the ruthenium centres are reduced stepwise from Ru(III)Ru(III) to Ru(III)Ru(II) to Ru(II)Ru(II).

The uv/vis spectroelectrochemistry of the two electron reduction of ["Bu$_4$N]$_2$[{Ru(dmso-S)Cl$_4$}$_2$(μ-pyz)] in DMF at 217 K is shown in Figure 8. The dianion has a maximum absorbance at 24350 cm^{-1} and the tetra-anion a maximum absorption at 19650 cm^{-1}. The spectra show clearly that the reduction proceeds *via* two steps as indicated by one isosbestic point at 22 000 cm^{-1} for the initial reduction process followed by a separate isosbestic point at 23 350 cm^{-1} for the second process. Thus the spectra can be separated into the first one electron reduction and then the second one electron reduction [80]. Figure 9 shows the spectra of [{Ru(dmso-S)Cl$_4$}$_2$(μ-pyz)]$^{2-}$, [{Ru(dmso-S)Cl$_4$}$_2$(μ-pyz)]$^{3-}$ and [{Ru(dmso-S)Cl$_4$}$_2$(μ-pyz)]$^{4-}$ in CH$_2$Cl$_2$ at 217 K. Table 3 gives the band maxima and electronic assignments.

The uv/vis spectrum of the fully oxidised species (formally Ru(III)/Ru(III)) is very similar to that of the monomeric Ru(III) complex (Fig.6, Table 2) and is dominated by a Cl$^-$ $\longrightarrow$ Ru t$_{2g}^5$ LMCT transition at 24 350 cm^{-1}

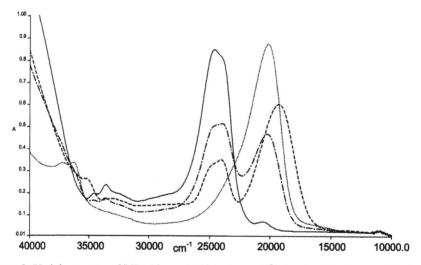

Figure 9. Uv/vis spectra of [{Ru(dmso-S)Cl$_4$}$_2$(μ-pyz)]$^{2-}$ (——), [{Ru(dmso-S)Cl$_4$}$_2$(μ-pyz)]$^{3-}$ (------), [{Ru(dmso-S)Cl$_4$}$_2$(μ-pyz)]$^{4-}$ (·····) in CH$_2$Cl$_2$/0.4 M ["Bu$_4$N]BF$_4$ at 217 K and calculated spectrum of the disproportionation reaction (·-·-·).

(ε = 18 600 M^{-1} cm^{-1}) compared to 24050 cm^{-1} (ε = 7 200 M^{-1} cm^{-1}) in the mononuclear species. The extinction coefficient of the band in the binuclear complex is much larger than the corresponding band in the mononuclear compound as expected since the binuclear species has two Ru(III)Cl$_4$ fragments compared to one in the mononuclear.

Table 3. Absorption band characteristics and electronic assignments for [^{n}Bu$_4$N]$_2$[{Ru(dmso-S)Cl$_4$}$_2$(μ-L)]$^{z-}$ at 217 K.

L	z	Cl$^-\rightarrow$ Ru(III)^a	Ru(II) $\rightarrow$ L	Solvent
pyz	2	24350 (18 600)		DMF
	3	24070	18600	DMF
	4		19650 (20600)	DMF
pyz	2	24600		C$_2$H$_5$CN
	3	24100	19150	C$_2$H$_5$CN
	4		20050	C$_2$H$_5$CN
pyz	2	24660		CH$_2$Cl$_2$
	3	24070	19350	CH$_2$Cl$_2$
	4		20250	CH$_2$Cl$_2$
pym	2	24200 (10350)		DMF
	4		24900 (6150)	DMF
bpy	2	24820 (12000)		DMF
	4		21300 (11900)	DMF
bpa	2	24420 (11000) 33350 (8900)		DMF
	4		27210 (10900)	DMF
bpe	2	25130 (12300)		DMF
	4		19780 (7300)	DMF

The uv/vis spectrum of the fully reduced species (formally Ru(II)/Ru(II)) resembles that of the dianionic Ru(II) mononuclear compound. It is dominated by a MLCT transition, from the Ru t$_{2g}$ set of orbitals to the antibonding π^* orbital of the pyrazine ligand at 19 650 cm^{-1} (ε = 20600 M^{-1} cm^{-1}) compared to 22300 cm^{-1} (ε = 7150 M^{-1} cm^{-1}) in [RuCl$_4$(dmso-S)(pyz)]$^{2-}$. The low energy shift of the MLCT band in the Ru$_2$ compound compared to the mononuclear complex is a result of the bridging pyz ligand bonding to two Ru(II) centres rather than one.

Similar spectral changes were observed in the chemical reduction of [{Ru(dmso-S)Cl$_4$}$_2$(μ-pyz)]$^{2-}$ to [{Ru(dmso-S)Cl$_4$}$_2$(μ-pyz)]$^{4-}$ by ascorbic acid [25]. The mixed valence state was not identified in the chemical reduction.

The spectrum of the mixed-valence species [{Ru(dmso-S)Cl$_4$}$_2$(μ-pyz)]$^{3-}$ is a combination of the spectrum of the individual Ru$_2$(II) and Ru$_2$(III) species with slightly perturbed features. The spectrum of [{Ru(dmso-S)Cl$_4$}$_2$(μ-pyz)]$^{3-}$ is not simply an average of the Ru$_2$(III) and Ru$_2$(II) spectra which would arise if the mixed valence compound disproportionates into the fully oxidised and reduced species (see Fig. 9), i.e., 2 Ru(III)Ru(II) $\rightleftharpoons$ Ru$_2$(III) + Ru$_2$(II). Note that in [{Ru(dmso-S)Cl$_4$}$_2$(μ-pyz)]$^{3-}$ the MLCT band is lower in energy than in the fully reduced Ru$_2$(II) complex since in the mixed valence compound the pyz ligand is bound to a Ru(II) and a Ru(III) centre whereas in [{Ru(dmso-S)Cl$_4$}$_2$(μ-pyz)]$^{4-}$ the pyz is bound to two t$_{2g}^6$ Ru centres which will both π-back bond to the bridging ligand raising the level of the π* antibonding orbital and hence cause the MLCT band to occur at higher energy.

All the evidence points to the mixed-valence compound having Class II character and therefore the spectrum of the mixed valence compound might be expected to contain an intervalence charge transfer transition. Repeating the spectroelectrochemical experiment on a more concentrated solution of [{Ru(dmso-S)Cl$_4$}$_2$(μ-pyz)]$^{2-}$ resulted in the spectra shown in Fig. 10.

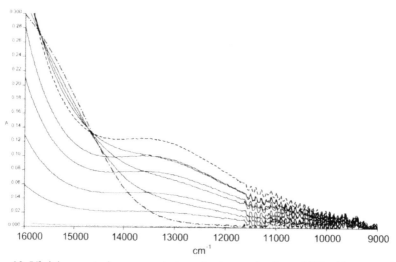

Figure 10. Vis/nir spectral response to two electron reduction of [{Ru(dmso-S)Cl$_4$}$_2$(μ-pyz)]$^{2-}$ in C$_2$H$_5$CN/0.1 M [^{n}Bu$_4$N]BF$_4$ at 217 K, E$_{app}$ = -0.75 V, [{Ru(dmso-S)Cl$_4$}$_2$(μ-pyz)]$^{2-}$ (······),[{Ru(dmso-S)Cl$_4$}$_2$(μ-pyz)]$^{3-}$ (– – – –),[{Ru(dmso-S)Cl$_4$}$_2$(μ-pyz)]$^{4-}$ (· – · –). Solid lines are spectra of intermediate solutions between 2-, 3- and 4- states.

A low energy band ε_{max} = 13500 cm^{-1} ($\varepsilon \sim$ 150 M^{-1} cm^{-1}) in C$_2$H$_5$CN is observed to grow in as the compound is reduced by one electron. Addition of a second electron leads to the disappearance of this band. The energy of the IVCT band depends on the solvent, see Table 4. It does not correlate with $1/D_{op}$ – $1/D_s$ as predicted by Hush [111] but does with the Gutmann acceptor number [112]. However it should be noted that the energy of the charge transfer bands in [{Ru(dmso-S)Cl$_4$}$_2$(μ-pyz)]$^{3-}$ are also solvent dependent, see Table 3.

Table 4. Position of IVCT transition in various solvents for [{Ru(dmso-S)Cl$_4$}$_2$(μ-pyz)]$^{3-}$.

Solvent	Acceptor Number	IVCT band max/cm^{-1}
pyridine	14.2	12 300
benzonitrile	15.5	12 700
dimethylformamide	16.0	12 500
acetonitrile	18.9	13 100
dichloromethane	20.4	13 200
chloroform	23.1	13 400

Thus [{Ru(dmso-S)Cl$_4$}$_2$(μ-pyz)]$^{2-}$ undergoes two separate reversible one-electron reductions. The separation between these two couples depends on both the solvent and the counter ion. A significant difference between the CT ion and the binuclear compound investigated above, apart from the net overall charges, is that the stable form of the CT ion contains Ru(II)Ru(II) centres and the mixed valence state is achieved by one electron oxidation whereas the ground state of the [{Ru(dmso-S)Cl$_4$}$_2$(μ-pyz)]$^{2-}$ dianion has Ru(III)Ru(III) centres and the d^5/d^6 configuration is gained *via* a one electron reduction. The mixed valence state [{Ru(dmso-S)Cl$_4$}$_2$(μ-pyz)]$^{3-}$ has Class II character ie a weaker metal-metal interaction than shown by the CT ion. The difference in the redox potentials (and hence the K_c value) is much smaller for the [{Ru(dmso-S)Cl$_4$}$_2$(μ-pyz)]$^{3-}$ species compared to the CT ion, see Table 5. Thus on replacement of the σ-only donor NH$_3$ ligands by π-donating chloride ligands the Ru-Ru coupling is reduced to a similar degree as found in [{Ru(bpy)$_2$Cl}$_2$(μ-pyz)]$^{3+}$ [19].

The complexes [{Ru(dmso-S)Cl$_4$}$_2$(μ-L)]$^{n-}$, L = pym, bpy, bpe, bpa, all exhibit significantly decreased Ru-Ru interactions, compared to the pyz bridged analogue, as evidenced by their electrochemical and spectroelectrochemical behaviour.

N N N N N N N—(CH$_2$)$_2$—N N—(CH)$_2$—N

pyz   pym   bpy   bpa   bpe

Thus their cyclic voltammograms only show one reduction wave which involves two electrons as determined by coulometry. Half wave potentials are given in Table 1. Comparison of the mono- and binuclear pym complexes shows that both species exhibit a single reduction wave with the half wave potential of the binuclear species being 25 mV more negative than the mononuclear compound. Thus formation of the binuclear complex has not resulted in significant communication between the metal centres. Previous studies with bridging pym has also shown that the transmission of electronic effects between groups bound to the *m*-nitrogen atoms through the π^* system of the diazine ring is relatively ineffective [115]. The half wave potentials of the binuclear complexes for L = pym, bpy, bpe and bpa are all similar which has also been reported for the fragment Ru(bpy)$_2$Cl connected by the same bridging ligands [114,116]. However in that instance L = pyz and pym both showed two separate oxidation processes with L = bpy, bpa and bpe having only a single two-electron redox process.

Table 5. Separation between the two redox couples ($\Delta E = E_1 - E_2$) and K$_c$ values for the pyz bridged metal fragments.

Fragment	ΔE/mV	K$_c$
Ru(NH$_3$)$_5$	430	$10^{7.3}$ [113]
Ru(bpy)$_2$Cl	120	10^2 [114]
Ru(py)$_4$Cl	280	$10^{4.7}$[114]
RuCl$_4$(dmso-S)	150	10^2

The uv/vis spectroelectrochemical results for the reductions of [{Ru(dmso-S)Cl$_4$}$_2$(μ-L)]$^{2-}$, L = pym, bpa are shown in Fig. 11 and the peak maxima are given in Table 3 which also includes the data for L = bpy, bpe. Close inspection of the spectra generated during the reduction of [{Ru(dmso-S)Cl$_4$}$_2$(μ-pym)]$^{2-/4-}$, Figure 11a, shows two different sets of isosbestic points. Thus there may be a small but real interaction between the two metal centres. However no features unique to the mixed valence state could be distinguished. The spectrum of the fully oxidised species is very similar to that of the mononuclear Ru(III)-pym complex, see Figure 6 and Table 2, and is dominated by a $Cl^- \to Ru\ t_{2g}^5$LMCT transition. The band positions are identical but more intense, as expected, for the binuclear species since there are two equivalent ruthenium centres contributing to the band compared to the monomer. The spectrum of the fully reduced species is also very similar to the spectrum of the reduced Ru(II)-pym mononuclear species with a MLCT (Ru(II) $\to$ pym π^*) transition centred at 24900 cm^{-1}. Thus the spectra of the fully oxidised Ru(III)/Ru(III) and the fully reduced Ru(II)/ Ru(II) compounds are very similar to the spectra of the monomer containing isoelectronic ruthenium centres. The Ru-Ru interaction must be minimal. Smaller Ru-Ru interactions with pyrimidine compared to pyrazine have also been noted for the fragments Ru(bpy)$_2$Cl and Ru(NH$_3$)$_5$ [116,117].

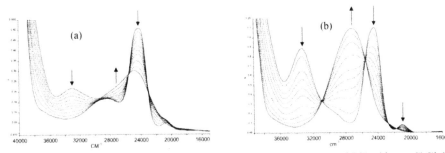

Figure 11. Uv/vis spectral response for two electron reduction of [{Ru(dmso-S)Cl$_4$}$_2$-(μ-L)]$^{2-}$ in DMF/0.1 M [^{n}Bu$_4$N]BF$_4$ at 217 K, E$_{app}$ = –0.75 V; (a) L = pym, (b) L = bpa.

The spectroelectrochemical study of the related complexes containing the bridging ligands bpy, bpa, bpe all showed similar results to the bpa case, Figure 11b and Table 3. A single set of isosbestic points is observed and the fully oxidised spectra exhibit $Cl^- \to Ru(III)$ charge transfer transitions at similar energies. The fully reduced spectra show MLCT bands, Ru(II) $t_{2g}^6 \to L\ \pi^*$, whose energy depends on the electronic character of the bridging ligand ie is at low energy for the fully conjugated π-ligand bpe (19780 cm^{-1}) and at considerably higher energy in bpa (27210 cm^{-1}). The complex [{Ru(dmso-S)Cl$_4$}$_2$(μ-bpa)]$^{2-}$ has a band at 33350 cm^{-1}, Figure 11b, which is assigned to a charge transfer transition involving the bpa ligand and collapses on reduction.

Thus the binuclear species containing the Ru(dmso-S)Cl$_4$ fragments bridged by pyz, pym, bpy, bpa and bpe ligands all show significantly decreased Ru-Ru interactions compared to the complexes with Ru(NH$_3$)$_5$ fragments. The metal-metal interaction must be bridging ligand mediated since the spacial distance between the metals precludes direct interaction. Consideration of the Ru-based t_{2g} orbitals in the Ru(dmso-S)Cl$_4$ ligated moieties indicates that the redox active metal based orbital is primarily located in the RuCl$_4$ plane with little direct overlap with the bridging ligand which is in agreement with the electrochemical and spectroelectrochemical results presented.

3.2.2 Biological Systems

The spectroelectrochemical experiments where the uv/vis spectrum is recorded as a function of applied potential is a useful method for measuring the half wave potential of a redox process and the number of electrons transferred in that process [118-120]. It is a particularly good method when the electron transfer process is slow at an electrode surface and the resultant cyclic voltammogram too broad to analyse. Biological molecules often fall into this category [121,122]. They are usually large and the redox active centre may be buried within the molecule making direct electrochemical study of the half-wave potential difficult. The solvent system containing the biological species of interest may or may not contain molecular mediators to aid electron transfer. The spectroelectrochemical technique using a gold minigrid OTE was first applied to biological molecules, in particular, to horse heart cyctochrome c in the presence of redox mediators in 1975 [123]. Other proteins investigated spectroelectrochemically include cytochrome c [124-127], myoglobin [127,128], blue Cu protein [127,129], cytochrome c_3 [130]. Two further examples from our laboratories, employing the OTE cell described in Fig.1, are discussed below.

The first example details the spectroelectrochemical study of the Co corrinoid [131], aquohydroxo-cobinamide (Cbi). Co corrinoids are derivatives of vitamin B$_{12}$ in which the central Co(III) metal is ligated to four nitrogen donor atoms (corrin ligand) in the equitorial plane and a further two axial ligands. A direct cyclic voltammetric study of diaquo-Co(III) Cbi in an aqueous solution/1 M Na$_2$SO$_4$/0.1 M NaNO$_3$ (pH = 7) at 293 K on a glassy carbon working electrode does not reveal the Co(III)/Co(II) couple but does show a very broad Co(II)/Co(I) couple centred at –0.73 V with a peak-to-peak separation of 190 mV. The reduction process Co(II) + e$^-$ → Co(I) appears to be considerably more sluggish than the reoxidation step Co(I) → Co(II) + e$^-$ and may be related to catalytic hydrogen discharge [132].

The spectroelectrochemical behaviour of aquohydroxo-Co(III) Cbi at pH = 7 was investigated using applied potentials from +0.504 V (Co(III)) to –1.008 V (Co(I)). Figure 12a shows the spectral changes accompanying the applied voltage changes from +0.504 to –0.112 V, that is Co(III) reducing to Co(II). Fig 12b shows the spectra associated with the applied potential varying from –0.608

to –1.008 V, that is, Co(II) reducing to Co(I). The spectral changes shown in Fig. 12a are very similar to those previously reported [132]. The fully oxidised Co(II) spectrum is dominated by absorption bands at 526, 495 and 350 nm and the Co(II) spectrum shows maximum absorbances at 471 and 313 nm. Further reduction to Co(I) results in the collapse of the characteristic Co(II) bands and the growth of Co(I) peaks at 387 and 288 nm.

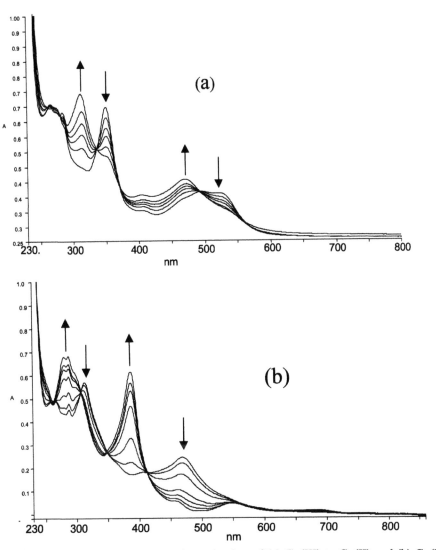

Figure 12. Uv/vis changes on stepwise reduction of (a) Co(III) to Co(II) and (b) Co(II) to Co(I) in aquohydroxo-Cbi at 293 K

The Nernst equation for the redox reaction O + ne $\rightleftharpoons$ R is given by Eq.1:

$$E_{app} = E^{o\prime} + (RT/nF)\ln\{[O]/[R]\} \tag{1}$$

Where E_{app} = applied potential, $E^{o\prime}$ = formal redox potential, $[O]$, $[R]$ = concentration of oxidised and reduced species respectively. The ratio of the concentrations of oxidised to reduced forms of the redox couple can be equated to absorbance values *via* the Beer-Lambert law as in Eq 2:

$$[O]/[R] = (A_R - A)/(A - A_O) \tag{2}$$

where A is the absorbance at a given applied potential and A_R and A_O are the absorbance values for the fully reduced and oxidised forms at the monitored wavelength. A plot of E_{app} against $\ln\{(A_R - A)/(A - A_O)\}$ will be linear with $E^{o\prime}$ occurring when $\ln\{(A_R - A)/(A - A_O)\} = 0$ and with a slope of 59 mV for a one electron transfer at 293 K.

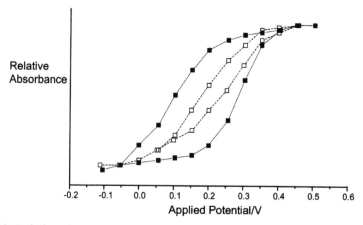

Figure 13. Relative absorbance *vs.* applied potential for aquohydroxo-Cbi at 530 nm, solid squares = 1×10^{-5} mol dm^{-3}, open squares = 2×10^{-5} mol dm^{-3}.

The aquohydroxo-Co(III) Cbi spectra were recorded as a function of applied potential. Note that equilibrium was established for every potential. Analysis of the Co(III)/Co(II) spectra at wavelengths 313, 350, 471 and 525 nm (chosen for maximum absorbance differences between the Co(III) and Co(II) spectra) gave a value for $E^{o\prime}$ of +240 mV *vs* sce and a slope of 58 mV. Similar plots at 387 and 470 nm for the Co(II)/Co(I) couple determines a value of –750 mV *vs* sce for this couple.

On reversing the applied potential back to +0.500 V to check the chemical reversibility of the system it was noted that the $E^{o'}$ potential for the Co(II) → Co(III) + e⁻ oxidation occurred at a significantly different potential than the forward reduction process Co(III) + e⁻ → Co(II) with $E^{o'}$ Co(III)/Co(II) measured at a less positive potential than the Co(II)/Co(III) process. Thus a plot of absorbance versus applied potential for the Co(III)/Co(II) reduction and Co(II)/Co(III) oxidation exhibits hysteresis, see Figure 13.

The width of the hysteresis loop is found to be concentration dependent and shows a separation of 200 mV at 1×10^{-5} mol dm⁻³ but a reduced separation of 70 mV at 2×10^{-5} mol dm⁻³. The observation of hysteresis indicates that the Cbi must exist in two forms with differing redox potentials but very small differences in their uv/vis spectra. The sequence $ox_A \rightarrow red_A \rightarrow red_B \rightarrow ox_B \rightarrow ox_A$, which involves a change of conformation (A, B) as well as oxidation state (ox, red: Co(III), Co(II)), where ox_A and red_B are the most stable forms, may explain the experimental hysteresis [133-135]. The decrease in width with increasing concentration suggests some sort of redox-catalysed interconversion may be occurring.

The second example discusses the spectroelectrochemical study of the two one-electron transfers associated with the flavin mononucleotide (FMN) domain of neuronal NO synthase (nNOS). The fully oxidised form of FMN is yellow and has a maximum absorbance around 450 nm. It is converted by a two-electron reduction to the colourless hydroquinone form FMNH⁻ or FMNH₂. In the FMN domain of nNO synthase the semi-quinone form is stabilised giving the blue free radical FMNH• with $\lambda_{max} \sim 592$ nm, see Figure 14, where R = $CH_2(CH(OH))_3(CH_2)OPO_3$.

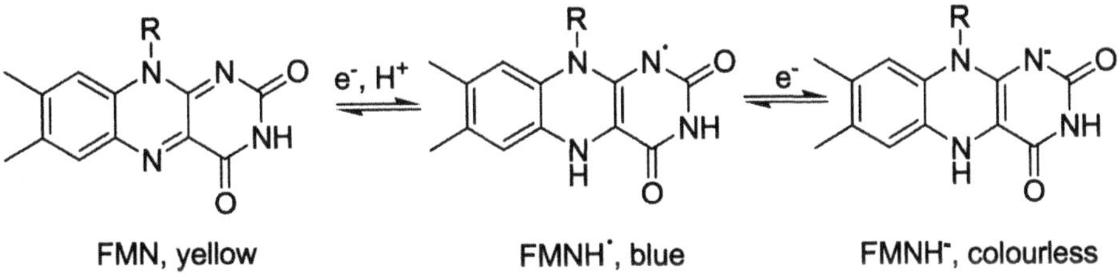

FMN, yellow   FMNH•, blue   FMNH⁻, colourless

Figure 14. Oxidation states of FMN

Previous methods of determining half wave potentials for flavoproteins include redox potentiometry and chemical reduction with dye indicators. The spectroelectrochemical conditions used in the experiments detailed below are: 250 μM nNOS FMN domain in 100 mM Tris/HCl, 0.5 M KCl at pH 7.5.

Mediators used were 2-hydroxy-1,4-naphthaquinone (13 µM), FMN (6 µM), pyocyanine (10 µM), benzyl viologen (6 µM) and methyl viologen (8 µM). The spectral changes accompanying the electron transfer processes in neuronal NO synthase FMN domain are shown in Figure 15. The spectrum of the fully oxidised FMN form has a maximum absorbance at 457 nm, the radical species, FMNH$^{\bullet}$, a maximum at 592 nm and the fully reduced form FMNH$^-$ has no absorbance maximum in the visible region of the spectrum. Thus the two one-electron transfer processes are accompanied by a steady collapse of the 457 nm band and the growth of the 592 nm band as the radical species FMNH$^{\bullet}$ is formed followed by its decay on further reduction to FMNH$^-$. Figure 16 shows the change in absorbance at 470 nm and 592 nm, O and Δ respectively, as a function of applied potential.

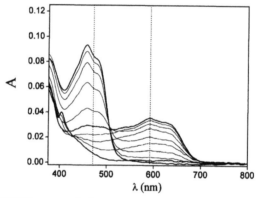

Figure 15. UV/Vis spectral changes on reduction of FMN domain in nNOS.

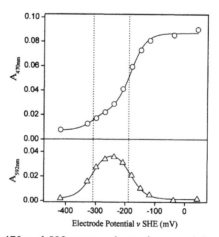

Figure 16. Absorbance at 470 and 592 nm vs. electrode potential

The degree of reduction, measured by the absorbance at either 470 or 592 nm, can be fitted to Eq. 3, derived by extension of the Nernst equation in combination with the Beer-Lambert law[136].

$$\text{flavin absorbance} = \frac{a10^{(E-E_1')/59} + b + c10^{(E_2'-E)/59}}{1+10^{(E-E_1')/59}+10^{(E_2'-E)/59}} \tag{3}$$

where a, b, c are the absorption coefficients for FMN, FMNH$^\bullet$ and FMNH$^-$ respectively, E is the electrode potential, E_1' and E_2' are the formal potentials for the couples below.

Analysis of the data in Figure 16 gives the formal potentials:

$$FMN + e^- + H^+ \rightarrow FMNH^\bullet \quad E_1' = -183 \text{ mV } \textit{vs.} \text{ she} \quad \text{electron transfer 1}$$
$$FMNH^\bullet + e^- \rightarrow FMNH^- \quad E_2' = -310 \text{ mV } \textit{vs,} \text{ she} \quad \text{electron transfer 2}$$

It is interesting to note that the electron transfer step 1 above is slow and hence the system takes a long time to equilibrate on alteration of the applied potential $E_{app} = -183 \pm 20$ mV, see Figure 17. Inspection of the FMN site in the related enzyme cytochrome P450 reductase, Fig. 18, reveals that N(5) is hydrogen bonded to the peptide amide nitrogen of Glycine 810 (of neuronal NO synthase) and this linkage must be broken on reduction. This may account for the slow equilibration time of electron transfer step 1 above. Reversal of the applied potential from −418 to 0 mV gives identical spectral results to those shown, that is, providing the system is given sufficient time to adjust there is no hysterises observed at either electron transfer step.

Spectroelectrochemical methods of determining electron transfer potentials and the number of electrons transferred in the process have several advantages over redox titration methods, particularly for biological systems. Firstly, the

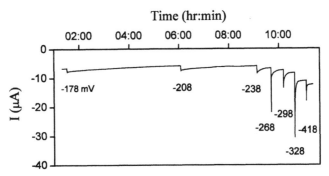

Figure 17. Current vs. time profile for given electrode potentials.

solution containing the redox active species is not contaminated by reducing/ oxidising agents and therefore may be reused indefinitely. In addition a more accurate measurement of the formal electrode potential may be measured as the applied potential can be continuously varied over a wide voltage range. Absorption problems from precipitating protein are minimised as stirring of the solution and addition of chemical reductants/oxidants is unnecessary. Finally, the limitations of the redox ranges available to redox dyes are circumvented.

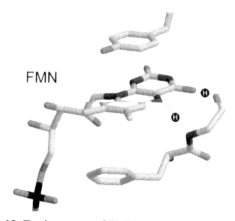

Figure 18. Environment of FMN in rat neuronal NO synthase.

4   EPR SPECTROELECTROCHEMISTRY

In situ epr spectroelectrochemistry experiments are not as frequently reported as uv/vis cases. This is somewhat surprising since one-electron redox processes should involve a paramagnetic species either initially or as the product of the one-electron transfer. There are several reports detailing the epr spectra of electron transfer processes where the electrogeneration step has been carried out *ex situ* and the paramagnetic solution transferred to an epr tube for study [61,82,137-141]. Such studies are not discussed further since this paper is concerned with *in situ* technologies [142-144]. Epr spectroelectrochemical experiments are most often used to determine the site of redox activity [145]. Examples of *in situ* epr cells from the recent literature are discussed below.

4.1   Recent Developments in epr spectroelectrochemistry

Hartl and Vlcek Jr. used a Pt minigrid electrode in an epr tube to study the redox chemistry of $[Mn(CO)_3(DBCat)]^{1-}$, DBCat = 3,5-di-tert-butyl catecholate in THF [146]. The solution epr spectrum of the one-electron reduction product exhibited features typical of a metal centred radical with a(Mn) = 53.4 G whereas the one-electron oxidation is centred on the DBCat ligand. The oxidation is followed by a chemical reaction resulting in formation of [Mn(CO)$_3$(THF)(DBSQ)] which exhibits an epr signal with a small coupling to Mn with a(Mn) = 3.7 G, that is, the unpaired electron is primarily based on the semi-quinone ligand [146].

Hartl *et al* describe a low temperature, air tight epr cell based on a round tube design [147]. The Au working electrode has a helical shape with the Pt counter electrode placed down the middle of the working electrode [148]. The electrochemical circuit is completed by a Ag wire pseudo-reference electrode. The small solution volume between the working electrode and the cell wall permits the use of solvents such as CH$_3$CN. The cell was used to study the one-electron oxidation of 3,6-diphenyl-1,2-dithiine in CH$_2$Cl$_2$ at 233 K whose epr signal consisted of a 1:2:1 triplet, a(H) = 3.5 G from coupling of the unpaired electron to two equivalent hydrogen nuclei. Other systems studied were the radical anions of 6-methyl-6-phenyl fulvene and *fac*-[Re(Bn)(CO)$_3$(dmb)]$^-$, where Bn = CH$_2$Ph, dmb = 4,4'-dimethyl-2,2'-bipyridine.

3,6-diphenyl-1,2-dithiine

A comparative study by Vlcek *et al* on [M(CO)$_4$(N,N)], M = Cr or W, N,N = 1,10-phenanthroline or 3,4,7,8-tetramethyl-1,10-phenanthroline, shows that on N,N-based reduction a ^{2}B$_1$ ground state arises for N,N = phen but a ^{2}A$_2$ ground state is found for the tetramethyl analogue [149].

Bond *et al* have developed a small volume (0.2 ml) variable temperature epr spectroelectrochemical cell which enables simultaneous rapid scan voltammetry and epr measurements to be made [150,151]. The performance of this cell is compared to that of a flow-through cell designed by Coles and Compton [152,153]. The small volume cell permits cyclic voltammetric studies at variable temperatures but has significantly lower sensitivity compared to the flow-through cell which is not amenable to low temperature work.

In 1990 Kaim *et al* detailed an *in situ* two electrode epr cell with a Pt tip working electrode and a Pt wire electrode contained in an epr tube with an internal diameter of 1 mm [154]. Radical species are generated by short electrolysis times. The epr signal for the *in situ* one-electron reduction of [(η^5-C$_5$Me$_5$)ClRh(μ-bptz)Re(CO)$_3$Cl]PF$_6$, bptz = 3,6-bis(2-pyridyl)-1,2,4,5-tetrazine is very similar to the spectrum from [(bptz)Re(CO)$_3$Cl]$^-$ and shows that the reduction electron enters a bptz ligand-based molecular orbital [155]. A similar study on [(OC)$_3$ClRe]$_n$(abpy), n = 1, 2 and abpy = 2,2'-azobispyridine concludes that the site of reduction is the abpy ligand [156].

A laminated platinum-mesh working electrode positioned in an epr flat cell combined with a Ag pseudo-reference electrode and a Pt counter electrode was used by Dunsch *et al* for the *in situ* study of $C_{120}O$ [157]. The same group also performed simultaneous variable temperature epr/uv/vis/nir experiments on Wurster's reagent and thianthrene [158,159]. Similar cells using laminated ITO and gold working electrodes have also been reported [160,161].

Wurster's reagent   Thianthrene

An epr microspectroelectrochemical low temperature cell has been detailed by Wilgocki and Rybak with a Pt working electrode and has been used to characterise the unstable cation $[O=Re(OEt)Cl_2(py)_2]^+$ [162].

4.2 Results and Discussion

The original low temperature epr cell used in this laboratory had a cylindrical design with a Pt tube working electrode, a Ag/AgCl reference electrode placed inside the tube and a Pt counter electrode [34]. This epr cell gave acceptable spectra but closely spaced signals could not be resolved. The epr cell therefore underwent a complete design change and the new cell based on a variable temperature flat cell was developed as shown in Fig. 2 in the experimental section [33]. Use of the flat cell enables high dielectric constant solvents, such as DMF or CH_3CN, which are good electrochemical but poor epr solvents, to be employed. The cell has been successfully used over the temperature range 100 – 350 K. Fig. 19 shows the growth of the epr signal as $[Pt(bpy)Cl_2]$, bpy = 2,2'-bipyridine, is reduced by one-electron in DMF/0.1 M $[^nBu_4N]BF_4$ at 253 K. The diamagnetic $[Pt(bpy)Cl_2]$ is epr silent whereas the one-electron reduction product $[Pt(bpy)Cl_2]^{1-}$ is paramagnetic and is responsible for the growth of the signal observed in Figure 19. The epr signal arises from coupling of the unpaired electron to ^{195}Pt ($I= \frac{1}{2}$, natural abundance 34%). The resolution of this cell is much improved over the cylindrical design as illustrated by the study of $[Pt(4,4'-^tBu_2-bpy)(C_2(C_6H_4-p-NO_2))_2]$, Figure 20.

The first spectroelectrochemical report of the two-electron reduction of $[Pt(4,4'-^tBu_2-bpy)(C_2(C_6H_4-p-NO_2))_2]$ in DMF/0.1 M $[^nBu_4N]BF_4$ at 293 K concluded that the sites of reduction were the two $(-C\equiv C(Ph-NO_2-p))$ ligands which were electronically non-communicating [30]. The experimental and simulated solution epr spectra of the dianion recorded in the cylindrical cell are

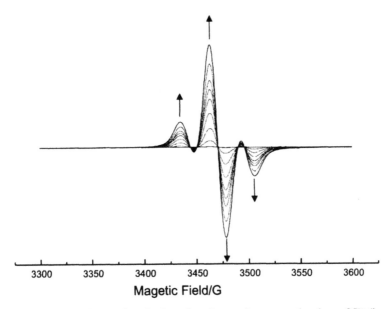

Figure 19. Epr spectroelectrochemical study of one-electron reduction of [Pt(bpy)Cl$_2$] (E$_{gen}$ = –1.6 V) in DMF/0.1 M [^{n}Bu$_4$N]BF$_4$ at 253 K.

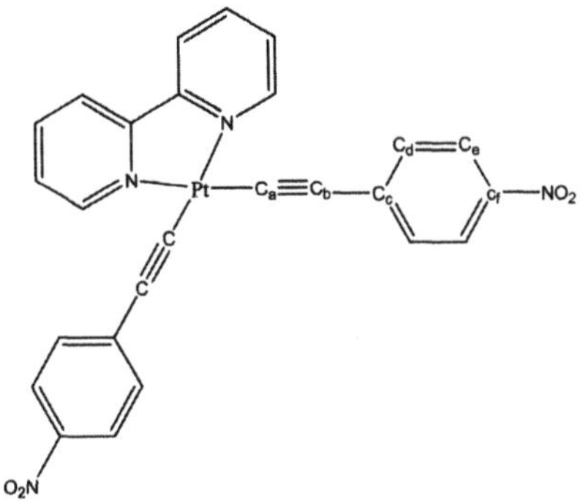

Figure 20. [Pt(4,4'-^{t}Bu$_2$-bpy)(C$_2$(C$_6$H$_4$-p-NO$_2$))$_2$]

shown in Figure 21a. Repeating the study in the variable temperature flat cell results in the significantly more resolved spectrum given in Fiure 21b.

Note that the spectra in Fig. 21a contain the whole spectral range of 100 G whereas the more resolved spectra in Figure 21b only contain the low field half of the signal. Both spectra in Figure 21 a and b can best be simulated to spin ½ systems. The spectrum in Figure 21b shows coupling of the unpaired electron to the Pt nucleus, one N atom and two sets of two equivalent protons and can be successfully simulated using the parameters noted in Figure 3b. Assignment of the coupling to specific nuclei was done using the results of semi-empirical molecular orbital calculations at the PM3 level. These calculations reveal that the LUMO of $[Pt(4,4'-{}^tBu_2-bpy)(C_2(C_6H_4-p-NO_2))_2]$ is based on one $(C_2(C_6H_4-p-NO_2))$ ligand and that the spin densities in the carbon $2p_z$ orbital contributing to the π system (ρ_C^π) on C_e and C_d, Figure 20, are 0.130 and 0.055 respectively. Thus the reduction electron is coupling to the Pt nucleus (15.20 G), the N atom of the NO_2 group (9.59 G), two equivalent protons attached to C_e (3.32 G) and two equivalent protons bonded to C_d (1.16 G). The McConnell equation, Eq. 4, which relates the hydrogen hyperfine coupling value to the spin density, ρ_C^π, [163], where Q is a semi-empirical parameter of the order of 24 G [145] can be used to check the above assignments.

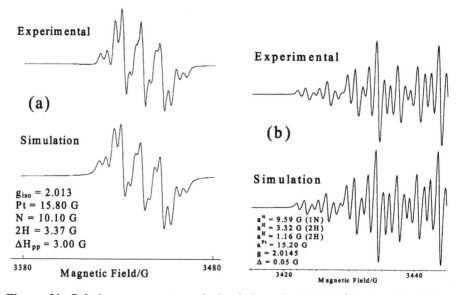

Figure 21. Solution epr spectra and simulations for $[Pt(4,4'-{}^tBu_2-bpy)(C_2(C_6H_4-p-NO_2))_2]^{2-}$ in DMF/0.1 M $[{}^nBu_4N]BF_4$ in (a) cylindrical cell, (b) flat cell.

$$a^H = Q\rho_C^\pi \tag{4}$$

Feeding the values obtained above for a^H and ρ_C^π into Eq. 4 gives values of Q of 25.5 and 21.1 G for C_e and C_d respectively.

The power of *in situ* generation of radical species is fully exploited in the study of a redox series. In such series several redox states can be characterised from the starting material which undergoes more than one reversible reduction/ oxidation process. The site of redox activity accompanying the stepwise reductions or oxidations is difficult to probe by means other than *in situ* spectroelectrochemical studies since it is imperative that complete generation of each of the redox partners is achieved and recorded for successful assignment. The use of a large surface area working electrode makes accomplishment of 100% electrogeneration of the sequential redox species possible. In addition, use of a reference electrode such as Ag/AgCl or sce rather than a pseudo-reference electrode is preferred so that the exact value of E_{app} is known.

The donor-acceptor (diad) molecule 5,10,15,20-[*N*-benzyl -*N'*-(4-benzyl-4,4'-bipyridinium-4-pyridyl)]-triphenylporphyrin tris(hexafluorophosphate, Figure 22, comprises a porphyrin ring (donor) covalently attached to a dibenzylviologen unit (acceptor). The diad has a one-electron reduction at −0.30 V followed by two closely spaced one-electron reductions centred at −0.78 V in $CH_3CN/0.1$ M[^{n}Bu$_4$N]BF$_4$ [31].

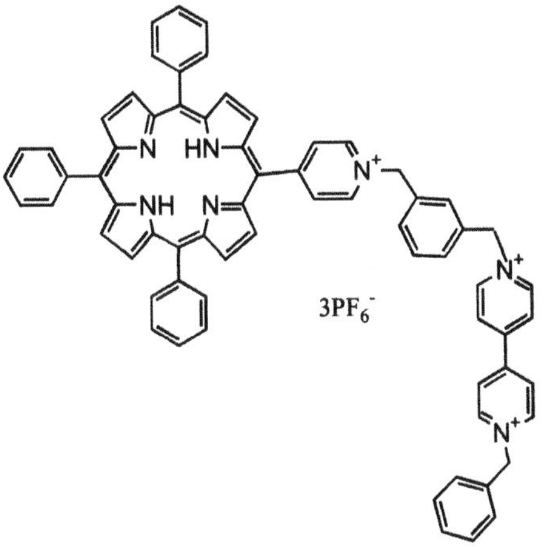

Figure 22. Structure of 5,10,15,20-[*N*-benzyl-*N'*-(4-benzyl-4,4'-bipyridinium-4-pyridyl)]-triphenylporphyrin tris(hexafluorophosphate.

Epr spectroelectrochemistry at 233 K elucidated which centres on the molecule were reduced. Reduction at −0.61 V results in the epr spectrum shown in Figure 23a which is best simulated by coupling of the unpaired electron to three equivalent nitrogens (a(N) = 4.08 G) and two sets of six equivalent protons (a(H) = 1.41, 1.00 G). Consideration of the structure of the diad in Fig. 22 suggests that the redox site at −0.30 V be attributed to the acceptor part of the molecule which has three equivalent nitrogen sites. It is likely that six H couplings observed

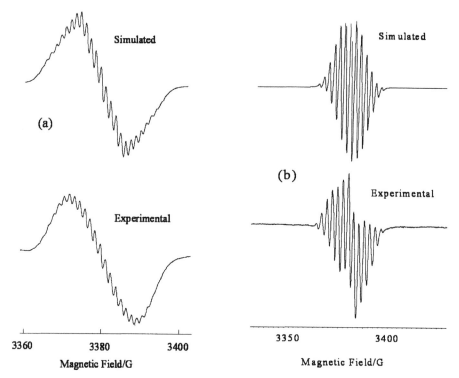

Figure 23. epr, experimental and simulated, spectra for reduced diad in DMF/0.2 M [^{n}Bu$_4$N]BF$_4$ at 233 K; (a) one-electron reduction product E_{gen} = –0.61 V, g = 2.012; (b) three-electron reduction product E_{gen} = –1.40 V, g = 2.011. Simulation parameters as in text and line width = 1.00 G.

in the epr spectrum are due to methylene protons (2 per nitrogen) and six are aromatic protons (again 2 per nitrogen) adjacent to N$^+$.

On application of a potential of –1.40 V the epr signal decays as the system becomes diamagnetic. Thus the second reduction electron probably enters the same orbital as the first on the acceptor part of the compound. A new signal subsequently develops as a third electron is added to the molecule, see Figure 23b. The third electron enters a molecular orbital based to the porphyrin ring since the epr signal is best simulated using four equivalent nitrogen nuclei (2.65 G) and two equivalent protons (5.40 G) and this fragment is clearly to be found at the heart of the porphyrin ring. Note that the epr spectra generated on addition of one and three reduction electrons are very different in form and are centred at differing magnetic fields. Reversal of the applied potential to –0.61 V shows collapse of the signal at g = 2.011 to give a featureless epr spectrum and then growth of the first signal at g = 2.012. The epr experiments show that at 233 K

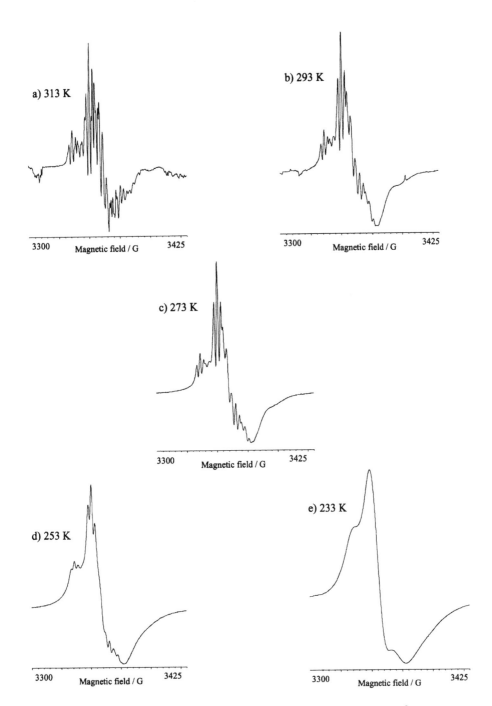

Figure 24. Variable temperature epr spectra of $[Pt(4,4'-(NO_2)_2\text{-bpy})Cl_2]^{2-}$ in DMF/0.5 M $[^nBu_4N]BF_4$.

the diad is stable in four redox states, two of which are diamagnetic and two are paramagnetic.

The ability to use *in situ* spectroelectrochemical cells at depressed temperatures can lead to the study of reactive redox products since their stability will be increased at low temperature, one example of this is shown in Figure 3. However decreased temperatures can also result in a loss of resolution particularly for epr signals. Figure 24 show the epr spectra for $[Pt(4,4'-(NO_2)_2-bpy)Cl_2]^{2-}$ generated at –0.75 V in DMF/0.5 M $[^nBu_4N]BF_4$ at various temperatures [32]. Note the significantly increased resolution at 313 K compared to 233 K.

As the temperature of the cell is decreased so the viscosity of the solution increases thereby hindering the tumbling of the paramagnetic species and resulting in loss of signal resolution. Thus the variable temperature facility of the *in situ* cell may well enable reactive intermediates to be generated and stabilised, however, low temperatures do not necessarily lead to well resolved epr spectra. The cell in Figure 2 may be cooled so that the solution glasses and a frozen spectrum recorded. Figure 25 shows the frozen solution spectrum of $[Pt(4,4'-(NO_2)_2-bpy)Cl_2]^{2-}$,whose solution spectra are given in Figure 24, at 150 K.

The epr signal can be recorded as a function of applied potential in an analogous experiment to the uv/vis spectroelectrochemical experiment described in section 3.2.2. Thus as a paramagnetic species is generated so the epr signal will increase in intensity, with the relative concentration of the epr active material obtained

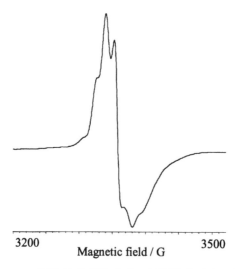

3200   3500
                        Magnetic field / G

Figure 25. Epr spectrum of $[Pt(4,4'-(NO_2)_2-bpy)Cl_2]^{2-}$ in DMF/0.5 M $[^nBu_4N]BF_4$ at 150 K.

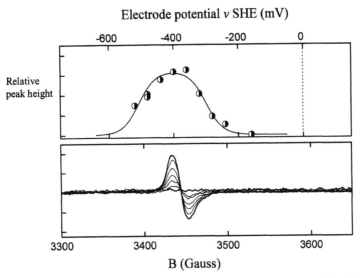

Figure 26. Epr signal of reduction of *E. coli* flavodoxin from –150 to –550 mV at 293 K

from the relative resonance intensity [164]. The epr response as a function of applied potential for a solution of *E. coli* flavodoxin from –150 to –550 mV is presented in Figure 26.

As previously described for a flavodoxin containing species (section 3.2.2) two redox processes occur, the first one-electron reduction yielding a paramagnetic radical and the second one-electron resulting in the diamagnetic form. Thus as the applied potential is stepped from –150 to –550 mV the epr signal increases in intensity to a maximum value and then decreases. Reversing the applied potential from –550 to –150 mV results in the growth of the epr signal and then its collapse with a profile which exactly mirrors that of the response in the forward direction. Analysis of the data in Fig. 26 gives the two formal potentials –300 and –506 mV *vs* she for the first and second electron transfer processes respectively. Note that using a large surface area working electrode considerably shortens the electrolysis time and that the solution is allowed to equilibrate fully at each applied potential before the epr signal is recorded.

ACKNOWLEDGEMENTS

This work has been partially supported by the EPSRC, the BBSRC and the University of Edinburgh. In addition, a contribution from the Edinburgh Protein Interaction Centre funded by the Wellcome Trust is gratefully acknowledged.

REFERENCES

1. Russell, J.W.; Overend, J.; Scanlon, K.; Severson, M.W.; Bewick, A. *J. Phys. Chem,* *86*, 3066 (1982).
2. Fiedler, J.; Nervi, C.; Osella, D.; Calhorda, M.J.; Godinho, S.S.M.C.; Merkel, R.; Wadepohl, H. *J. Chem. Soc., Dalton Trans.* 3705 (2002).
3. Krejcik, M.; Danek, M.; Hartl, F. *J. Electroanal. Chem. 317*, 179 (1991).
4. Best, S.P.; Clark, R.J.H.; McQueen, R.C.S.; Clooney, R.P. *Rev. Sci. Instrum. 58*, 20711 (1987).
5. Best, S.P.; Ciniawskii, S.A.; Humphrey, D.G. *J. Chem. Soc., Dalton Trans.* 2945 (1996).
6. al Dulaimi, J.P.; Bond, A.M.; Clark, R.J.H.; Harden, N.C.; Humphrey, D.G. *J. Chem. Soc., Dalton Trans.* 2541 (2002).
7. Bellec, V.; De Backer, M.G.; Levillain, E.; Sauvage, F.X.; Sombret, B.; Wartelle, C. *Electrochem. Commun. 3*, 483 (2001).
8. Rosenberg, E.; Abedin, M.J.; Rokhsana, D.; Osella, D.; Milone, L.; Nervi, C.; Fiedler, J. *Inorg. Chim. Acta 300-302*, 769 (2000).
9. Büschel, M.; Stadler, C.; Lambert, C.; Beck, M.; Daub, J. *J. Electroanal. Chem. 484*, 24 (2000).
10. Kardash, D.; Huang, J.; Korzeniewski, C. *J. Electroanal. Chem. 476*, 95 (1999).
11. Mosier-Boss, P.A.; Newbery, R.; Szpak, S.; Lieberman, S.H.; Rovang, J.W. *Anal. Chem. 68*, 3277 (1996).
12. Shaw, M.J.; Geiger, W.E. *Organometallics 15*, 13 (1996).
13. Atwood, C.G.; Geiger, W.E.; Bitterwolf, T.E. *J. Electroanal. Chem. 397*, 279 (1995).
14. Yu, J.S.; Zhou, T.Y. *J. Electroanal. Chem. 504*, 89 (2001).
15. Prenzler, P.D.; Bramley, R.; Downing, S.R.; Heath, G.A. *Electrochem. Commun. 2*, 516 (2000).
16. Zhu, Y.; Cheng, G.; Dong, S. *Biophys. Chem. 97*, 129 (2002)
17. Jasperson, S.N.; Burge, D.K.; O'Handley, R.C. *Surf. Sci. 37*, 548 (1973).
18. Salbeck, J. *J. Electroanal. Chem. 340*, 169 (1992).
19. Hartl, F.; Stufkens, D.J.; Vlcek, A. *Inorg. Chem. 31*, 2869 (1992).
20. Lin, C-Y.; Spiro, T.G. *Inorg. Chem. 35*, 5237 (1996).
21. Farley, N.R.S.; Gurman, S.J.; Hillman, A.R. *Electrochem. Commun. 1*, 449 (1999).
22. Soderholm, L.; Antonio, M.R.; Williams, C.; Wasserman, S.R. *Anal. Chem. 71*, 4622 (1999).
23. Plieth, W.; Wilson, G.S.; Gutiérrez de la Fe, C. *Pure and Appl. Chem. 70*, 1395 (1998),
24. Plieth, W.; Wilson, G.S.; Gutiérrez de la Fe, C. *Pure and Appl. Chem. 70*, 2409 (1998).
25. Iengo, E.; Mestroni, G.; Geremia, S.; Calligaris, M.; Alessio, E. *J. Chem. Soc., Dalton Trans.* 3361 (1999).
26. Baldwin, D.A.; Betterton, E.A.; Pratt, J.M. *J. Chem. Soc., Dalton Trans.* 217 (1983).
27. Ford, S.H.; Nichols, A.; Shambee, M.J. *Inorg. Biochem. 41*, 235 (1991).
28. Zou, X.; Evans, D.R.; Brown, K.L. *Inorg. Chem. 34*, 1634 (1995).

29.　Koetsier, M.J.; Craig, D.H.; Mackay, F.; Daff, S.; Chapman, S.K. Flavins and Flavoproteins (Rudolf Weber), Chapman,S.K.; Perham, R.N.; Scrutton, N.S. eds., 587 (2002).

30.　Adams, C.J.; James, S.L.; Liu, X.M.; Raithby, P.R.; Yellowlees, L.J. *J. Chem. Soc., Dalton Trans.* 63 (2000).

31.　Barton, M.T.; Rowley, N.M.; Ashton, P.R.; Jones, C.J.; Spencer, N.; Tolley, M.S.; Yellowlees, L.J. *New J. Chem.24*, 555 (2000).

32.　McInnes, E.J.L.; Welch, A.J.; Yellowlees, L.J. *J. Chem. Soc., Chem. Commun.* 2393 (1996).

33.　McInnes,E.J.L.; Farley, R.D.; Rowlands, C.C.; Welch, A.J.; Rovatti, L.; Yellowlees, L.J. *J. Chem. Soc., Dalton Trans.* 4203 (1999).

34.　Macgregor, S.A.; McInnes, E.J.L.; Sorbie, R.J.; Yellowlees, L.J. Molecular Electrochemistry of Inorganic, Bioinorganic and Organometallic Compounds (Kluwer, Netherlands), Pombeiro, A.J.L.; McCleverty, J.A. eds. 530 (1993).

35.　Morse, P. Scientific Software Services, Bloomington, USA (1994).

36.　Quantum CAChe version 3.2, Oxford Molecular Ltd. (1999).

37.　WIN-EPR SimFonia, version 1.25, Bruker, Billerica, USA (1996).

38.　Kuwana, T.; Heineman, W.R. *Acc. Chem. Res. 9*, 241 (1976).

39.　Heineman, W.R.; Hawkridge, F.M.; Blount, H.N. *Electroanalytical Chemistry,* Bard, A.J. (ed.) Marcel dekker, New York, *13*, 1 (1984).

40.　Szentrimay, R.; Yeh, P.; Kuwana, T. *Electrochemical Studies of Biological Systems,* Sawyer, D.T. (ed.) ACS Symp. Ser. *38*, 143 (1977).

41.　Murray, R.W.; Heineman, R.W. O'Dom, G.W. *Anal. Chem. 39*, 1666 (1967).

42.　De Angelis, T.P.; Heineman, W.R. *J. Chem. Educ. 53*, 594 (1976).

43.　Kirchhoff, J.R.; Allen, M.R.; Cheesman, B.V.; Okamoto, K.; Heinemaan, W.R.; Deutsch, E. *Inorg. Chim. Acta 262*, 195 (1997).

44.　Toma, H.E.; Sernaglia, R.L. *Talanta 42*, 1867 (1995).

45.　Bagatin, I.A.; Toma, H.E. *Spectrosc. Letts. 29,* 1409 (1996).

46.　Toma, H.E.; Alexiou, A.D.P.; Dovidauskas, S. *Eur. J. Inorg. Che.* 3010 (2002).

47.　Enemark, J.H.; Feltham, R.D. *Coord. Chem. Rev. 13*, 339 (1974).

48.　Eisenberg, R. Meyer, C.D. *Acc. Chem. Res. 8*, 26 (1975).

49.　Bottomley, F. *Coord. Chem. Rev. 26*, 7 (1978).

50.　Westcott, B.L.; Enemark, J.H. *Transition Metal Nitrosyls*, John Wiley & Sons, Inc., Vol. II: Applications and Case Studies (1999).

51.　Nunes, F.S.; Toma, H.E. *J. Coord. Chem. 36*, 33 (1995).

52.　López-Palacios, J.; Colina, A.; Herras, A.; Ruiz, V.; Fuente, L. *Anal. Chem. 73*, 2883 (2001).

53.　Qian, W.; Zhu, S-M.; Luo, Q-H.; Hu, X-L.; Wang, Z-L *Bioelectrochem, 58*, 197 (2002).

54.　Losada, J.; del Peso, I.; Beyer, L. *Inorg. Chim. Acta 321*, 107 (2001).

55.　Dag, O.; Yaman, S.O.; Önal, A.M.; Isci, H. *J. Chem. Soc., Dalton Trans.* 2819 (2001).

56.　Hill, M.G.; Mann, K.R. *Inorg. Chim. Acta 243*, 219 (1996).

57.　Hill, M.G.; Bailey, J.A.; Miskowski, V.M.; Gray, H.B. *norg. Chem. 35*, 4585 (1996).

58. Stinner, C.; Wightman, M.D.; Kelly, S.O.; Hill, M.G.; Barton, J.K. *Inorg. Chem. 40*, 5245 (2001).

59. Jain, S.L.; Crayston, J.A.; Richens, D.T.; Woollins, J.D. *Inorg. Chem. Commun. 5*, 853 (2002).

60. Hartl, F.; Luyten, H.; Nieuwenhuis, H.A.; Schoemaker, G.C. *Appl. Spectroscopy 48*, 1522 (1994).

61. Knödler, A.; Wanner, M.; Fiedler, J.; Kaim, W. *J. Chem. Soc., Dalton Trans.* 3079 (2002).

62. Glöckle, M.; Kaim, W.; Katz, N.E.; Posse, M.G.; Cutin, E.H.; Fiedler, J. *Inorg. Chem. 38*, 3270 (1999).

63. Kaim, W.; Berger, S.; Greulich, S.; Reinhardt, R.; Fiedler, J. *J. Organomet. Chem. 582*, 1533 (1999).

64. Baumann, F.; Kaim, W.; Olabe, J.A.; Parise, A.R.; Jordanov, J. *J. Chem. Soc., Dalton Trans.* 4455 (1997).

65. Baumann, F.; Kaim, W.; Posse, M.G.; Katz, N.E. *Inorg. Chem. 37*, 658 (1998).

66. Berger, S.; Hartmann, H.; Wanner, M.; Fiedler, J.; Kaim, W. *Inorg. Chim. Acta 314*, 22 (2001).

67. Kaim, W.; Sixt, T.; Weber, M.; Fiedler, J. *J. Organomet. Chem. 637-639*, 167 (2001).

68. Sixt, T.; Fiedler, J.; Kaim, W. *Inorg. Chem. Commun.3*, 80 (2000).

69. Farrell, I.R.; Hartl, F.; Záliš, S.; Wanner, M.; Kaim, W.; Vlcek Jr., A. *Inorg. Chim. Acta 318*, 143 (2001).

70. Kaim, W.; Schwederski, B.; Dogan, A.; Fiedler, J.; Kuehl, C.J.; Stang, P.J. *Inorg. Chem. 41*, 4025 (2002).

71. Crespo, M.; Grande, C.; Klein, A. *J. Chem.Soc., Dalton Trans.* 1629 (1999).

72. Lin, X.Q.; Kadish, K.M. *Anal. Chem. 57*, 1498 (1985).

73. Wei, Z.; Ryan, M.D. *Inorg. Chim. Acta 314*, 49 (2001).

74. Akiba, K.; Nadano, R.; Satoh, W.; Yamamoto, Y.; Hagase, S.; Ou, Z.; Tan, X.; Kadish, K.M. *Inorg. Chem. 40*, 5533 (2001).

75. Chaudhuri, P.; Verani, C.N.; Bill, E.; Bothe, E.; Weyhermüller, T.; Wieghardt, K. *J. Am. Chem. Soc. 123*, 2213 (2001).

76. Sun, X.; Chun, H.; Hildenbrand, K.; Bothe, E.; Weyhermüller, T.; Neese, F.; Wieghardt, K. *Inorg. Chem. 41*, 4295 (2002).

77. Chun, H.; Verani, C.N.; Chaudhuri, P.; Bothe, E.; Bill, E.; Weyhermüller, T.; Wieghardt, K. *Inorg. Chem. 40*, 4157 (2001).

78. Barthram, A.M.; Ward, M.D. *New J. Chem.* 501 (2000).

79. Behrendt, A.; Couchman, S.M.; Jeffrey, J.C.; McCleverty, J.A.; Ward, M.D. *J. Chem. Soc., Dalton Trans.* 4349 (1999).

80. Lee, S-M.; Kowallick, R.; Marcaccio, M.; McCleverty, J.A.; Ward, M.D. *J. Chem. Soc., Dalton Trans.* 3443 (1998).

81. Lee, S-M.; Marcaccio, M.; McCleverty, J.A.; Ward, M.D. *Chem. Mater. 10*, 3272 (1998).

82. Chakraborty, S.; Laye, R.H.; Paul, R.L.; Gonnade, R.G.; Puranik, V.G.; Ward, M.D.; Lahiri, G.K. *J. Chem. Soc., Dalton Trans.* 1172 (2002).

83. Sakar, B.; Laye, R.H.; Mondal, B.; Chakraborty, S.; Paul, R.L.; Jeffrey, J.C.; Puranik, V.G.; Ward, M.D.; Lahiri, G.K. *J. Chem. Soc., Dalton Trans.* 2097 (2002).

84. Duff, C.M.; Heath, G.A. *Inorg. Chem. 30*, 2528 (1991).
85. Arnold, D.P.; Heath, G.A. *J. Am. Chem. Soc. 115*, 12197 (1993).
86. Arnold, D.P.; Heath, G.A.; James, D.A. *J. Porphyrins Phthalocyanines 3*, 5 (1999).
87. Arnold, D.P.; Hartnell, R.D.; Heath, G.A.; Newby, L.; Webster, R.D. *J. Chem. Soc., Chem. Commun.* 754 (2002).
88. Bruce, M.I.; Low, P.J.; Costuas, K.; Halet, J-F.; Best, S.P.; Heath, G.A. *J. Am. Chem. Soc. 122*, 1949 (2000).
89. Webster, R.D.; Heath, G.A.; Bond, A.M. *J. Chem. Soc., Dalton Trans,* 3189 (2001).
90. Marcaccio, M.; Paolucci, F.; Paradisi, C.; Roffia, S.; Fontanesi, C.; Yellowlees, L.J.; Serroni, S.; Campagna, S.; Denti, G.; Balzani, V. *J. Am. Chem. Soc. 121*, 10081 (1999).
91. Robertson, N.; Liu, X.; Yellowlees, L.J. *Inorg. Chem. Commun. 3*, 424 (2000).
92. Araujo, C.S.; Drew, M.G.B.; Jack, L.; Madureira, J.; Newell, M.; Roche, S.; Santos, T.M.; Thomas, J.A.; Yellowlees, L.J. *Inorg. Chem. 41*, 2250 (2002).
93. Nairn, A.K.; Bhalla, R.; Foxon, S.P.; Liu, X.; Yellowlees, L.J.; Gilbert, B.C.; Walton, P.H. *J. Chem. Soc., Dalton Trans.* 1253 (2002).
94. Marcaccio, M.; Paolucci, F.; Paradisi, C.; Carano, M.; Roffia, S.; Fontanesi, C.; Yellowlees, L.J.; Serroni, S.; Campagna, S.; Balzani, V. *J. Electroanal. Chem. 532*, 99 (2002).
95. Brown, A.R.; Taylor, K.J.; Yellowlees, L.J. *J. Chem. Soc., Dalton Trans.* 2401 (1998).
96. Kennedy, B.J.; Heath, G.A. *Inorg. Chim. Acta 195*, 101 (1992).
97. Creutz, C.; Taube, H. *J. Am. Chem. Soc. 91*, 3988 (1969).
98. Robin, M.B.; Day, P. *Adv. Inorg. Radiochem. 10*, 247 (1967).
99. Fürholz, U.; Bürgi, H-B.; Wagner, F.E.; Stebler, A.; Ammeter, H.H.; Krausz, E.; Clark, R.J.H.; Stead, M.J.; Ludi, A. *J. Am. Chem. Soc. 106*, 121 (1984).
100. Richardson, D.E.; Taube, H. *J. Am. Chem. Soc. 105*, 40 (1983).
101. Sutton, J.E.; Taube, H. *Inorg. Chem. 20*, 3125 (1981).
102. Callahan, R.W.; Keene, F.R.; Meyer, T.J.; Salmon, D.J. *J. Am. Chem. Soc. 99*, 1064 (1977).
103. Creutz, C.; Chou, M.H. *Inorg. Chem. 26*, 2995 (1987).
104. Callahan, R.W. PhD Dissertation, University of North Carolina, Chapel Hill, 1975.
105. Coe, B.J.; Meyer, T.J.; White, P.S. *Inorg. Chem. 34*, 593 (1995).
106. Hornung, F.M.; Baumann, F.; Kaim, W.; Olabe, J.A.; Slep, L.D.; Fiedler, J. *Inorg. Chem. 37*, 311 (1998).
107. Lay, P.A.; Magnuson, R.H.; Taube, H. *Inorg. Chem. 37*, 2364 (1988).
108. Balducci, G.; Mestroni, G.; Alessio, E. *Current Topics in Electrochemistry, 2*, 323 (1993).
109. Alessio, E.; Balducci, G.; Calligaris, M.; Costa, G.; Attia, W.M.; Mestroni, G. *Inorg. Chem. 30*, 609 (1991).
110. Gutmann, V. The Donor-Acceptor Approach to Molecular Interactions, Plenum Press, New York (1980).

111. Hush, N.S. *Prog. Inorg. Chem. 8*, 391 (1967).
112. Gutmann, V.; Resch, G.; Linert, W. *Coord. Chem. Rev. 43*, 133 (1982).
113. Creutz, C.; Chou, M.H. *Inorg. Chem. 31*, 3170 (1992).
114. Creutz, C. *Prog. Inorg. Chem. 30*, 29 (1983).
115. Powers, M.J.; Meyer, T.J. *Inorg. Chem. 17*, 2955 (1978).
116. Powers, M.J.; Meyer, T.J. *J. Am. Chem. Soc. 102*, 1289 (1980).
117. Richardson, D.E.; Taube, H. *Inorg. Chem. 20*, 1278 (1981).
118. Taboy, C.H.; Bonaventura, C.; Crumbliss, A.L. *Bioelectrochem. Bioenerg, 48*, 79 (1999).
119. Ding, X.D.; Weichsel, A.; Andersen, J.F.; Shokhieeva, T.K.; Balfour, C.; Pierik, A.J.; Averill, B.A.; Montfort, W.R.; Walker, F.A. *J. Am. Che. Soc. 121*, 128 (1999).
120. Borrsari, M.; Benini, S.; Marchesi, D.; Ciurli, S. *Inorg. Chim. Acta 263*, 379 (1997).
121. Niki, K.; Vrána, O.; Brabec, V. *Experimental Techniques in Bioelectrochemistry* (Birkhäuserr Verlag) Brabec, V.; Walz, D.; Milazzo, G. (eds.) Basel (1996).
122. Comtat, M.; Durliat, H. *Biosensors Bioelectronics 9*, 663 (1994).
123. Heineman, W.R.; Norris, B.J.; Goelz, J.F. *Anal. Chem. 47*, 79 (1975).
124. Norris, B.J.; Meckstroth, M.L.; Heineman, W.R. *Anal. Chem. 48*, 630 (1976).
125. Kreishman, G.P.; Anderson, C.W.; Su, C-H.; Halshall, H.B.; Heineman, W.R. *Bioelectrochem. Bioenerg. 5*, 196 (1978).
126. Anderson, C.W.; Halsall, H.B.; Heineman, W.R. *Anal. Biochem. 93*, 366 (1979).
127. Taniguchi, V.T.; Sailasuta-Scott, N.; Anson, F.C., Gray, H.B. *Pure Appl. Chem. 52*, 2275 (1980).
128. Heineman, W.R.; Meckstroth, M.L.; Norris, B.J.; Su, C-H. *Bioelectrochem. Bioenerg. 6*, 577 (1979).
129. Sailasuta, N.; Anson, F.C.; Gray, H.B. *J. Am. Chem. Soc. 101*, 455 (1979).
130. Fan, K.J.; Akutsu, H.; Niki, K.; Higuchi, N.; Kyogoku, Y. *J. Electroanal. Chem. 278*, 295 (1990).
131. Pratt, J.M. *Chemistry and Biochemistry of B_{12}* (Wiley) Banerjee, R. (ed.) New York, 73 (1999).
132. Lexa, D.; Saveant, J-M.; Zickler, J. *J. Am. Chem. Soc.102*, 4851 (1980).
133. Sano, M.; Taube, H. *J. Am. Chem. Soc. 113*, 2327 (1991).
134. Sano, M.; Taube, H. *Inorg. Chem. 33*, 705 (1994).
135. Koppenhofer, A.; Turner, K.L.; Allen, J.W.A.; Chapman, S.K.; Ferguson, S.J. *Biochemistry 39*, 4243 (2000).
136. Daff, S. N.; Chapman, S. K.; Turner, K.L.; Holt, R.A.; Govindaraj, S.; Poulos, T.L.; Munro, A.W. *Biochemistry, 36*, 13816 (1997).
137. Webster, R.D.; Heath, G.A.; Bond, A.M. *J. Chem. Soc. Dalton Trans.* 31189 (2001).
138. van Staveren, D.R.; Bothe, E.; Weyhermüller, T.; Metzler-Nolte,N. *Eur. J. Inorg. Chem.* 1518 (2002).
139. Winter, R.F. *Eur. J. Inorg. Chem.* 2121 (1999).
140. Lomoth, R.; Huang, P.; Zheng, J.; Sun, L.; Hammarström, L.; Akermark, N.; Styring, S. *Eur. J. Inorg. Chem.* 2965 (2002).
141. Robben, M.P.; Reiger, P.H.; Geiger, W.E. *J. Am. Chem. Soc. 121*, 367 (1999).

142. McKinney, T.M. *Electroanalytical Chemistry Vol. 10: Electron Spin Resonance and Electrochemistry* Bard, A.J. (ed) (Marcel Dekker) New York, 97 (1977).

143. Bard, A.J.; Faulkner, L.R. *Electrochemical Methods: Fundamentals and Applications* (John Wiley and Sons) New York, 615 (1980).

144. Compton, R.G.; Waller, A.M. *Spectroelectrochemistry: Theory and Practise* Gale, R.J. (ed.) (Plenum Press) New York 349 (1988).

145. Goldberg, I.B.; McKinney, T.M. *Laboratory techniques in Electroanalytical Chemistry*, Kissinger, P.T.; Heineman, W.R. (eds) (Marcel Dekker) New York, 901 (1996).

146. Hartl, F.; Vlcek, Jr. A. *Inorg. Chem. 30*, 3048 (1991).

147. Hartl, F.; Groenestein, R.P.; Mahabiersing, T. *Collect. Czech. Chem. Commun. 66*, 52 (2001).

148. Allendoerfer, R.D.; Martinchek, G.A.; Bruckenstein, S. *Anal. Chem. 47*, 890 (1975).

149. Farrell, I.R,; Hartl, F.; Zalis, S.; Mahabiersing, T.; Vlcek Jr. A. *J. Chem. Soc. Dalton Trans.* 4323 (2000).

150. Fiedler, D.A.; Koppenol, M.; Bond, A.M. *J. Electrochem Soc. 142*, 863 (1995).

151. Bond, A.M.; Dyson, P.J.; Humphrey, D.G.; Lazarev, G.; Suman, P. *J. Chem. Soc., Dalton Trans.* 443 (1999).

152. Coles, B.A.; Compton, R.G. *J. Electroanal. Chem. 144*, 87 (1983).

153. Webster, R.D.; Bond, A.M.; Coles, B.A.; Compton, R.G. *J. Electroanal. Chem. 404*, (1996).

154. Kaim, W.; Ernst, S.; Kasack, V. *J. Am. Chem. Soc. 112*, 173 (1990).

155. Scheiring, T.; Fiedler, J.; Kaim, W. *Organometallics 20*, 1437 (2001).

156. Hartmann, H.; Scheiring, T.; Fiedler, J.; Kaim, W. *J. Organomet. Chem. 604*, 267 (2000).

157. Rapta, P.; Staško, A.; Gromov, A.V.; Bartl, A.; Dunsch, L. *Electrochem. Soc. Proceed.* 10 (2000).

158. Rapta, P.; Kress, L.; Hapiot, P.; Dunsch, L. *Phys. Chem. Chem. Phys. 4*, 4181 (2002).

159. Rapta, P.; Dunsch, L. *J. Electroanal. Chem. 507*, 287 (2001).

160. Rapta, P.; Fáber, R.; Dunsch, L.; Neudeck, A.; Nuyken, O. *Spectrochim. Acta Part A 56*, 357 (2000).

161. Neudeck, A.; Petrr, A.; Dunsch, L. *Synth. Mets. 107*, 143 (1999).

162. Wilgocki, M.; Rybak, W.K. *Portugalie Electrochim. Acta 13*, 211 (1995).

163. McConnell, H.M. *J. Chem. Phys. 24*, 764 (1958).

164. Long, Y-T.; Yu, Z-H.; Chen, H-Y. *Electrochem. Commun. 1*, 194 (1999).

Part 6
Unconventional Electrochemistry

Chapter 12

Electrochemistry at Ultramicroelectrodes: Small and Fast May Be Useful

Christian Amatore, Stéphane Arbault, Emmanuel Maisonhaute, Sabine Szunerits, and Laurent Thouin

Ecole Normale Supérieure, Département de Chimie
UMR CNRS-ENS-UPMC 8640 "PASTEUR"
24 rue Lhomond, 75231 Paris Cedex 05, France.
(http://www.chimie.ens.fr/w3amatore/)

1   INTRODUCTION

Electrochemical techniques have long been used to investigate the kinetics of various types of reactions [1]. In this case, the time scale, θ, of the experiment is determined by the length of time that it takes for a molecule to move from the surface of the electrode to the outer edge of the diffusion layer. The nature of the observed response depends on the relative magnitude of the time scale, θ, compared to the half life of the chemical reaction, $\tau_{1/2}$, with the former being controlled by the sweep rate. For the two limiting cases, $\theta \ll \tau_{1/2}$ and $\theta \gg \tau_{1/2}$, the system behaves either as if entirely under diffusion control or under control of the chemical kinetics, respectively. Each of these two limiting cases has very recognizable manifestations and when θ is comparable to $\tau_{1/2}$, the observed response is a convolution of these two limiting cases [1].

Increasing the scan rate, υ, in cyclic voltammetry allows one to decrease the duration of diffusion, *i.e.*:

$$\theta \simeq (RT/F\upsilon) \tag{1}$$

so that, *formally*, the characteristic time of the experiment, θ, may be adjusted at will, at least within the experimental limits within which the scan rate may be varied. Since the earliest times of transient electrochemical methods, this possibility of decreasing and adjusting at will the experimental time scale has been used mostly with the aim of investigating faster and faster phenomena. Yet, diffusion of molecules over a duration θ implies that the solution domain adjacent to the electrode surface which is explored electrochemically has a thickness of the order of:

$$\delta \propto (D\theta)^{1/2} = (DRT/F\upsilon)^{1/2} \qquad (2)$$

where D is the diffusion coefficient of the species considered. Increasing the scan rate, υ, in cyclic voltammetry thus allows one to decrease the thickness of the diffusion layer, and adjust this thickness to extremely small dimensions provided that ultrafast scan rates may be achieved experimentally with minimal instrumental distortions.

However, when υ becomes increasingly large, two phenomena alter the quality of the pure electrochemical information [2]. First, the electrochemical perturbation, E(t), cannot be applied to the electrode faster than the cell time constant. Second, a time-dependent fraction of the energy, i(t)E(t), applied to the cell is lost through ohmic heating occurring while the current flux is passing through the solution. It has been recognized since the early eighties that these two problems are minimized when performing electrochemistry at disk ultramicroelectrodes [2].

Albeit using ultramicroelectrodes considerably reduces problems associated to time constants and ohmic drop, this is not enough to allow a decrease of δ to nanometric dimensions in transient electrochemical experiments. To reach these limits, so that nanometric dimensions may be explored voltammetrically, *on-line* and *real-time* electronic compensation of the cell resistance must be used in addition to ultramicroelectrodes.

In this chapter we wish therefore to begin by explaining the physical origin of these distortions and show how one can eliminate them adequately by designing proper instrumentations. In the second part we will present three series of applications of fast electrochemistry at ultramicroelectrodes. The first one deals with the classical kinetic views associated with ultrafast voltammetry. The second one will show that access at these nanotime scales allows to probe electron transfers within a nanoscale redox object taking advantage that the diffusion layer may then be adjusted precisely to the geometrical dimensions of the object. The third example will be used to demonstrate that the combined use of very small electrodes and very small diffusion layers may serve to probe the local composition of a solution so that dynamic concentration profiles may be monitored.

2 THEORETICAL AND EXPERIMENTAL LIMITATION OF ULTRAFAST CYCLIC VOLTAMMETRY AT ULTRAMICROELECTRODES

2.1 Theoretical Considerations

In the electrochemical kinetic investigations the electrochemical perturbation must be applied to the Faradaic impedance with minimal distortion. Reaching scan rates in the megavolt-*per*-second range requires this to be performed within a sub-nanosecond time scale [3]. Although such requirements are rather easy to fulfill in usual modern electronic equipments owing to today's fast integrated components, the specificities of an electrochemical cell make them a real challenge.

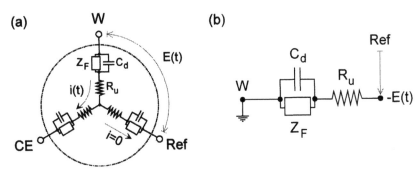

Figure 1. Equivalent circuit for a three-electrode electrochemical cell (see text). (a) Complete cell (W: Working electrode, Ref: Reference electrode, CE: Counter-electrode). (b) Apparent electrical circuit of the cell when using a potentiostat.

The electrical circuit of an electrochemical cell is represented in Figure 1a for a three-electrode configuration. Formally, such a circuit depends on three Faradaic impedances, three capacitances and three resistances. However, using a potentiostat and an adequate reference electrode allows the application of the desired electrochemical perturbation to the working electrode circuit element alone [1,4] (see Figure 1b; Z_F, C_d and R_u indicate respectively the Faradaic impedance, the double layer capacitance at the working electrode and the so-called uncompensated resistance). However, one should really be able to apply the electrochemical perturbation to the Faradaic impedance (*viz.*, Z_F in Fig. 1b) and measure simultaneously the Faradaic current i_F only. In cyclic voltammetry or in chronoamperometric methods the electrochemical perturbation consists of an imposed potential variation, E(t), that is applied to the circuit in Fig. 1b.

Thus one has:

$$E(t) = Z_F\, i_F + R_u\, i(t) \tag{3}$$

where $i = i_F + i_C$, is the measured current, and:

$$i_C(t) = C_d\,[d(Z_F i_F)/dt] = C_d\,\{d\,[E(t) - R_u\, i(t)]/dt\} \tag{4}$$

Extraction of the true Faradaic current, *viz.*, of $i_F(t) = f[E(t) - R_u\, i(t)]$, *i.e.*, of the *undistorted voltammogram*, seems formally possible by application of equations (3,4), provided that R_u and C_d are known. However, this requires also the independent knowledge (or assumption) of the Faradaic impedance $Z_F(t)$, *viz.*, of the final result. The validation may then be performed *a posteriori* either by simulation of the distorted voltammogram [5] or based on the reconstructed Faradaic information which may be obtained through convolution integral procedures [6]. However, either approach is tantamount to a verification that the hypothesized $Z_F(t)$ function (*viz.*, the hypothesized mechanism) is adequate, since a given mechanism, *i.e.*, a given $Z_F(t)$ is postulated to solve equations (3,4), which allows ultimately to verify whether the $i_F(t)$ experimental function agrees with the initial mechanistic hypothesis or not. There is absolutely nothing criticizable *per se* in such "hypothesis-verification" procedures, but in our view they present an important disadvantage for the study of complex unknown kinetics.

Because of our specific interest in using electrochemistry for investigating complex chemical reactions we developed another strategy to eliminate the problem borne by equations (3,4). This is based on positive feedback: R_u may be known and $i(t)$ measured sufficiently fast, so that one may apply $E(t) + (1 - \varepsilon) \times R_u\, i(t)$ to the cell instead of $E(t)$ (note that ε may be made small enough for $|E(t)| \gg \varepsilon \times |R_u\, i(t)|$, while $(1 - \varepsilon)$ must remain positive in order to avoid severe electronic instabilities [7,8]; *vide infra*). Equations (3,4) become then:

$$E(t) = Z_F\, i_F + \varepsilon\, R_u\, i(t) \approx Z_F\, i_F \tag{5}$$

$$i_F(t) = i(t) - i_C(t) \tag{6}$$

where:

$$i_C(t) = C_d\,\{d\,[E(t) - \varepsilon\, R_u\, i(t)]/dt\} \approx C_d\,[dE(t)/dt] \tag{7}$$

Equations (6-7) establish that one may then measure the experimental values of $i_F(t)$ and therefore deduce $Z_F(t)$ without supposing *a priori* any specific mechanism. Furthermore, $i_C(t)$ in equation (7) is now independent of the Faradaic impedance, and may be recorded independently. It can thus be experimentally subtracted from the experimental current $i(t)$ (eqn. 6) to afford *on-line* a display of the ohmic-free and capacitive-free Faradaic voltammogram.

(a)   (b)

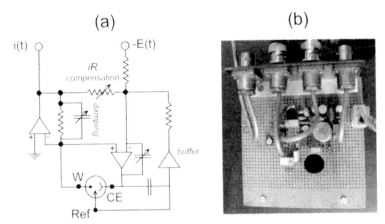

Figure 2. (a) Electronic diagram of a potentiostat allowing on-line ohmic drop and time constant electronic compensations up to 2,500,000 Vs^{-1} (adapted from refs. [9,10]). **(b)** Photograph of the actual potentiostat. The cell is placed in the hole made in the circuit board (dark circle) so that the lengths of contacts are minimized.

The experimental cost associated to this strategy is that it requires the continuous *on-time* re-injection of a potential increment being as close as possible to R_u i(t), *viz.*, with ε being as close as possible to zero (eqn. 5), into the signal applied to the working electrode. This becomes increasingly difficult when the duration of the whole voltammetric scan becomes excessively short. Figure 2 presents the electronic circuit of a potentiostat with on-line electronic compensation of ohmic drop and time constant which has been recently designed in our laboratory and allows undistorted voltammograms to be recorded up to scan rates of 2,500,000 Vs^{-1} [9,10].

2.2   *Experimental Tests of the Potentiostat*

Reduction of anthracene in acetonitrile is a paramount of redox systems affording among the highest measurable heterogeneous rate constants of electron transfer. Such rate constants are too high to be measurable by standard cyclic voltammetry. However, the use of scan rates in the megavolt per second range allowed its facile determination: $k_0 = 5.1$ cm s^{-1} ($\alpha = 0.45$), as evidenced in Figure 3 [10]. Furthermore, the validity of the procedure for ohmic drop and time constant on-line compensations is evidenced by the "rectangular" aspect of the capacitive current in Figure 3a. The experimental validity of this concept, as well as of the subtraction procedure when ohmic drop is compensated, is established by the extremely good agreement between the experimental subtracted voltammogram in Figure 3b and the theoretical one which was simulated without considering ohmic drop, time constant or any instrumental distortions [10].

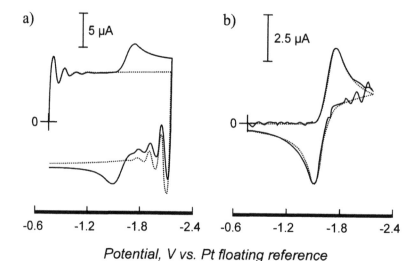

Potential, V vs. Pt floating reference

Figure 3. Voltammetry of anthracene (14.3 mM) in acetonitrile (0.9 M NBu$_4$BF$_4$) at a gold disk electrode (2.5 μm diameter) at a scan rate of 960,000 Vs^{-1}. 20°C. (a) Experimental voltammogram in the absence of anthracene (dashed trace) or in the presence of anthracene (solid trace) while keeping the same settings and the same electrode. (b) Solid trace: Faradaic voltammogram as obtained by subtracting the traces shown in (a); Dashed trace: simulated voltammogram for a pure Faradaic process at this scan rate (E^0 = −1.61 V *vs.* Pt, k_0 = 5.1 cm s^{-1}, α = 0.45, D = 1.6 × 10^{-5} cm^{2}s^{-1}). (adapted from ref. [10]).

3   KINETIC APPLICATIONS OF ULTRAFAST CYCLIC VOLTAMMETRY AT ULTRAMICROELECTRODES

In this section we wish to illustrate the interest of our strategy for the monitoring of ultrafast kinetic processes which are associated to the one-electron loss or one-electron uptake of a molecule to the ultramicroelectrode surface.

3.1   *Oxidation of Methylbenzenes: Determination of Fast Decay Rate Constants*

Using the same methodology but resorting "*only*" to scan rates in the range of a few ten thousand volts per second, we were able to measure the deprotonation rate constants of a series of polymethylbenzenes cation radicals by pyridine bases (eqn. 8) [11]:

These rate constants are reported in Table 1 and are compared to those estimated previously through laser piscosecond spectroscopy [12]. Besides the electrochemical interest of such data *per se*, their excellent agreement with those determined by the much more sophisticated picosecond laser spectroscopic method clearly demonstrate again the validity and great usefulness of this electrochemical approach.

Table 1. Rate constants for the deprotonation of a series of methylbenzenes by pyridine bases according to the reaction in equation (7) in acetonitrile. 20°C.

Pyridine (Z)	Methylbenzene[a]	k_H (M^{-1}s^{-1})	
		Cyclic voltammetry[b]	Picosecond spectroscopy[c]
2,6-Me$_2$	HMB	1.1×10^7	2.1×10^7
	d_{18}-HMB	2.9×10^6	7.5×10^6
	PMB	3.6×10^7	4.0×10^7
	DUR	1.1×10^8	5.8×10^7
2,4,6-Me$_2$	HMB	2.5×10^7	3.5×10^7

[a] HMB: hexamethylbenzene; PMB: pentamethylbenzene; DUR: durene.
[b] From ref. [11]. [c]From ref. [12].

3.2   Dimerization of Pyrylium Radical: Determination of a Bimolecular Rate Constant Close to the Diffusion Limit

A second example illustrating the great kinetic interest of ultrafast cyclic voltammetry at ultramicroelectrodes is shown in Figure 4 for reduction of a pyrylium cation (eqn. 9) [13,14]. Indeed, the *on-line* corrected voltammogram in Fig. 4b clearly shows a reoxidation wave for the fraction of pyrylium radicals which have been generated during the forward scan and have survived their rapid dimerization in equation (9) [13-16]:

(9)

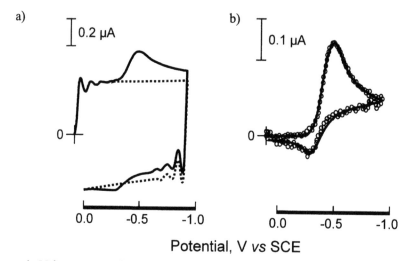

Figure 4. Voltammetry of 5 mM pyrylium cation (Pyr⁺,BF₄⁻) in acetonitrile (0.3 M NBu₄BF₄) at a platinum disk electrode (5 μm diameter) at a scan rate of 207,000 Vs⁻¹. 20°C. (a): Voltammograms obtained in the absence (i_C, dashed line) or in the presence of pyrylium cation (i_C+i_F, solid line). (b): Open circles: Faradaic voltammogram (i_F) obtained by subtracting the two current traces shown in (a); solid line: theoretical voltammogram simulated for the sequence in equation (8) (k_{dim} = 0.9×10⁹ M⁻¹ s⁻¹, E⁰ = –0.38 V *vs.* SCE, k_0 = 2 cm s⁻¹, α = 0.4). (Adapted from ref. [14]).

Here again, the experimental strategy based on *on-line* ohmic drop compensation procedure is validated by the "rectangular" aspect of the capacitive current in Figure 4a. Note that the oscillations which are apparent in Figure 4a (as well as in the previous Figure 3a) almost cancel in the subtracted voltammogram shown in Figure 4b (as it occurred in Figure 3b). This happens because these oscillations pertain mostly to the capacitive current since they arise from very small instabilities in the potential applied at the electrode capacitance (see Figure 1b) due to the electronic compensation procedure. These very small amplitude instabilities of the potential are greatly magnified by the capacitive current due to the derivative effect of the electrode capacitance while they perturb only slightly the Faradaic current [17].

The adequacy of the method is demonstrated again by the extremely good agreement between the experimental voltammogram in Figure 4b with the theoretical voltammogram simulated without considering any ohmic drop nor time constant or any other instrumental distortions. It is also noteworthy that the rate constant k_{dim} = (0.9±0.3)×10⁹ M⁻¹ s⁻¹ of the dimerization reaction in equation (9) [13,14] thus determined is near the diffusion limit and is almost equal to that which could be estimated previously by flash-photolysis techniques [15].

4   MONITORING ELECTROCHEMICAL PROCESSES WITHIN A NANOMETRIC REDOX OBJECT

These above examples clearly illustrate the validity and the great experimental interest of ultrafast cyclic voltammetry at ultramicroelectrodes. In this section we wish to elaborate about different topics, still related to ultrafast cyclic voltammetry but in which the voltammetric scan is used to adjust the extent of the diffusion layer thickness to the dimension of a nanometric object, so that diffusion may be monitored at the molecular level within a nanometric molecule. Indeed, voltammetric experiments within nanosecond time-scales correspond to the development of diffusion layer whose sizes approach those of single molecules (eqn 2).

Dendrimers with pendant redox centers are well suited for investigation by ultrafast cyclic voltammetry due to their size. The specific molecule studied here is a fourth-generation polyamidoamide (PAMAM) dendrimer capped with 64 ruthenium(II) bis-terpyridine moieties, $Ru(tpy)_2$, which was synthesized in Prof. Abruña's group at Cornell, Ithaca, USA [18]. The molecule has a spheroidal shape in solution with a diameter of approximately 10 nm [18], with its 64 redox sites distributed over the surface of the sphere. Additionally, in acetonitrile solution, these dendrimer molecules adsorb strongly to platinum surfaces to form a monolayer which may thus be probed electrochemically [18].

If cyclic voltammetry is carried out at scan rates such that δ, the extent of the diffusion layer, see equation (2), would be much greater than the dimension of the adsorbed dendrimer, all the Ru^{II} metal centers of each adsorbed molecule are oxidized into Ru^{III} within one voltammetric scan, so that the voltammogram is that of an adsorbed monolayer [18].

On the other hand, if the scan rate may be made fast enough so that δ results smaller than the dimension of the molecule, a semi-infinite diffusion response is expected. By careful consideration of this sort of diffusionnal response, information can thus be gained about the topology of the space in which the diffusive process occurs, giving insight into whatever deformation of the dendrimer, if any, takes place upon adsorption to an electrode surface. Spatial information on this type of redox dendrimer has already been obtained by STM techniques [19], but those techniques only serve to image the "top" side of the dendrimer and provide no information about the side of the molecule in contact with the surface. In other words, STM allows a geometrical description of the top face of the electrode-molecule assembly, so that it is impossible to examine at which extent the dendrimer molecules "squash" onto the electrode surface when they adsorb.

For very fast scan rates, the voltammograms should exhibit characteristics due to semi-infinite diffusion. However, the redox sites under consideration are covalently fixed at the surface of the spherical dendrimer by the dendritic polyamidoamide branches and are thus unable to diffuse *"physically"* towards the electrode surface. Yet, if one considers the occurrence of electron transfer between adjacent redox sites, it becomes clear that one might expect to observe an apparent diffusion coefficient (D_{hop}) that is due to *"electron hopping"* between the redox centers [20]. None of the redox sites will be displaced from its rest position, except for small motions around this position, but the electrochemical perturbation wave will sweep across the surface of the molecule as if the ruthenium centers had physically moved in the sense of Nernst-Einstein [1]. A similar problem was addressed several years ago for diffusion in macroscopic polymeric films containing covalently linked redox sites [20-27].

Extensive theoretical work has been performed in our laboratory in order to fully characterize the electrochemical behavior of this system, and these results have been reported in detail elsewhere [28]. Our aim here is only to summarize the results obtained through these complete analyses. However, in order to understand what follows, several definitions need to be presented and some of the theoretical aspects need to be mentioned with some specificity. First, we consider a spherical dendrimer molecule of radius R_0 which has been cut at one end to provide a flat surface that is in contact with the electrode (see Figure 5).

The disk that is in contact with the electrode can be considered to be the base of a cone, whose apex is at the center of the sphere, with half angle ϕ_0. The 64 redox sites are distributed in a shell of thickness d (*viz.*, the diameter of one

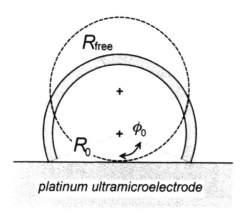

Figure 5. Schematized shape of a PAMAM dendrimer molecule adsorbed onto a platinum ultramicroelectrode surface. The shaded area represents the thin shell into which the 64 ruthenium redox centers are dispersed, while the white zone inside features the dendrimer covalent tethers. The circle shown in dashed line represents the size of the free dendrimer molecule in solution for comparison.

Ru(tpy)$_2$ moiety) at the surface of the sphere. Within this shell, the apparent surfacic diffusion coefficient resulting from electron hopping (D_{hop}) is given by equation (10) [20-28]:

$$D_{hop} = (4/\pi)\,(k/d) \tag{10}$$

where k is the apparent rate of electron transfer between a pair of vicinal Ru^{II}/Ru^{III} sites in the dendrimer shell. In order to be able to compare this 2-dimension rate constant with 3-dimension homogeneous ones which are more familiar to chemists, one should note that [28]:

$$k \approx k_{act}^{soln}/6N_A \tag{11}$$

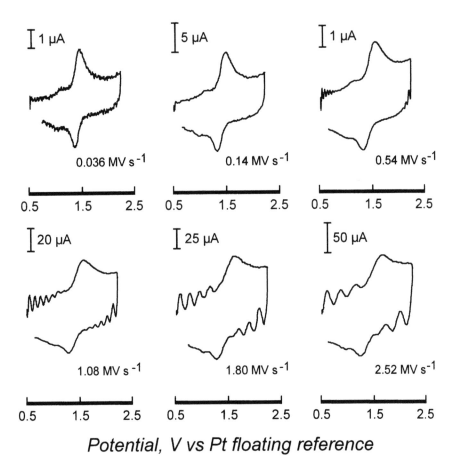

Potential, V vs Pt floating reference

Figure 6. Ultrafast cyclic voltammetry of a monolayer of 4th generation PAMAM dendrimer with 64 Ru(tpy)$_2$ redox centers adsorbed onto a 5 μm radius Pt ultramicroelectrode (acetonitrile, 0.6 M NEt$_4$BF$_4$; 20°C). The scan rates are indicated on each panel in MVs[-1]. (Adapted from refs. [28,29]).

where N_A is Avogadro's constant and k_{act}^{soln} is the molar isotopic activation rate constant determined for the same redox centers in a homogeneous bulk solution giving rise to a similar environment of the redox centers. A factor of 1/6 is considered because in bulk solution reactions can occur in either direction along any of three coordinates whereas in the spherical shell, only one of the six possibilities produces a net apparent site displacement.

This analysis shows that oxidation of the ruthenium(II) centers dispersed in the dendrimer shell should obey an apparent diffusive behavior, so that in principle one should be able to investigate the cross-talking between Ru(II) and Ru(III) moieties within the dendrimer molecule by voltammetric methods. There are however two difficulties for that. One is obvious, in the sense that such voltammetric analysis requires that one is able to control the time-extension of the apparent diffusion layer within the nanometric range, and this obviously demands access to ultrafast scan rates. The other difficulty stems from the fact that the 64 redox centers present in a single dendrimer molecule are not in sufficient number to lead to a statistical behavior. This second difficulty is however solved by using a 2-dimension array of of dendrimer molecules adsorbed onto the ultramicroelectrode surface. Indeed, the respective sizes of the PAMAM dendrimers (*ca.*, 5 nm radius) *vs.* that of the disk ultramicroelectrode (5 μm radius), ensures that about one million of dendrimer molecules are present on the electrode surface, so that the

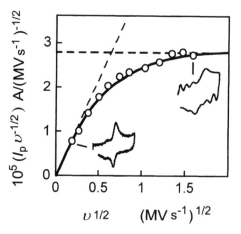

Figure 7. Ultrafast cyclic voltammetry of a monolayer of 4th generation PAMAM dendrimer with 64 Ru(tpy)$_2$ redox centers (see Figure 6). Symbols: variations of the experimental scan rate-normalized voltammetric current peak intensity as a function of square root of scan rate. Solid line: predicted behavior for $\phi_0 = 1.2$ rd and $D_{hop}/R_0^2 = 1.65 \ 10^7 \ s^{-1}$. Same conditions as in Figure 6. (Adapted from ref. [28]).

measurements are satisfactorily averaged statistically as was tested by performing random-walk simulations of diffusion (C. Amatore, F. Grün, E. Maisonhaute, *Angew. Chem.*, in press).

Statistically significant voltammograms could thus be obtained for these nano-metric redox objects upon using scan rates ranging from 36 kV/sec to 2.52 MV/sec. A representative set of voltammograms is shown in Figure 6 [28,29], which exemplifies the voltammetric gradual changes observed upon increasing the scan rate up to the megavolt-*per*-second range.

The variations of the scan-rate-normalized peak current intensity, $I_p/v^{1/2}$, are presented in Figure 7 as a function of the scan rate v. At "slow" scan rates, *i.e.*, within the kVs^{-1} range, where $\delta \gg 2R_0$, the peak current intensity I_p is propor-tional to the scan rate (*viz.*, $I_p/v^{1/2} \propto v^{1/2}$, in Fig. 7) as expected for the voltam-metry of a molecularly-thin redox film adsorbed on a surface [1]. At higher scan rates, *i.e.*, within the MVs^{-1} range, where $\delta \ll 2R_0$, I_p tends to be proportional to the square root of scan rate (*viz.*, $I_p/v^{1/2}$ = constant, in Figure 7) as expected for a system undergoing semi-infinite diffusion [1]. This shift between the two expected limiting behaviors is clearly apparent in Figure 7.

The exact transition between these two limits (*viz.*, the exact shape of the solid curve in Figure 7) is strongly dependent on the topology of the space avail-able to diffusion (which is determined by R_0 and ϕ_0) as well as on the apparent diffusion coefficient, D_{hop}. The diffusion equations for electron hopping on the surface of a sphere were formulated and solved with the appropriate boundary conditions. This afforded expressions for the peak current intensity in terms of the two parameters D_{hop}/R_0^2 and ϕ_0 [28]. In order to extract useful information from the voltammograms, simulations were carried out to determine the param-eters which best matched the experimentally observed responses [28]. The first parameter of interest is ϕ_0, which is a topological measure of the 2-dimension space shape since it determines how the dendrimer deforms as it "squashes" down on the electrode surface (see Figure 5). A series of working curves which characterize the variations of $\log(I_p/v^{1/2})$ with $\log v$, for several values of ϕ_0, were calculated and then the experimental data were examined to identify which working curve best fit the data. From this analysis, a value of $\phi_0 = 1.2\pm0.1$ radi-ans was determined [28] and the sketch in Figure 5 is drawn for this value of ϕ_0. The best fit to the experimental data in real time scales (*viz.*, $I_p/v^{1/2}$ *vs.* $v^{1/2}$ as shown in Figure 7) then afforded the value of $D_{hop}/R_0^2 = 1.65 \times 10^7$ s^{-1} which allows to reconcile the "topological" time scale, *viz.*, R_0^2/D_{hop}, to the voltam-metric one, *viz.*, $\theta = RT/Fv$ (eqn. 1).

From the ϕ_0 value determined above it is deduced that the PAMAM dendrimer molecules do not rest on the electrode surface by retaining their total spherical solution shape, but that they adsorb by forcing a significant fraction of their chain linkers to be in close contact with the electrode surface (see sketch in Figure 5). In fact, within the framework of this model, the disk of contact with the electrode has a radius of $R_0 \sin\phi_0 \approx 0.93 R_0$. This shows that the adsorbed dendrimer molecule more closely resembles a hemisphere than a sphere as in solution [30]. For example, assuming that the dendrimer molecule retains its inner volume when it "squashes" on the electrode surface, one estimates that $R_0 \approx 1.1 \times R_{free}$ where $R_{free} \approx 5$ nm is the estimated radius of a free dendrimer molecule in solution [30]. Such data compares extremely well with the result of STM investigations of related dendrimers on platinum electrode surfaces in which it was noted that the radius of the adsorbed globule was *ca.* 10% larger than that evaluated for the free molecule in solution [30].

Based on the estimation of $R_0 \approx 1.1 \times R_{free} \approx 5.5$ nm and the independent determination of $D_{hop}/R_0^2 = 1.65 \times 10^7$ s^{-1}, $D_{hop} = 5 \times 10^{-6}$ cm^{2}s^{-1} is obtained [28]. Despite there is obviously some crudeness in the above evaluation of R_0, the ensuing value of D_{hop} is not affected by more than 10-20%. Using equation (10) and noting that the thickness of the spherical shell of redox centers is $d = 1.4$ nm (*i.e.*, is equal to the molecular diameter of one Ru(tpy)$_2$ redox center) yields a value of k of *ca.* 4.8×10^{-16} L s^{-1}. When converted into a homogeneous molar self-exchange rate constant using equation (11), this 2-dimension rate constant k would correspond to a value of $k_{act}^{soln} = 1.7 \times 10^9$ M^{-1}s^{-1} for the rate of electron transfer between a Ru^{II} and a Ru^{III} metal center in solution.

To the best of our knowledge, k_{act}^{soln} has never been reported for Ru$^{II/III}$(tpy)$_2$ complexes, but the self-exchange rate constant $k_{act}^{soln} \approx 10^9$ M^{-1}s^{-1} has been determined for the related Ru$^{II/III}$(bpy)$_3$ system in water [31]. In other words, the measured D_{hop} value corresponds exactly to what is expected for a diffusion-limited electron self-exchange, *viz.*, involving two redox centers which are adjacent in the transition state.

However, the redox sites distributed on the surface of the dendrimer molecule are neither adjacent, nor are they able to move to achieve permanent close contact. It can be estimated that, to achieve full coverage of the dendrimer surface with all redox centers in close contact, 140 redox sites would be necessary. Since there are only 64 on each dendrimer molecule, the Ru$^{II/III}$ centers cannot be in close contact, the average center-to-center distance being *ca.* 2 nm. Since the radius of one ruthenium center is *ca.* 0.7 nm, this implies that a pair of two nearby centers should move cumulatively by 0.6 nm from their equilibrium position to reach close contact.

Several studies of long-range electron transfer have shown that the rate of electron transfer drops exponentially with the distance λ between the two centers at the transition state, with β being the attenuation factor per unit of distance such that $k \propto \exp(-\beta\lambda)$ [32-34]. A typical value of β is 1 Å^{-1}, *i.e.*, 10 nm^{-1} [33,34].

Considering the above average distance $\lambda = 2$ nm between the tethered redox centers and using $\beta = 10$ nm^{-1} predicts that the rate of electron transfer observed between centers on the surface of the dendrimer should be approximately one thousandth of the observed one. The experimental value of k, however, is comparable to that expected for free centers in solution which come into close contact upon electron transfer. That leads to the conclusion that, even though the redox centers are immobilized by dendritic branches, they must possess considerable mobility at the scale of a few nanometers inside their potential wells, allowing them to reach easily close contact in the transition state for electron transfer, without a high energetic cost nor severe viscous drag.

Using the experimentally determined k value, it is possible to minimize the force constant attracting each redox center in its potential well and to subsequently estimate the maximum tolerable surfacic viscosity of a site when it moves within its potential well [28]. The result of that calculation is that the surfacic viscosity, η_S, of one Ru(tpy)$_2$ molecule in its potential well is considerably less than 10^{-20} N s m^{-1}. For comparison, application of the Stokes-Einstein relationship to $D_{soln} = 5 \times 10^{-6}$ cm^{2}s^{-1}, taken as a typical value for the diffusion coefficient of a free Ru(tpy)$_2$ molecule in solution, would predict a surfacic kinematic viscosity $\eta_s^{pred} = \eta_{soln}d = 6\eta(d/2)D_{soln}/k_BT = 7 \times 10^{-13}$ N s m^{-1}. Such a value predicted on the basis of free solution behavior exceeds considerably the maximum allowable value of η_S for the dendrimer system. In other words, the viscosity of the redox centers on the spherical dendritic shell is at least 10^6 times smaller than that in a free solution.

At first glance, this is counterintuitive since in a dendrimer the Ru(tpy)$_2$ centers are linked to radial chains which ought to dissipate energy to overcome the sterical rubbing against their neighbors during any Ru(tpy)$_2$ center motion. Our results shows that these frictions between linkers are minimal when a nearby pair of Ru^{II} and Ru^{III} centers move over a distance of *ca.* half a nanometer towards each other. Therefore, even if these friction exist necessarily, they are not the important phenomenon which dissipate energy during the motion of ruthenium centers, at least for the sub-nanometric displacements which are required here. Most important therefore is the fact that the dendrimer structure prevents the building-up of a tight solvent and ionic atmosphere around each redox center, thus eliminating the severe electrophoretic drag which occur in any solution. This is expected to enhance considerably the apparent mobility compared to what is reflected by D values and prevails when diffusion occurs into a solution over longer distances. To the best of our knowledge, this important effect has never been observed nor considered before. However, this unsuspected facile mobility of dendrimer-borne active centers certainly needs to be taken into account in designing dendrimer-supported catalysts. Indeed, this may induce considerable chemical cross-talk between vicinal catalytic centers and thus enforce mechanistic paths which may differ from those known for the same catalysts when they are homogeneously dispersed or covalently bound to a solid support.

5   MONITORING DYNAMIC CONCENTRATION PROFILES NEAR AN ACTIVE INTERFACE

We have illustrated through the above series of examples how access to extremely fast scan rates allows ultrafast kinetics to be monitored under classical homogeneous conditions or within a nanometric redox object. In this final section we wish to illustrate another advantage which follows from the ability of applying fast electrochemical perturbations at an ultramicroelectrode.

Indeed, taking advantage of the intrinsic small size of an ultramicroelectrode and the fact that the extent of solution probed by this electrode, *viz.*, the thickness of its diffusion layer, can be made extremely small (eqn. 2), one understand that an ultramicroelectrode may be converted into a very local sensor, able to report with a micrometric resolution onto the concentration of any redox active material dispersed into a solution in which a gradient of this material exists. In other words, such a dispositive may be used to track concentration profiles within a solution in a similar way as this is performed along microtome analysis in solid materials. However, the challenge here is that a concentration profile within a solution has no intrinsic existence *per se* since it vanishes immediately in the absence of the time-dependent perturbation which is imposed by the generating interface.

5.1   *Principle*

In electrochemistry, concentration gradients are central since they command the transport of molecules to-and-from the electrode surface, and in particular impose the current at the electrode. Engstrom [35,36] first showed that an ultra-microelectrode could be used as a local probe to map amperometrically concentration profiles created by a large active surface. Concentration *vs.* distance profiles with a resolution better than 5 μm were also reported using absorption spectrometry [37]. At the same time, SECM (*viz.*, scanning electrochemical microscopy) [38-40] was introduced, allowing the investigation of sample surfaces topologies together with the monitoring of their electrochemical activities. Due to its high spatial resolution numerous applications of SECM have been reported to provide quantitative insights into liquid / liquid, solid / liquid and liquid / gas interfaces [40-44], or to monitor potential distributions near active surfaces [45,46]. Schultze *et al.* [47-49] developed the Scanning diffusion Limited Current Probe-method (SLCP) for spatially resolved concentration measurements during cathodic alloy deposition in microstructures. Our group contributed recently by showing that an unprecedented precision could be achieved through the use of confocal Raman microscopy together with spectroscopic information on the electrogenerated species [50,51]. These works demonstrate the intense effort made to perform such measurements since it is the paragon of many other chemically or biologically important situations where concentration profiles near an active surface or any interface play a key role.

One of our interests was to probe the molecular composition of samples of micrometric sizes where complex kinetic situations were initiated upon electrochemical stimulation. Thus we have been able to demonstrate that amperometric and potentiometric detections [52,53] could be performed with an ultramicroelectrode probe to map concentration profiles generated by another electrode. The principle of the experiment is summarized in Figure 8.

In the following, we want to illustrate the great interest of such approaches, but first we need to establish and discuss in more detail the principle of such measurements. For this we will use the example of a canonical electrochemical system consisting in the oxidation of an aqueous solution of $Fe(CN)_6^{4-}$ [54].

Figure 9a-A shows the chronoamperometric current observed for a solution of 10 mM $Fe(CN)_6^{4-}$ when the potential of the generator millimetric electrode was set on the oxidation plateau (+0.6 V *vs.* SCE $\gg E^0_{Fe(II/III)}$) of the $Fe(CN)_6^{4-}$ wave. While the generator electrode is active, an ultramicroelectrode probe could be positioned with a micromanipulator in the solution at a micrometric distance z from the generator surface (see sketch in Figure 8). In a typical experiment, the generator is allowed to perform during a time span Δt during

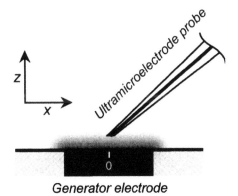

Generator electrode

Figure 8. Experimental principle of the set-up used for monitoring concentration profiles extending near a working electrode with an ultramicroelectrode probe.

which the ultramicroelectrode is disconnected electrically. After this time delay Δt the ultramicroelectrode probe was connected electrically during a few millisecond duration θ (so that $(D\theta)^{1/2} \ll z$) at a potential selected so that either the local concentration $Fe(CN)_6^{4-}$ could be oxidized ($E_{probe} \gg E^0_{Fe(II/III)}$), or the local concentration of $Fe(CN)_6^{3-}$ be reduced ($E_{probe} \ll E^0_{Fe(II/III)}$). The chronoamperometric current recorded at the ultramicroelectrode probe at the end of the period θ (Figure 9a-B) is thus a direct measurement of the local concentration of the selected species at the,time $\Delta t + \theta$ and at the point z of the solution where the tip of the probe is placed. Upon varying z while keeping Δt constant one can thus monitor the concentration profile of either species as a function of the distance from the electrode surface. The same procedure may be repeated at different times Δt so that the variations of the concentration profiles developing at the generator surface can be monitored as a function of time [54].

a)   b)

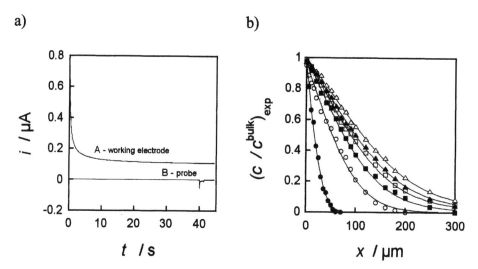

Figure 9. The principle of monitoring concentration profiles near a working electrode is exemplified by the oxidation of an aqueous solution of $Fe(CN)_6^{4-}$, 10 mM in 1 M KCl. (a) A: Variations of the working millimetric electrode (generator) current as a function of time when its potential was set at +0.6 V *vs.* SCE; B: the trace labeled "probe" shows the current monitored at the ultramicroelectrode probe when this was connected at $\Delta t = 40$ s. (b) Variations of the experimental (symbols) concentration of $Fe(CN)_6^{3-}$ monitored with the ultramicroelectrode probe during the chronoamperometric oxidation of $Fe(CN)_6^{4-}$ at different times Δt (from left to right, $\Delta t = 0.5, 5, 10, 15, 20$ and 40 s). Solid curves: theoretical predictions [54] for diffusion perturbed by spontaneous convection for $D = 6 \times 10^{-6}$ cm^{2}s^{-1} and a steady state Nernst diffusion layer thickness of $\delta = 230$ μm. (Adapted from ref. [54]).

The result of this procedure is shown in Figure 9b under the form of a series of reconstructed concentration profiles (symbols) measured for $Fe(CN)_6^{3-}$ at different times Δt. As expected for planar diffusion [1], it is observed that the concentration profile of $Fe(CN)_6^{3-}$ initially propagates as a function of $(\Delta t)^{1/2}$. However, it is soon observed that this propagation rapidly breaks down when Δt approaches the range of a few tens of a second, so that the concentration profile becomes more or less time independent of the time duration Δt. This progressive phenomenon occurs in phase with the fact that the current monitored at the generator electrode deviates from its Cottrellian behavior (*viz.*, $i \propto (\Delta t)^{-1/2}$, observable when Δt is less than a few seconds) to reach its steady state limit when Δt is larger than a few tens of a second (Figure 9a).

This behavior reflects the progressive interference of "spontaneous" convection into the diffusional transport of molecules to-and-from the electrode. This progressive dual transport process can be modeled, so that the predicted concentration profiles shown as solid curves in Figure 9b take into account the increasing interference of convection when the distance z from the generator electrode surface increases [54]. It is of interest to note that the limiting concentration profile at very long times is very close from that which could be deduced if the hypothetical Nernst layer simplification would be applicable. In other words, these results validate *a posteriori* this classical text-book approximation and show that despite its apparent crudeness, it is not unrealistic at all as soon as the generator electrode current has reached steady state.

5.2   *Investigation of Redox-Catalysis*

Redox catalysis provides the mean of activating electrochemically a substrate S at a potential located much before (*i.e.*, positive to, in reduction, or negative to, in oxidation) its electrochemical wave [55-57]. This is performed through the use of an electron transfer mediator, M, whose chemically stable reduced or oxidized form, noted $M^{\bullet\pm}$ accordingly, see equation (12), is able to exchange endergonically an electron with the substrate S, equation (13). When the resulting activated substrate $S^{\bullet\pm}$ is sufficiently unstable chemically (eqn. 14), the up-hill equilibrium in equation (13) is pulled continuously to its right-hand-side so that $S^{\bullet\pm}$ is continuously generated at a maximum rate imposed by the forward rate constant, k_{ET}, of equation (13):

$$M \pm e^- \longrightarrow M^{\bullet\pm} \tag{12}$$

$$M^{\bullet\pm} + S \longrightarrow M + S^{\bullet\pm} \quad (k_{ET}) \tag{13}$$

$$S^{\bullet\pm} \longrightarrow etc. \tag{14}$$

The fundamental kinetic reasons explaining why the rate of the follow-up reaction (14) may be sufficient to pull the homogeneous electron transfer at a significant rate while the same reaction does not succeed in doing so when the substrate is reduced or oxidized directly at the potential of the redox mediator, have been extensively discussed previously [55,56].

When one uses redox catalysis for preparative purposes, a large turn-over is desired in order that a small fraction of redox mediator affords a current density comparable to that which would be obtained for a direct electrolysis of the substrate. As a consequence, the overall kinetics of the process in equations (13,14) needs to be much larger than the rate of diffusion of $M^{\cdot\pm}$ from the electrode surface to the solution bulk. Under such conditions $M^{\cdot\pm}$ cannot survive enough to escape from the diffusion layer towards the solution bulk. Similarly, S cannot survive its penetration into the diffusion layer up to the electrode surface. It follows that whenever k_{ET} (eqn. 16) has a significant value, $M^{\cdot\pm}$ and S cannot co-exist in the diffusion layer except within a narrow strip of solution located at $y = \mu$, which separates the diffusion layer into two distinct compartments with different compositions (Figure 10a) [58,59]. One zone ($0 < y < \mu$), adjacent to the electrode surface contains M and $M^{\cdot\pm}$ and

a)   b)

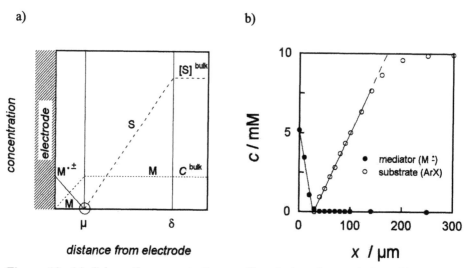

Figure 10. (a) Schematic concentration profiles during the catalytic oxidation (or reduction) of a substrate S by the radical cation (or radical anion) of a redox mediator $M^{\cdot\pm}$ generated at the working electrode surface. **(b)** Experimental concentration profiles determined during the reduction of iodobenzene (10 mM) catalyzed by benzophenone (5 mM) in 0.1 M nBu_4NBF_4/DMF. Benzophenone radical anion: solid circles; iodobenzene: open circles. (Adapted from ref. [59]).

not S or S$^{\cdot\pm}$. The other, located between this narrow strip and the solution bulk ($\mu < y < \mu$) contains only M and S. M$^{\cdot\pm}$ and S may then react together (eqn. 13) only within the excessively narrow domain located at $y = \mu$ which separates the two above zones. In this narrow strip of solution M$^{\cdot\pm}$ and S have then necessarily extremely small concentrations [58,59].

We wish to show hereafter that thanks to the development of in-situ mapping techniques developed recently in our laboratory (see above and refs [58,59]), the precise concentration profiles of M$^{\cdot\pm}$ and S can be monitored so as to establish definitively the existence of the above diffusion layer structuring when redox catalysis is performed under conditions which are similar to those occurring within a realistic preparative electrolysis [59]. Because of this latter concern, all the experiments reported here are performed under steady state conditions as imposed by "spontaneous" convection (*viz.*, as discussed above; see Figure 9b).

We wish to illustrate this phenomenon through the *in situ* investigation of the catalytic reduction of an aromatic halide (ArX) [60,62]:

$$M + e^- \rightleftarrows M^{\cdot-} \tag{15}$$

$$M^{\cdot-} + ArX \rightleftarrows M + ArX^{\cdot-} \quad (k_{ET}) \tag{16}$$

$$ArX^{\cdot-} \longrightarrow Ar^{\cdot} + X^- \tag{17}$$

$$A^{\cdot} + e^- \rightleftarrows Ar^- \tag{18}$$

$$Ar^{\cdot} + ArX^{\cdot-} \longrightarrow Ar^- + ArX \quad \text{and/or} \quad Ar^{\cdot} + M^{\cdot-} \longrightarrow Ar^- + M \tag{19}$$

$$Ar^- + H^+ \longrightarrow ArH \tag{20}$$

The neutral radical Ar$^{\cdot}$ thus formed may then be reduced at the electrode (eqn 18) or in solution (eqn 19) leading to the strong base Ar$^-$. Protonation of the later in the medium affords ArH (eqn 20). Under such conditions, the overall process is tantamount to a two-electron reduction of ArX into ArH provided that the lifetimes of the anion radicals M$^{\cdot-}$ and ArX$^{\cdot-}$ are sufficiently short [56,60,62].

A millimetric disk electrode (generator) was used to generate locally the concentration gradients. Before any measurement, this electrode was polarized on the plateau of the mediator reduction wave during several seconds until the steady state regime imposed by the natural convection of the solution was reached (see above and Figure 9b). The resulting concentration profiles were then monitored by placing a platinum-disk ultramicroelectrode (probe) at selected positions within the diffusion layer of the generator electrode as discussed above (Figures 8,9b). An amperometric detection was performed, which allowed a direct evaluation of the concentrations of the target species [59].

Figure 10b reports the concentration profiles recorded by the probe when the reduction of 10 mM iodobenzene (ArX; $E^o_{ArX} = -2.10$ V *vs.* SCE) is catalyzed by the reduction of 5 mM benzophenone (M; $E^o_M = -1.75$ V *vs.* SCE). The millimetric generator electrode was poised at a potential $E_{gen} = -1.95$ V *vs.* SCE, located in between the half-wave potentials of benzophenone and iodobenzene. For each measurement, the ultramicroelectrode probe was not connected during a time delay Δt of 50 s after application of the potential step to the generator. This ensured that the generator current reached its steady state (see above, Figure 9a). After this time delay, the probe potential was set during a sampling time θ (see above) either at potential $E_{oxdn} = 0$ V *vs.* SCE, *viz.*, on the plateau of the oxidation wave of the benzophenone radical anion, or at potential $E_{redn} = E_{gen} = -1.95$ V *vs.* SCE. In the first case (*viz.*, when $E_{probe} = E_{oxdn}$), the only process which can be monitored at the probe is the oxidation of the mediator anion radical if any is present:

at any y: $c_{M^-} = [i(\theta,E_{oxdn},y)/i_0] \times (D_M/D_{M^-})^{1/2} c^{bulk}$ (21)

where i_0 is the current measured for the same sampling time θ when the probe is placed in the bulk solution containing benzophenone only at concentration c^{bulk}, and D_M and D_{M^-} the diffusion coefficients of M and M·⁻ respectively.

In the second case (*viz.*, when $E_{probe} = E_{redn}$), the probe reduces the local benzophenone concentration. Therefore, when this latter is alone ($y < \mu$, in Figure 9a), the probe current reports only on the benzophenone concentration, so that:

$0 < y < \mu$ $c_M = [i(\theta,E_{redn},y)/i_0] \times c^{bulk}$ (22)

However, when the solution domain which is investigated contains both benzophenone and iodobenzene (*viz.*, when $\mu < y < \delta$; see Figure 9a), one observes at the probe also the catalytic two-electron reduction of iodobenzene in addition to that of benzophenone provided that k_{ET} (eqn. 16) is sufficiently large [56,60]. The probe current is then:

$y > \mu$ $i(\theta,E_{redn}, y) = FA/(\pi\theta)^{1/2}[D_M^{1/2}c_M + 2D_{ArX}^{1/2}c_{ArX}]$ (23)

where D_{ArX} is the diffusion coefficient of ArX and c_{ArX} its concentration at the point y tested in the solution. Since by definition, at $y > \mu$ the concentration c_M of benzophenone in the diffusion layer of the working electrode is equal to c^{bulk} (see Figure 10a), one obtains:

$y > \mu$ $c_{ArX} = [(i(\theta,E_{redn}, y) - i_0)/2i_0] \times (D_M/D_{ArX})^{1/2} c^{bulk}$ (24)

Based on the probe current measurements performed at either potential E_{oxdn} or E_{redn}, all the three concentration profiles (*viz.*, of M, M·⁻ and ArX) can be determined based on the application of equation (21,22 or 24) accordingly. The result of this procedure is represented in Figure 10b for a particular set of the mediator (5 mM) and substrate (10 mM) concentrations. One can verify that these profiles are in good agreement with the predictions in Figure 10a.

The thickness δ of the diffusion layer can be calculated by extrapolating the linear dependence of the concentrations observed near the electrode surface to a zero intercept when benzophenone alone is present, as was observed above (see Figure 9b at long durations Δt). In those experimental conditions, δ is estimated at 175 μm (see Figure 10a, for [PhI] = 0 M). One must stress that the same thickness is obtained upon extrapolating the concentrations of iodobenzene to the point where it reaches its bulk value (Figure 10b).

Within the framework of redox catalysis [59], it is expected that increasing the concentration of ArX should "push" the boundary at $y = \mu$ towards the electrode

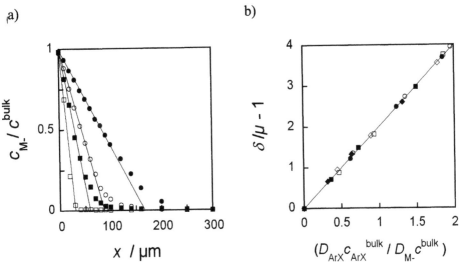

Figure 11. Influence of the addition of iodobenzene on the concentration profiles of the benzophenone radical anion during the catalytic reduction of iodobenzene. (a) Experimental concentration profiles of the benzophenone anion radical (benzophenone: 10 mM in 0.1 M nBu$_4$NBF$_4$/DMF) monitored in the presence of different concentrations of iodobenzene: [PhI] = 0 mM (closed circles), 5 mM (open circles), 10 mM (closed squares), 10 mM (open squares). (b) Experimental validation of eqn. (25), for a series of substrates and mediators, see text (slope = 2.004, R^2 = 0.9995): iodobenzene + perylene (open circles), iodobenzene + benzophenone (closed circles), 2-bromonaphthalene + 1,3-diphenylisobenzofuran (open squares), 2-bromonaphthalene + 1-cyanonaphthalene (closed squares), 2-bromopyridine + 1-cyanonaphthalene (open diamond), 2-bromopyridine + methyl benzoate (closed diamonds). (Adapted from ref. [59]).

surface because the flux demands increases upon increasing the forward rate of reaction (16). In fact, it can be predicted that the position $y = \mu$ of the interfacial domain must vary with ArX concentration as in eqn. (25) [59]:

$$(\delta/\mu)-1 = 2\times(D_{ArX}/D_{M^-})\times(c_{ArX}^{bulk}/c^{bulk}) \qquad (25)$$

where all parameters have been defined above. Interestingly, this law does not depends on the particular mediator / substrate couple except through the ratio of the diffusion coefficients D_{ArX}/D_{M^-} which can be measured independently.

   This effect is observed to be qualitatively correct through the series of concentration profiles displayed in Figure 11a for the mediator anion radical when the concentration of iodobenzene is increased from 0 to 10 mM, since it is seen that μ evidently decreases upon increasing c_{ArX}^{bulk}. The same effect is better established on Figure 11b, upon plotting $[(\delta/\mu)-1]$ versus the concentration ratio of substrate over mediator for several mediators (benzophenone, $E^0 = -1.75$ V/SCE; 1-cyanonaphthalene, $E^0 = -1.78$ V/SCE;   1,3-diphenylisobenzofuran, $E^0 = -1.82$ V/SCE; methyl benzoate, $E^0 = -2.17$ V/SCE; perylene, $E^0 = -1.63$ V/SCE) and substrates (iodobenzene, $E^0 = -2.10$ V/SCE; 2-bromonaphthalene, $E^0 = -2.20$ V/SCE; 2-bromopyridine, $E^0 = -2.34$ V/SCE). As expected from eqn. (25), a linear correlation is obtained provided only that the data are corrected by the ratio on the corresponding diffusion coefficients.

6   FINAL REMARKS

   By taking advantage of the capability to obtain undistorted cyclic voltammograms up to scan rates in excess of 2.5 MV s^{-1}, we have been able to demonstrate that voltammetric methods can be used with great kinetic profit to monitor very fast chemical processes occurring in the nanosecond time domain.

   Furthermore, since these ultrafast scan rates correspond to diffusion over sub-nanometric distances, voltammetric methods can be elaborated into "molecular microtomes" offering new experimental tools through which the topology of nanoscopic redox objects adsorbed on surfaces may be investigated. Thus, topology and dynamics of nanometric objects can be studied by a careful adjustment of the size of the diffusion layer, which allows one to progressively span the geometrical dimensions of the object under investigation.

   Finally, the ability of exploring very small domains of a solution located at the very tip of an ultramicroelectrode affords an experimental mean of investigating the concentration profiles which develop near an active interface. We have established the usefulness and validity of this method by considering the case of concentration profiles generated by an electrode, but this may be extended to any other interface provided that a redox marker is available to report on the local concentration. Also, due to the present stage of development of the method, we relied on conditions in which the concentration profiles investigated extended over several tens of a micrometer in order that the micrometric size (about 5 μm) of the electrode remained small *vis-à-vis* the size of the sampled domains. In this case sampling times θ of a few milliseconds were enough to ensure the desired precision of the measurements. However, ultramicroelectrodes of truly nanometric dimensions may now be produced and be well characterized [63] so that the method will be applicable for the investigation of sub-nanometric domains. In this case, much smaller sampling times will be required to ensure that the diffusion layer thicknesses extending at the probe surface will remain small enough.

ACKNOWLEDGEMENTS

This work was supported in parts by CNRS (UMR CNRS-ENS-UPMC 8640 "PASTEUR"), Ecole Normale Supérieure and the French Ministry of Research. Eng. Gérard Simonneau is more than gratefully acknowledged for his considerable involvement in the design and construction of the potentiostats used in this studies. The work on the PAMAM dendrimer has been performed within the framework of a collaborative research [28,29] with the group of Prof. Hectór Abruña at Cornell University (Ithaca, USA) whom we are glad to acknowledge gratefully.

REFERENCES

1. Bard, A.J.; Faulkner, L.R. *"Electrochemical Methods"*, J. Wiley & Sons, New York (1980).
2. Amatore C., in *"Physical Electrochemistry"*, Rubinstein I., Ed., M. Dekker, New York, Chapter 4 (1995).
3. Amatore C.; Lefrou C. *J. Electroanal. Chem.*, *296*, 335 (1990).
4. P.T. Kissinger; W.R. Heineman, Eds., *Laboratory Techniques in Electroanalytical Chemistry*, M. Dekker, New York (1984).
5. Wipf D.O.; Wightman R.M. *Anal. Chem.*, *60*, 2460 (1988).
6. Andrieux C.P.; Garreau D.; Hapiot P.; Pinson J.; Savéant J.M. *J. Electroanal. Chem.*, *243*, 321 (1988).
7. Amatore C.; Lefrou C.; Pflüger F. *J. Electroanal. Chem.*, *270*, 43 (1989).
8. Amatore C.; Lefrou C. *J. Electroanal. Chem.*, *324*, 33 (1992).
9. Amatore C.; Maisonhaute E.; Simonneau G. *Electrochem. Commun.*, *2*, 81 (2000).
10. Amatore C.; Maisonhaute E.; Simonneau G. *J. Electroanal. Chem.*, *486*, 141 (2000).
11. Amatore C.; Lefrou C. *J.Electroanal. Chem.*, *325*, 239 (1992).
12. Masnovi J.M.; Sankararaman S.; Kochi J.K. *J. Phys. Chem.*, *111*, 2263 (1989).

13. Amatore C.; Jutand A.; Pflüger F. *J. Electroanal. Chem.*, *218*, 361 (1987).
14. Amatore C.; Arbault S., *unpublished results*.
15. Kawata H.; Suzuki Y.; Nuzuma S. *Tetrahedron Lett.*, *27*, 4489 (1986).
16. Amatore C.; Jutand A.; Pflüger F.; Jallabert C.; Strzelecka H.; Veber M. *Tetrahedron Lett.*, *30*, 1383 (1989).
17. Amatore C.; Lefrou C. *J. Electroanal. Chem.*, *324*, 33 (1992).
18. Takada K.; Storrier G.D.; Morán M.; Abruña H.D. *Langmuir*, *15*, 7333 (1999).
19. Díaz D. J.; Storrier G.D.; Bernhard S.; Takada K.; Abruña H.D. *Langmuir*, *15*, 7351 (1999).
20. Murray R. W. *Phil. Trans. R. Soc. London A*, *302*, 253 (1981).
21. Dahms H.J. *J. Phys. Chem.*, *72*, 362 (1968).
22. Ruff I.; Friedrich V.J. *J. Phys. Chem.*, *75*, 3297 (1971).
23. Ruff I.; Friedrich V.J.; Demeter K.; Csillag K. *J. Phys. Chem.*, *75*, 3303 (1971).
24. Andrieux C.P.; Savéant J.M. *J. Electroanal. Chem.*, *111*, 377 (1980).
25. Laviron E. *J. Electroanal. Chem.*, *112*, 1 (1980).
26. Facci J.; Murray R.W. *J. Phys. Chem.*, *85*, 2870 (1981).
27. Blauch D.N.; Savéant J.M. *J. Phys. Chem.*, *97*, 6444 (1993).
28. Amatore C.; Bouret Y.; Maisonhaute E.; Goldsmith J.I.; Abruña H.D. *Chem. Eur. J.*, *7*, 2206 (2001).
29. Amatore C.; Bouret Y.; Maisonhaute E.; Goldsmith J.I.; Abruña H.D. *ChemPhysChem Eur. J.*, *2*, , 130 (2001).
30. Storrier G.D.; Takada K.; Abruña H.D. *Langmuir*, *15* , 872 (1999).
31. Sutin N.; Creutz C., *Pure Appl. Chem.*, *52*, 2717 (1980).
32. Chance B.; DeVault D.; Frauenfelder H.; Marcus R.A.; Schrieffer J.R.; Sutin N., Eds., *"Tunneling in Biological Systems"*, Acad. Press, New York (1979).
33. Mayo S.; Ellis W.R.; Crutchley R.J.; Gray H.B. *Science*, *233*, , 948 (1986).
34. Miller C.J., in *"Physical Electrochemistry"*, Rubinstein, I., Ed., M. Dekker: New York, Chapter 2 (1995), and refs. therein.
35. Engstrom R.C.; Weber M.; Wunder D.J.; Burgess R.; Winquist S. *Anal. Chem.*, *58*, 844 (1986).
36. Engstrom R.C.; Meaney T.; Tople R.; Wightman R.M. *Anal. Chem.*, *59*, 2005 (1987).
37. Jan C.C.; McCreery R.L. *Anal. Chem.*, *58*, 2771 (1986).
38. Bard A.J.; Fan F.F.; Kwak J.; Lev O. *Anal. Chem.*, *61*,132 (1989).
39. For a recent book on SECM, see: Bard, A.J.; Mirkin, M.V., Eds., *"Scanning Electrochemical Microscopy"*, Marcel Dekker, New York (2001).
40. Barker A.L.; Gonsalves M.; Macpherson J.V.; Slevin C.J.; Unwin P.R. *Anal. Chim. Acta*, *385*, 223 (1999) and references therein.
41. Slevin C.J.; Unwin P.R. *Langmuir*, *13*, 4799 (1997).
42. Slevin C.J.; Macpherson J.V.; Unwin P.R. *J. Phys.Chem. B*, *101*, 10851 (1997).
43. Zhang J.; Slevin C.J.; Unwin P. *Chem. Com.*, 1501 (1999).
44. Slevin C.J.; Unwin P.R. *Langmuir*, *15*, 7361 (1999).
45. Horrocks B.R.; Mirkin M.V.; Pierce D.T.; Bard A.J. *Anal. Chem.*, *65*, 1213 (1993).
46. Wei C.; Bard A.J.; Nagy G.; Toth K. *Anal. Chem.*, *67*, 1346 (1995).
47. Kupper M.; Schultze J.W. *Electrochimica Acta*, *42*, 3023 (1997).

48. Kupper M.; Schultze J.W. *Electrochimica Acta, 42,* 3085 (1997).

49. Klusmann E.; Schultze J.W *Electrochimica Acta, 42,* 3123 (1997).

50. Amatore C.; Bonhomme F.; Bruneel J.-L.; Servant L.; Thouin L. *Electrochem. Commun., 2,* 235 (2000).

51. Amatore C.; Bonhomme F.; Bruneel J.-L.; Servant L.; Thouin L. *J. Electroanal. Chem., 484,* 1 (2000).

52. Amatore C.; Szunerits S.; Thouin L. *Electrochem. Commun., 2,* 248 (2000).

53. Amatore C.; Szunerits S.; Thouin L.; Warkocz J.S. *Electrochem. Commun., 2,* 353 (2000).

54. Amatore C.; Szunerits S.; Thouin L.; Warkocz J.S. *J. Electroanal. Chem., 500,* 62 (2001).

55. Lund H.; Michel M.A.; Simonet J. *Acta. Chem. Scand., B28,* 901 (1974).

56. Andrieux C.P.; Savéant J.M., in *"Investigations of rates and mechanisms of reactions"*, Bernasconi F., Ed., *Vol. 6, 4/E, Part 2,* John Wiley & Sons, Inc., p. 305 (1986).

57. Lund H.; Daasbjerg K.; Lund T.; Pedersen S. *Acc. Chem. Res., 28,* 313 (1995).

58. Amatore C.; Szunerits S.; Thouin L.; Warkocz J.-S. *Electroanalysis., 13,* 646 (2001).

59. Amatore C., Pebay C., Scialdone O., Szunerits S., Thouin L.. *Chem. Eur. J., 7,* 2933 (2001).

60   Andrieux C.P.; Blocman C.; Dumas-Bouchiat J.M.; Savéant J.M. *J. Am. Chem. Soc., 101,* 3431 (1979).

61. Andrieux C. P.; Savéant J.M.; Su K.B. *J. Phys. Chem., 90,* 3815 (1986).

62. Andrieux C. P.; Savéant J.M. *J. Am. Chem. Soc., 115,* 8044 (1993).

63. Watkins J.J.; Chen J.; White H.S.; Abruna H.D.; Maisonhaute E.; Amatore C. *Anal. Chem., 75,* 3962 (2003).

Chapter 13

Electrochemical Advances Using Fluoroarylborate Anion Supporting Electrolytes

Frédéric Barrière,[1] Robert J. LeSuer, William E. Geiger*[2]

Department of Chemistry, University of Vermont
Burlington, VT 05405, USA

[1]*Present address: Université de Rennes 1, Institut de Chimie, UMR CNRS 6510*
35042 Rennes Cedex, France

[2]*e-mail: wgeiger@zoo.uvm.edu*

1   INTRODUCTION

1.1   Role of Supporting Electrolytes

A supporting electrolyte is required in virtually every experiment involving molecular electrochemistry [1-3]. Exceptions to this (e.g., through use of an ultra-microelectrode [4]) may be found when the voltammetric currents are sufficiently low to approach potentiostatic conditions, but even in those cases it is assumed that adventitious ions are present to carry the current. In the vast majority of cases, an electrochemically suitable electrolyte [5] is prepared by adding to the solvent a salt in 50-100 fold excess over that of the electroactive compound (i.e., the analyte). Within the context of nonaqueous solvents, the most important supporting electrolytes are those comprised of anions such as $[ClO_4]^-$, $[BF_4]^-$, $[PF_6]^-$ and $[CF_3SO_3]^-$ as tetraalkylammonium salts. The use of these anions became standard practice over four decades ago [3], so that their $[NR_4]^+$ salts may now be referred to as 'traditional' supporting electrolytes. It is fair to say that the choice of a supporting electrolyte had become a pragmatic, rather than an intellectual, matter. The purpose of this Chapter is to discuss very recent developments involving a new family of salts which promise to supplant the traditional electrolytes under many experimental conditions, especially those involving lower-polarity solvents having dielectric constants (ε) of about 10 or less. The crucial characteristic of the new electrolyte family is the inclusion of relatively large anions of very low nucleophilicity.

Since comprehensive reviews of the general role of supporting electrolytes in electrochemistry are available elsewhere [1-3], we restrict our introductory comments to factors most relevant to molecular electrochemistry. For example, although the supporting electrolyte may significantly influence electrode double-layer properties [6], that subject is not covered here, nor do we discuss certain mass-transfer properties such as migration effects [7].

As pointed out by Fry [1], certain considerations of electrolyte choice are practical ones: the salt should be readily obtained in pure form, dissolve sufficiently in the solvent of choice, be non-hygroscopic (for applications in organic solvents), and be easy to separate from solvent and/or product(s) if a macro-scale electrolysis is performed. In addition to these physical properties, the electrolyte should have an electrochemical 'window' sufficiently broad to allow the investigation of analytes which may have very positive or negative $E_{1/2}$ values. Sawyer *et al.* emphasize the importance of the electrolyte in regulating cell resistance and add a list of specific chemical effects such as ion-pairing interactions between an electrolyte ion and at least one form of the redox couple [2]. The points addressed by both sets of authors provide a foundation upon which we will base our discussion. We will also include two additional chemical factors, namely the increased solubility and kinetic stabilities of analyte cations, which allow for widened electrochemical studies in lower-polarity solvents.

1.2   What are Fluoroarylborate (FAB and FAB-type) Anions ?

This Chapter focuses on two tetraarylborate anions which are substituted either by perfluorination of the phenyl ring in tetrakis(pentafluorophenyl)borate, $[B(C_6F_5)_4]^-$ (**1**), or by trifluoromethylation at the 3 and 5 positions of the phenyl ring in tetrakis[3,5-bis(trifluoromethyl)phenyl]borate, $[B\{C_6H_3(CF_3)_2\}_4]^-$ (**2**). Although the latter is formally a trifluoromethyl-aryl rather than fluoro-aryl derivative, we take the liberty of referring to both as **fluoroarylborate**, or FAB-type [8], anions. The properties of these anions are best-understood within the larger chemical context of a search for anions of decreased coordinating ability, nucleophilicity, and basicity, and increased size and lipophilicity. This search was fueled by the desire to find anions capable of stabilizing reactive cations or promoting desired (especially, catalytic) reactions of such cations. The interested reader is referred to a number of excellent reviews on this subject [9], the details of which are beyond the scope of this Chapter. The preparations of anions **1** and **2** were first reported in 1962 [10] and 1984 [11], respectively. They, along with other large anions, have been shown to be an important factor in early transition metal "single-site" catalysis for the polymerization of olefins [12-14].

The earliest published use of either anion in electrochemical experiments appears to be the 1991 paper by Mann and co-workers [15] which utilized $[B\{C_6H_3(CF_3)_2\}_4]^-$, and to which we will refer later. There were few subsequent publications [16] building on this groundbreaking work, and almost a decade passed before the first description of $[B(C_6F_5)_4]^-$ as an electrolyte anion for molecular electrochemistry appeared [17]. This slow development can be traced to a point made in Section 1.1, namely that of easy availability. These anions, which are not trivial to prepare, are now commercially available in quantities allowing systematic electrochemical work [18].

(1)   (2)

1.3 Electrochemical Advantages of FAB Anions

The most important electrochemically relevant physical and chemical properties of large 'inert' anions are given in Table 1, where the implied comparison is with electrolytes containing the traditional anions.

Property 1, that of increased electrolyte solubilities, lessens the resistance effects routinely encountered. Large-amplitude controlled-potential techniques such as cyclic voltammetry (CV) are beset by measurement errors owing to ohmic effects that arise from uncompensated resistance, R_u, between the working and reference electrodes. Although a number of approaches may be used to diminish these errors, ohmic loss (iR_u) still limits the accuracy and utility of both voltammetric and bulk electrolytic experiments. In describing ways to overcome solution resistance, Roe lists three experimental approaches, one being the use of highly conducting solutions, which is "often impractical for organic solvents" [19]. As shown in sections 2.1 and 2.2, the new electrolytes provide the opportunity to capitalize on this experimental approach (see points

1 and 3 in Table 1). The salts we describe result in more conductive solutions through their increased solubilities and decreased association constants in lower-polarity solvents.

Table 1. Electrochemical Consequences of Chosen Properties of 'Inert' Electrolyte Anions

Physical or Chemical Property	Consequences for Electrochemisry
1) Increased electrolyte solubilities	(i) Lower medium resistance
	(ii) Better use of low polarity solvents
2) Increased solubilities of Ox/Red	Less electrode adsorption
3) Reduced ion pairing	(i) Lower medium resistance
	(ii) Positive $E_{1/2}$ shift for redox reactions having products of increased positive charge
4) Reduced nucleophilicity	Kinetic stabilization of reactive cations

Poor solubility of an electrode *product* often comes into play in voltammetric experiments by leading to adsorption effects. While interesting in themselves [20], adsorption effects are often not welcome because of complications to analytical, mechanistic and electro-synthetic applications. As will be apparent from the examples below, the increased solubilities of cationic redox products (Point 2, Table 1) are an important advantage of the new family of electrolytes.

The effects of the solvent/electrolyte medium on $E_{1/2}$ values are an important aspect of experimental design in electrochemistry. If the one-electron process of Eq 1 has a formal potential of $E_{1/2(0)}$, an ion-pairing interaction between Ox^+ and the anion A^- (Eq 2) leads to a modified $E_{1/2}$ value given by Eq 3 [21, 22].

$$Ox^+ + e^- \rightleftharpoons Red \qquad\qquad E_{1/2(0)} \qquad\qquad (1)$$

$$Ox^+ + A^- \rightleftharpoons OxA \qquad\qquad K_A \qquad\qquad (2)$$

$$E_{1/2} = E_{1/2(0)} - (RT/F) \ln (1 + K_A[A^-]) \qquad\qquad (3)$$

The qualitative effect of Eq 3 is that ion-pairing of A^- with Ox^+ leads to a negative shift in the redox potential. That is, ion pairing leads to thermodynamic stabilization of Ox^+. Conversely, if ion-pairing is turned *off*, the oxidized state is *de*stabilized. The fact that this thermodynamic effect is larger in more highly charged products [24], e.g., for dications compared to monocations, allows one to exploit ion pairing to control comproportionation equilibria in multi-electron-transfer systems (Section 2.3.2).

Quantitatively, the measured shifts in $E_{1/2}$ upon addition of A^- may be used in Eq 3 to determine the association constant, K_A, provided that $E_{1/2(0)}$ is known. This raises an important point having to do with the commonly made assumption that ion pairing between cations and the traditional anions such as $[PF_6]^-$ can be ignored when conducting such experiments. Consider the following.

Suppose that it is desired to measure K_A for an associative interaction between a cationic product Ox^+ and a strong ion-pairing anion such as Cl^-. The traditional experimental approach has been to obtain $E_{1/2(0)}$ by measuring the potential in a medium containing a 'non-ion-pairing' anion such as $[PF_6]^-$, followed by addition of Cl^- to obtain the required $E_{1/2}$ shifts. The problem here is that the measured value for K_A is, in fact, only a <u>ratio</u> of K_A values for Ox^+ with Cl^- compared to $[PF_6]^-$ (see the 'incremental' model mechanism in Miller *et al* [23]). Comparative $E_{1/2}$ effects (section 2.3.2) using **1** or **2** in place of $[PF_6]^-$, $[BF_4]^-$ or $[ClO_4]^-$ establish that these earlier assumptions are seriously flawed and are likely to lead to large errors in reported association constants.

Point 4 of Table 1 concerns the nucleophilicity of the anion. Since cationic oxidation products are commonly susceptible to nucleophilic attack [25], an anion of low nucleophilicity is of obvious benefit to the stabilization of these products. The past 10-15 years have seen an intense search for ever less nucleophilic anions [9, 26, 27]. Although $[B(C_6F_5)_4]^-$ and $[B\{C_6H_3(CF_3)_2\}_4]^-$ are not the least nucleophilic anions known [27], they are vastly superior to the traditional anions (see Section 2.4).

1.4   Metathesis and Purification of FAB-type Electrolytes

The general procedure for metathesizing salts containing a fluorinated arylborate anion consists of mixing solutions of water or alcohol-soluble precursors. Owing to the increased solubility of FAB-containing salts in a variety of solvents, attention must be paid to the selection of starting salts and solvents. This is especially true for the $[B\{C_6H_3(CF_3)_2\}_4]^-$ anion. We have successfully used the aqueous approach (see below) to metathesize $[B(C_6F_5)_4]^-$ with the following cations: $[N(CH_3)_4]^+$, $[N(C_nH_{2n+1})_4]^+$ (n = 2, 4), $[N(PPh_2)_2]^+$, ethylmethyl-imidazolium and butylmethyl-imidazolium. Although salts of the $[B\{C_6H_3(CF_3)_2\}_4]^-$ anion have also been prepared, in this case with tetramethyl- or tetrabutylammonium cations, as well as with several imidazolium cations, the extremely high solubility of these salts in a variety of solvents leads to lower efficiency purification steps. We offer two procedures which are representative of the metathesis methods used in our laboratory.

$[N(C_4H_9)_4] [B(C_6F_5)_4]$ from $Li[B(C_6F_5)_4]$:

The method most frequently used by the authors is performed on a 10 g scale under ambient atmosphere. The yields given represent the average of six replicates. 10.0 g of $Li[B(C_6F_5)_4]/$ n Et_2O (Boulder Scientific, 11.0 mmol assuming n = 3) is dissolved in 20 mL of HPLC-grade methanol. To this solution, 5.1 g (15.8 mmol) of $[N(C_4H_9)_4]Br$ in 10 ml of methanol is added drop-wise with stirring over a 15 minute period, after which 3 mL of nanopure water is added dropwise. The solution is cooled to *ca.* 0° C for 30 minutes and then stored at -25° C overnight. The resulting precipitate is separated by filtration through a medium frit and washed with 10 mL of cold (-25° C) methanol. The solid is dissolved in excess (30 mL) dichloromethane and stirred over $MgSO_4$ for two hours. Filtration and evaporation yields 9.56 g (94%) of the off-white crude product.

The sample is then recrystallized from dichloromethane / diethyl ether or dichloromethane / hexane. Here we describe the former. The crude salt is dissolved in 11 mL of dichloromethane. This typically requires slight warming of the solution. A quantity (typically 50 to 60 mL) of anhydrous diethyl ether is then added dropwise with stirring over 20 minutes until a precipitate starts to form. On some occasions the salt appears to supersaturate in the dichloromethane/ether solution. Once a precipitate starts to form, the solution is stored at *circa* 0° C for one hour followed by -25° C overnight. The resulting precipitate is filtered and rinsed with 30 mL of room temperature hexanes. A fine white crystalline material is obtained, yield 9.09 g (89% overall), m.p. 147-148.5°. The filtrate can be evaporated to yield additional crude salt suitable for further purification.

In order to obtain a salt of electrochemical-grade purity, the recrystallization step typically needs to be repeated at least once. We have observed that when water is not used in the metathesis, a second recrystallization may not be necessary. The water-free metathesis is identical to the one described above, except that instead of adding 3 mL of nanopure water, the methanolic solution is boiled to reduce the volume. $[N(C_4H_9)_4][B(C_6F_5)_4]$ is fairly soluble in hot methanol and the volume can be reduced significantly prior to precipitate formation. When a white solid forms, the solution is removed from heat, cooled in an ice bath with stirring, and filtered. Although a smaller quantity of product is obtained than in the procedure described earlier, it is electrochemically more pure and requires only one recrystallization. More product may be recovered by cooling the filtrate to -25° C and going through more recrystallizations. The overall yields of the two different metathesis procedures are similar.

We note that $K[B(C_6F_5)_4]$ has recently become commercially available. It is ether-free and non- hygroscopic, both of which are advantageous in storing the sample and in performing the metathesis. The potassium salt has also been used to prepare $[N(C_4H_9)_4][B(C_6F_5)_4]$ in our laboratory.

$[NEt_4]$ $[B\{C_6H_3(CF_3)_2\}_4]$ from $Na[B\{C_6H_3(CF_3)_2\}_4]$

This procedure is taken directly from the original metathesis described by Nishida *et al.* [11]. We have used it in our laboratory and re-produce it here for the convenience of the reader.

Hot aqueous tetraethylammonium iodide was added to a saturated hot aqueous solution of $Na[B\{C_6H_3(CF_3)_2\}_4]$ to yield colorless precipitates, which were successively eluted from a silica-gel column with hexane, benzene, ether and methanol. The solid residue, which was obtained from the ether and methanol fractions by evaporation, was washed with cold chloroform to give the tetraethylammonium salt, mp 181-182 °C [11].

2   ADVANTAGES OF NEW SUPPORTING ELECTROLYTES

2.1   Increased Solubilities for Salts of FAB Anions

Fluoroarylborate salts are generally very soluble in organic solvents owing to the weakening of cation-anion coulombic energies compared to the solvation energies of the individual ions. Salts of $[B\{C_6H_3(CF_3)_2\}_4]^-$, in particular, are known to exhibit extraordinary solubilities [28]. This characteristic provides for the increased use of *lower-polarity solvents* and leads to *decreased analyte adsorption effects* in molecular electrochemistry. Regarding the first point, the desire to probe redox reactions in solvents having different physical and chemical (e.g., donor and acceptor) properties is often thwarted by the lack of an appropriate supporting electrolyte [29]. For this reason, aliphatic ethers have been underutilized in electrochemistry in spite of their importance in synthetic chemistry. Although some lithium salts are reasonably soluble in ethers, the poor solubility of salts having organic cations, along with the low dielectric constants of aliphatic ethers (e.g., 4.3 for OEt_2) has up to now made it difficult to conduct serious electrochemical studies in these solvents. The salts $[NR_4][1]$ and $[NR_4][2]$ are found, however, to allow standard CV and electrolysis experiments in aliphatic ethers. Acceptable CV scans have been reported for anthraquinone in methyl tert-butyl ether containing 0.075 M $[NBu_4][1]$ (Figure 1). Experiments in diethyl ether were less successful using $[NBu_4][1]$, but acceptable steady state voltammetry was obtained at a 125 μm electrode at 210 K for this electrolyte medium [17]. Improved results have been obtained using $OEt_2/[NBu_4][2]$, wherein resistivities comparable to those of CH_2Cl_2 / $[NR_4][PF_6]$ are possible (present authors, unpublished work). $Na[B\{C_6H_3(CF_3)_2\}_4]$ has also proven effective in diethylether [16(b)].

Fluoroarylborate anions also generally increase the solubilities of cationic electrode reactants and products, thereby leading to fewer electrode adsorption effects. Unless one is specifically probing analyte/surface interactions, it is advantageous that both members of the redox couple be soluble in order to simplify mass transport effects and to avoid problems of electrode history and passivation. In anodic processes, lower solubilities are often encountered for products with charges greater than +1, particularly when using solvents such as dichloromethane. Halocarbons and haloarenes such as CH_2Cl_2, dichlorobenzene, and difluorobenzene are often used for anodic electrochemistry of organic and organometallic systems owing to the relatively low nucleophilicity and coordinating ability of these solvents, but their rather low polarities (ε: CH_2Cl_2, 8.9; 1,2-$C_6H_4Cl_2$, 9.9; $C_6H_4F_2$, 11.2) often lead to poorly soluble mono-, di- or tri-cations when using the traditional anions.

Figure 2 compares the CV scan of a triferrocenyl-compound in CH_2Cl_2 / 0.1 M $[NBu_4][PF_6]$. with that in CH_2Cl_2 / 0.1 M $[NBu_4][1]$. Two differences are immediately seen. The first is that the $E_{1/2}$ values, particularly of the second

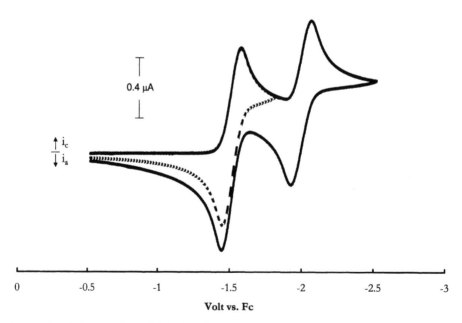

0.4 µA

i_c
i_a

| 0 | -0.5 | -1 | -1.5 | -2 | -2.5 | -3 |

Volt vs. Fc

Figure 1. CV scans (v = 0.2 V/s) of 1 mM anthraquinone in methyltertbutyl ether containing 0.075 M $[NBu_4][B(C_6F_5)_4]$. Reprinted with permission from reference 17. Copyright Wiley.

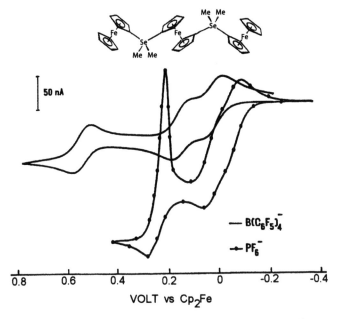

Figure 2. CV scans (v= 0.2 V/s) of a triferrocenyl complex in CH_2Cl_2 containing two different electrolyte salts.—●— contains 0.1 M [NBu$_4$] [PF$_6$]; —— contains 0.1 M [NBu$_4$] [B(C$_6$F$_5$)$_4$]. Reprinted with permission from reference 17. Copyright Wiley.

and third oxidation processes, are shifted to more positive potentials when $[B(C_6F_5)_4]^-$ is the electrolyte anion. Aspects of potential shifts are discussed in detail in section 2.3. Second, and relevant to the discussion at hand, is that all three oxidation processes give diffusion-controlled behavior in the presence of $[B(C_6F_5)_4]^-$, whereas the third anodic wave displays a stripping peak, diagnostic of adsorption of the trication, with $[PF_6]^-$ in solution. Similar results were obtained by Camire and co-workers [30] on a number of other multi-ferrocenyl compounds. Figure 3 shows the CV behavior of terferrocene (**4**) in several

(3)   (4)

different nonaqueous media. In CH_2Cl_2 / 0.1 M [NBu$_4$][PF$_6$] [Fig 3(c)] it is again evident that the trication, in this case [**4**][PF$_6$]$_3$, adsorbs onto the glassy carbon electrode. Electrochemistry in CH_3CN / [NBu$_4$][PF$_6$] [Figs 3(a) and 3(b)] avoids the adsorption problem, but leads to decomposition of the trication. The increased stability and solubility of **4**$^{3+}$ in CH_2Cl_2 / 0.1 M [NBu$_4$] [B(C$_6$F$_5$)$_4$] is evident from the well-behaved waves of Fig 3(d). The E$_{1/2}$ shifts manifested under the medium changes will be discussed more broadly in Section 2.3.

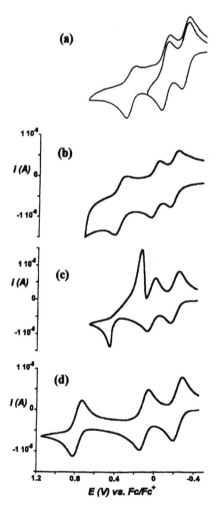

Figure 3. CV scans (v = 0.2 V/s) of terferrocene (**4**) in different nonaqueous media: (a) 1:1 CH_3CN:CH_2Cl_2/0.1 M [NBu$_4$][PF$_6$] (b) CH_3CN/0.1 M [NBu$_4$] [PF$_6$] (c) CH_2Cl_2/ 0.1 M [NBu$_4$] [PF$_6$] (d) CH_2Cl_2/0.1 M [NBu$_4$][B(C$_6$F$_5$)$_4$]. Reprinted with permission from reference 30. Copyright Elsevier.

Additional insight into the solubility benefits of $[B(C_6F_5)_4]^-$ and $[B\{C_6H_3\text{-}(CF_3)_2\}_4]^-$ comes from consideration of "piano-stool"-shaped metal complexes containing as ligands only a pi-bonded polyolefin (such as arene or cyclopentadienyl) and a number of carbonyl groups. This family includes the important first-row $18e^-$ transition metal complexes $(C_5H_5)Mn(CO)_3$, $(C_6H_6)Cr(CO)_3$, and $(C_5H_5)Co(CO)_2$ [31]. It is well known that weakening of the M-CO bond must be taken into account in attempts to synthesize the corresponding $17e^-$ cations [32]. Less well appreciated is how the poor solubilities of the cations have restricted electrochemical inquiries. A neutral metal carbonyl compound has electronegative CO groups which interact well with an acceptor solvent such as CH_2Cl_2, aiding its solubility. In their cationic relatives, this carbonyl-solvent interaction is weakened and may be insufficient to overcome

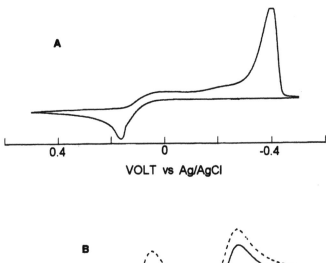

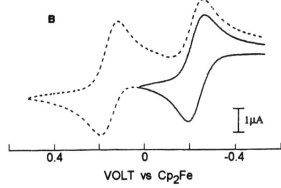

Figure 4. CV scans ($v = 0.1$ V/s) of $(C_9H_7)CpMo(CO)_2$ (**3**) under two different sets of conditions in CH_2Cl_2: (a) 0.1 M $[NBu_4]$ $[PF_6]$ at Pt electrode (b) 0.1 M $[NBu_4][B(C_6F_5)_4]$ at glassy carbon electrode. Source: Michael E. Stoll, Ph.D. dissertation, University of Vermont, 2000.

the ion-pairing coulombic interactions between the cation and the electrolyte anion, resulting in precipitation of the organometallic salt onto the electrode surface. A stronger donor solvent such as THF may increase the solubility of the cation, but at the cost of a decrease in its stability [32]. The coulombic interactions may be minimized with relatively non-coordinating anions, favoring increases in solubilities that lower or eliminate adsorption effects for the solvents in which organometallic cations are most stable. Two examples follow.

The indenyl complex $(\eta^3\text{-Ind})(C_5H_5)Mo(CO)_2$, **3**, Ind = (C_9H_7) is the parent compound of a series being studied as models for redox-induced hapticity changes in metal-ligand π–bonding [33]. When at least one carbonyl is replaced by a phosphine or other donor ligand, these compounds exhibit well-behaved

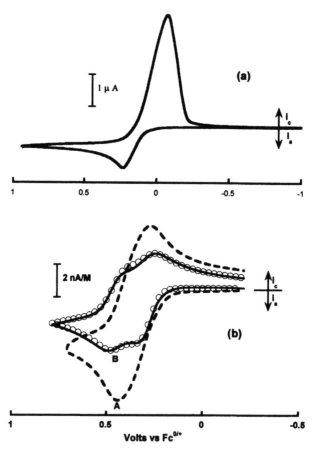

Figure 5. CV scans of $CpCo(CO)_2$ (**5**) in CH_2Cl_2 using two different electrolyte anions: (i) 0.1 M $[NBu_4][PF_6]$, $v = 0.2$ V/s (ii) 0.1 M $[NBu_4][B(C_6F_5)_4]$, $v = 0.5$ V/s. The two different scans under (ii) are taken at different concentrations of **5** and the y-axis is normalized for [**5**]: ----- = 0.3 mM **5**, —O—O = 2.3 mM **5**. Reprinted with permission from reference 38. Copyright Am. Chem. Soc.

electrochemistry [34]. Oxidation of the parent dicarbonyl compound in CH_2Cl_2 / 0.1 M $[NBu_4][PF_6]$, however, produces only a stripping wave for the anodic product [Figure 4(a)]. When the electrolyte is changed to $[NBu_4][B(C_6F_5)_4]$, the anodic behavior of **3** is experimentally accessible [Figure 4(b)], revealing that **3** undergoes two consecutive one-electron processes, first to 3^+ and then to 3^{2+}, at $E_{1/2}$ potentials (vs Ag/AgCl) of -0.21 V and 0.13 V. Bulk electrolysis at 0 V produced 3^+, which was characterized by esr spectroscopy [35].

A related example concerns the oxidation of $CpCo(CO)_2$, **5**, $(Cp = C_5H_5)$, a starting material for the preparation of $CpCoL_2$ compounds [36]. The oxidative electrochemistry of **5** is poorly developed owing to severe electrode adsorption of the anodic product(s) in lower-polarity solvents and their rapid decomposition in more polar but also more nucleophilic solvents [37]. Overcoming the solubility problems with $[NBu_4][B(C_6F_5)_4]$ enables one to generate 5^+ and to probe its chemistry [38]. Figure 5 shows comparative CVs of **5** in CH_2Cl_2 with either $[NBu_4][PF_6]$ or $[NBu_4][B(C_6F_5)_4]$. This system is described more fully in section 2.4.

Because of our personal areas of research emphasis, the examples in this section predominantly involved organometallic compounds. The same basic principles can be applied, however, to organic and inorganic redox processes.

2.2   Decreased Solution Resistance

As pointed out in Section 1.3, ohmic loss (i.e., iR_u 'drop') is a significant source of voltammetric potential control error. Although a number of compensation and analysis methods have been introduced for the purpose of minimizing the measurement errors [19, 39], few would dispute the intrinsic benefit of obtaining one's data from more highly conductive (i.e., less resistive) media. In Table 2 we list specific resistances for a number of supporting electrolytes in nonaqueous solvents in order to emphasize the considerably

Table 2. Selected solvent resistivities (in Ω cm) of selected tetrabutylammonium salts at ambient conditions.

Anion	THF	CH_2Cl_2	BTF[a]	MTBE[a]	diethylether
$[PF_6]^-$	2300	720	--[d]	--[d]	--[d]
$[BF_4]^-$	3400	800	4300	--[d]	--[d]
$[TFAB]^-$	490	470	790	2500[b]	22 K[c]

[a] MTBE = methyltertbutylether, BTF = benotrifluoride
[b] 0.075 M
[c] saturated solution < 0.025 M at 235 K
[d] either sparingly or not soluble

smaller resistivities of FAB-type salts in lower dielectric constant solvents. Of special relevance are the data in methyltertbutylether (MTBE). The resistivity of MTBE with 0.075 M $[NBu_4][B(C_6F_5)_4]$ (2.5 KΩcm) is comparable to those observed for traditional media (e.g., resistivity of THF/0.1 M $[NBu_4][PF_6]$ is 2.3 KΩcm). The solubility, and more importantly the dissociation, of FAB-type salts in ethers opens new possibilities for voltammetric investigations in these very low polarity solvents.

We draw particular attention to three solvents: dichloromethane, THF, and benzotrifluoride [BTF, $C_6H_5(CF_3)$]. The first two have become arguably the solvents of choice for oxidations and reductions, respectively, involving reactive electrode products. The third is a newly-introduced solvent suitable for both nonaqueous and fluorous-phase electrochemistry [40]. A significant drawback to these solvents as far as electrochemical investigations are concerned is their polarity. The low dielectric constants (CH_2Cl_2, 8.9; THF, 7.5; BTF, 9.2) give rise to significant ohmic distortions. In CV scans, solution resistance manifests itself as an increase in ΔE_p, the difference between the forward and reverse peak potentials. The peak separation is an important diagnostic for determining the rate of heterogeneous electron transfer, information that is subject to serious error when measured under conditions of large uncompensated resistance.

Table 3 provides an example of the effect of ohmic drop under typical experimental conditions for an essentially nernstian couple, ferrocene / ferrocenium in THF. The data were acquired with a 1.5 mm glassy carbon electrode using as a supporting electrolyte either 0.1 M $[NBu_4][PF_6]$ or 0.1 M $[NBu_4][B(C_6F_5)_4]$. An experimental value of about 60 mV is expected, but when the $[PF_6]^-$ salt is used as the electrolyte anion, the much larger value of 123 mV

Table 3. Change in peak separation (ΔE_p) with scan rate (v) for the oxidation of 1.0 mM ferrocene in THF at a 1.5 mm diameter glassy carbon electrode. Electrolyte concentration is 0.1 M.

v	$[NBu_4][PF_6]$	$[NBu_4][TFAB]$	$[NBu_4][TFAB]^a$
		ΔE_p (mV)	
0.1	123	62	
0.5	189	90	
1	234	105	
5	398	167	90
10			107
50			155

^{a} Positive feedback compensation used to correct for 85% of the ohmic drop.

was measured at $v = 0.1$ V/s. At the still modest scan rate of 1 V/s, the ΔE_p value had increased to such an extent that it was now diagnostic of an irreversible charge-transfer mechanism, in clear disagreement with the well-known reversibility of this couple. By way of contrast, the behavior with $[B(C_6F_5)_4]^-$ approaches the expected result, with $\Delta E_p = 62$ mV at 0.1 V/s. At $v = 5$ V/s the error in ΔE_p is only about one-quarter with $[B(C_6F_5)_4]^-$ what it is with $[PF_6]^-$. These results serve to demonstrate that the use of $[NBu_4][B(C_6F_5)_4]$ allows for more accurate measurement of higher electron-transfer rates and assists the study of reaction mechanisms by increasing the accuracy of diagnostic parameters based on shifts and shapes of voltammetry curves.

Despite the significant decrease in solution resistance upon changing the supporting electrolyte anion from $[PF_6]^-$ to $[B(C_6F_5)_4]^-$, ohmic drop is not eliminated. The third column of Table 3 presents the observed ΔE_p values obtained when positive feedback was used to compensate for the residual ohmic loss in a $[B(C_6F_5)_4]^-$-containing solution. The data indicate that in the presence of 0.1 M $[NBu_4][B(C_6F_5)_4]$, semi-quantitative information regarding the kinetics of heterogeneous electron-transfer is accessible in lower-polarity solvents at scan rates upwards of 50 V s^{-1} at an r = 1.5 mm disk.

2.2.1 Implications of Conductance Properties

Turning from utilitarian factors to more fundamental properties, we now present some conductivity results that are informative about two important physical properties of the salts under discussion, namely their ionic associations (K_A) and molecular conductivities at infinite dilution (Λ_o). Present-day theory utilizes the two-step reaction sequence of Eq 4 to describe the dissociation of a 1:1 electrolyte:

$$A^+ + B^- \overset{K_r}{\rightleftharpoons} A^+ \text{---} B \overset{K_s}{\rightleftharpoons} A^+B^- \tag{4}$$

Free ions are those that are unaffected by the presence of other ionic species in solution and will interact with ions of opposite charge through diffusion-controlled processes (K_r) to form solvent separated ion-pairs (SSIP). The further formation of contact ion pairs (CIP) is likely dominated by electrostatic interactions and Brownian motion although ill-defined specific (chemical) interactions may play a significant role as well. The association constant K_A is related to the two individual equilibria of Eq 4 as $K_A = K_r (1 + K_s)$. One outcome of this model is that the different entities-free ions, SSIPs and CIPs -contribute differently to the conductance of a solution. For the present discussion it is sufficient to note that free ions and SSIPs increase the conductivity of a solution and CIPs do not. In low polarity solvents, the presence of CIPs cannot be dismissed and as will be shown below, the assumption that the concentration of conducting ions is equal to the concentration of added solute is grossly inadequate.

This paired-ion model was used extensively by Fuoss to extract meaningful physical properties of ions from conductivity data [41]. We have used the Fuoss method to analyze conductance data for a number of the salts under discussion here and some of the results involving [NBu$_4$][PF$_6$] and [NBu$_4$][B(C$_6$F$_5$)$_4$] are collected in Table 4. Three points are worthy of note. First, based on Λ_0 values the inherent conductivity of the [PF$_6$]$^-$ ion is greater than that of the [B(C$_6$F$_5$)$_4$]$^-$ ion. This is due primarily to the relative size of the ions. Since conductivity is a measure of the *mobility* of an ion in solution, the smaller ion will move more freely in a viscous medium. Second, the association constant for the [B(C$_6$F$_5$)$_4$]$^-$ containing salt is one-to-two orders of magnitude smaller than when the salt contains the [PF$_6$]$^-$ ion in low-polarity solvents such as dichloromethane or tetrahydrofuran. Thirdly, there is no significant difference between the association constants of the two salts in a high-polarity solvent such as acetonitrile.

Table 4. Association constants (K$_A$) and limiting molar conductivities (Λ_0 in Ω^{-1} cm^2 mol^{-1}) for tetrabutylammonium salts in various electrochemical solvents.

Electrolyte	Solvent	K$_A$	Λ_0
[TFAB]$^-$	CH$_2$Cl$_2$	1.8×10^3	87
[PF$_6$]$^-$	CH$_2$Cl$_2$	1.7×10^4	109
[TFAB]$^-$	THF	6.2×10^3	86
[PF$_6$]$^-$	THF	3.7×10^5	121
[TFAB]$^-$	BTF^a	3.4×10^3	72
[PF$_6$]$^-$	BTF^a	5.6×10^4	58
[BF$_4$]$^-$	BTF^a	2.6×10^5	83
[TFAB]$^-$	CH$_3$CN	42	118
[PF$_6$]$^-$	CH$_3$CN	36	167

^{a} BTF = benzotrifluoride

These data demonstrate that the observed decrease in resistance between [PF$_6$]$^-$ and [B(C$_6$F$_5$)$_4$]$^-$ solutions is due to the significant increase in concentration of charge-carrying species in the [B(C$_6$F$_5$)$_4$]$^-$ solution. They suggest that in

CH_2Cl_2, for example, the fractional concentration of free ions in a $[B(C_6F_5)_4]^-$ solution is about ten times that of a $[PF_6]^-$ solution. Because the inherent conductivity of the $[B(C_6F_5)_4]^-$ anion is approximately 20% less than the $[PF_6]^-$ ion, solutions of the latter would be more conductive if ion-pairing were not a factor. Indeed, in the high-polarity solvent acetonitrile ($\varepsilon = 36$), where the association constants of the two salts are similar and small, the $[B(C_6F_5)_4]^-$ salt performs poorly as an electrolyte. These conductivity results help to rationalize and quantify the experimental observation that FAB-type electrolytes give superior (lower) resistivities only in lower-polarity solvents.

2.3   Modification of $E_{1/2}$ Values

2.3.1   General Considerations

As discussed in section 1.3, reduced ion pairing with electrogenerated cations or polycations produces a positive $E_{1/2}$ shift which may in principle be easily measured by polarography or cyclic voltammetry. When the potential shifts are not large, however, their accurate measurement is flawed owing to changes in liquid junction potentials which accompany those in the electrolyte medium [42]. Liquid junction potentials are usually not known for the types of reference electrodes and low-polarity solvents commonly employed in modern molecular electrochemistry. The use of an internal reference redox couple, most often that of ferrocene/ferrocenium [43, 44], overcomes part of the measurement problem. The accuracy is still limited, however, by unknown subtleties in solvation and ion pairing for the two forms of the reference couple.

The choice of a more 'protected' redox site such as decamethylferrocene$^{0/+}$ may be preferred, and the goal of a 'medium-independent' reference couple has still not been achieved [45].

In part for this reason, we chose to focus on the effects of ion-pairing on the *differences* between $E_{1/2}$ shifts for successive one-electron processes involving formation of first a monocation and then a dication (Eq 5). Based on electrostatic effects, ion-pairing with the latter will be much stronger than the former, owing to the quadratic dependence of electrostatic attraction on molecular charge. The electrostatic effect may be expected to dominate medium shifts involving a single redox site (e.g., $Cp_2Ni^{0/+/2+}$) in lower polarity solvents, but play a diminished role either in more polar solvents or for systems having multiple redox sites with limited electronic interactions (e.g., ferrocenyl-labeled dendrimers).

$$(ML_n)_2 \underset{}{\overset{E_{1/2}(1)}{\rightleftharpoons}} [(ML_n)_2]^+ \underset{}{\overset{E_{1/2}(2)}{\rightleftharpoons}} [(ML_n)_2]^{2+} \qquad (5)$$

Baring formation of higher aggregates, the ion-pairing interactions of the cations generated in Eq 5 are described in the reactions of equations 6-8. Defining $\Delta E_{1/2}$ as

$$[(ML_n)_2]^+ \; + \; A^- \; \rightleftharpoons \; (ML_n)_2A \qquad\qquad K_{A(11)} \qquad\qquad (6)$$

$$[(ML_n)_2]^{2+} \; + \; A^- \; \rightleftharpoons \; [(ML_n)_2A]^+ \qquad\qquad K_{A(21)} \qquad\qquad (7)$$

$$[(ML_n)_2]^{2+} \; + \; 2A^- \; \rightleftharpoons \; (ML_n)_2A_2 \qquad\qquad K_{A(22)} \qquad\qquad (8)$$

$E_{1/2}(2)$-$E_{1/2}(1)$ and the change in $\Delta E_{1/2}$ with added ion-pairing anion as $\Delta\Delta E_{1/2}$, the latter is described quantitatively by Eq 9, where C_{A-} is the concentration of A^-:

$$\Delta\Delta E_{1/2} \; = \; \frac{RT \; \ln}{F} \left[\frac{(1 + K_{A(11)} \, C_{A-})^2}{1 + K_{A(21)} C_{A-} + K_{A(21)} \, K_{A(22)} C^2_{A-}} \right] \qquad (9)$$

Since it is generally true that $K_{A(22)}$, $K_{A(21)} \gg K_{A(11)}$, Eq 9 implies that increased ion pairing leads to smaller $\Delta E_{1/2}$ values, i.e. less separation of the two one-electron processes. The effect of the concentration of A^- on $\Delta\Delta E_{1/2}$ is relatively small within the context of our discussion. For example, a decrease in $\Delta E_{1/2}$ from 762 mV to 704 mV is measured for the two oxidations of bisfulvalene dinickel, **8** in CH_2Cl_2 as the concentration of $[NBu_4][B(C_6F_5)_4]$ is increased from 0.02 M to 0.67 M, a change that is much less than that brought about by an alteration of the nature of the solvent and electrolyte. We therefore de-emphasize the effects of electrolyte concentration on $\Delta E_{1/2}$ in the discussion below, although brief mention of concentration effects is found in section 2.3.2.5.

One area of application in which $\Delta\Delta E_{1/2}$ effects are quite important is that of mixed-valence chemistry. Consider a dinuclear complex with two equivalent redox sites capable of significant electronic interactions [46]. We start with a conceptual model in which the two E^o values in weakly ion-pairing media are $E_{1/2}(1)$ and $E_{1/2}(2)$ (eq 5). In a stronger ion-pairing environment, the first oxidation potential, $E_{1/2}(1)$, is only slightly shifted if it involves a neutral-to-monocation couple. The second electron-transfer potential, $E_{1/2}(2)$, involving the monocation-to-dication couple, is much more sensitive to medium effects because of the presence of a positive charge in the immediate vicinity of the redox center. Mann and co-workers showed that the ion-pairing induced changes in $E_{1/2}(2)$- $E_{1/2}(1)$ can be very large under these circumstances [15, 47]. We therefore chose to measure $\Delta E_{1/2}$ values of a dinuclear redox complex as part of a systematic investigation of ion-pairing and solvation effects on comproportionation phenomena. Although the present model is directed at mixed-valence systems, the conclusions are likely to be pertinent to a broad range of multi-electron-transfer processes. These results offer a view of the ion-pairing properties of a given anion, or more accurately of a given electrolyte (anion and cation of the supporting salt at a given concentration in a given solvent) towards model cationic receptors.

A broad investigation of medium effects is made feasible by the high lipophilicity of FAB anions, which as noted earlier increases the solubilities of their salts in lower-polarity solvents. Keeping in mind also the increased solubilities of analyte oxidation products, FAB anions are nearly ideal for systematic medium probes of oxidation reactions. As shown in the following sections, ion-pairing effects are indeed very strong in solvents of low dielectric constant, accentuating differences in the electrostatic effects of large and small anions in low-polarity media.

The role of solvent in determining the medium effects cannot be downplayed. The donor and acceptor properties of the solvent, as well as their dielectric properties, play a very important role in medium effects [48]. For example, solvents of low dielectric constant such as ethers can have their electrical properties overtaken by the rather strong electron-donating properties of the oxygen lone pair towards an electrophile or positively charged center. Dichloromethane, by contrast, even if more polar, will show larger potential shifts due to its very poor electron-donating abilities. In turn, it becomes evident that one should also consider the acceptor properties of the cations with respect to these donating effects, i.e., how open/protected is the cationic site to/from interactions with the medium (solvent and/or counter ion). We now illustrate these points by selected examples based on polyferrocenyl or bisfulvalenyl metal complexes.

2.3.2 Control of Comproportionation

The thermodynamic stability of a mixed-valent species ($[(ML_n)_2]^+$ in Eq 5) is usually expressed in terms of its comproportionation constant ($K_{comp} = \{[(ML_n)_2]^+\}^2 / [(ML_n)_2] [(ML_n)_2]^{2+}$), a quantity that is directly related to $\Delta E_{1/2}$ by $\Delta G = -RT \ln K_{comp} = -nF \Delta E_{1/2}$. Different contributions of statistical origin, or arising from electrostatic or electronic coupling, enter this energetic term [49]. The use of medium effects to tune the $\Delta E_{1/2}$ value will essentially affect the electrostatic component in $\Delta E_{1/2}$. The effect of the structure of the analyte and the nature of the medium (supporting electrolyte and solvent) are now illustrated.

2.3.2.1 Role of Analyte Structures

The structural aspects of mixed-valent comproportionation effects have been reviewed elsewhere [46]. We restrict ourselves to a point about the spread in $\Delta E_{1/2}$

values that might be expected for $(ML_n)_2$ systems having linkages of varying length and delocalization abilities.

Biferrocene (**6**) and bis(ferrocenyl)ethane (**7**) are closely related dinuclear complexes differing structurally by the length separating the equivalent ferrocenyl redox centers. Since the redox centers are further apart in bis(ferrocenyl)ethane, the electrostatic effect of a positive charge of one center on the other (after the first electron transfer) will be more weakly felt. Hence the available tuning potential window will be narrower. For instance, in dichloromethane with 0.1 M $[Bu_4N][B(C_6F_5)_4]$, a $\Delta E_{1/2}$ value of 530 mV is found for biferrocene but a $\Delta E_{1/2}$ of only 180 mV is found for bis(ferrocenyl)ethane[30]. In the latter complex, the distance between the redox centers considerably diminishes their mutual electrostatic influence.

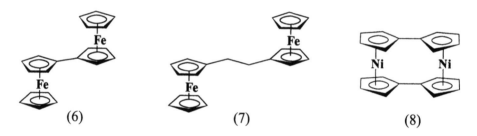

$\quad\quad$ (6) $\quad\quad\quad\quad\quad\quad\quad\quad$ (7) $\quad\quad\quad\quad\quad\quad\quad\quad$ (8)

In both cases, however, the $\Delta E_{1/2}$ values are significantly increased over the values when using the traditional anions. A $\Delta E_{1/2}$ value of about 330 mV has been reported for biferrocene in CH_2Cl_2 / 0.1 M $[NBu_4][PF_6]$, and the $\Delta E_{1/2}$ value of bis(ferrocenyl)ethane was reported as too small to measure under the same conditions [50]. The increased $\Delta E_{1/2}$ value for bis(ferrocenyl)ethane is particularly significant because of its effect on the chemistry of the mixed-valent monocation, which is now seen to be accessible when $[B(C_6F_5)_4]^-$ is the counter-ion.

A logical outgrowth of these ideas is to make use of electrolyte effects to manipulate redox systems having close to "inverted" $\Delta E_{1/2}$ values [51] that may favor single two-electron CV behavior. The goal would be to use medium effects systematically to favor either two separate one-electron CV waves or a single two-electron wave, depending on the experimental needs. Such investigations are in progress.

2.3.2.2 Control of $\Delta E_{1/2}$ by Electrolyte Anion

Di- or poly-nuclear metal complexes generally show an appreciable increase in $\Delta E_{1/2}$ as the electrolyte anion in organic solvents is increased in size from halide to traditional to large. For example, the two consecutive one-electron oxidations of bisfulvalene dinickel (**8**) in dichloromethane with 0.1 M $[Bu_4N][A]$, A = Cl$^-$,

$[PF_6]^-$ or $[B(C_6F_5)_4]^-$ are separated by $\Delta E_{1/2} = 273, 480$ and 753 mV respectively (Figure 6). Hence, a large window of at least 480 mV is available for tuning the thermodynamic stability of the Ni(II)Ni(III) mixed-valent species depending on the choice of the anion. This is qualitatively explained by a model in which ion-pairing increases in the order $[B(C_6F_5)_4]^- < [PF_6]^- < Cl^-$, and that the ion-pairing effect increases dramatically in the dication compared to the monocation. This trend has been shown to hold for a number of halides, traditional ions, and large anions for bisfulvalene dinickel, a partial description of which has been published [52]. One finding of interest is that in a lower-dielectric solvent like dichloromethane the $\Delta E_{1/2}$ values for media with the traditional anions tend to be closer to those measured for halides than to those seen for either $[B(C_6F_5)_4]^-$ or $[B\{C_6H_3(CF_3)_2\}_4]^-$.

Among all the media studied, a maximum $\Delta E_{1/2}$ spread of 850 mV was achieved using $CH_2Cl_2 / Na[B\{C_6H_3(CF_3)_2\}_4]$. This very large spread is due in part to the electronic nature of bisfulvalene dinickel, the monocation of which

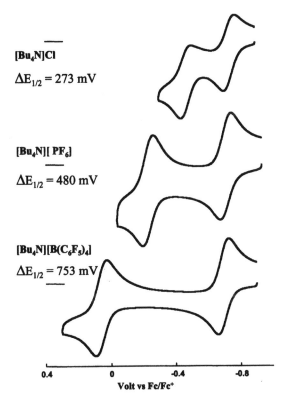

Figure 6. CV scans ($v = 0.1$ V/s) of bisfulvalenedinickel (**8**) in $CH_2Cl_2/0.1$ M $[NBu_4]Cl$, $[NBu_4][PF_6]$, or $[NBu_4][B(C_6F_5)_4]$.

is intrinsically delocalized [53]. The less drastic changes in $\Delta E_{1/2}$ values for the two ferrocenyl complexes discussed in section 2.3.2.1 arise because their metal centers are less strongly interacting.

2.3.2.3 Effect of Solvent Polarity on $\Delta E_{1/2}$ Control

Solvent electrical properties play an important role on the extent of the available tuning potential range of mixed-valent complexes. As shown above, a difference in $\Delta E_{1/2}$ values, $\Delta\Delta E_{1/2}$, of 480 mV was observed for **8** in dichloromethane. If the studies are conducted in a much more polar solvent like DMSO ($\varepsilon = 47.2$) the tuning potential window drops considerably to a mere 56 mV as is evident from comparison of Figures 6 and 7. Good dissociation and solvation of ions in more polar solvent considerably diminishes the electrostatic interaction between ions and therefore levels the changes in $\Delta E_{1/2}$ values as a function of the supporting salt. $\Delta E_{1/2}$ is respectively 337, 371 and 393 in DMSO with 0.1 M $[Bu_4N]^+$ salts of Cl^-, $[PF_6]^-$ or $[B(C_6F_5)_4]^-$.

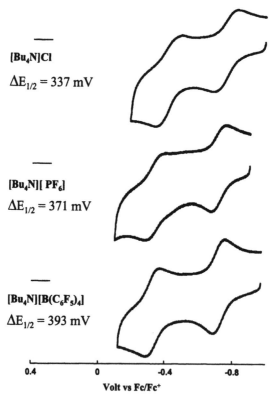

[Bu₄N]Cl

$\Delta E_{1/2} = 337$ mV

[Bu₄N][PF₆]

$\Delta E_{1/2} = 371$ mV

[Bu₄N][B(C₆F₅)₄]

$\Delta E_{1/2} = 393$ mV

0.4   0   -0.4   -0.8

Volt vs Fc/Fc⁺

Figure 7. CV scans ($v = 0.1$ V/s) of bisfulvalenedinickel (**8**) in DMSO/0.1 M $[NBu_4]Cl$, $[NBu_4][PF_6]$, or $[NBu_4][B(C_6F_5)_4]$.

2.3.2.4 *Effect of Solvent Donor Properties on* $\Delta E_{1/2}$ *control*

For solvents with similar electrical properties, appreciable differences on the range of accessible $\Delta E_{1/2}$ values can be obtained by carefully choosing the donor-acceptor properties of the lower-polarity solvent [52]. For example, dichloromethane is a very weak donor (donor number, DN, = 0 by definition on the Gutmann scale [48]). In this solvent with 0.1 M [NBu$_4$][B(C$_6$F$_5$)$_4$], a $\Delta E_{1/2}$ of 753 mV is measured for bisfulvalene(dinickel). However, in the even less polar tetrahydrofuran solvent ($\varepsilon = 7.5$) with the same supporting electrolyte, $\Delta E_{1/2}$ does not increase as one would expect when considering only the polarity parameters, but rather decreases to 595 mV. This 158 mV difference for $\Delta E_{1/2}$ can be directly accounted for by the higher donating abilities of tetrahydrofuran (DN = 20).

2.3.2.5 *Role of Electrolyte Cation in* $\Delta E_{1/2}$ *control*

The examples given so far have allowed discussion of the two most important requirements of the solvents in promoting a large tuning potential window of a given analyte: a low polarity and poor donor property. In addition to these factors and the obvious importance of the anion, varying the properties of the cation in the supporting electrolyte can add another point of control. Since the equilibria of Eq 6-8 are secondarily affected by competitive equilibria with other counterions, the $\Delta E_{1/2}$ values will depend somewhat on the nature of the cation of the supporting electrolyte, even if no anionic analytes are involved in the electron transfer reaction. This is confirmed by experiments on **8** in dichloromethane, for which we have compared the $\Delta E_{1/2}$ values using two different [B{C$_6$H$_3$(CF$_3$)$_2$}$_4$]$^-$ salts. The value of 744 mV measured using 0.1M [NBu$_4$]-[B{C$_6$H$_3$(CF$_3$)$_2$}$_4$] is increased to 850 mV for a saturated solution (*ca* 0.02 M of Na[B{C$_6$H$_3$(CF$_3$)$_2$}$_4$] in the same solvent. Certainly, the concentration effects mentioned above and described in Eq 9 contribute to part of the effect. However, the full increase in $\Delta E_{1/2}$ must arise in part because of efficient competitive ion pairing between the small sodium cation (poorly solvated by dichloromethane) and the fluoroarylborate anion with respect to the electrogenerated analyte cations.

(9)

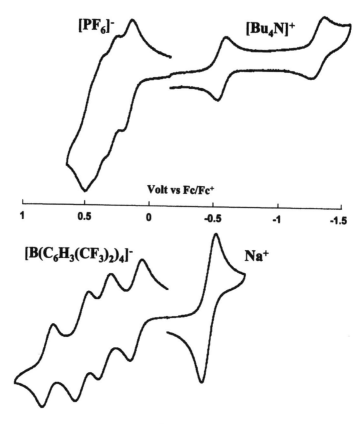

Figure 8. CV scans ($v = 0.1$ V/s) of tetraferrocenyl(nickel dithiolene) (**9**) in CH_2Cl_2 /0.1 M [NBu$_4$][PF$_6$], or 0.02 M Na[B{C$_6$H$_3$(CF$_3$)$_2$}$_4$].

Another example of a secondary cationic effect is found for tetraferrocenyl-(nickeldithiolene), **9**, which can be oxidized to the tetracation in four consecutive reversible steps. The potential difference between the first and the fourth oxidation process is 292 mV in dichloromethane/0.1 M [Bu$_4$N][PF$_6$] whereas it increases to 510 mV and 682 mV with [NBu$_4$][B(C$_6$F$_5$)$_4$] or Na[B{C$_6$H$_3$(CF$_3$)$_2$}$_4$] respectively (Figure 8). Dichloromethane saturated with Na[B{C$_6$H$_3$(CF$_3$)$_2$}$_4$] (ca 0.02 M) is thus seen to maximize the $\Delta E_{1/2}$ values of polycationic redox species since it combines favorable solvent properties (low polarity and weak donicity) with favorable electrolyte properties (a large and weakly donating anion and a small, unsolvated, and charge-localized cation). This refined set of solvent and ionic properties has a symmetrically reverse effect on electrogenerated anions as is discussed next.

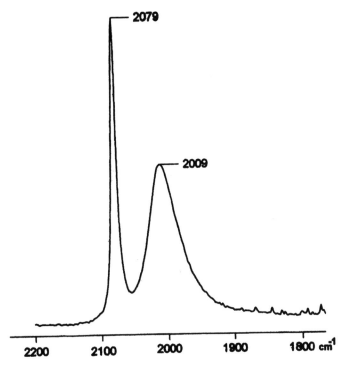

Figure 9. IR spectrum of $[(C_6H_6)Cr(CO)_3]^+$ generated by anodic oxidation of **10** in $CH_2Cl_2/0.1$ M $[NBu_4][B(C_6F_5)_4]$.

2.3.2.6 *Mirror Image Effect for Electrogenerated Anions*

The tetraferrocenyl(nickeldithiolene) complex can also be reversibly reduced in two consecutive steps to the di-anion 9^{2-}, with $\Delta E_{1/2}$ of 770 mV in dichloromethane 0.1 M $[Bu_4N][PF_6]$. This value can be gradually decreased by carefully selecting the solvent and supporting salt properties along the lines exemplified in the previous sections. Hence, $\Delta E_{1/2}$ decreases to 472 mV and 183 mV in solutions of $Na[B\{C_6H_3(CF_3)_2\}_4]$ in tetrahydrofuran and anisole respectively. The $\Delta E_{1/2}$ can be decreased to essentially zero in dichloromethane saturated with $Na[B\{C_6H_3(CF_3)_2\}_4]$. Thus, the electrolyte shown to maximize $\Delta E_{1/2}$ with polycations, logically minimizes it for polyanions. In the case of tetraferrocenyl(nickeldithiolene) a reversible two-electron reduction wave is obtained (Figure **8**), which opens the way to inverted electron transfer reactions [51, 52]. Medium effects on $\Delta E_{1/2}$ for formation of cations therefore behave in a mirror-image way for the $\Delta E_{1/2}$ values involving formation of anions.

2.3.3 Summary of $\Delta E_{1/2}$ Control Section

In conclusion, access to large, lipophilic, and weakly coordinating ions, combined with the array of available organic solvents, opens the way to a fine tuning of the thermodynamic stability of mixed-valent species and other ionic intermediates of EE mechanisms. As discussed and illustrated in this section, the tuning effect is fully understood only if one takes into account all the parameters of ion size, solvent electrical properties, donor-acceptor characteristics of both ions and solvent, and receptivity of the analyte to these interactions, in addition to the respective concentrations of both analyte and electrolyte. Depending on the experimental choices, these properties add to or diminish each other in the determination of $\Delta E_{1/2}$ values.

2.4   Kinetic Stabilization of Oxidation Products

It might be anticipated that the low nucleophilicities of fluoroarylborates would make them attractive as counterions for the preparation of relatively electrophilic positively-charged compounds. This expectation has been confirmed by a number of reports detailing the synthesis of highly electron-deficient cations that had previously escaped isolation with traditional anions [9,26,27,54]. A similar advantage is anticipated when using large electrolyte anions in nonaqueous anodic electrochemistry.

Indeed, this was a motivation in the first reported study employing a FAB-type electrolyte [15]. In that work Hill *et al.* showed that the oxidation of ruthenocene was a reversible one-electron process in CH_2Cl_2 / $[NBu_4][B\{C_6H_3(CF_3)_2\}_4]$, whereas similar conditions utilizing the traditional anions gave irreversible two-electron processes.

Certainly, a frequent limitation when trying to use molecular electrochemistry for preparative work is an electrode product which is stable on the CV time scale (a maximum of 0.10 s), but subject to decomposition on a bulk electrolysis time scale of 10 minutes or more. This finding is often ascribed to 'adventitious impurities' such as moisture or oxygen, or to reactions with solvent. A generally overlooked possibility is that the cationic product reacts slowly with the somewhat nucleophilic electrolyte anion. Future studies will tell just how frequently this occurs, but we chose to probe this possibility using the oxidation of the oft-studied $(C_6H_6)Cr(CO)_3$, **10**, as a test model.

The anodic behavior of this compound has been intensively studied in an effort to generate and to prepare the 17-electron radical cation **10$^+$**. In a dozen or more published studies [55,56], an almost exhaustive list of experimental conditions were attempted, including changes in solvent or solvent dryness, electrolyte, and temperature. Invariably, the long-term electrolysis products were free benzene and a usually unspecified Cr(II) or Cr(III) moiety, both apparently arising from nucleophilic attack on **10⁺**. Most relevant to the present discussion are the results of Stone *et al.* [56(a)] and Zoski *et al.* [56(b)], who showed that the chemical reversibility of the couple **10/10⁺**, as measured by CV, increased as the supporting electrolyte anion was changed in the order $[ClO_4]^- < [BF_4]^-$, $[CF_3SO_3]^- < [PF_6]^-$. Although these experiments still did not allow the long-term generation of **10⁺** (even the CV reversibility in CH_2Cl_2/ [NBu₄][PF₆] required low temperatures), they suggested to us the possible benefit of an anion even-less nucleophilic than $[PF_6]^-$.

When the anodic oxidation of **10** is carried out in CH_2Cl_2/ [NBu₄][B(C₆F₅)₄], the desired **10⁺** exhibits a remarkable increase in lifetime, being persistent in solution for many hours at room temperature [38]. This enabled us to report, for the first time, the ESR and IR (Figure 10) spectra of **10⁺** and to conduct controlled reactions of this cation. It must be emphasized that these spectra were not obtained by any of the spectro-electrochemistry techniques that have been developed to deal with short-lived radicals [57]. After ambient temperature electrolysis in a drybox or under Schlenk conditions, samples were simply removed by syringe, transferred to a spectroscopy cell, and taken to the spectrometer for analysis.

Substitution reactions of **10⁺** are carried out easily under these conditions. Oxidation of **10** in the presence of PPh₃ or treatment of solutions of **10⁺** with PPh₃ gave the substitution product $[(C_6H_6)Cr(CO)_2PPh_3]^+$, **11⁺**. Subsequent reduction of **11⁺** gave neutral **11** in high yield. The overall electron-transfer chemistry (Scheme 1), which constitutes an electrochemical switch mechanism, provides a useful alternative to photochemically-induced processes for the substitution in **10** of CO by donor ligands [38].

Oxidation of another half-sandwich organometallic carbonyl compound, namely CpCo(CO)₂, **5**, provides further insights into the role of the electrolyte

$(C_6H_6)Cr(CO)_3$ $\xrightarrow{-e^-}$ $[(C_6H_6)Cr(CO)_3]^+$

10   **10⁺**

$\downarrow$ PPh₃

$(C_6H_6)Cr(CO)_2PPh_3$ $\xrightarrow[+e^-]{}$ $[(C_6H_6)Cr(CO)_2PPh_3]^+$

11   **11⁺**

Scheme 1

anion. Although **5** is the parent for a family of very important $CpCoL_2$ compounds [36], its oxidation to **5**$^+$ had proven unsuccessful. Analogues having derivatized Cp rings such as $(C_5Me_5)Co(CO)_2$ or substitution of one carbonyl ligand {e.g., $CpCo(CO)(PPh_3)$} have been shown to display reversible one-electron oxidations to the corresponding cations [58]. **5**$^+$ itself, however, is reported to lose CO when formed by chemical oxidations [59] and to form a mercurous complex when **5** is oxidized at a mercury electrode [58].

The anodic electrochemistry of **5** is beset by problems having to do not only with the susceptibility of **5**$^+$ to nucleophilic attack, but also with the poor solubility of **5**$^+$ in low-donor, but also low-polarity, solvents such as dichloromethane (see obvious electrode adsorption in Figure 5(a). Both of these problems are obviated using fluoroarylborate anion electrolytes, with evidence of the 17-electron radical **5**$^+$ being clearly obtained in CV scans (Figure 5(b)). Contrary to the findings for $(C_6H_6)Cr(CO)_3$, however, the oxidation of $CpCo(CO)_2$ is not an uncomplicated one-electron process. Rather, the CV behavior is a function of temperature, scan rate, and the concentration of **5**. At low concentrations the CV behavior approximates that of a simple one-electron couple (dashed line A of Figure 5(b)), whereas at higher concentrations a more complex pair of anodic features are observed (solid line B of Figure 5(b)). This behavior has been interpreted in terms of reversible formation of a metal-metal bonded dimer monocation, $[CpCo(CO)_2]_2^+$, which is responsible for the additional one-electron wave as the more positive anodic feature of Figure 5(b) [38]. At the time of this writing, the chemistry of **5**$^+$ and its dimeric derivatives are under investigation. Pertinent to the theme of this Chapter is that the use of a $[B(C_6F_5)_4]^-$-based electrolyte allows investigation of a potentially valuable new area of organometallic redox chemistry.

3   FUTURE DIRECTIONS

Advantages of supporting electrolytes containing FAB-type anions are demonstrated in the work discussed and quoted in this report. It is clear that important gains are possible for applications in lower-polarity solvents, for broadening of ion-pairing studies, for regulation of comproportionation reactions, and for increased use of electrochemistry for synthetic purposes, to name only a few.

A number of other anions which are being explored for the same purposes may prove superior in some ways to FAB-type anions. These include carborane anions, which are arguably the least coordinating anions known at this time [27]. One of these, as $Li[CB_{11}Me_{12}]$, has been reported to easily support voltammetry in benzene [60], and a family of partially-halogenated carboranes such as

$[NBu_4][CB_{11}HMe_5Br_6]$ also has promising electrochemical properties [61]. Yet other possibilities are highly-fluorinated aluminates such as $[Al\{OC(H)(CF_3)_2\}_4]^-$ [62]. Some of these anions, as well as others not mentioned or still unknown, will certainly prove to be of sufficient synthetic accessibility to warrant their consideration in analytical and synthetic aspects of electrochemistry. Ultimately one might hope to have available families of anions differing in both physical and chemical properties in order to meet a broad range of experimental needs.

ACKNOWLEDGEMENTS

The authors are grateful to the National Science Foundation (CHE-0092702) for support of the research described in this Chapter.

REFERENCES

1. Fry A. in Kissinger, P.T.; Heineman, W.R. (eds) "Laboratory Techniques in Electro-analytical Chemistry", Marcel Dekker, New York, 2nd Ed., 1996, Chapter 15
2. Sawyer, D.T.; Sobkowiak, A.; Roberts, Jr., J.L. "Electrochemistry for Chemists", John Wiley & Sons, New York, 1995, 2nd ed, Chapter 7
3. Mann, C.K.; Barnes, K.K. "Electrochemical Reactions in Nonaqueous Systems", Marcel Dekker, New York, 1970, p 23
4. Amatore, C. in Rubinstein, I. (ed) "Physical Electrochemistry", Marcel Dekker, New York, 1995, Chapter 4
5. The term "electrolyte" refers properly to the combined solvent and supporting electrolyte. In the present treatment, however, we will use "electrolyte medium" to convey the same idea and will frequently substitute "electrolyte" for "supporting electrolyte"
6. Bard, A.J.; Faulkner, L.R. "Electrochemical Methods", John Wiley & Sons, New York, 2001, 2nd ed., Chapter 13
7. Bond, A.M. "Modern Polarographic Methods in Analytical Chemistry", Marcel Dekker, New York, 1980, p 76
8. The organometallic literature has frequently used the acronym FAB for the trisubstituted borane $(C_6F_5)_3B$. Within that context, **tetrakis**-substituted $[(C_6F_5)_4B]^-$ is labeled TFAB. The anion $[\{C_6H_3(CF_3)_2\}_4B]^-$, sometimes referred to as BArF, is a close relative and is referred to here as a FAB-type anion
9. (a) Strauss, S. *Chem. Reviews*, *93*, 927 (1993) (b) Reed, C.A. *Accounts Chem. Res.*, *31*, 133 (1998) (c) Bochmann, M. *Angew. Chem. Int. Ed.*, *31*, 1181 (1992) (d) Seppelt, K. *Angew. Chem. Int. Ed.*, *32*, 1025 (1993)
10. Massey, A.G.; Park, A.J. *J. Organometal. Chem.*, *2*, 245 (1964)
11. Nishida, H.; Takada, N.; Yoshimura, M.; Sonoda, T.; Kobayashi, H. *Bull. Chem. Soc. Jpn.*, *57*, 2600 (1984). An improved synthesis of $Na[\{C_6H_3(CF_3)_2\}_4B]$ has recently appeared: Reger, D.L.; Wright, T.D.; Little, C.A.; Lamba, J.J.S.; Smith, M.D. *Inorg. Chem.*, *40*, 3810 (2001)

12. Turner, H. *European Patent* No. 277,004 (1988); *Ibid.,* U.S. Patent No. 4,752,597 (1988)

13. Stevens, J.C.; Neithamer, D.R., U.S. Patent No 5,064,802 (1991)

14. Chen, E.Y-X.; Marks, T.J. *Chem. Reviews, 100,* 1391 (2000)

15. Hill, M.G.; Lamanna, W.M.; Mann, K.R. *Inorg. Chem., 30,* 4687 (1991)

16. (a) Gassman, P.G.; Sowa, Jr., J.R.; Hill, M.G.; Mann, K.R. *Organometallics, 14,* 4879 (1995) (b) Chávez, I.; Alvarez-Carena, A.; Molins, E.; Roig, A.; Maniukiewicz, W.; Arancibia, A.; Arancibia, V.; Brand, H.; Manriquez, J.M. *J. Organometal. Chem., 601,* 126 (2000) (c) Shao, Y.; Stewart, A.A.; Girault, H.H. *J. Chem. Soc., Faraday Trans., 8,* 2593 (1991)

17. LeSuer, R.J.; Geiger, W.E. . *Angew. Chem. Int. Ed., 39,* 248 (2000) 18. The authors have obtained TFAB-like salts from Boulder Scientific Company, P.O. Box 548; 598 Third St., Mead CO 80542 U.S.A.

19. Roe, D.K. in Ref 1, Chapter 7

20. (a) Reference 6, Chapter 13 (b) Oldham, K.B.; Myland, J.C. *Fundamentals of Electrochemical Science,* Academic Press, New York, 1994, pp 338-352

21. Galus, Z. "Fundamentals of Electrochemical Analysis", Ellis Horwood Limited, Chichester, 1976, Chapter 14

22. The designation of a positive charge for Ox^+ is intended to emphasize its tendency to ion-pair with available anions

23. Miller, S.R.; Gustowski, D.A.; Chen, Z-H; Gokel, G.W.; Echegoyen, L.; Kaifer, A.E. *Anal. Chem., 60,* 2021 (1988)

24. Bockris, J. O'M.; Reddy, A.K.M. *Modern Electrochemistry,* 2nd ed., *Ionics,* Plenum Press, New York, 1998, Volume 1, pp 304 ff

25. Astruc, D. *Electron Transfer and Radical Processes in Transition-Metal Chemistry,* VCH Publishers, New York, 1995 contains many references to nucleophilic chemical follow-up reactions

26. (a) Reed, C.A.; Kim, K-C.; Bolskar, R.D.; Mueller, L.J. *Science, 289,* 101 (2000) (b) LaPointe, R.E.; Roof, G.R.; Abboud, K.A.; Klosin, J. *J. Am. Chem. Soc., 122,* 9560 (2000)

27. Stasko, D.; Reed, C.A. *J. Am. Chem. Soc., 124,* 1148 (2002)

28. Brookhart, M.; Liu, Y.; Goldman, E.W.; Timmers, D.A.; Williams, G.D. *J. Am. Chem. Soc., 113,* 927 (1991)

29. Abbott, A. *Chem. Soc. Rev. 22,* 435 (1993)

30. Camire, N.; Mueller-Westerhoff, U.T.; Geiger, W.E. *J. Organometal. Chem., 637-639,* 823 (2001)

31. Collman, J.P.; Hegedus, L.S.; Norton, J.R.; Finke, R.G. *Principles and Applications of Organotransition Metal Chemistry,* University Science Books, Mill Valley, CA,1987, Chapter 3

32. Broadley, K.; Connelly, N.G.; Geiger, W.E. *J. Chem. Soc., Dalton Trans.,* 121 (1983)

33. Gamelas, C.A.; Herdtweck, E.; Lopes, J.P.; Romão, C.C. *Organometallics 18,* 506 (1999) and references therein

34. Stoll, M.E.; Belanzoni, P.; Calhorda, M.J.; Drew, M.G.B.; Félix, V.; Geiger, W.E.; Gamelas, C.A.; Goncalves, I.S.; Romão, C.C.; Veiros, L.F. *J. Am. Chem. Soc., 123,* 10595 (2001)

35. Stoll, M.; Ph.D. Dissertation, University of Vermont, 2000 36. Kemmitt, R.D.W.; Russell, D.R. in Wilkinson, G.; Stone, F.G.A.; Abel, E.W. (eds) *Comprehensive Organometallic Chemistry*, Pergamon Press, Oxford, Vol 5, p 248 (1982)

37. Gennett, T.; Grzeszczyk, E.; Jefferson, A.; Sidur, K.M. *Inorg. Chem.*, *26*, 1856 (1987)

38. Camire, N.; Nafady, A.; Geiger, W.E. *J. Am. Chem. Soc.*, *124,* 7260 (2002)

39. Ref 6, pp 648-650

40. Ohrenberg, C.; Geiger, W.E. *Inorg. Chem. 39,* 2948 (2000)

41. Fuoss, R.M. *J. Phys. Chem., 82,* 2427 (1978); *Ibid. 79,* 525 (1975)

42. Ref 6, pp 63-74

43. Gagne, R.R.; Koval, C.A.; Lisensky, G.C. *Inorg. Chem.*, *19,* 2854 (1980)

44. Gritzner, G.; Kuta, J. *Pure Appl. Chem.*, *56,* 461 (1984)

45. (a) Bashkin, J.K.; Kinlen, P.J. *Inorg. Chem.*, *29,* 4507 (1990) (b) Ruiz, J.; Astruc, D. *C.R. Acad. Sc. Ser.II*, 21(1998) (c) Noviandri, I.; Brown, K.N.; Fleming, D.S.; Gulyas, P.T.; Lay, P.A.; Masters, A.F.; Phillips, L. *J. Phys.Chem. B, 103*, 6713 (1999) (d) Noviandri, I.; Bolskar, R.D.; Lay, P.A.; Reed, C.A. *J. Phys.Chem. B, 101,* 6350 (1997)

46. (a) Demadis, K.D.; Hartshorn, C.M.; Meyer, T.J. *Chem. Rev.*, *101,* 2655 (2001) (b) Chen, P.; Meyer, T.J. *Chem. Rev.*, *98,* 1439 (1998) (c) Brunschwig, B.S.; Creutz, C.; Sutin, N. *Chem. Soc. Rev.*, *31,* 168 (2002) (d) Launay, J-P. *Chem. Soc. Rev., 30,* 386 (2001) (e) Paul, F.; Lapinte, C. *Coord. Chem. Rev., 178-180,* 431 (1998) (f) McCleverty, J.A.; Ward, M.D. *Accounts Chem. Res., 31,* 842 (1998) (g) Ward, M.D. *Chem. Soc. Reviews 24,*121 (1995)

47. Siedle, A.R.; Hanggi, B.; Newmark, R.A.; Mann, K.R.; Wilson, T. *Macromol. Symp., 89,* 299 (1995)

48. Linert, W.; Fukuda, Y.; Camard, A. *Coord. Chem. Rev., 218,* 113 (2001)

49. (a) Evans, C.E.B.; Naklicki, M.L.; Rezvani, A.R.; White, C.A.; Kondratiev, V.V.; Crutchley, R.J. *J. Am. Chem. Soc.*, *120,* 13096 (1998) (b) Crutchley, R.J. *Coord. Chem. Rev., 219-221,* 125 (2001)

50. Morrison, Jr.; W.H.; Krogsrud, S.; Hendrickson, D.N. *Inorg. Chem.*, *12,* 1998 (1973)

51. Evans, D.H.; Hu, K. *J. Chem. Soc., Faraday Trans., 92,* 3983 (1996)

52. Barrière, F.; Camire, N.; Geiger, W.E.; Mueller-Westerhoff, U.T.; Sanders, R. *J. Am. Chem. Soc.*, *124,* 7262 (2002)

53. Smart, J.C.; Pinsky, B.L. . *J. Am. Chem. Soc.*, 99, 956 (1977)

54. (a) Harlan, C.J.; Hascall, T.; Fujita, E.; Norton, J.R. *J. Am. Chem. Soc., 121,* 7274 (1999) (b) Priego, J.L.; Doerrer, L.H.; Rees, L.H.; Green, M.L.H. *J. Chem. Soc., Chem. Commun.* 779 (2000) (c) Alvarez, M.A.; Anaya, Y.; García, M.E.; Riera, V.; Ruiz, M.A. *Organometallics, 22,* 456 (2003)

55. See leading references in (a) Howell, J.O.; Goncalves, J.M.; Amatore, C.; Klasnic, L. Wightman, R.M.; Kochi, J.K. *J. Am. Chem. Soc.*, *106,* 3968 (1984) (b) Hunter, A.D.; Mozol, V.; Tsai, S.D. *Organometallics, 11,* 2251(1992)

56. (a) Stone, N.J.; Sweigart, D.A.; Bond, A.M. *Organometallics, 5,* 2553 (1986) (b) Zoski, C.G.; Sweigart, D.A.; Stone, N.J.; Rieger, P.H.; Mocellin, E.; Mann, T.F.; Mann, D.R.; Gosser, D.K.; Doeff, M.M.; Bond, A.M. *J. Am. Chem. Soc., 110,* 2109 (1988)

57. For leading references see: (a) Heineman, W.R.; Hawkridge, F.M.; Blount, H. in Bard, A.J. (ed) *Electroanalytical Chemistry*, Marcel Dekker, New York, 1984, vol 13, pp 1-113 (b) Handzlik, J.; Hartl, F.; Szymańska-Buzar, *New J. Chem.*, *26*, 145 (2002) (c) Shaw, M.J.; Geiger, W.E. *Organometallics*, *15*, 13 (1996) (d) Krejčik, M.; Daněk, M.; Hartl, F. *J. Electroanal. Chem. 317*, 179 (1991)

58. (a) Gennett, T.; Grzeszczyk, E.; Jefferson, A.; Sidur, K.M. *Inorg. Chem.*, *26*, 1856 (1987) (b) Broadley, K.; Connelly, N.G.; Geiger, W.E. *J. Chem. Soc., Dalton Trans.*, 121 (1983)

59. McKinney, R.J. *Inorg. Chem.*, *21*, 2051 (1982). For leading references to the reaction chemistry of cations of $[CpML_2]^+$, M = Co, Rh, Ir see Carano, M.; Cicogna, F.; D'Ambra, I.; Gaddi, B.; Ingrosso, G.; Marcaccio, M.; Paolucci, D.; Paolucci, F.; Pinzino, C.; Roffia, S. *Organometallics*, *21*, 5583 (2002)

60. Pospíšil, L.; King, B.T.; Michl, J. *Electrochim. Acta*, *44*, 103 (1998)

61. Reed, C.; Nafady, A.; Geiger, W.E., unpublished results

62. (i) Preparation: (a) Ivanova, S.M.; Nolan, B.G.; Kobayashi, Y.; Miller, S.M.; Anderson, O.P.; Strauss, S.H. *Chem. Eur. J.*, *7*, 503 (2001) (c) Krossing, I. *Chem. Eur. J.*, *7*, 490 (2001) (ii) Voltammetry: DeWitte, R.; Krossing, I.; Geiger, W.E., work in progress

Chapter 14

Illustration of Experimental and Theoretical Problems Encountered in Cyclic Voltammetric Studies of Charged Species Without Added Supporting Electrolyte

Alan M. Bond

School of Chemistry, PO Box 23, Monash University
Victoria 3800, Australia

1   INTRODUCTION

Voltammetric measurements are usually undertaken under a fairly standard set of conditions. Thus, 0.1 M or higher concentrations of electrolyte are usually present, a single time domain-single technique (cyclic, square wave, differential pulse) is employed in any given experiment, a three-electrode potentiostatted cell configuration is used and conditions close to ambient temperature and pressure are employed. However, in each of these areas it may be argued that a generally conservative approach has been adopted by users of electrochemical methods and that the frontiers of the discipline may be expanded with careful consideration to any experimental limitations that are actually imposed on the basis of fundamental limitations rather than common practice considerations.

In this chapter of the book, the commonly perceived need to routinely add high concentrations of supporting electrolyte is critically assessed in the context of voltammetric studies of charged electroactive species at macrodisc electrodes. The conclusion is reached that whilst significant obstacles are sometimes encountered under these dilute electrolyte conditions, studies are frequently tractable with careful attention to experimental design. Importantly, rigorous theory is now often available to assist with the understanding of additional complexities associated with migration and uncompensated resistance, although the presence of a very diffuse double layer that may be of significant concern in highly dilute electrolyte media.

1.1   Circumstances where undertaking voltammetric measurements in the absence of added electrolyte may be desirable

Figure 1 contains cyclic voltammograms obtained at a macrodisc electrode for the oxidation of uncharged ferrocene $[Fe(\eta^5\text{-}C_5H_5)_2]$ with and without added supporting electrolyte. On the basis of the result presented without electrolyte, it would quite reasonably be concluded that electrolyte addition is mandatory. The need to avoid experiencing such poor quality, highly distorted voltammetry with a significant uncompensated resistance problem, is of course a major reason why high concentrations of added electrolyte are routinely included in most publications that describe cyclic voltammetric experiments undertaken with macrodisc electrodes.

However, in some of the earliest voltammetric experiments by Heyrovský and others [1-4] describing the reduction of charged metal ions and complexes at a dropping mercury electrode, pure water was routinely employed as the solvent without added electrolyte. Somewhat remarkably, since that time, except for

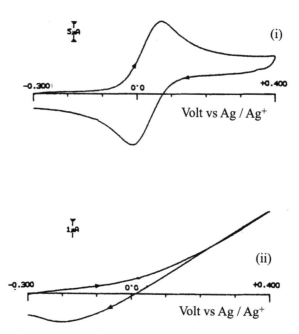

Figure 1. Cyclic voltammograms obtained at a 0.8 mm radius gold macrodisc electrode for the oxidation of 1 mM ferrocene in acetonitrile at 20°C using a scan rate of 250 mV s^{-1}. (i) with 0.1 M Et_4NClO_4 supporting electrolyte and (ii) without any deliberately added electrolyte. Reproduced by courtesy: *Anal. Chem.* **57**, 2764 (1985). Copyright, American Chemical Society.

specialized studies at microelectrodes (see later), it has become customary to use excess supporting electrolyte irrespective of whether or not electroactive species are uncharged or charged. Reasons usually given for addition of electrolyte include [4-6]:

- By increasing the cell's electrical conductivity, problems with uncompensated resistance (R_u) or ohmic (IR_u) drop are diminished.
- When excess supporting electrolyte is present, migration is an unimportant transport mechanism, massively aiding the modeling of the voltammetric experiment.
- The double layer at the working electrode is beneficially affected by the addition of supporting electrolyte, becoming narrower and less influenced by the presence or absence of the electroactive species, and thereby facilitating "background corrections". "Frumkin effects" [2] (see later) also are minimized.
- Gradients of concentration of electroactive species occur in the vicinity of the working electrode. In the absence of supporting electrolyte, these concentration gradients can give rise to appreciable density gradients and so engender unwanted convection. When other solution components are present in large excess, the dependence of the solution density on the concentration of a minor component is greatly suppressed.
- At least one of the electroactive species (reactant or product) must be an ion. The activity coefficient of that ion, in the absence of excess electrolyte, is a function of its concentration and therefore changes unwelcomely with time and/or distance. A high ionic strength enforces a constant activity coefficient.

These advantages are so strong that it would appear that nowadays few experimentalists ever consider the prospect of dispensing with electrolyte. However, there is also a downside to employing solutions that contain large concentrations of ionic species other than the electroreactants. These include:

- It becomes difficult to compare electrochemical evidence from solutions of high ionic strength with data obtained from spectroscopic or other non-electrochemical techniques, which are customarily conducted without added electrolyte.
- Many organic solvents of low permittivity do not permit the dissolution of large concentrations of electrolytes. Even when special electrolytes are found that do dissolve significantly in such solvents, it is likely that they do so by forming ion pairs, and so do not achieve the desired effect of significantly increasing the ionic strength.
- Some analytical applications of voltammetry, such as *in situ* analysis of river waters, preclude the addition of salts.

- The speciation of the electroreactant is always in doubt when other solutes are present in excess. In the case of ions, for example, ion pairs or even complex ions may be formed. This prevents or impedes the reliable measurement of data of thermodynamic significance (standard potentials, for example), or other data such as diffusion coefficients and rate constants.

- The need to have electrolyte present at much higher concentrations imposes an unwelcome upper limit on the concentrations of electroactive species that can be studied.

Whether the advantages of adding supporting electrolyte outweigh the disadvantages depends, of course, on the circumstances and on the purpose of the voltammetric experiment. However, there will be occasions when they do not.

Whether or not it is desirable to exclude electrolyte, there is no inherent reason why addition of supporting electrolyte is necessary when undertaking voltammetric studies on the majority of charged electroactive species. This is so because the electroreactant ion and its counter ion that is necessarily present will usually provide the means whereby current may be carried through the cell. The product of the electrode reaction also will assist in this process, if it is an ion. An exception to this situation is the unusual case in which the product of the electrode reaction is an ion of opposite sign to that of the ionic reactant. In such cases, special complications [7] arise which are not addressed here. Thus, the focus of this Chapter will be on the oxidation or reduction of an ion of either sign which generates another ion of the same sign but of either a smaller or a larger magnitude. Thus, the electrode reaction type to be addressed is confined to the electron transfer process

$$A(soln) + (z_A\text{-}z_B)e^- \rightleftharpoons B(soln) \tag{1}$$

In equation 1, the charges z_A and z_B of the reactant A and the product B have the same sign, which means that the counter-ion **X** has a charge z_X of opposite sign.

Recent reviews that have contained discussion of voltammetry in dilute electrolyte media [e.g. 4,8,9] have almost exclusively addressed microelectrode studies under the near steady-state conditions achievable at low scan rates where IR_u drop is minimal, relative to that encountered under transient conditions that commonly apply with macrodisc electrode studies. Data summarised for electrode reactions in these articles demonstrate that useful results often, but not always [4,10], can be obtained with minimal difficulty from near steady-state voltammetric experiments conducted without added electrolyte. However, as will be demonstrated in this Chapter of the book, with care, the technique of cyclic voltammetry at a macrodisc electrode also can be used without added supporting electrolyte under transient conditions. This methodology is invariably the one preferred by organic and inorganic chemists who employ voltammetry to elucidate the nature of an electrode process. In order to provide a didactic form of presentation, detailed consideration of two case studies will be provided in the initial part of this article that specifically encompass problems that are likely to be encountered when employing cyclic voltammetry with conventional macrodisc electrodes under transient conditions in dilute electrolyte media. However, this article necessarily also contains reference to selected studies with microdisc electrodes under near steady-state conditions in order to convey the authors' perspective of where the subject of voltammetry without added electrolyte could develop in the next few years, now that it is apparent that the full arsenal of both transient and steady-state methodologies can be applied to a given problem.

2　CASE STUDY I: THE REVERSIBLE REDUCTION OF $[S_2Mo_{18}O_{62}]^{4-}$ UNDER CONDITIONS OF CYCLIC VOLTAMMETRY IN ACETONITRILE IN THE ABSENCE OF ADDED SUPPORTING ELECTROLYTE

The voltammetric reduction of $[S_2Mo_{18}O_{62}]^{4-}$ in acetonitrile and other organic solvents with added Bu_4NClO_4 or other electrolytes is exceptionally rich. Eight essentially reversible one-electron reduction steps may be detected under strictly aprotic conditions. The extensive voltammetry of this polyoxometallate compound in acetonitrile has been reviewed in reference 11 under conditions where at least 0.1 M concentrations of supporting electrolyte is present. In this initial case study, the first two one-electron reduction processes associated with this system, *viz*,

$$[S_2Mo_{18}O_{62}]^{4-} + e^- \rightleftharpoons [S_2Mo_{18}O_{62}]^{5-} \tag{2}$$

and

$$[S_2Mo_{18}O_{62}]^{5-} + e^- \rightleftharpoons [S_2Mo_{18}O_{62}]^{6-} \tag{3}$$

process will be examined in the absence of added supporting electrolyte in order to convey an idea of the experimental and theoretical complexities that need to be considered in media where only the dissolved $[Bu_4N]_4[S_2Mo_{18}O_{62}]$ or other salt provides the electrolyte needed to carry the current.

Figure 2 contains examples of a cyclic voltammograms obtained as a function of scan rate for the two processes of interest in the presence of 0.2 M Bu_4NClO_4 supporting electrolyte, and provides a reference point for data reported in the absence of added supporting electrolyte. In the particular case of these essentially reversible cyclic voltammograms (uncompensated resistance problems distort voltammograms at high scan rate) obtained in the presence of 0.2M electrolyte, the average of the reduction (E_p^{red}) and oxidation (E_p^{ox}) peak potentials, $(E_p^{red} + E_p^{ox})/2$, gives the reversible midpoint potential or $E_{1/2}$-value which approximates to the reversible formed potential or E_f^0-value. Furthermore [11,12], the Randles-Sevcik relationship (equation 4) is valid

$$I_p = \pm 0.4463 \; nF(nF/RT)^{1/2}AD^{1/2}v^{1/2}C \tag{4}$$

where I_p is the peak current for oxidation (positive) or reduction (negative) (I_p^{red} or I_p^{ox} respectively), A is the electrode area, D is the diffusion coefficient, v is the magnitude of the scan rate, C is the bulk solution concentration, n is the number of electrons transferred, T is the temperature, R is the Universal Gas Constant and F is Faraday's Constant. Consequently, I_p^{red}, should depend on the square root of scan rate and be linearly related to concentration. Other expectations that apply in this situation include $\Delta E_p \; (= E_p^{red} - E_p^{ox})$ should be independent of scan rate and have a value of about 56/n mV at 25°C. These and other well known relationships that are available for the reversible processes under conditions of cyclic voltammetry at a macrodisc electrode, as well as for other voltammetric techniques, are summarized in references 11 and 12.

2.1 *Experimental Considerations*

In this first case study, derived in the main data presented in reference 13, and unpublished data provided by D.C. Coomber, the working electrode refers to the use of ≤ 1.5 mm radii glassy carbon macrodisc electrodes and a platinum wire auxiliary electrode. Both these working and auxiliary electrodes are typical of those used in conventional studies with added supporting electrolyte. However, instead of a classical Ag/AgCl, saturate calomel (SCE) or Ag/Ag$^+$ (0.01 M Ag$^+$, 0.1 M Bu$_4$NPF$_6$ in CH$_3$CN) reference electrode separated from the test solution by a salt bridge [11], the reference electrode used to provide data presented in this case study frequently was a platinum wire configured in the shape of a hook, such that the tip represents the closest point of approach of the reference electrode to the working electrode surface [13]. Furthermore, the reference electrode casing is attached to a thread, which can be screwed, in order to provide a variable but known distance between the centre of the working electrode surface and the tip of the hook. This platinum wire quasi-reference electrode was used in order to minimize the risk of introduction of ions via leakage from the very high electrolyte concentration present in a conventional reference electrode (e.g. SCE) configuration. The disadvantage of this quasi-reference electrode is that the potential drifts with time. However, if it is the effect of electrolyte, rather than the electrode potentials that are being probed, then a quasi-reference electrode is appropriate. Alternatively, if highly accurate values of potential are required, then *in situ* use of the usual uncharged ferrocene (Fc) which gives a reversible $Fc^{0/+}$ process (see Figure 1) or another non-interacting neutral compound which gives a reversible process whose potential against the $Fc^{0/+}$ standard known, can be recommended [11]. Obviously, addition of a charged reference potential standard such as the commonly employed cobaltocenium cation (Cc^+) and the $Cc^{+/0}$ process must be avoided because this cationic reference material would provide a source of electrolyte. With great care, a conventional reference electrode, with a double salt bridge arrangement may be used, but the nature of the junction between high (reference electrode) and very dilute electrolyte concentration (test solution) conditions makes problems with electrolyte leakage almost unavoidable. Additionally, calculation of the junction potential established under these conditions is most problematical.

A cyclic voltammogram obtained for the first reduction process of $[n\text{-}Hex_4N]_4[S_2Mo_{18}O_{62}]$ in acetonitrile containing no deliberately added supporting electrolyte is shown in Figure 3 at a scan rate of 20 mV s^{-1}. In this experiment, the distance from the tip of the platinum quasi-reference electrode to the center of the glassy carbon working electrode was set at 2 mm and the direction of the potential scan was reversed from negative (reduction) to positive (oxidation) just prior to the commencement of the second reduction process. Clearly, well defined voltammograms akin to those observed in the presence of electrolyte (Figure 2) are obtained at slow scan rates (Figure 3). Rather broader processes are detected at faster scan rates (Figure 4), but despite this distortion in wave shape it is apparent that cyclic voltammograms can be obtained in an organic solvent such as acetonitrile over a reasonably wide scan rate range in the absence of added supporting electrolyte for a highly charged species such as $[S_2Mo_8O_{62}]^{4-}$.

The cyclic voltammogram displayed for the reduction of $1 \times 10^{-3}M$ $[S_2Mo_{18}O_{62}]^{4-}$ in Figure 3 at a scan rate of 20 mV s^{-1} exhibits an $E_{1/2}$-value (E_f^0-value) of 125 ± 5 mV *vs* the Pt wire quasi-reference electrode and an ΔE_p-value of 160 ± 5 mV. Solutions of the $[S_2Mo_{18}O_{62}]^{4-}$ anion in acetonitrile, containing 0.1 M Bu_4NPF_6 as the added supporting electrolyte give rise to cyclic voltammograms having an $E_{1/2}$-values of -120 mV ± 10 mV vs Pt wire (after stabilization of the reference) and ΔE_p value of 70 ± 5 mV when using the same scan rate and electrode placement. Thus, whilst the $E_{1/2}$-value is similar, the ΔE_p-value is significantly enhanced in the absence of added electrolyte. The difference in ΔE_p values at a scan rate of 20 mVs^{-1} observed in the absence and presence of supporting electrolyte may be attributed predominantly to the enhanced IR_u drop encountered in the absence of added supporting electrolyte

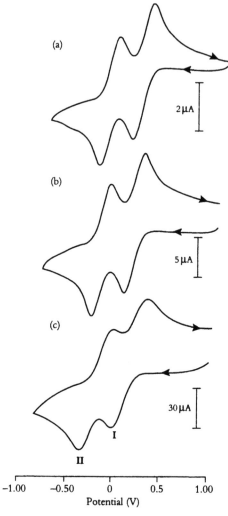

Figure 2. Cyclic voltammograms obtained at a 0.5 mm radius glassy carbon macrodisc electrode for the reduction of 2.0×10^{-3} M $[S_2Mo_{18}O_{62}]^{4-}$ at 22°C in acetonitrile containing 0.2 M Bu_4NClO_4 added as the supporting electrolyte. Potential scale is V vs Fc/Fc$^+$. Scan rates are (a) 0.05; (b) 0.50 and (c) 10 V s^{-1}. Reproduced by courtesy: *Inorg. Chem.* **36**, 2826 (1997). Copyright, American Chemical Society.

[13]. The IR_u distortion then becomes enhanced as the scan rate increases. That is, as the scan rate increases, the current values and hence IR_u is enhanced which leads to E_p^{red} being shifted to more negative values, E_p^{ox} to more positive potentials, to produce a negative shift in the apparent $E_{1/2}$ value and an increase in the value of ΔE_p.

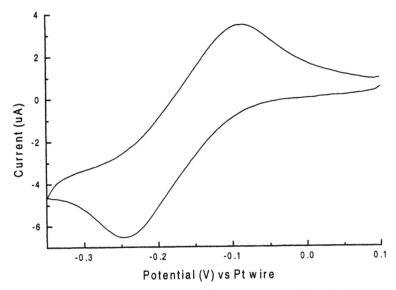

Figure 3. Cyclic voltammograms obtained using a scan rate of 20 mV s^{-1} at 22°C for the reduction of 1.0×10^{-3} M $[S_2Mo_{18}O_{62}]^{4-}$ with a 1.5 mm radius glassy carbon macrodisc electrode in acetonitrile in the absence of added supporting electrolyte. Provided by courtesy of D.C. Coomber.

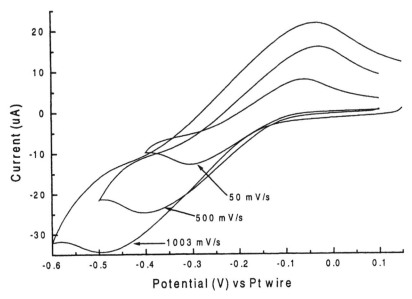

Figure 4. Cyclic voltammograms obtained as a function of designated scan rate for reduction of 1×10^{-3} M $[S_2Mo_{18}O_{62}]^{4-}$ in acetonitrile in the absence of added supporting electrolyte. Other experimental conditions as in Figure 3. Provided by courtesy of D.C. Coomber.

Table 1. Peak potential, peak current and resistance data obtained at a scan rate of 20 mV s^{-1} for the reduction of 1.0×10^{-3} M [$S_2Mo_{18}O_{62}$]$^{4-}$ in acetonitrile under conditions of cyclic voltammetry when the platinum quasi-reference electrode is placed at a specified distance from the 1.5 mm radius glassy carbon working electrode surface.

Electrochemical Parameter			Scan rate of 20 mVs^{-1}					Scan rate of 100 mVs^{-1}				
Working-Reference Electrode Distance (mm)	R^a (ohm)	$R_u{}^b$ (ohm)	E_p^{red} (mV)	E_p^{ox} (mV)	I_p^{red} (µA)	$E_{1/2}{}^c$ (mV)	$\Delta E_p{}^d$ (mV)	E_p^{red} (mV)	E_p^{ox} (mV)	I_p^{red} (µA)	$E_{1/2}{}^c$ (mV)	$\Delta E_p{}^d$ (mV)
8	5970	5970	-229	-66	5.49	-148	163	-296	-37	7.44	-167	259
8 (corr)	5970	460	-174	-82	6.25	-128	92	-201	-85	12.5	-145	119
5	5040	5040	-215	-64	5.77	-140	151	-268	-35	8.88	-152	233
5 (corr)	5040	150	-164	-81	6.35	-123	83	-187	-84	13.1	-136	103
2	4465	4465	-209	-64	5.73	-137	145	-257	-34	9.33	-146	223
2 (corr)	4465	0	-172	-88	6.28	-130	84	-188	-85	13.3	-137	103

[a] Data provided by D.C. Coomber. Also see reference 13 for further details.
[b] Measured at a potential of 0 V vs Pt
[c] Measured as $(E_p^{red} + E_p^{ox})/2$
[d] Measured as $E_p^{ox} - E_p^{red}$

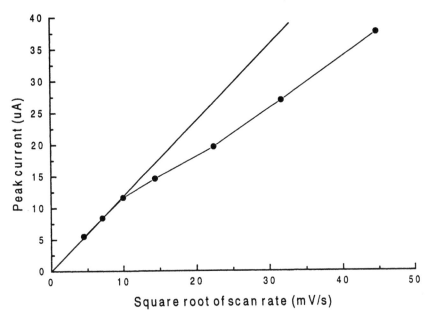

Figure 5. Dependence on the peak current for reduction of 1×10^{-3} M $[S_2Mo_{18}O_{62}]^{4-}$ on the square root of scan rate under transient voltammetric conditions in acetonitrile in the absence of added supporting electrolyte. Other experimental conditions as in Figure 3. Provided by courtesy of D.C. Coomber.

In the absence of uncompensated resistance effects, I_p values for the $[S_2Mo_{18}O_{62}]^{4-}$ anion should be similar in the presence and absence of added electrolyte as migration effects are predicted to be small for highly charged ions [14]. Furthermore, equation 4 should still also apply to a good approximation [15]. In practice, a plot of the I_p^{red} versus the square root of the scan rate in the absence of added supporting electrolyte is close to linear (Figure 5) at slow scan rates (20 to 100 mV s^{-1}) where the IR_u drop is not too severe. However, at a scan rate of 2000 mV s^{-1}, I_p^{red} should theoretically [15] be 10 times greater than that found at a scan rate of 20 mVs^{-1}. Experimentally, it is only 7 times greater (Figure 5) so that IR_u 'distortion' at fast scan rates is significant in acetonitrile in the absence of added supporting electrolyte.

2.1.1 Factors Contributing to the Magnitude of the Uncompensated Resistance

Reference electrode position

A crucial feature of transient cyclic voltammetry without added electrolyte clearly is the presence of uncompensated resistance. Thus, the position of the reference electrode must be far more carefully considered than it usually is in the case with experiments containing high concentrations of added supporting electrolyte. To illustrate the importance of this aspect of the experiment, uncompensated resistance data are presented in Table 1 for the reduction of 1×10^{-3} M $[S_2Mo_{18}O_{62}]^{4-}$ in acetonitrile when the tip of the platinum quasi-reference electrode was placed at a known distance from the 1.5 mm radius glassy carbon working electrode. Voltammetric E_p and I_p data, fully uncompensated and residual uncompensated resistance (R_u) values remaining after partial correction of the IR_u drop (measured at an applied potential of 0 mV vs Pt) are contained in Table 1, at scan rates of 20 and 100 mV s^{-1}. The technique used for obtaining these uncompensated data are contained in reference 16. As theoretically expected, the R_u value decreases as the tip of the quasi-reference electrode is placed closer to the working electrode. However, close to nominally 100% IR_u instrumental compensation could only be achieved routinely when the tip of the reference was 2 mm from the working electrode. At closer distances, significant variation in the uncompensated resistance was evident from experiment to experiment. This inconsistency may be associated with the shielding of the current or disturbance of the current distribution [17], but not from working electrode products reaching the quasi-reference electrode on these cyclic voltammetric time scales [13]. With instrumental IR_u correction, the $E_{1/2}$ value is −126 ± 5 mV vs Pt and the ΔE_p approaches 80 mV when the reference electrode is located 2 mm from the working electrode. These values now agree well with data obtained in the presence of 0.1 M $[Bu_4N][PF_6]$ supporting electrolyte.

IR_u instrumentally corrected and uncorrected voltammograms for the first reduction process of the $[S_2Mo_{18}O_{62}]^{4-}$ anion at scan rates of 20 and 100 mVs^{-1} when the reference electrode tip is placed 2 mm from the working electrode surface are shown in Figures 6 and 7 respectively. Despite the fact that instrumental correction for IR_u may seem to be attractive, in practice, perfect resistance compensation is very unlikely to be achievable [18]. For example, the value of R_u in data presented in Table 1 was obtained at 0V vs Pt and assumed to be independent of potential. In practice, R_u may vary during the course of an experiment. Consequently, a dynamic rather than a static form of instrumental compensation should be applied. Furthermore, the presence of a potential dependent IR_u value, effectively means that a potential dependent rather than constant scan rate is being employed. As a consequence, it may be prudent to not attempt instrumental IR_u compensation and rather include the measured value of R_u in the theory and experimental comparisons (see Section 2.2).

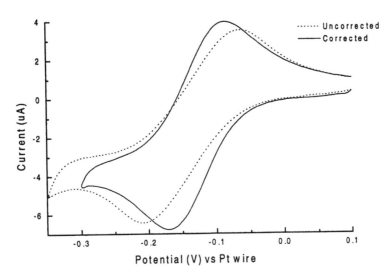

Figure 6. A comparison of cyclic voltammograms obtained at a scan rate of 20 mV s^{-1} with and without instrumental correction for uncompensated resistance for reduction of 1×10^{-3} M $[S_2Mo_{18}O_{62}]^{4-}$ in acetonitrile without added supporting electrolyte. Other experimental conditions are as for Figure 3. Provided by courtesy of D.C. Coomber.

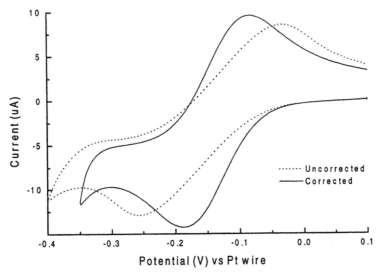

Figure 7. A comparison of cyclic voltammograms obtained at a scan rate of 100 mV s^{-1} with and without instrumental correction for uncompensated resistance for reduction of 1×10^{-3} M $[S_2Mo_{18}O_{62}]^{4-}$ in acetonitrile without added supporting electrolyte. Other experimental conditions are as for Figure 3. Provided by courtesy of D.C. Coomber.

Dependence on [S$_2$Mo$_{18}$O$_{62}$]$^{4-}$ concentration

As expected, the value of the resistance between the reference and working electrode decreases with an increase in [S$_2$Mo$_{18}$O$_{62}$]$^{4-}$ concentration, (Figure 8) but not quite as the reciprocal of concentration, as would be expected theoretically if complete dissociation of the tetra-n-hexyl ammonium salt occurred in acetonitrile. These data imply that concentration dependent ion pairing occurs between the [n-Hex$_4$N]$^+$ cation and the [S$_2$Mo$_{18}$O$_{62}$]$^{4-}$ anion.

Cyclic voltammograms obtained for the [S$_2$Mo$_{18}$O$_{62}$]$^{4-/5-}$ and [S$_2$Mo$_{18}$O$_{62}$]$^{5-/6-}$ reduction processes over the concentration range of 5×10^{-4} M to 3×10^{-2} M in acetonitrile at a scan rate of 20 mV s^{-1}, without deliberately added electrolyte, with the reference electrode tip positioned 2 mm from the working electrode surface, but without instrumental IR_u compensation, are shown in Figure 9. Ideally, in the absence of added electrolyte, the peak height increases linearly with concentration of [S$_2$Mo$_{18}$O$_{62}$]$^{4-}$ and the value of R_u decreases linearly with concentration so that IR_u in the absence of ion pairing may be expected to be independent of concentration. The voltammograms reveal that a slight increase in the contribution from IR_u drop actually occurs as the [S$_2$Mo$_{18}$O$_{62}$]$^{4-}$ concentration increases, as might be expected if ion pairing is present.

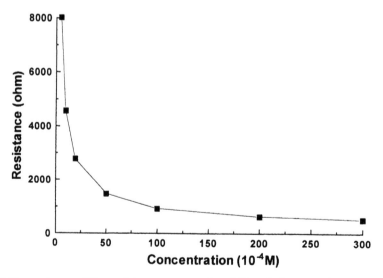

Figure 8. Dependence of the resistance measured at 0 V vs Pt in acetonitrile at 22°C on [n-Hex$_4$N][S$_2$Mo$_{18}$O$_{62}$] concentration when the distance between the working 1.5 mm radius glassy carbon electrode and a Pt quasi-reference electrode is 2 mm. Provided by courtesy of D.C. Coomber. Also see reference 13 for additional details.

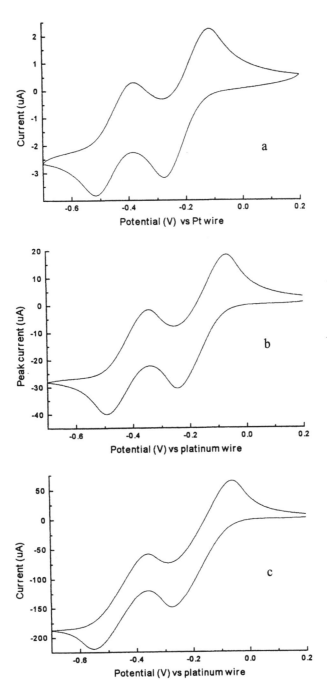

Figure 9. Cyclic voltammograms obtained in acetonitrile for the reduction of (a) 4.9×10^{-4} M, (b) 5.0×10^{-3} M and (c) 3.0×10^{-2} M [n-Hex$_4$N]$_4$[S$_2$Mo$_{18}$O$_{62}$] at a scan rate of 20 mV s^{-1} and at 22°C in the absence of either added supporting electrolyte or instrumental IR_u correction when a glassy carbon macrodisc working electrode is separated from a Pt quasi-reference electrode by 2 mm. Adapted from reference 13.

Dependence on the added supporting electrolyte concentration

Voltammograms obtained at a range of electrolyte concentrations are shown in Figure 10 whilst Figure 11 illustrates the dependence of R_u on the added $[Bu_4N][PF_6]$ supporting electrolyte concentration. Comparison of data in Figure 11 with data in Figure 9 where $[n\text{-}Hex_4N]_4[S_2Mo_{18}O_{62}]$ provides the electrolyte, indicate that the resistance in this latter case is higher for a given electrolyte concentration than when the more completely dissociated $[Bu_4N][PF_6]$ electrolyte is used. Comparison of data obtained (Figure 10, Table 2) with or without added supporting electrolyte reveals that the peak current for the first process is not significantly altered by the addition of electrolyte, implying that the migration current terms are not very large for highly charged complexes under conditions of cyclic voltammetry (also see Section 2.2).

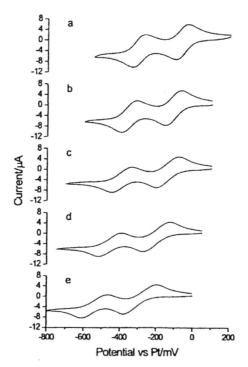

Figure 10. Cyclic voltammograms obtained at a scan rate of 20 mV s^{-1} and at 22°C without instrumental IR_u compensation for the reduction of 1.0×10^{-3} M $[n\text{-}Hex_4N]_4$ $[S_2Mo_{18}O_{62}]$ when a glassy carbon macrodisc working electrode is separated by a distance of 2 mm for a Pt quasi-reference electrode in acetonitrile containing (a) 1.00×10^{-1} (b) 1.0×10^{-2} (c) 1.0×10^{-3} (d) 1.0×10^{-4} and (e) 0 M $[n\text{-}Hex_4N][PF_6]$ as the added supporting electrolyte. Reproduced by courtesy: *Anal. Chem.* **73**, 352 (2001). Copyright, American Chemical Society.

Table 2. Voltammetric data obtained[a] at a scan rate of 20 mV s^{-1} for the reduction of 1.0×10^{-3} M $[S_2Mo_{18}O_{62}]^{4-}$ at 22°C using a glassy carbon electrode in acetonitrile as a function of electrolyte concentration ($[Bu_4N][PF_6]$)

		Uncorrected			IR Corrected		
$[Bu_4N][PF_6]$ (mM)	R (ohm)	$E_{1/2}{}^b$ (mV)	$\Delta E_p{}^c$ (mV)	I_p^{red} (μA)	$E_{1/2}{}^b$ (mV)	$\Delta E_p{}^c$ (mV)	I_p^{red} (μA)
0	4711	-140	144	5.49	-134	71	6.15
1	4214	-129	141	5.76	-125	74	6.43
5	3593	-123	133	6.02	-119	70	6.65
10	3055	-116	123	6.35	-112	68	6.90
10	3291	-138	128	5.85	-134	70	6.50
20	2593	-130	115	6.02	-126	68	6.49
50	1558	-123	97	6.18	-120	66	6.53
91	990	-114	86	6.44	-114	63	6.69
200	514	-106	76	6.62	-104	66	6.79
500	283	-100	71	6.58	-98	75	6.73
1000	181	-92	67	6.32	-90	72	6.48

[a] Data provided by D.C. Coomber. See reference 13 for further details.
[b] Measured as $(E_p^{red} + E_p^{ox})/2$
[c] Measured as $E_p^{ox} - E_p^{red}$

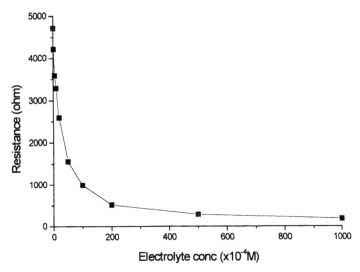

Figure 11. Dependence of the resistance measured at 100 mV vs Pt on added [n-Hex$_4$N] [ClO$_4$] supporting electrolyte concentration at 22°C when the distance between the 1.5 mm radius glassy carbon macrodisc working electrode and the Pt quasi-reference electrode is 2 mm in acetonitrile solution. Provided by courtesy of D.C. Coomber. Also see reference 13 for additional details.

Experimental Overview

Provided adequate care is taken with respect to the positioning of the reference and working electrodes, cyclic voltammograms under transient conditions at macrodisc electrodes may be obtained for a highly charged complexes such as $[S_2Mo_{18}O_{62}]^{4-}$ without addition of electrolyte in solvents such as acetonitrile [13]. In particular, for these highly charged species, the migration current is not a highly significant term. Nevertheless, quantitative comparisons with theory would appear to require detailed attention to uncompensated resistance and also possibly to ion pairing. It is interesting to note that Nicholson [19,20] first drew attention to the problem of uncompensated resistance in cyclic voltammetry in the presence of supporting electrolyte almost 40 years ago. However, only small deviations in IR_u of up to 0.1 V were considered in these [19,20] pioneering studies. In the absence of added electrolyte, far larger IR_u drop effects (e.g. 300 mV) may need to be considered from both experimental and theoretical perspectives.

2.2   The Theoretical Perspective for a Reversible Process under Conditions of Cyclic Voltammetry

The theoretical perspective relevant to cyclic voltammetric experimental data presented for the reversible $[S_2Mo_{18}O_{62}]^{4-/5-}$ and $[S_2Mo_{18}O_{62}]^{5-/6-}$ processes in Section 2.1 are based on concepts presented in reference 14. Only analyses of cyclic voltammetric responses at a (planar) macrodisc electrode with semi-infinite linear diffusion are considered in any detail. As noted in the introduction, cyclic voltammetry undertaken in the presence of excess supporting electrolyte is widely employed by organic and inorganic chemists for evaluating the reversibility of an electrode process, for confirming the stability of electrode products and for establishing the reversible potential of the redox couple of interest. However, as deduced from experimental data (Section 2.1) in the absence of added supporting electrolyte, the resistance between the working electrode and the reference electrode, R_u can produce a large potential drop, IR_u such that E_{eff}, the effective potential across the double layer of the working electrode, is significantly different from E_{CV}, the potential specified by the cyclic voltammetric protocol [14] which is

$$E_{eff} = E_{CV} - IR_u \qquad\qquad (5)$$

As noted in Section 2.1.1 good electrochemical practice dictates [14,18] that the reference electrode should be placed as close as possible to the working electrode, thereby reducing R_u (and therefore IR_u) as much as possible. Also as noted in Figures 6 and 7, when a potentiostated three electrode system is employed with positive feedback forms of instrumental IR_u compensation, the effect of R_u can be further diminished. Positive feedback [14,18] increments E_{CV} by a voltage component that is proportional to the current, I, and R_u^0, where R_u^0 is the value of R_u measured at the initial potential in the cyclic voltammetric perturbation. If R_u^0 is assumed to remain constant throughout the course of the experiment, then, equation 5 can be rewritten [14] as:

$$E_{eff} = E_{CV} - IR_u + \gamma IR_u^0 \tag{6}$$

where $0 \leq \gamma \leq 1$. When $\gamma = 1$, perfect compensation has been achieved as long as R_u is constant (independent of potential). Under this circumstance $E_{eff} = E_{CV}$, but the system can also become unstable when $\gamma \geq 1$, although in common [14,18] experimental practice, $\gamma < 1$.

An interesting and important complication arises in systems without added supporting electrolyte in that variations in the composition of the depletion layer that occur during the perturbation will induce a change in R_u from the initial value, R_u^0. The role of that change, ΔR_u, could be critical in some circumstances. If equation 5 is rewritten as:

$$E_{eff} = E_{CV} - I(IR_u^0 + \Delta R_u) + \gamma IR_u^0 = E_{CV} - (1 - \gamma) IR_u^0 - I\Delta R_u \tag{7}$$

and the term

$$\Delta\gamma = \frac{\Delta R_u}{R_u^0} \tag{8}$$

is introduced, then equation 7 becomes [14]

$$E_{eff} = E_{CV} - (1 - \gamma + \Delta\gamma)IR_u^0 \tag{9}$$

and it is clear that a negative $\Delta\gamma$ produced by a negative ΔR_u (equation 8) could introduce over compensation and instability. On this basis, the term "perfect compensation" means that the entire resistance is compensated at all times, including any changes in resistance introduced by changes in the composition of the depletion layer. "Partial compensation" therefore means that γ is fixed and compensation is based on a constant value of R_u^0 with no adjustment being made for changes in resistance that occur during the course of a cyclic voltammetric experiment. Note that even when $\gamma = 1$ the system is still only partially compensated, since no adjustments are made for the changes in resistance caused by changes in the composition of the depletion layer. In the analysis presented below, which is based on that presented in reference 14, capacitive currents are ignored, which is a reasonable assumption for many cases of interest. However, in practice there is no simple way to execute a background correction. Consequently, it could be appropriate to introduce a capacitive current component into the simulation.

The reversible system of interest in the example to be considered theoretically is a one-electron charge transfer process at a temperature of 25°C (298.2 K). As in the experimental case, the system initially comprises only the solvent and an ionic redox moiety **A** or **B** (equation 1) with its counter-ion **X**. Thus, the voltammetry is described (see equation 1) by the reversible one electron transfer reaction

$$\mathbf{A}^{z_A} + e^- \rightleftharpoons \mathbf{B}^{z_A - 1} \tag{10}$$

where $\mathbf{A}^{z_A}$ or $\mathbf{B}^{z_A - 1}$ is the charged redox species initially present in solution and z_X is the charge on the counter ion. Thus, the $[S_2Mo_{18}O_{62}]^{4-/5-}$ process considered experimentally in Section 2.1, is an example where $z_A = -4$, $z_A - 1 = -5$ and $z_X = 1$

The fundamental assumptions for computations of problems involving migration and diffusion are that mass transport is described by the Nernst-Planck equation coupled with the electroneutrality constraint. The Nernst-Planck relationship is [12,14]

$$f_j = -D_j \frac{dc_j}{dx} + \frac{F}{RT} E D_j c_j z_j \tag{11}$$

where f_j, c_j and E are dependent upon time (t) and distance (x), and correspond to the flux and concentration of the jth species and the electric field, respectively and D_j and z_j are the diffusion coefficient and charge of the jth species respectively. Invoking the electroneutrality constraint, which can be expressed mathematically as:

$$\sum_j c_j z_j = 0 \tag{12}$$

greatly simplifies the analysis [14]. A planar electrode with semi-infinite linear diffusion is assumed (i.e. no edge or radial diffusion effects [11]). The reference electrode is assumed to be well outside the depletion layer [17]. E_{eff}, the potential across the compact and diffuse double layers (which is referred to as the interfacial region [18]) is assumed to be exactly equal to E_{CV}, which is easy to do in a simulation, but very difficult in an experiment. However, the simulation therefore does at least allow an estimate of exactly what compensation would be required to achieve such a goal. The resistance, R_u, between the working and reference electrodes cannot be constant since the concentrations of all species, and therefore the resistance, will be changing within the depletion layer.

Computations and discussions reproduced from reference 14 below, focus on systems where z_A is the charge on the redox active species counter ion, $\mathbf{X}$, z_A is the charge of the redox species initially present in solution and the reduction product has the charge $z_A - 1$ (see equation 10). Thus, the analyses presume that there is no complexation and no ion-pairing. The latter condition becomes increasingly unlikely with increasing $|z_A|$, and is certainly present in the voltammetry of $[S_2Mo_{18}O_{62}]^{4-}$ considered from an experimental perspective in Section 2.1.

Simple analysis presented in reference 14 confirms that the magnitude of $I\Delta R_u$, when perfect compensation is presented, as would be expected, is independent of scan rate, initial concentration, and the diffusion coefficient. However, in order to understand experimental data of the kind described in Section 2.1, the behaviour of uncompensated IR_{ref} needs to be analysed.

For convenience of presentation of theoretical cyclic voltammograms it is useful [14] to define an uncompensated resistance parameter, P_u:

$$P_u \approx \frac{F}{RT} I_p^{red} R_u \qquad (13)$$

which provides a convenient basis for characterizing the effect of uncompensated resistance on the shape of the cyclic voltammograms. In the discussion presented below, which is adapted from reference 14, an explicit finite difference algorithm was used to simulate cyclic voltammograms with either *perfect compensation* (i.e., $E_{eff} = E_{CV}$) or with partial compensation (i.e., $E_{eff} = E_{CV} - I(1-\gamma) R_u^0$).

2.2.1 Simulations with Perfect Compensation

Simulation of $I\Delta R_u$ during the course of a cyclic voltammograms

The objective of these simulations, presented in reference 14, and executed with "perfect compensation" (i.e., $E_{eff} = E_{CV}$ at all times), is to ascertain the magnitude of $I\Delta R_u$, the change in the compensating potential effected by the change in resistance within the depletion layer. Figure 12 shows the computed values of $(F/RT) I\Delta R_u$ during the course of a cyclic voltammogram as a function of $(F/RT)(E_{CV}-E^0)$. The curves in Figure 12a were computed for $z_A = -1$, -2, -3, and -4 with $z_X = -1$. Clearly, $|I\Delta R_u|$ decreases dramatically with increasing values of z_A whilst as noted above and in more detail in reference 14, the results are independent of scan rate, diffusion coefficients (assuming relative values are fixed), and the initial concentration of the redox active salt. When $z_A \leq -1$ (see Figure 12a), the reduction product, $\mathbf{B}^{z_A-1}$ produced in the depletion layer is more highly charged. The net result is a decrease in the resistance within the

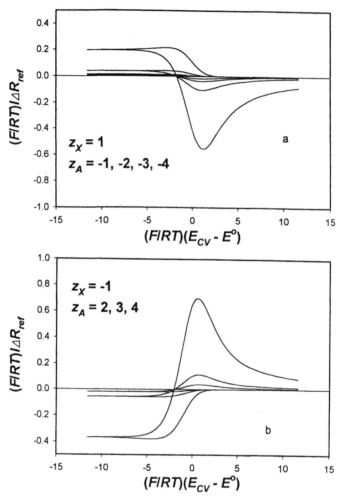

Figure 12. Plots of $(F/RT)I\Delta R_u$ vs $F/RT(E_{CV} - E_f^0)$ (a, top) for $z_X = 1$ and for curves from highest to lowest: $z_A = -1, -2, -3,$ and -4 respectively (b, bottom) for $z_X = -1$ and for curves from highest to lowest: $z_A = 2, 3,$ and 4 respectively. Reproduced by courtesy: *J. Phys. Chem. B*, **102**, 9966 (1998). Copyright, American Chemical Society.

depletion layer thus providing a negative value ΔR_{ref} throughout the course of the cyclic voltammogram. Since the reductive currents during the initial half-cycle of the experimental are negative (IUPAC convention) $I\Delta R_u$ will be positive during this phase of the cyclic voltammogram. When oxidation occurs during the second half cycle, I becomes positive, ΔR_u remains negative, and therefore $I\Delta R_u$ becomes negative. When $z_A \geq 2$, ΔR_u will be 0 or positive (Figure12b) throughout the course of the cyclic voltammogram and $I\Delta R_u$ will be negative during the initial (reductive) half-cycle and positive during the second half cycle of the cyclic voltammogram. Note also that at potentials well negative of the reduction peak, $I\Delta R_u$ is virtually constant which is a consequence of the fact that the change in the thickness of the depletion layer $|\Delta R_u|$ increases with $t^{1/2}$ while $|I|$ decreases with $t^{1/2}$ [6].

Data in Figure 12 and reference 14 suggest that as long as $z_A \leq -3$ or ≥ 3 as applies in the reduction of $[S_2Mo_{18}O_{12}]^{4-}$, the maximum values of $[I\Delta R_u|$ at 25°C will be less than 0.001 V, and not a major problem.

The shape of cyclic voltammograms simulated with perfect compensation

In Figure 13, a cyclic voltammograms with perfect compensation, for $z_A = -1, -2, -3, -4, -1000, 4, 3$ and 2 with $z_X = \pm 1$ as required, are presented as the dimensionless current function [12,14,19-21], χ plotted (equation 14 represents the maximum value of χ or χ_p) as a function of the normalized potential (F/RT) $(E_{CV} - E_f^0)$.

$$\chi_p = \frac{I_p}{FAc_A\sqrt{\frac{F}{RT}D_A|v|}} = 0.4463 \tag{14}$$

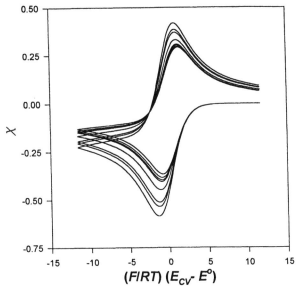

Figure 13. Plot of the current function χ (see Equation 14 and related discussion) vs $F/RT(E_{CV} - E_f^0)$ with $z_X = \pm 1$ as required for increasing values of $z_A = -1, -2, -3, -4, -1000, 4, 3,$ and 2 corresponding to increasing peak currents. Reproduced by courtesy: *J. Phys. Chem. B*, **102**, 9966 (1998). Copyright, American Chemical Society.

The shape of the voltammogram for a very high charged species ($z_A = -1000$) is indistinguishable from the classical curves computed for a system with excess inert electrolyte [21]. The normalized difference $(F/RT)(E_p^{ox} - E_p^{red})$ or $(F/RT\Delta E_p)$ does not change significantly from 2.22, the value obtained when there is excess electrolyte, although as usual [12] there are, small changes induced into this value by changing the switching potential. The effect of migration on the normalized peak potentials $(F/RT)(E_p - E_f^0)$ for the forward and reverse peaks is shown in Figures 14 as a function of $1/z_A$. Note that there is a smooth transition from negative to positive values of $1/z_A$ and the interpolated value of $(F/RT)(E_p - E_f^0)$ corresponding to $1/z_A = 0$ is within 0.004 of 1.109 (equals 0.0285 V at T = 298.2) the value predicted for a reversible cyclic voltammogram with excess electrolyte [12]. The effect of migration on the current has also been evaluated in reference 14 as a function of $1/z_A$ and the ratio of χ_p^{red} and χ_p^{ox} in the absence and presence of excess added supporting electrolyte. Again, there is a smooth transition from negative to positive values of $1/z_A$ [14] and the interpolated value of ratio corresponding to $1/z_A = 0$ is 1.000. All these data lead to the conclusion that the influence of migration becomes small for large values of z_A.

While the simulations of systems with perfect compensation are helpful for understanding of migration effects, executing an experiment with perfect

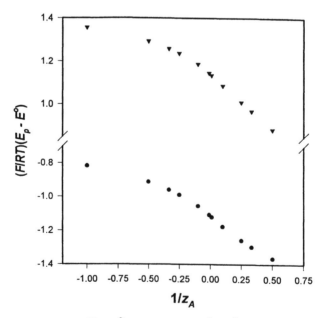

Figure 14. Plots of F/RT $(E_p^{ox} - E_f^0)$ and F/RT $(E_p^{red} - E_f^0)$ values of z_A given in Figure 13. Reproduced by courtesy: *J. Phys. Chem. B*, **102**, 9966 (1998). Copyright, American Chemical Society.

compensation is not possible. Thus, it is essential to examine the simulations of cyclic voltammogram with partial compensation under conditions encountered in experimental situations.

2.2.2 Simulations with partial compensation

Effect of uncompensated resistance on the shape of a cyclic voltammogram.

A sense of how uncompensated resistance modifies the shapes of cyclic voltammograms can be gleaned from examination of Figure 15 which shows normalized cyclic voltammograms (χ vs. $(F/RT)(E_{CV}-E_f^0)$ for $z_A = -1$ and $z_X = 1$ with P_u (equation 13) = 1.973, 4.932, 9.86, 14.80 and 19.73. All curves were initiated at a normalized potential of $(F/RT)(E_{CV}- E_f^0)$ of 0.3(F/RT) and all except one, were reversed at $(F/RT)(E_{CV}- E_f^0) = -0.3(F/RT)$. The voltammogram reversed at $(R/RT)(E_{CV}-E_f^0) = -0.5(F/RT)$ corresponds to $P_u = 19.73$. Note that the oxidation peak potential for this switching potential is less positive than the oxidation peak of the corresponding curve reversed at $-0.3(F/RT)$ for $P_u = 19.73$

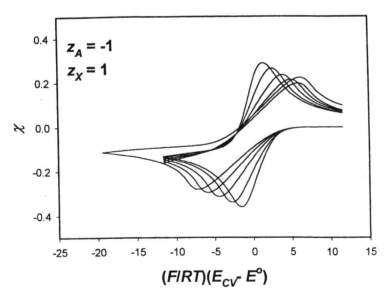

Figure 15. Plot of χ vs $F/RT (E_{CV} - E_f^0)$ for $z_A = -1$, $z_X = 1$ with $P_u = 1.973, 4.932, 9.86, 14.80$ and 19.73. The vertical normalised potential in all simulated voltammograms is $F/RT (E - E_f^0) = 0.3 (F/RT)$ and the normalized switching potential except one was $F/RT (E_{CV} - E_f^0) = 0.3 (F/RT)$. The cyclic voltammogram reversed at $F/RT (E_{CV} - E_f^0) = 0.5 (F/RT)$ corresponds to $P_u = 19.73$. Reproduced by courtesy: *J. Phys. Chem. B*, **102**, 9966 (1998). Copyright, American Chemical Society.

because the current is significantly lower and therefore IR_u, is smaller. This dependence on switching potential is a major reason for the difficulty in using the average value of the peak potentials, $(E_p^{ox} + E_p^{red})/2$, as an estimate of E_f^0.

Determining E_f^0 from cyclic voltammograms with partial compensation.

The best way to determine the E_f^0- value is to fit the simulated response to the experimentally obtained cyclic voltammogram. That is, set the values of the known input parameters for the simulation and then adjust the unknown parameters until the simulated and experimental voltammograms match. Other inherently more complex approaches are contained in reference 14.

2.2.3 Conclusions based on theoretical studies

The analysis of the cyclic voltammetric behaviour of a reversible one-electron reduction process without added supporting electrolyte systems suggests that:

- ΔIR_u (the change in the solution resistance between the working and reference electrodes effected by the cyclic voltammetric perturbation) is negligible for $z_A \le -3$ and $z_A \ge 4$. However, interestingly, when z_A is negative, ΔR_{ref} is negative. This decrease in solution resistance could destabilize the system when instrumental compensation using a potentiostat with positive feedback, is close to unity.

- Data analysed with the use of a three electrode system, a potentiostat, and positive feedback type methods to compensate for ohmic drop, IR_u, between the working and reference electrodes must be treated with caution as only partial compensation is likely to be achieved under experimental conditions.

- Information obtained from a cyclic voltammogram without added supporting electrolyte can be used to confirm the overall reversibility of the electrochemical process. Furthermore, E_f^0 can be evaluated approximately and with caution by averaging E_p^{red} and E_p^{ox}.

- E_f^0 is best determined by direct matching of simulated and experimental cyclic voltammogram.

- Simulations available in commercial packages such as DigiSim [22] which consider uncompensated resistance but do not consider the effects of migration on mass transport or the effects of changes in the depletion layer, provide adequate approximations to experimental data if the value of $|z_A|$ is large (see Figure 13).

- Whilst a one electron reversible *reduction* that mimics the $[S_2Mo_{18}O_{62}]^{4-/5/6-}$ has been considered as the model system processes considered experimentally in Section 2.1, the analysis of a one electron *oxidation* is analogous. In order to solve this problem changes to be made include the following: signs of z_A and z_X must be changed, excluded values of z_A will be –1 and 0; the initial species will be B and its counter ion will be X; the initial (normalized) potential will be –0.3 (F/RT), and the initial sign of v will be positive.

2.3 A Comparison of Experimental and Theoretical Data Obtained for the Reduction of $[S_2Mo_{18}O_{62}]^{4-}$ in the Absence of Added Supporting Electrolyte

Qualitatively, the experimental data presented in Section 2.1 are consistent with the response obtained by simulations of the kind considered in Section 2.2. However, new features associated with voltammetry without added supporting electrolyte emerge when quantitative comparisons are made. To assess to what extent mass transport by diffusion plus migration, and also uncompensated resistance account for the experimental observations reported in Section 2.1, cyclic voltammograms obtained for the reduction of 1.05×10^{-3} M $[(n\text{-Hex})_4N]_4[S_2Mo_{18}O_{62}]$ in the absence of added supporting electrolyte at a scan rate of 20 mV s^{-1} and with a reference-to-working electrode distance of 2 mm are compared with those obtained theoretically using the DigiSim software package [22] with mass transport solely by diffusion as well as the numerical method reviewed in Section 2.3 which incorporates both diffusion and migration. Both forms of simulation can include uncompensated resistance effects. The reported diffusion coefficient (D-value) of $[S_2Mo_{18}O_{62}]^{4-}$ in acetonitrile of 6.2×10^{-6} cm^2 s^{-1} obtained in the presence of supporting electrolyte [23] was used for all species in the simulations. It has been shown that the D-values of both the one- and two-electron-reduced polyoxometalate species are essentially identical [23]. The dependence of diffusion coefficients on ionic strength [2] was neglected. D for the $[(n\text{-Hex}_4)_4N]^+$ counterion is unknown, although variation by a factor of 5 using the simulation method discussed in Section 2.2 led to only small changes in the theoretically predicted voltammograms [14]. Both the $[S_2Mo_{12}O_{62}]^{4-/5-}$ and $[S_2Mo_{12}O_{62}]^{5-/6-}$ reduction processes were assumed to be electrochemically reversible in all simulations.

A comparison of simulated voltammograms incorporating only diffusion and resistance with those obtained experimentally at a scan rate of 20 mV s^{-1} in the absence of electrolyte, with and without resistance compensation, is shown in Figure 16. Simulation without migration provides reasonable agreement with experiment only in the potential region associated with the initial reduction process. As expected, incorporation of migration into the theory enables superior

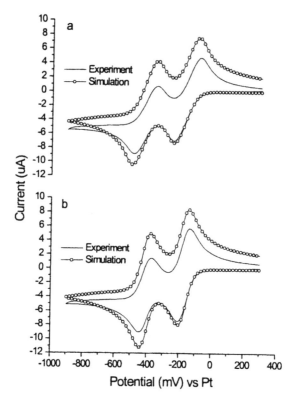

Figure 16. Comparison of simulated (DigiSim software package [22]) and experimental cyclic voltammograms obtained at a scan rate of 20 mV s^{-1} and 22°C with a 0.5 mm radius glassy carbon working-Pt quasi-reference electrode separation of 2 mm for the reduction of 1.05×10^{-3} M $[S_2Mo_{18}O_{62}]^{4-}$ in acetonitrile in the absence of added supporting electrolyte. (a) Experiment without IR_u compensation and simulation incorporating an uncompensated resistance value of 4900 ohm. (b) Experiment with nominally 100% IR_u compensation and simulation with zero uncompensated resistance. Reproduced by courtesy: *Anal. Chem.* **73**, 352 (2001). Copyright, American Chemical Society.

agreement to be obtained, at least for the reduction component of the second process (Figure 17). Furthermore, agreement between theory and experiment improves as the scan rate increases to 100 mV s^{-1} (Figure 17). Incorporation of spherical diffusion into the simulations made only a small difference and certainly does not account for more than a minor fraction of the discrepancy between experiment and theory. The variation of resistance as a function of potential also was included in this simulation but found to have a negligible influence for the highly charged 4-/5- and 5-/6- systems.

The differences between experimental and simulated voltammograms that remains even after inclusion of migration and R_u may be explained by a combination of factors that include the following: the $[S_2Mo_{18}O_{62}]^{4-}$ anion having a slightly different diffusion coefficient in the absence of excess added supporting electrolyte; the absence of charging current in the simulation; small variations in reference-working electrode distance from the simulated value of 2.0 mm; a small components of spherical diffusion, small departures from reversibility and concomitant introduction of diffuse double-layer Frumkin effects (see Section 4.1). However, most critically, the diffusion plus migration simulation predicts a significantly different mass transport-limited current from that obtained experimentally at potentials beyond the peak potential for

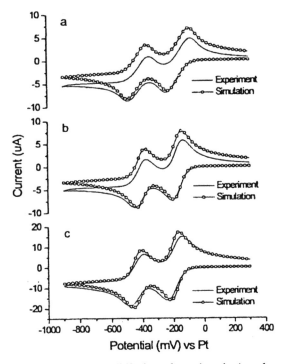

Figure 17. Comparison of simulated (diffusion plus migration) and experimental cyclic voltammograms obtained at 22°C in acetonitrile in the absence of added supporting electrolyte using a 0.5 mm radius glassy carbon working-reference electrode separation of 2 mm for the reduction of 1.05×10^{-3} M $[S_2Mo_{18}O_{62}]^{4-}$: (a) experimental (without IR_u compensation) and simulated (migration, without IR_u compensation) voltammograms at a scan rate of 20 mV s^{-1}, (b) experimental (with IR_u compensation) and simulated (migration, with 100% IR compensation) voltammograms at a scan rate of 20 mV s^{-1}, and (c) experimental (with IR_u compensation) and simulated (migration, with 100% IR_u compensation) at a scan rate of 100 mV s^{-1}. Reproduced by courtesy: *Anal. Chem.* **73**, 352 (2001). Copyright, American Chemical Society.

the second process and at all potentials on the entire reverse scan, regardless of the extent of IR_u compensation. This discrepancy is most readily observed in voltammograms obtained at slow scan rate as a large current "offset" when comparing experimental and simulated voltammograms (Figure 17a). Thus, it is concluded that incorporation of migration and uncompensated resistance into the theory fails to provide a completely adequate description of the cyclic voltammetry of $[S_2Mo_{18}O_{62}]^{4-}$ in the absence of added supporting electrolyte. Simulations incorporating a very fast (diffusion-controlled) cross-redox reaction for comproportionation of $[S_2Mo_{18}O_{62}]^{4-}$ and $[S_2Mo_{18}O_{62}]^{6-}$ with both diffusion and migration were shown [14] to have virtually no effect upon the shape of the voltammogram, so that this does not account for the discrepancy. Additionally, it should be noted that ion pairing would not be predicted to give an increased current at potentials beyond more negative than the second reduction process.

The major feature excluded from the simulations is a possible contribution from convection, which appears to be enhanced by both the absence of added supporting electrolyte and by high $[S_2Mo_{18}O_{62}]^{4-}$ concentrations. That is, the cyclic voltammogram for a 3.0×10^{-2} M solution of $[S_2Mo_{18}O_{62}]^{4-}$ shown in Figure 9c, exhibits such pronounced evidence of convection that characteristics of a steady-state response are exhibited. If the origin of the problem is convection, then use of higher scan rates should minimize the relative importance of convection. Indeed, as noted above, significantly superior agreement between experiment and simulation (diffusion plus migration) is obtained at a scan rate of 100 mV s^{-1} (Figure 17c). However, a slightly enhanced current is still observed in the experimental voltammogram at potentials beyond the peak potential for the second process, and despite the employment of nominally 100% instrumental IR_u compensation, fitting the peak potentials at these faster scan rates still requires the inclusion of significant resistance in the simulation.

2.3.1 Dependence of Cyclic Voltammograms on Electrode Angle

If the occurrence of a redox process at an electrode surface leads to density gradients within the depletion layer resulting in natural convective mass transfer, this phenomenon should cause current values to be dependent on the angle of the electrode, because of the subsequent influence of gravity. The situation in the complete absence of added electrolyte is more complex than when excess added supporting electrolyte is present as density gradients may be generated by transport of the counterion as well as by the oxidized and reduced forms of the electroactive moieties.

Cyclic voltammograms have been obtained [14] for the reduction of [(n-Hex)$_4$N]$_4$[S$_2$Mo$_{18}$O$_{62}$] in acetonitrile in the absence of added electrolyte at glassy carbon working electrode angles from vertical (electrode downward, defined as 0°) to horizontal (defined as 90°) at a scan rate of 20 mV s^{-1}. Inspection of data in Figure 18a reveals that the peak potential and current values of the reduction processes are essentially independent of the electrode angle. In contrast, it is obvious that currents in the mass transport-limited, more negative potential region, are strongly electrode-angle dependent and have many of the characteristics of a steady-state response. Figure 19 contains plots of the ratio of the current measured at a potential of –800 mV vs Pt for a range of electrode angles against that at an angle of 0°. It may be deduced from data contained in this

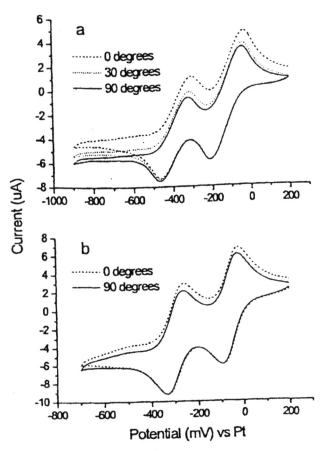

Figure 18. Cyclic voltammogram obtained at 22°C in acetonitrile at a scan rate of 20 mV s^{-1} using specified 0.5 mm radius (glassy carbon) electrode angles for the reduction of 1.0×10^{-3} [(n-Hex)$_4$N][S$_2$Mo$_{18}$O$_{62}$] (a) without and (b) with 0.10 M [(n-Hex)$_4$N][ClO$_4$] added supporting electrolyte. Reproduced by courtesy: *Anal. Chem.* **73**, 352 (2001). Copyright, American Chemical Society.

plot that for electrode angles greater than 45°, the value of the current at −800 mV vs Pt remains constant, within experimental error, but at a significantly greater value than obtained at 0°. Angles above 0° and below 45° show increased, but intermediate, current values. On the reverse potential sweep, the peak potential values are shifted slightly to more negative potentials as the angle increases. Although the peak current values (corrected for the convection current) are similar in magnitude on the reverse and forward scans, the absolute peak current values on the reverse scan direction are markedly different, with lower values observed at larger angles. Thus, density-induced convection provides a credible explanation for the major part of the current offset observed when comparing experimental and theoretical voltammograms. At the faster scan rate of 100 mV s⁻¹, minimal dependence on electrode angle is observed in cyclic voltammograms at a $[S_2Mo_{18}O_{62}]^{4-}$ concentration of 1.0 mM in the absence of added supporting electrolyte, indicating that the duration of the voltammetry is now brief enough to inhibit the extent of the influence of the natural convection process.

Voltammograms obtained for the reduction of $[S_2Mo_{18}O_{62}]^{4-}$ in the presence

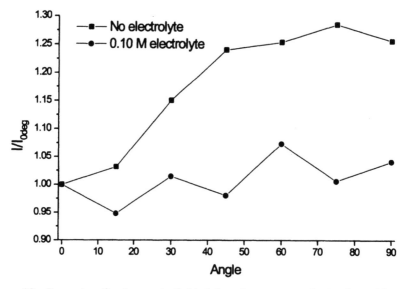

Figure 19. Current ratio (current divided by the current obtained at 0°) against (0.5 mm radius glassy carbon) electrode angle for the reduction of 1.0 mM $[(n\text{-}Hex)_4N][S_2Mo_{18}O_{62}]$ at 22°C in the absence of added supporting electrolyte (measured at −800 mV vs Pt) and in the presence of 0.10 M $[(n\text{-}Hex)_4N][ClO_4]$ supporting electrolyte (measured at −650 mV vs Pt). Reproduced by courtesy: *Anal. Chem.* **73**, 352 (2001). Copyright, American Chemical Society.

of 0.10 M $[(n\text{-Hex})_4N]_4[ClO_4]$ electrolyte show a much lower dependence on electrode angle over the range 0-90°, when a scan rate of 20 mV s^{-1} is employed, (Figure 18b). In the presence of 0.1 M added electrolyte, peak potentials are independent of angle for both processes. A plot of reduction current measured at –650 mV vs Pt, (Figure 19) which is more negative than the peak potential for the second process shows no dependence on electrode angle, within experimental error. However, the magnitude of current values associated with the oxidative sweep scan direction decrease slightly as the electrode angle is increased over the range 0-90° (Figure 18) so that the presence of added supporting electrolyte minimizes rather than eliminates convection associated with density gradient. At a scan rate of 100 mV s^{-1}, the entire voltammetric response is independent of the electrode angle, when 0.10 M $[(n\text{-Hex})_4N]_4[ClO_4]$ is present as the added electrolyte.

2.3.2 Conclusions Derived from Theory-Experimental Comparisons for a Reversible Process

The quantitative evaluation of cyclic voltammogram obtained for the reversible reduction of $[(n\text{-Hex})_4]N_4[S_2Mo_{18}O_{62}]$ at a glassy carbon macrodisc electrode in acetonitrile in the absence of added supporting electrolyte requires care. For example, voltammograms obtained at slow scan rates are less affected by IR_u drop and hence are more readily resistance compensated, but they are more severely influenced by convection. The reverse situation applies at high scan rates where connection problems are minimal, but IR_u problems may become reverse. As expected, careful placement of the reference electrode assists in lowering the uncompensated resistance.

Studies on the related polyoxometalate system $[Bu_4N]_2[SMo_{12}O_{40}]$ also have been reported [24] in acetonitrile in the presence and absence of added supporting electrolyte for the processes

$$[SMo_{12}O_{40}]^{2-} + e^- \;\rightleftharpoons\; [SMo_{12}O_{40}]^{3-} \tag{15}$$

$$[SMo_{12}O_{40}]^{3-} + e^- \;\rightleftharpoons\; [SMo_{12}O_{40}]^{4-} \tag{16}$$

Figure 20 shows that excellent comparison of experimental and simulated data exist for the $[SMo_{12}O_{40}]^{2-/3-}$ and $[SMo_{12}O_{40}]^{3-/4-}$ process when migration and IR_u drop are included in the simulations. This example represents a case where precipitation of $[Bu_4N]_2[SMo_{12}O_{40}]$ occurs when $[Bu_4N][PF_6]$ is added as the electrolyte, so that only studies are possible in the absence of added supporting electrolyte. Consequently, if one-electron reduced $[Bu_4N]_3[SMo_{12}O_{40}]$, which in contrast to $[Bu_4N]_2[SMo_{12}O_{40}]$, is soluble in both the presence and absence

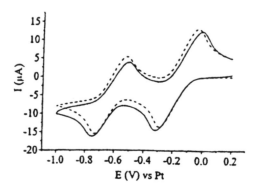

Figure 20. Comparison of (—) experimental cyclic voltammogram (obtained at a glassy carbon electrode in acetonitrile with added supporting electrolyte) and simulated (····) data (accounting for migration and IR_u drop). Experimental conditions and parameters: 1 mM [(n-Hex$_4$)N]$_2$[SMo$_{12}$O$_{40}$] scan rate = 100 mV s^{-1}, $D = 1 \times 10^{-5}$ cm^2 s^{-1} for all solution species. Reproduced by courtesy: *Electroanalysis* **13**, 1475 (2001). Copyright, Wiley-VCH Verlag GmbH.

of added supporting electrolyte is present in bulk solution, then the oxidation process in the absence of electrolyte (Equation 17)

$$[SMo_{12}O_{40}]^{3-} \rightleftharpoons [SMo_{12}O_{40}]^{2-} + e^- \tag{17}$$

becomes complicated (Figure 21a) as [Bu$_4$N][PF$_6$] electrolyte is added because of precipitation of [Bu$_4$N]$_2$[SMo$_{12}$O$_{40}$] at the electrode surface. Consequently, the voltammetry of [SMo$_{12}$O$_{40}$]$^{3-}$ at slow scan rates is more complicated when added supporting electrolyte is present (Figure 21b) because the process corresponds to the reaction scheme given by the combination of equations 18a and b.

$$[SMo_{12}O_{40}]^{3-}_{solution} \rightleftharpoons [SMo_{12}O_{40}]^{2-}_{solution} + e^- \tag{18a}$$

$$\downarrow [Bu_4N]^+$$

$$[Bu_4N]_2[SMo_{12}O_{40}]_{(solid)} \tag{18b}$$

Of course, employment of faster scan rates than used in the absence of added supporting electrolyte (≥ 500 mV s^{-1}) enables the precipitation reaction (equation 18b) to be outrun, and as shown in Figure 21b, cyclic voltammograms obtained at fast scan rate simply correspond to the reaction in equation 18a.

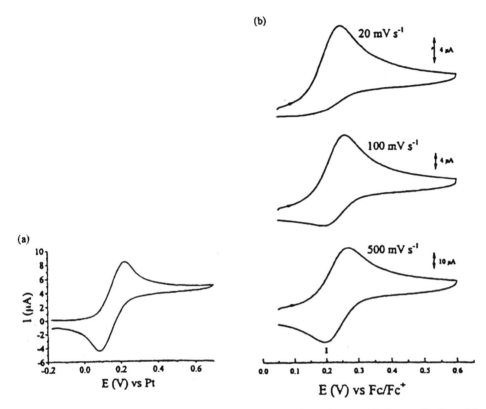

Figure 21. Cyclic voltammograms obtained for the oxidation of 1 mM $[(Bu_4)N]_3[SMo_{12}O_{40}]$ at a 1.5 mm radius glassy carbon electrode in CH_3CN (a) no added supporting electrolyte with a scan rate of 20 mV s^{-1}; (b) in the presence of 0.1 M $(Bu_4)NClO_4$ at indicated scan rates. Reproduced by courtesy: *Electroanalysis* **13**, 1475 (2001). Copyright, Wiley-VCH Verlag GmbH.

3   CASE STUDY II: CYCLIC VOLTAMMETRIC STUDIES IN WATER ON THE QUASI-REVERSIBLE $[Fe(CN)_6]^{3-/4-}$ PROCESS IN THE PRESENCE AND ABSENCE OF ADDED SUPPORTING ELECTROLYTE

The quasi-reversible voltammetric processes (19) and (20)

$$[Fe^{III}(CN)_6]^{3-} + e^- \rightleftharpoons [Fe^{II}(CN)_6]^{4-} \tag{19}$$

$$[Fe^{II}(CN)_6]^{4-} \rightleftharpoons [Fe^{III}(CN)_6]^{3-} + e^- \tag{20}$$

have been studied in aqueous media almost since the dawn of the era of modern electrochemistry [2] because of the ready availability of highly stable $K_3[Fe^{III}(CN)_6]$

and $K_4[Fe^{II}(CN)_6]$ and related salts. Inherently, it may be anticipated that these processes should be simple one electron charge transfer process, in which case the voltammetric studies involving reduction (Equation 19) and oxidation (Equation 20) of highly charged species should be expected to be ideal with or without added supporting electrolyte. Data reported below are adapted from studies in references 25 and 26, and reveal significant additional complexity relative to the voltammetry associated with the reversible reduction of $[S_2Mo_{18}O_{62}]^{4-}$.

3.1   Cyclic Voltammetry at High $[Fe(CN)_6]^{3-}$ and $[Fe(CN)_6]^{4-}$ Concentrations

Figure 22 shows cyclic voltammograms obtained at a 1.46 mm radius glassy carbon electrode for the reduction of 50 mM $[Fe(CN)_6]^{3-}$ (as the potassium salt) in water in the absence of added supporting electrolyte over a wide range of scan rates. For this high concentration of $[Fe(CN)_6]^{3-}$ situation, generally excellent agreement between theory [26] and experiment is achieved at intermediate scan rates in the range of about 1 V s^{-1} to 100 mV s^{-1}. In this case, the simulation is based on a quasi-reversible electron transfer rather than a reversible process, as considered in Case Study I. At a scan rate of 5 V s^{-1} the IR_u term becomes very large and in the region of the peaks, agreement of simulated and experimental voltammograms become marginally less satisfactory. The influence of natural convection again can clearly be seen at low scan rates where the experimental current exceeds the simulated current at potentials following the peaks.

Cyclic voltammograms for the reduction of 50 mM $[Fe(CN)_6]^{3-}$ at scan rates between 5 V s^{-1} and 20 mV s^{-1} in the presence 1 M KCl as the added supporting electrolyte are presented in Figure 23. In all cases, except at very slow scan rates of 20 mV s^{-1}, where natural convection effects remain important, agreement between theory and experiment is excellent, suggesting that the problems observed at very high scan rates in the absence of added supporting electrolyte are associated with imperfect consideration of the large IR_u drop.

Although a conventional reference electrode system can result in leakage of ions that can have a significant effect on electrochemical studies undertaken in the absence of deliberately added supporting electrolyte (Section 2.1), the use of Ag/AgCl (saturated KCl) reference electrode was found to have a negligible effect [26] on voltammetric measurements undertaken with high concentrations (≥ 50 mM) of $K_3[Fe(CN)_6]$ and $K_4[Fe(CN)_6]$ in the absence of added supporting electrolyte. For this reason, a Ag/AgCl rather than a Pt quasi- reference electrode was used [26] when high concentrations (50 mM) of electroactive species were being studied. In contrast, for low concentration studies [25], a platinum quasi-reference electrode system was employed to study the processes described by equation 19 and 20.

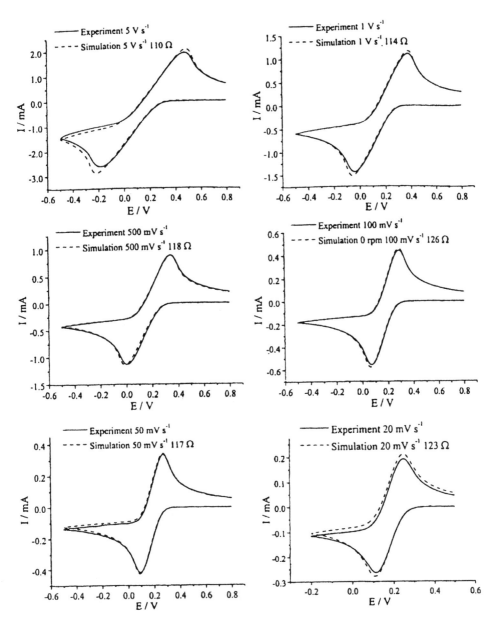

Figure 22. Comparison of simulated (—) and experimental (····) voltammograms obtained at a 1.46 mm diameter glassy carbon macrodisc electrode for the quasi-reversible reduction of 50 mM $[Fe(CN)_6]^{3-}$ without added supporting electrolyte using the designated scan rates and R_u values. Simulation parameters are defined in the text. Reproduced by courtesy: *J. Phys. Chem. A* **105**, 9085 (2001). Copyright, American Chemical Society.

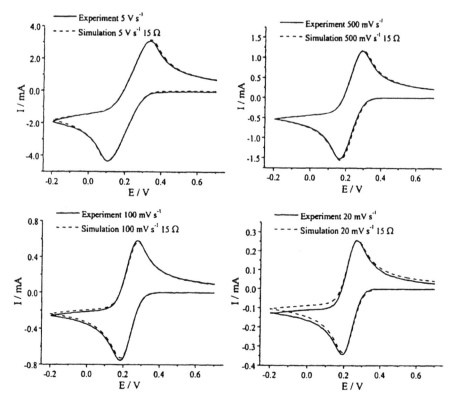

Figure 23. Comparison of simulated (—) and experimental (····) voltammograms obtained at a 1.46 mm radius glassy carbon electrode with designated scan rates and R_u values for the quasi-reversible reduction of 50 mM $[Fe(CN)_6]^{3-}$ in the presence of 1 M KCl as added supporting electrolyte. Simulation parameters are defined in the text. Reproduced by courtesy: *J. Phys. Chem. A* **105**, 9085 (2001). Copyright, American Chemical Society.

In the simulations, presented for the reduction of 50 mM $K_3[Fe(CN)_6]$ in the absence (Figure 22) and presence (Figure 23) of 1 M KCl added supporting electrolyte the parameter used were: $D = 7.6 \times 10^{-6}$ for $[Fe(CN)_6]^{3-}$, $D = 6.3 \times 10^{-6}$ for $[Fe(CN)_6]^{4-}$, and for both K^+ and Cl^-, $D = 2 \times 10^{-5}$. For the electron transfer process, values for the charge-transfer coefficient (α) of 0.5 and heterogeneous charge-transfer rate constant (k^0) of 0.04 cm s^{-1} were used. No attempts to provide small improvements in particular theory-experiment comparisons by varying these parameters were undertaken, as the fit, in the global sense, over a wide range of conditions was excellent. As described in reference 25, simulated and experimental cyclic voltammograms obtained for oxidation of 50 mM $K_4[Fe(CN)_6]^{3-}$ also show excellent agreement (Figure 24).

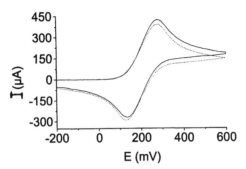

Figure 24. Comparison of experimental (—) and simulated (····) cyclic voltammograms obtained at a 1.5 mm radius glassy carbon electrode and using a scan rate of 20 mV s^{-1} for the oxidation of 50 mM K$_4$[Fe(CN)$_6$] in water in the absence of added supporting electrolyte. Reproduced by courtesy: *Anal. Chem.* **72**, 3486 (2000). Copyright, American Chemical Society.

3.2   Cyclic Voltammetry in the Normal Concentration Range

The almost ideal response obtained at (voltammetrically) very high 50 mM concentrations is deceptively simple. When operating with the usual mM or less, concentration range then far from ideal cyclic voltammograms are detected for both reduction of [Fe(CN)$_6$]$^{3-}$ and oxidation of [Fe(CN)$_6$]$^{4-}$ in the absence

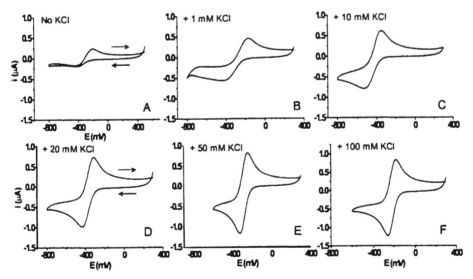

Figure 25. Cyclic voltammograms obtained at a 0.9 mm radius glassy carbon electrode with a scan rate of 2.0 mV s^{-1} for the reduction of 0.5 mM [Fe(CN)$_6$]$^{3-}$ in the absence and in the presence of incremental concentrations of KCl as the added supporting electrolyte. Provided by courtesy of M.B. Rooney.

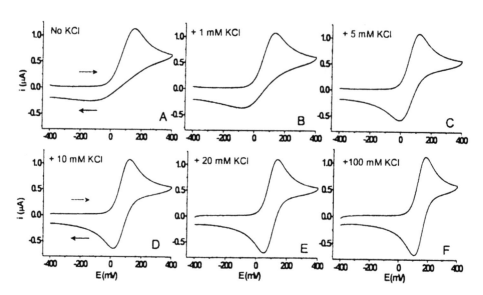

Figure 26. Cyclic voltammograms obtained at a 0.9 mm radius glassy carbon electrode for oxidation of 0.5 mM $[Fe(CN)_6]^{4-}$ in the absence and in the presence of incremental concentrations of KCl as the added supporting electrolyte. Provided by courtesy of M.B. Rooney.

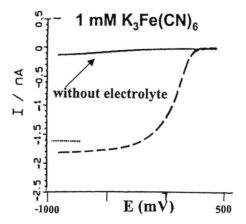

Figure 27. Near steady-state voltammogram obtained at a carbon fibre microdisc electrode for reduction of 1 mM $K_3[Fe(CN)_6]$ in the presence (····) and absence (——) of added supporting electrolyte. The theoretical limiting current obtained in the absence of supporting electrolyte is represented by (····). The potential axis is mV vs a saturated calomel reference electrode. Reproduced by courtesy: *J. Electroanal. Chem.* **323**, 381 (1992). Copyright, Elsevier.

or even in the presence of small KCl electrolyte concentrations (Figures 25 and 26). Analogously anomalous behaviour was reported in polarographic studies at the dropping mercury electrode [2] where the limiting current in the negative potential region for reduction of mM $[Fe(CN)_6]^{3-}$ in the absence of KCl electrolyte is far lower than predicted on the basis of diffusion/migration theory. Furthermore, data reported by Lee and Anson [27] (Figure 27) reveal that at a carbon fibre microdisc electrode, no reduction current is detected in the absence of electrolyte, and in contrast to theoretical predictions based on diffusion/migration theory [8-10]. Clearly, the concentration of $[Fe(CN)_6]^{3-}$ and $[Fe(CN)_6]^{4-}$ used is crucial in the absence of added supporting electrolyte.

3.2.1 Cyclic Voltammetry at Macrodisc Electrodes: 5 and 0.5 mM $K_4[Fe(CN)_6]$

In principle, the shapes of cyclic voltammograms should be independent of the concentration of $[Fe(CN)_6)]^{4-}$ with respect to the influence of IR_u drop [14]. This follows because, on decreasing the concentration, the effects of increased resistance and decreased current magnitudes should cancel each other with respect to the IR_u term, assuming the absence of ion-pairing. In accordance with this concept, the shapes of cyclic voltammograms obtained with 5 mM solutions of $[Fe(CN)_6]^{4-}$ (not shown) are similar to the 50 mM case shown in Figure 24, with both oxidative and reductive processes being well-defined. However, as the concentration of $K_4[Fe(CN)_6]$ is lowered to 0.5 mM, far more irregular voltammetry is observed at all electrode materials (glassy carbon, gold, platinum) considered in reference 24 (Figure 28). For glassy carbon and gold electrodes, the oxidative component is well-defined, but clearly broadened relative to that obtained when KCl supporting electrolyte is present. More obviously, the reductive component is virtually eliminated. The voltammetric behaviour at the platinum electrode is even more complex.

The non-idealities observed when utilizing low $[Fe(CN)_6]^{4-}$ concentrations in the absence of added electrolyte have been attributed to a number of factors. Surface adsorption and blockage by the oxidation product, ferricyanide, or a reaction product was proposed by Beriet and Pletcher [28] to explain the complex steady-state voltammetric behavior observed for low mM concentrations of $K_4[Fe(CN)_6]$ at a platinum microdisc electrode.

A significant contribution to the irregular electrochemistry observed at low concentrations is most likely a diffuse double layer effect (see Section 4.1), as suggested by Lee and Anson [27]. The inhibited reduction current may very well occur at potentials negative of the point of zero charge, which give rise to repulsion of the negatively charged ferricyanide anion at the electrode surface in

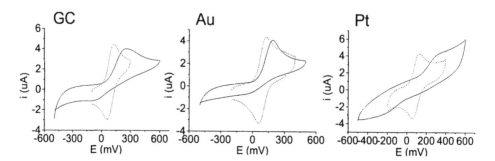

Figure 28. Cyclic voltammograms obtained at a scan rate of 20 mV s^{-1} with 0.5 mM K$_4$[Fe(CN)$_6$] in aqueous solution using 1.5 mm radius carbon, gold and platinum macrodisc electrode. [(—) no added supporting electrolyte, (····) 1 M KCl supporting electrolyte]. Reproduced by courtesy: *Anal. Chem.* **72**, 3486 (2000). Copyright, American Chemical Society.

this potential region. In the presence of added supporting electrolyte, screening provided by the modified compact double layer should avoid this effect. It follows that the well-defined voltammetry of high K$_4$[Fe(CN)$_6$] concentrations in the absence of KCl may also occur because of the increased availability of charge carriers and the resulting attainment of a relatively well-defined compact double layer.

As noted above, the electrochemistry of 0.5 mM K$_4$[Fe(CN)$_6$] in the absence of electrolyte at the platinum macroelectrode is qualitatively different from that at glassy carbon and gold electrodes. On platinum, both the oxidative and reductive components of cyclic voltammograms are broad, so that the value of ΔE_p is significantly greater than that observed with the glassy carbon or gold surfaces. Thus, features associated with both a decrease in the rate of electron transfer kinetics (in a diffuse double layer) and electrode blockage are indicated. However, while exhibiting some agreement with the steady-state results obtained by Beriet and Pletcher [28], transient voltammetric responses obtained at the platinum macrodisc electrode [25] are still significantly better defined than under near steady-state conditions at a platinum microdisc electrode.

3.2.2 Cyclic Voltammetry at Macrodisc Electrodes: 5 and 0.5 mM K$_3$[Fe(CN)$_6$]

As with K$_4$[Fe(CN)$_6$], the cyclic voltammograms obtained for K$_3$[Fe(CN)$_6$] in the absence of added electrolyte reveal apparently anomalous behavior when present at low concentrations. Data for 5 mM K$_3$[Fe(CN)$_6$] is qualitatively similar to that obtained for 5 mM K$_4$[Fe(CN)$_6$], but significantly different behaviour [25] is found at 0.5 mM for the two oxidation states as shown by

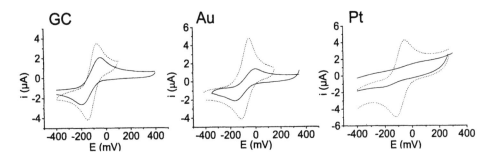

Figure 29. Cyclic voltammograms obtained at a scan rate of 20 mV s^{-1} with 0.5 mM K$_3$[Fe(CN)$_6$] in aqueous solution using 1.5 mm radius carbon, gold and platinum macrodisc electrode. [(—) no added supporting electrolyte, (····) 1 M KCl supporting electrolyte]. Reproduced by courtesy: *Anal. Chem.* **72**, 3486 (2000). Copyright, American Chemical Society.

comparing voltammograms displayed in Figures 28 and 29. The results at the glassy carbon macroelectrode [25] are in contrast to those reported by Lee and Anson at carbon fibre microelectrodes, where essentially no reduction wave was detected (Figure 27) at low concentrations of K$_3$[Fe(CN)$_6$].

Consistent with results obtained for K$_4$[Fe(CN)$_6$] oxidation, cyclic voltammograms for 0.5 mM K$_3$[Fe(CN)$_6$] reduction at the Pt electrode are most detrimentally affected by the absence of excess supporting electrolyte. In contrast, at all 3 electrode surfaces, well-defined cyclic voltammograms are observed when 1 M KCl is the added supporting electrolyte. Voltammetric results at Pt macrodisc electrodes parallel those of Beriet and Pletcher, where no reduction wave was observed at a platinum microelectrode for K$_3$[Fe(CN)$_6$] concentrations <0.5 mM in the absence of added inert electrolyte [28]. Surface blockage/passivation effects at platinum electrode surfaces, appears to be far more significant at platinum than at gold and glassy carbon electrode surfaces. (For a detailed discussion of Pt surface blockage by the [Fe(CN)$_6$]$^{3-/4-}$ redox couple, see references 28 and 29).

3.2.3 Near Steady-State Voltammetry at Microdisc Electrodes as a Function of Concentration

Almost ideal voltammetric data also are obtained with high 50 mM concentrations of [Fe(CN)$_6$]$^{4-}$ and [Fe(CN)$_6$]$^{3-}$ under microelectrode steady-state conditions where an essentially normal double layer conditions is available. The oxidation of 50 mM K$_4$[Fe(CN)$_6$] (Figure 30) and the reduction of 50 mM K$_3$[Fe(CN)$_6$] (Figure 31) at microdisc electrodes (radius = 5-6 μm)

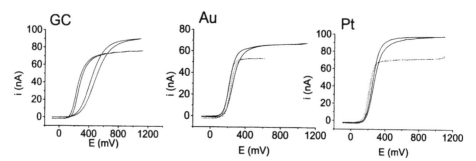

Figure 30. Steady-state voltammogram for oxidation of 50 mM $K_4[Fe(CN)_6]$ glassy carbon (radius = 6 μm), gold (radius = 10 μm) and platinum (radius = 5 μm) microdisc electrodes in the absence (—) and presence (····) of 1 M KCl added supporting electrolyte. Provided by courtesy of D.C. Coomber.

in the absence of added supporting electrolyte has been investigated at carbon, gold and platinum macrodisc electrode [25]. For the oxidative process, the steady-state currents at the 3 electrode surfaces are reported [25] to be 1.25 ± 0.03 times greater than that obtained in the presence of electrolyte [25], a value slightly larger than that predicted by diffusion plus migration theory (ratio of limiting currents in absence of added electrolyte (I_{lim}) and with inert electrolyte (I_d), or $I_{lim}/I_d = 1.2$) [10,30]. For the reductive process, the limiting current [25] was reduced by 10 % ($I_{lim}/I_d = 0.9$) in the absence of KCl, a value consistent with that predicted by diffusion-migration theory [10,30]. For all three electrode materials, the $[Fe(CN)_6]^{3-}$ reductive current was virtually eliminated for 0.5 mM $K_3Fe(CN)_6$ in the absence of added electrolyte (shown on carbon for $K_3([Fe(CN)_6]$ in Figure 31, consistent with the results of Lee and Anson [27] and Beriet and Pletcher [28].

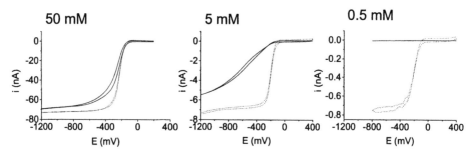

Figure 31. Near steady-state voltammogram for reduction of 50 mM $K_3[Fe(CN)_6]$ at a 6 μm radius glassy carbon microdisc electrode in the absence (—) and presence (····) of 1 M KCl added supporting electrolyte. Reproduced by courtesy: *Anal. Chem.* **72**, 3486 (2000). Copyright, American Chemical Society.

4   THE DIFFUSE DOUBLE LAYER EFFECT

Figure 32 schematically illustrates the double-layer that is present in a voltammetric experiment at a negatively charged working electrode. The double-layer develops when a working electrode is placed in contact with a solution containing ions, present either as charged redox active species such as $[Fe(CN)_6]^3$, and their counterions, added supporting electrolyte (e.g. KCl), charged species present in the solvent (e.g. H^+ or OH^-), and/or charged contaminants. In order to balance the charge on the working electrode, two capacitive components form between the electrode and the bulk solution. The first, C_i, is due to specifically and nonspecifically adsorbed species within the Outer Helmholz Plane (OHP), and is relatively unaffected by the concentration of added supporting electrolyte. The second, C_{diff}, results from charge carriers present in the diffusion layer at distances beyond the OHP, which is assumed to be the plane of closest approach of the electroactive species. Thus, the total potential applied to the working electrode surface (V_T) is balanced via two potential drops in series, ϕ_1 immediately adjacent to the electrode surface, and ϕ_2 which extends through the diffusion layer. These potentials are related to the capacitances, C_i and C_{diff} by the equations in Figure 32. Ordinarily, when added supporting electrolyte is present in excess, the concentration of inert charge carriers in the diffusion layer is very

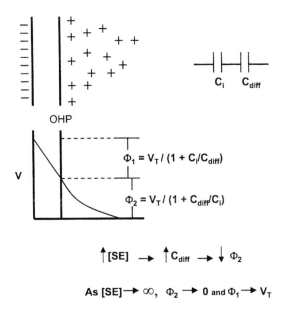

$$\Phi_1 = V_T / (1 + C_i/C_{diff})$$

$$\Phi_2 = V_T / (1 + C_{diff}/C_i)$$

$$\uparrow [SE] \;\rightarrow\; \uparrow C_{diff} \;\rightarrow\; \downarrow \Phi_2$$

$$\text{As } [SE] \rightarrow \infty, \; \Phi_2 \rightarrow 0 \text{ and } \Phi_1 \rightarrow V_T$$

Figure 32. Schematic diagram of the electrochemical double layer. Provided by courtesy of M.B. Rooney.

large, so that C_{diff} is also very large, resulting in a negligible value of ϕ_2. Thus, the entire voltage drop in solution occurs via ϕ_1, within the OHP, so that charged electroactive species "feel" no potential changes within the diffuse double-layer, and mass transport occurs via diffusion and migration alone. However, when the supporting electrolyte concentration is very dilute (i.e. in situations of low/no added supporting electrolyte and low concentrations of charged electroactive species/counterion), the concentration of charge carriers in solution is very low. Under these diffuse double layer conditions, C_{diff} becomes very small, resulting in the presence of a significantly enhanced value of ϕ_2. In this case, there is a potential gradient within the diffuse double-layer, and charged electroactive analyte species are attracted or repelled, depending on their charge, by this potential gradient. Thus, mass transport is no longer necessarily determined by diffusion and migration alone, but is now subject to diffuse double-layer effects. If, in the case depicted in Figure 32, a negatively charged redox active species would be repelled from the electrode surface, resulting in inhibition of the reductive current detected at the working electrode. Further details of the diffuse layer are available in references 2, 12 and 31.

4.1   The Frumkin Double Layer Effect

It follows from the above discussion that the double layer potential at an electrode/solution interface can be considered as the sum of two potentials: ϕ_1, the potential drop across the inner layer and ϕ_2, the potential drop across the diffuse layer. Thus

$$E - E_{\text{pzc}} = \phi_1 + \phi_2 = \frac{\sigma_M}{C_1} + \phi_2 \tag{21}$$

where E is the applied interfacial potential, E_{pzc} is the potential of zero charge, C_1 is the capacitance of the inner layer (the region between the electrode surface and the plane of closest approach (PCA) of solution ions) and σ_M is the charge on the metal. The Frumkin effect [12,31] arises partly because, c_j^{PCA}, the concentration of a charged species (including the redox active species, at least one of which *must* be charged) at the plane of closest approach will differ from c_j^δ, the concentration just outside the diffuse layer, according to the Boltzmann expression:

$$\frac{c_j^{\text{PCA}}}{c_j^\delta} = \exp\left[-\frac{F}{RT}\phi_2 z_j\right] \tag{22}$$

where z_j is the charge on the jth species. A fundamental assumption is that only the redox species at the PCA can participate in the electron transfer process.

Under the usual voltammetric conditions with excess added supporting electrolyte, the current for a one electron charge transfer process

$$A + e^- \rightleftharpoons B \tag{23}$$

is described by the Butler-Volmer relationship (Equation 24)

$$-I = AFk^0 \left(c_{A,x=0} \exp\left[-\alpha(F/RT)(E-E_f^0)\right] - c_{B,x=0} \exp\left[(1-\alpha)(F/RT)(E-E_0^f)\right]\right) \tag{24}$$

where k^0 is the heterogeneous charge transfer rate constant at E_0^f, α is the charge transfer coefficient, and $c_{A,\,x=0}$ and $c_{B,\,x=0}$ are the surface concentrations of the oxidized (A) and reduced (B) species, respectively.

However, where the diffuse double-layer model must be invoked, the potential difference $(E-E_f^0,)$ is replaced by $(E-E_f^0,-\phi_2)$ in the Butler-Volmer relationship (Equation 24) and the redox concentrations at distance x = 0, which correspond to the surface concentrations of the redox active components in the presence of excess added supporting electrolyte, are replaced by the concentrations at the OHP, which are a function of ϕ_2 (Equation 22). Once these substitutions are made, the resulting equation is the operative Butler-Volmer expression (written to be consistent with IUPAC convention):

$$-I = AFk_{eff}^0 \left(c_A^\delta\left[-\alpha(F/RT)(E-E_f^0)\right] - c_B^\delta \exp\left[(1-\alpha)(F/RT)(E-E_0^f)\right]\right) \tag{25}$$

where k_{eff}^0, the effective standard rate constant for electron transfer is defined [31] by:

$$k_{eff}^0 = k^0 \exp\left[\frac{F}{RT}\phi_2\left(\alpha - z_A\right)\right] \tag{26}$$

where k^0 is the true standard rate constant for the electron transfer. Thus, only the rate of electron transfer is altered by ϕ_2. The value of ϕ_2 can be determined from Equation 27, which was originally derived by Gouy [32,33]:

$$\sigma_M = \frac{E - E_{pzc}}{|E - E_{pzc}|}\left(2RT\varepsilon\varepsilon_0 \sum_j c_j^\delta \left(\exp\left[-\frac{F}{RT}\phi_2 z_j\right] - 1\right)\right)^{1/2} \tag{27}$$

where ε_0 (8.85 x 10^{-14} C^2 J^{-1} cm^{-1}) is the (permittivity of a vacuum), ε is the dielectric constant of the solvent (78 for water) and z_j is the charge on species j.

Equations 21 and 27 are combined to deduce a value of ϕ_2 for a given applied E and any combination of c_j^δ values. The value of ϕ_2, thus computed, is used to deduce the value of k_{eff} from Equation 22. The full simulation again may be executed using explicit finite difference routines based on the Nernst-Planck expression with the constraint of electroneutrality (Section 2.2).

The more familiar Gouy-Chapman expression used in double layer theory [12] represents a special case of Equation 27 that applies only for a symmetrical binary electrolyte.

In this case,

$$\phi_1 = \left(\frac{K}{C_i}\right)\sinh\left(\frac{F\phi_2}{2RT}\right) \tag{28}$$

where $K = (8RT\varepsilon\varepsilon_0 c_Z)^{1/2}$ (29)

and C_Z is the concentration of the binary supporting electrolyte, Z. Combining Equations 21 and 28 gives the non-linear equation

$$\left(\frac{K}{C_i}\right)\sinh\left(\frac{F\phi_2}{2RT}\right) + \phi_2 = E - E_f^0 - E_{pzc} \tag{30}$$

which can be solved for ϕ_2 using a simple bisection search algorithm.

For simplicity, the theoretical simulations presented below are based on the simpler Guoy-Chapman model which are necessarily undertaken with an added binary electrolyte present and migration or IR_u drop are not incorporated. Generally speaking, there are three distinct cases of diffuse-double layer effects, which depend on the variance of $E_f^0 - E_{PZC}$, and which lead to vastly different voltammetric behavior: (1) $E_f^0 \gg E_{PZC}$, (2) $E_f^0 = E_{PZC}$, and (3) $E_f^0 \ll E_{PZC}$.

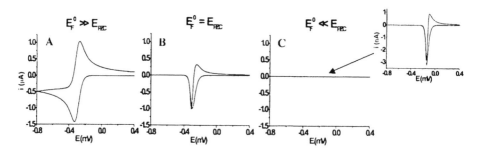

Figure 33. Simulation of the Frumkin effect for the one-electron reduction of a 3-charged anion (e.g. the $K_3[Fe(CN)_6]$ case) as a function of $E_f^0 - E_{PZC}$: electrode radius = 0.09 cm, v = 20 mV s^{-1}, concentration of electroactive species = 0.5 mM, z_A = -3 (counterion z = +1), added supporting electrolyte (e.g. KCl) concentration = 1.0 mM ($z = \pm 1$), E_f^0 = 0.3 V, k^0 = 0.01 cm s^{-1}, D = 7.6 × 10^{-6} cm^2 s^{-1}, α = 0.5. **A:** $E_f^0 > E_{PZC}$ ($E_f^0 - E_{PZC}$ = 0.5 V). **B:** $E_f^0 = E_{PZC}$. **C:** $E_f^0 < E_{PZC}$ ($E_f^0 - E_{PZC}$ = 0.2 V). Provided by courtesy of D.J. Gavaghan.

Figure 33 displays the diffuse double-layer simulations obtained in these three cases under conditions of transient cyclic voltammetry for one electron reduction of a 0.5 mM solution of redox active species such as $K_3[Fe(CN)_6]$ in water, with $z_B = -3$ (charge on the counterion, $z = +1$ e.g. K^+), $k^0 = 0.01$ cm s^{-1}, $v = 20$ mV s^{-1}, and D = 7.6×10^{-6} cm^2 s^{-1}. Because binary supporting electrolyte such as KCl must be present to employ the Gouy Chapman theory, the added supporting electrolyte concentration was taken to be 1.0 mM (electroactive cation = counterion, and charge on the counterion, $z = -1$ e.g. Cl$^-$). In the case of anion reduction and when $E_f^0 \gg E_{PZC}$ (ϕ_2 is positive, Figure 33a), there is little effect of the diffuse double-layer on a quasi-reversible electron transfer process, as expected. In the case of $E_f^0 = E_{PZC}$ (Figure 33b), the voltammogram proceeds as expected for a quasi-reversible process at potentials more positive than E_{PZC}. In contrast, at potentials more negative than E_{PZC}, the reduction current is inhibited as a result of repulsion within the diffuse double-layer. When the scan direction is reversed, the reduction current is resumed at potentials positive of E_{PZC}, resulting in an overlap of the forward and reverse reductive currents, whilst at potentials beyond this point, quasi-reversible behavior is resumed. The most extreme case is when $E_f^0 \ll E_{PZC}$, in which case the reductive current is dramatically reduced by a factor of 1000 (note different current scale of inset to Figure 33c). The current response reveals that the *shape* of the voltammogram is identical to the case where $E_f^0 = E_{PZC}$ (Figure 33b), but the reduction current is "turned off" far before E_f^0 is reached, resulting in the presence of a very small current signal.

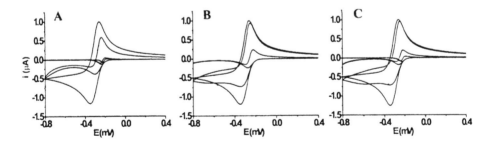

Figure 34. Simulations of the reduction of 0.5 mM solution of a 3– charged anion (e.g. the $K_3[Fe(CN)_6]$ case) with incremental additions of added supporting electrolyte (e.g. 1:1 electrolyte such as KCl). $E_f^0 - E_{PZC} = -0.1$ V; electrode radius = 0.09 cm; $v = 20$ mV s^{-1}; $E_f^0 = -0.3$ V; $k^0 = 0.01$ cm s^{-1}; D = 7.6×10^{-6} cm s^{-1}, $\alpha = 0.5$. **A:** $z_A = -3$; concentration of added supporting electrolyte = 1, 10, 100, 1000 mM, in order of increasing current. **B:** $z_A = -3$; concentration of added supporting electrolyte = 50, 250 mM, in order of increasing current. **C:** $z_A = -2$; concentration of added supporting electrolyte = 1, 10, 100, 1000 mM in order of increasing current. Provided by courtesy of D.J. Gavaghan.

Figures 34a and 34b show the theoretical effects of increasing the added 1:1 (e.g. KCl) supporting electrolyte concentration under transient cyclic voltammetric conditions when $E_f^0 - E_{PZC}$ is – 0.1 V. Under these conditions, the simulated data obtained in the presence of added supporting electrolyte concentrations ≤ 10 mM (Figures 34a and 34b) bear some but certainly not perfect resemblance to experimental data obtained for $[Fe(CN)_6]^{3-}$ reduction in the absence of added supporting electrolyte (Figure 35), while concentrations of added supporting electrolyte higher than those employed experimentally are required for the simulation to achieve voltammograms resembling experimental data obtained in the presence of higher concentrations added supporting electrolyte concentrations (Figure 25).

The double-layer effect is essentially removed in both experimental and simulated data when the added supporting electrolyte concentration greatly exceeds the electroactive analyte concentration, as is the case when the concentration of a binary electrolyte such as KCl is 1M. Under these conditions, and with $v = 20$ mV s^{-1} and $k^0 = 0.01$ cm s^{-1}, the $[Fe(CN)_6]^{3-/4-}$ process is close to diffusion controlled (reversible). Thus, whilst the simulated and experimental voltammograms have some features in common, as observed by Frumkin in his studies on mercury electrodes [2,34], the theoretical variation with added supporting electrolyte concentration is diminished relative to that observed experimentally. The importance of having a high concentration of electrolyte and hence avoiding a diffuse double layer also is demonstrated clearly by studies with 50 mM $K_3[Fe(CN)_6]$, where nearly ideally behaved reversible voltammograms

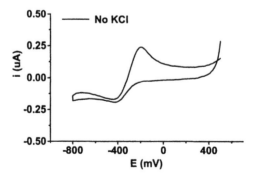

Figure 35. Cyclic voltammograms obtained at a scan rate of 20 mV s^{-1} at a 0.9 mm radius glassy carbon macrodisc electrode for the reduction of 0.5 mM $K_3[Fe(CN)_6]$ in water without added supporting electrolyte. Potential vs Pt quasi-reference electrode. Provided by courtesy of M.B. Rooney.

(Figure 23) are achieved in the absence of added supporting electrolyte [25,26].

The effects of decreasing the charge number of the species being reduced from $z_B = -3$, to $z_B = -2$ and hence decreasing repulsion in the double-layer is shown in Figure 34c. As suggested by Lee and Anson [27] it is quite likely that the effective charge of $[Fe(CN)_6]^{3-}$ is lowered from 3– towards 2– as a result of ion pairing with K^+. Heyrovsky [35-37] considered the formation of ion pairs to be important in bulk solution even in the absence of added supporting electrolyte, and Frumkin presumed their presence to be significant in his studies in double-layer effect [34,35,38]. It has long been realized that both $[Fe(CN)_6]^{3-}$ and $[Fe(CN)_6]^{4-}$ pair significantly with K^+, even in the most dilute aqueous solutions [39], resulting in a $[K^+]$-dependent redox potential [40] and a heterogeneous electron transfer rate constant (k^0) with a first-order dependence on potassium cation concentration [41,42]. Therefore, ion-pairing may contribute significantly to experimental, but not simulated results.

Significantly, distinctly different cyclic voltammograms have been observed for the oxidation of 0.5 mM $[Fe(CN)_6]^{4-}$ (Figure 26), relative to reduction of 0.5 mM $[Fe(CN)_6]^{3-}$ (Figure 25). Although somewhat broadened as a result of IR_u drop, a well-defined $[Fe(CN)_6]^{4-} \rightleftharpoons [Fe(CN)_6]^{3-} + e^-$ oxidation process is observed even in the absence of added supporting electrolyte (Figure 26a). However, almost complete elimination of the corresponding reductive wave is observed on reversing the potential scan direction, a feature that cannot be attributed to migration and IR_u alone. The expected reductive component of the

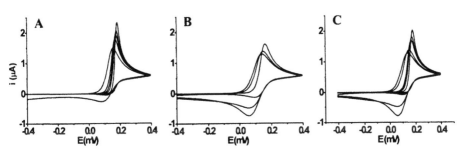

Figure 36. Simulations of the oxidation of 0.5 mM solution of a 4– anion (e.g. the $K_4[Fe(CN)_6]$ case) with incremental additions of added supporting electrolyte (e.g. 1:1 electrolyte such as KCl). $E_f^0 - E_{PZC} = -0.1$ V; $r = 0.09$ cm; $v = 20$ mV s^{-1}; $E_f^0 = 0.1$ V; $k^0 = 0.01$ cm s^{-1}; $D = 6.3 \times 10^{-6}$ cm s^{-1}, $\alpha = 0.5$ **A**: $z_B = -4$; concentration of added supporting electrolyte = 1, 5, 10, 20, 100 mM, in order of increasing reductive current. **C**: $z_{red} = -3$; concentration of added supporting electrolyte = 1, 5, 10, 20, 100, 1000 mM, in order of increasing reductive current. Provided by courtesy of D.J. Gavaghan.

cyclic voltammogram emerges as the concentration of KCl is increased (Figures 26b-f), although the initial oxidative wave is little affected except for a sharpening due to the lowering of solution resistance resulting from increased ionic strength. Consequently, these data support the concept that the reduction inhibition is a direct consequence of the combination of a diffuse double-layer in the absence of added supporting electrolyte and the fact that the $[Fe(CN)_6]^{4-/3-}$ process occurs at potentials negative of the PZC when a glassy carbon working electrode is used.

Simulations of cyclic voltammogram for oxidation of a species with a 4– charge in the presence of 1:1 electrolyte are shown in Figure 36 using a value of -0.1 V for $E_f^0 - E_{PZC}$. These simulations could approximately mimic the cyclic voltammetric oxidation of $K_4[Fe(CN)_6]$ with increasing KCl concentration because they reveal the same drastic inhibition of the reductive wave seen experimentally at low KCl concentrations and the disinhibition (return) of the reductive current with addition of higher added supporting electrolyte concentrations. The oxidative wave is sharper (exhibits higher peak current) with low concentrations of KCl, as potentials positive of the PZC effectively increase the electron transfer rate. Experimentally, this effect is opposed by the broadening effects of IR_u, for which the simulation does not account. However, again the simulated current values are far more diminished than the experimental current values and better agreement with the experimental data is obtained when higher added supporting electrolyte concentrations (> 50 times experimental values) and/or lower charge numbers ($z_A = -3$ or -2) are used in the simulation (Figure 36b and 36c).

5   METHODS OF AVOIDANCE OR MINIMIZATION OF FRUMKIN DIFFUSE DOUBLE LAYER EFFECTS

The above theory versus experiment comparison implies that only qualitative comparison between our simulated and experimental data under dilute electrolyte and hence diffuse double-layer conditions is appropriate. Quantitative comparison is impractical as a result of numerous phenomena neglected in the simulations, presented above. Thus, although the significant effects of migration and IR_u and also use of the more accurate Guoy equations could readily be included in the simulation, allowance must be made for: concentration polarization [35,43,44]. Local rather than mean values of ϕ_2 [35]; lack of extended electron transfer ("electron tunelling") [45] surface adsorption of electroactive species; and specific adsorption of the added supporting electrolyte ions. Other parameters which are dependent on added supporting electrolyte composition and concentration, and which are difficult to quantify, include diffusion coefficients (D), the charge trans-fer coefficient (α), E_f^0, and E_{PZC} [2]. Furthermore, ϕ_2 depends on the specific capacity (C') of the electrode, which is assumed to be constant but which, in actuality, depends somewhat on potential and solution species' concentration [35,45]. Thus, the number of variables changing in complicated and, in some cases, unpredictable manners, are so numerous that quantitative simulation of the voltammetry of low concentrations of redox active species in the absence of added supporting electrolyte is difficult to the point of being impractical. Nonetheless, the simulations presented, provided persuasive qualitative evidence that voltammetric experiments on quasi-reversible processes performed under dilute electrolyte conditions may show significant effects of the diffuse double-layer.

Because of the inherent difficulty of achieving quantitative simulations, which incorporate all aspects of the diffuse double-layer effect, it is desirable to ascertain experimental conditions in which the effect is negligible, so that only migration, diffusion and IR_u (and, in relevant cases, convection) need be considered. Clearly, one strategy is to restrict quantitative studies to systems that are decidedly reversible. Under this circumstance k^0 is very large, so the diffuse double-layer effect will be negligible (Equation 26). Calculation of reversible potentials and diffusion coefficients should be tractable under these circumstances. Alternatively, experiments can be conducted in the absence of added supporting electrolyte by utilizing higher than usual concentrations (i.e. 50 mM) of charged redox active analyte, such that a well-defined double-layer is maintained by the electroactive analyte, itself. However, the actual concentration of electroactive species that is adequate to minimize the diffuse double-layer effect probably will need to be determined on a case-by-case basis.

The concepts presented in this work indicate that mM or lower concentrations of charged redox active analytes exhibiting quasi-reversible electron transfer behavior could be studied by cyclic voltammetry in the absence of added supporting electrolyte, while avoiding significant diffuse double-layer effects, if the compound employed has a reversible potential which is significantly removed from the PZC of the electrode material, provided the sign of $(E_f^0 - E_{PZC})$ is opposite that of the reactant and product charge (Figure 33a). Thus, if cyclic voltammetry of a negatively charged species is to be considered, then a combination of a relatively positive E_f^0 value and an electrode with a PZC as negative as possible would be required to minimize the double-layer effect. Such may be the case for the oxidation of $[SMo_{12}O_{40}]^{3-}$ (1mM) at a glassy carbon electrode in acetonitrile (Figure 22), or indeed reduction of $[SMo_{12}O_{40}]^{2-}$ (1mM) (Figure 20) [24].

The above discussion implies that it should be possible on some occasions to deliberately select an electrode material whose PZC is significantly removed (in the required direction) from the E_f^0 value of the redox couple of interest in order to minimize the diffuse double-layer effect. This would mean that, under diffuse double-layer conditions of low ionic strength, the variable PZC values of working electrode materials may result in a redox process that exhibits a significant degree of electrode dependence.

Chemical modification of electrode surfaces with suitably charged materials also may provide a possible route to minimizing electrode-kinetic barriers. Furthermore, it is noteworthy that neutral compounds should be unaffected by ϕ_2, although in this case there is the significant difficulty of achieving charge balance upon electrolysis, so that some added supporting electrolyte or another source of electrolyte (e.g. adventitious impurity) will in all probability be required.

5.1   Enhanced Sensitivity of Near Steady-State Voltammetry at a Microdisc Electrode to the Diffuse Double-Layer Effect.

The Frumkin effect can be attributed to the influence of an effective rate of electron transfer (k_{eff}^0) adjusted for the influence of ϕ_2. Because of their effectively shorter time domains [46,47], steady-state experiments at rotating disk electrodes and microelectrodes are more subject to kinetic effects and, thus, demonstrate greater departures from reversibility than the very slow scan rate (20 mV s^{-1}) cyclic voltammetric data obtained in transient experimental studies at macroelectrodes.

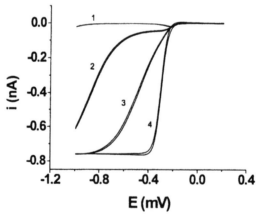

Figure 37. Simulations of the reduction of 0.5 mM K$_3$[Fe(CN)$_6$] with incremental additions of added supporting electrolyte concentration (e.g. a 1:1 electrolyte such as KCl). $E_f^0 - E_{PZC} = -0.1$ V; r = 0.5 μm; $v = 20$ mV s^{-1}; $E_f^0 = -0.3$ V; $z_{ox} = -3$; D = 7.6 × 10^{-6} cm s^{-1}, $\alpha = 0.5$. For curves 1-3: $k^0 = 0.01$ cm s^{-1}, concentration of added supporting electrolyte = 50 mM (curve 1), 250 mM (curve 2), and 1500 mM (curve 3), in order of increasing current. For curve 4, $k^0 \geq 1 \times 10^{-8}$ cm s^{-1}, concentration of added supporting electrolyte = 50 mM. Provided by courtesy of D.J. Govaghan

Thus, these steady-state techniques are expected to be even more severely affected by the diffuse double-layer (ϕ_2) present in solutions of low ionic strength. The theoretical curves shown in Figure 37 emphasize this feature. This expectation also is realized experimentally, in that the $[Fe(CN)_6]^{3-/4-}$ couple exhibits significantly more current inhibition in the absence of added supporting electrolyte at a microdisc electrode under steady-state conditions, than detected by transient voltammetry at stationary macrodisc electrodes under identical conditions (see references 25, 27, 28 and Figures 25-28 and 31). Also shown in Figure 37 (curve 4) is the simulation obtained with a very high rate of electron transfer. The voltammetry under this condition is reversible, even at low added supporting electrolyte concentrations, emphasizing that redox couples with sufficiently high rates of electron transfer will not exhibit detrimental diffuse double-layer effects.

On the basis of the above analysis it may be concluded therefore that whilst microdisc voltammetry under steady-state conditions minimizes the IR_u drop, slow scan rate transient voltammetric studies utilizing macrodisc electrodes appear to be slightly less prone to diffuse double-layer effects.

6   CONCLUSIONS

Comparison of theoretical and experimental data presented in this article reveal that studies in the absence of added electrolyte are tractable under the right conditions, but also treacherous under many circumstances. Qualitative comparison of theory based on Gouy-Chapman double-layer theory and Frumkin electron transfer models and experimental results implies that the current inhibition observed for the negatively charged $[Fe(CN)_6]^{3-/4-}$ processes in the absence of added supporting electrolyte is at least partially attributable to a diffuse double-layer effect. Cooper, Bond, and Oldham [10] also have reported a number of positively charged compounds that exhibit "anomalous" current inhibition, which is well beyond that attributable to migration, for studies at the ≤ 1 mM concentration level in the absence of added supporting electrolyte. These latter studies were conducted under near steady-state conditions at a platinum microdisc electrode. It may now be suggested that it is probably no coincidence that, of these anomalous cases, the oxidation processes of cations ($\{[Ru(bpy)_2]_2(\mu\text{-dpq})\}^{4+/6+}$; $Ru(COOEtbpy)_3^{2+/3+}$; $\{[Ru(bpy)_2]_2(\mu\text{-dpq})\}^{4+/5+}$; $[*Cp_3Ru_3Cl_3CH]^{1+/3+}$) exhibit very positive $E_{1/2}$ values (>1.3 V vs. Ag/AgCl), thereby in all probability requiring that they occur at a potential much more positive than the PZC, where a positive ϕ_2, and hence a positive charge at the Outer Helmholtz Plane, should repel the cationic species.

Since quantitative comparison with theory is impractical with the present state of knowledge, it is desirable to minimize the effect of the diffuse double-layer.

Suggested ways in which this may be accomplished, while still operating in the absence of added supporting electrolyte include:

- employing much higher than normal concentrations of charged redox active species (which may likely be the most generally reliable method, assuming compounds are sufficiently soluble).

- selecting a redox process/electrode material combination that results in an E_f^0 value that is far removed from the PZC, so that the charge associated with ϕ_2 is of opposite sign to the charge associated with the redox couple.

- use of redox couples exhibiting highly reversible electrochemical behavior (very large k^0 value).

- employment of slow scan rate cyclic voltammetry at macroelectrodes, rather than shorter time domain techniques such as near steady-state microdisc voltammetry and rotating macrodisc voltammetry.

Of course, these conclusions simply extend those that could be deduced from polarographic studies at a dropping mercury electrode half a century ago.

The advantages and disadvantages of working with and without added supporting electrolyte have been summarized in the introduction. Each electrochemical system must be evaluated in terms of these attributes, in order to decide whether voltammetric studies in the absence of added supporting electrolyte is worthwhile.

7   ACKNOWLEDGEMENTS

The content of this paper has extensively relied on concepts developed over the last decade during extensive discussions with Keith Oldham, Stephen Feldberg, Melissa Rooney, Darren Coomber, Nicholas Stevens and David Gavaghan. Particular thanks must be extended to these individuals for sharing of their insights and wisdom as well as generous provision of much of the experimental and theoretical data contained in Figures presented in the article. The author also acknowledges that much material has been adapted from collaborative publications with these individuals (see references 13, 14, 24-26 and 48 particularly). Finally, the Australian Research Council is acknowledged for generous financial support of our voltammetric studies in the absence of added supporting electrolyte as is Liza Verdan for technical assistance in preparation of this Chapter of the book.

REFERENCES

1. Heyrovský, J. *Chem. Listy*, *16*, 256 (1922).
2. Heyrovský, J.; Kůta, J. Principles of Pology, Academic Press, New York, 1966.
3. Slendyk, I. *Coll. Czeck. Chem. Commun.*, *3*, 385 (1931).
4. Ciskowska, M.; Stojek, Z. *Anal. Chem.*, *72*, 755A (2000) and references cited therein.
5. Oldham, K.B. Microelectrodes; Theory and Applications. Montenegro, M.I.; Queiros, M.A.; Daschbach, J.L. (eds) Kluwer Academic Publishers, Dordrecht, 1991, p 83.
6. Oldham, K.B.; Feldberg, S.W. *J. Phys. Chem. B.*, *103*, 1699 (1999).
7. Oldham, K.B. *J. Electroanal. Chem.*, *337*, 91 (1992).
8. Microelectrodes: Theory and Applications. Montenegro, M.I.; Queiros, M.A.; Daschbach, J.L. (eds) Kluwer Academic Publishers, Dordrecht, 1991.
9. Wightman, R.M.; Wipf, D.O. *Electroanal. Chem.*, *15*, 267 (1989).
10. Cooper, J.B.; Bond, A.M.; Oldham, K.B. *J. Electroanal. Chem.*, *33*, 887 (1992).
11. Bond, A.M. Broadening Electrochemical Horizons, Oxford University Press, 2002.
12. Bard, A.J.; Faulkner, L.R. Electrochemical Methods: Fundamentals and Applications, John Wiley, New York, 1st edition 1980, 2nd edition 2001.
13. Bond, A.M.; Coomber, D.C.; Feldberg, S.W.; Oldham, K.B.; Vu, T. *Anal. Chem.*, *73*, 352 (2001).
14. Bond, A.M.; Feldberg, S.W. *J. Phys. Chem. B.*, *102*, 9966 (1998).
15. Stevens, N.S.C.; Bond, A.M. *J. Electroanal. Chem.*, *538-539*, 25 (2002).
16. He, P.; Avery, J.P.; Faulkner, L.R. *Anal. Chem.*, *54*, 1313A (1982).
17. Hawkridge, F.M. in Laboratory Techniques in Electroanalytical Chemistry, Kissinger, P.T.; Heineman, W.R. (eds) Marcel Dekker, New York, 1984, Chapter 12.
18. Roe, D.K. in Laboratory Techniques in Electroanalytical Chemistry, Kissinger, P.T.; Heineman, W.R. (eds) Marcel Dekker, New York, 1986, Chapter 7.
19. Nicholson, R.S. *Anal. Chem.*, *37*, 667 (1965).
20. Nicholson, R.S. *Anal. Chem.*, *37*, 1351 (1965).
21. Nicholson, R.S.; Shain, I. *Anal. Chem.*, *36*, 706 (1964).
22. Rudolph, M.; Reddy, D.P.; Feldberg, S.W. *Anal. Chem.*, *66*, 589A (1994).
23. Way, D.M.; Bond, A.M.; Wedd, A.G. *Inorg. Chem.*, *36*, 2826 (1997).
24. Bond, A.M.; Coomber, D.C.; Harika, R.; Hultgren, V.M.; Rooney, M.B.; Vu, T.; Wedd, A.G. *Electroanalysis*, *13*, 1475 (2001).
25. Rooney, M.B.; Coomber, D.C.; Bond, A.M. *Anal. Chem.*, *72*, 3486 (2000).
26. Stevens, N.P.C., Rooney, M.B.; Bond, A.M.; Feldberg, S.W. *J. Phys. Chem. A*, *105*, 9085 (2001).
27. Lee, C.; Anson, F.C. *J. Electronanal. Chem.*, *323*, 381 (1992).
28. Beriet, C.; Pletcher, D. *J. Electroanal. Chem.*, *323*, 381 (1992).
29. Kawiak, J.; Kulesza, P.J.; Galus, Z. *J. Electroanal. Chem.*, *226*, 305 (1987).
30. Amatore, C.; Fosset, B.; Bartlet, J.; Deakin, M.R.; Lightman, R.M. *J. Electroanal. Chem.*, *256*, 255 (1988).

31. Oldham, K.B.; Myland, J.C. Fundamentals of Electrochemical Science, Academic Press, San Diego, 1994.
32. Gouy, M. *J. Phys. Radium*, *9*, 457 (1910).
33. Hunter, R.J. Fundamentals of Colloid Science, Clarendon Press, Oxford, 1987, p 341.
34. Frumkin, A.N.; Florianovich, G.M. *Doklady Akad. Nauk SSSR*, *80*, 907 (1951).
35. Heyrovsky, J.; Kuta, J. Principles of Polarography, Academic Press, New York, 1966, pp 20-21, pp229-266 and references therein.
36. Heyrovsky, J. Actualits scientifigues et industrielles, Herman, Paris, 1934, p 90.
37. Heyrovsky, J. Polarographie, Springer, Wien, 1941, p 74.
38. Frumkin, A.N. *Trans. Faraday Soc.*, *55*, 156 (1959).
39. Eaton, W.A.; George, P.; Hanania, G.I.H. *J. Phys. Chem.*, *71*, 2016 (1967).
40. Hanania, G.H.I.; Irvine, D.H.; Eaton, W.A.; George, P. *J. Phys. Chem.*, *71*, 2022 (1967).
41. Campbell, S.A.; Peter, L.M. *J. Electroanal. Chem.*, *364*, 257 (1994).
42. Peter, L.M.; Durr, W.; Bindra, P.; Gerischer, H. *J. Electroanal. Chem.*, *71*, 31 (1976).
43. Mejman, N. *Zh. Fiz. Khim.*, *22*, 1454, (1948).
44. Koutecky, J. *Chem. Listy*, *47*, 323 (1953).
45. Gavaghan, D.J.; Feldberg, S.W. *J. Electroanal. Chem.*, *491*, 103 (2000).
46. Aoki, K. *Electroanalysis*, *5*, 627 (1993).
47. Eklund, J.C.; Bond, A.M.; Alden, J.A.; Compton, R.G. *Advances in Physical Organic Chemistry*, *32*, 1 (1999).
48. Rooney, M.B.; Gavaghan, D.J.; Coomber, D.C.; Feldberg, S.W.; Bond, A.M. Unpublished work containing data and concepts that were significantly used in preparation of Sections 4 and 5 of this Chapter.

Chapter 15

Electron Transfer at the Liquid-Liquid Interface

Maurice L'Her and Vladimir Sladkov

UMR-CNRS 6521, Faculté des sciences et techniques, Université de Bretagne Occidentale
C.S. 93837, 29238 Brest CEDEX 3, France.

The aim of this chapter is to review the present state of knowledge about the thermodynamics and the kinetics of electron transfer (ET) through the liquid-liquid (LL) interface. A few reviews have already been written about the topic or some of its aspects [1-7], but the objective of this chapter is to describe the subject mainly from the point of view of an experimentalist, for chemists not familiar with the subject of the electron transfer through the liquid-liquid interface or even with electrochemistry at the interface between two immiscible electrolyte solutions (ITIES). This, of course, requires a preliminary brief description of the principles of charge transfer across the interface, of the electron in particular.

1 INTRODUCTION

Life is highly dependent on interfacial exchanges, particularly those of electrically charged species across biological membranes; all the respiration processes involve proton and electron transfers through membranes, between both the "aqueous" sides of the lipidic phase. The mechanisms of life rely on the kinetic control of thermodynamically possible reactions, and membranes are useful tools for that, being physical barriers to the diffusion of species as well as gates which opening can be monitored by the membrane potential. This explains the interest for the liquid-liquid interface and why research on membrane phenomena is a field where chemists frequently encounter electrophysiologists. These two groups of scientists worked close together at the beginning of the development of electrochemistry, which is not surprising as the contributions of Galvani and Volta are at the origin of that scientific domain. An electrophysiologist, Emil du Bois Reymond, has shown in 1848 the parallel between electrodes and the biological surfaces called membranes; Gibbs established in 1875 the equilibrium conditions for a membrane [8]. In 1902, Nernst and Riesenfeld described an experiment and the elements of the theory for ion transport at a liquid/liquid cell [9]. These facts prove that the two disciplines were still close; even the "technological" developments ensued and Waller published the first electrocardiogram in 1887. One of the earliest experimental studies involving an electron transfer through the liquid/liquid interface is the oxidation, by an aqueous solution of potassium permanganate, of the methyl group of N-*o*-tolyl-benzamide in benzene, described by Bell in 1928 [10]. However, the method for the preparation of colloidal gold by Faraday, recently cited by Schiffrin [11], is actually a biphasic electron transfer: an aqueous $AuCl_4^-$ solution is reduced by phosphorus in CS_2 or ether [12].

If the theory of membrane electrochemistry was clearly established at the beginning of the 20th century, the developments of research on that field were modest until the 1960s and 1970s, when Guastalla, Dupeyrat and their colleagues developed the study of the liquid-liquid interface [13-17]; they have been followed by other groups and particularly by the electrochemists of the Heyrovsky Institute in Prague [18,19]. Along these years, the collaboration between electrophysiologists and electrochemists has not been as effective as desirable; it has to be mentioned that the former were using micropipettes and working with microcurrents since a long time, when the electrochemists realized that working with microelectrodes would be interesting.

The contact surfaces between a biological membrane and the external or the internal aqueous phases of a cell has certain similarities with the interface which establishes between two immiscible media, such as between some organic solvents and solutions in water. A membrane is a bilayer of amphiphilic molecules, the phospholipids: their polar, ionizable heads point towards the aqueous media and the fatty long chains form the inner lipidic phase of the membrane through non covalent interactions. These membranes, which thickness is about 60 −100 Å, are permeable only to small molecules and not to ions which can however accumulate at the surface, in the vicinity of the polar heads. The transport of charged species, such as ions or electrons, from outside to the interior of a cell, through the membranes, is mediated by proteines (ionic pumps, ion channels, redox proteins ...) which are activated by potential differences or chemical gradients.

The frontier between the lipidic phase of membranes and the aqueous media appears to be quite sharp. This is not the case for the interface between two immiscible liquid media. Most solvents reckoned to be immiscible are in fact mutually soluble to a certain extend; *e.g.,* the solubility of chloroform in water is about 0.07 M, that of water in the solvent being 0.05 M. The nature of such an interface has been much debated and different models have been proposed along years. The interface between two immiscible electrolyte solutions (ITIES) is very complex, obviously more than the situation described in the first works dealing with that subject, where it was presented as a very sharp contact zone between two solvents considered as totally immiscible; two diffuse layers of electrolytes extend on each side of the contact zone. The structure of the interface has been the subject of numerous studies which are reported and discussed in many reviews. It is now accepted that the main features of the interface are: (i) a mixed solvent layer about 1nm thick in which the composition varies without discontinuity, (ii) two electrolyte diffuse layers that extend a few nm from both sides of the interface, (iii) concentration gradients which appear when charged species are transferred through the interface or when a molecule reacts in a phase [2,3,20,21]. Moreover, the interface is not rigid and thermal fluctuations will induce surface tension variations and, as a consequence, oscillations called capillary waves which roughen the surface. These phenomena, which can extend over ranges larger than the solvent layers, can greatly disturb the distribution of molecules and ions and of course the potential profiles too; the apparent roughness should vary along an experiment with the conditions affecting the surface tension and density of the liquid phases [5,22,23].

Many experimental methods are available for the study of the interface, some being used since a long time, like the electrochemical techniques and surface tension or electrocapillary measurements, as well as others which appeared more recently like the spectroscopic methods very useful for *in situ* investigations: ellipsometry, non-linear optical spectroscopy and other surface spectroscopic techniques. Molecular dynamics and Monte Carlo simulations are frequently used to model liquid-liquid interfaces [7,22,24].

Obviously, ITIES is not a simple interface and is much more complicated than the contact zone between a metal and a solution. That is the reason why a good model of the interface does not really exist. W. Schmickler noted *"In fact, this is one of the few areas of electrochemistry in which theory is leading experiment. Despite considerable progress in instrumentation, kinetic studies of reactions at liquid-liquid interfaces remain difficult ..."* [5,22,23]. Needless to say that in such circumstances theoricists will have grain to mill for a long time.

The situation at the interface is actually much more complicated than described above. As most solvents used for these studies have rather low dielectric constants, ion-pairing of the electrolytes plays a major role, at least in the organic medium and also in the aqueous phase under certain circumstances. That is mentioned quite often, but rarely is the association of ions considered as part of the interfacial reaction mechanism. Supporting electrolytes must be used for the electrochemical investigations, usually in a rather high concentration and this not only increases ion pairing, but also could create more complex situations by "cross ion-pairing". Transfer Gibbs energies for some ions, from water to solvents, have been measured from the partition of salts; in the presence of electrolytes added to lower the IR drop, as usually required for electrochemistry, the distribution of the salt under investigation could be different, as well as its apparent Gibbs energy of transfer. As voltammetry at microinterfaces can be done at lower concentrations of supporting electrolytes, such devices should help examining the effect of these salts, not really "indifferent" or "spectator" electrolytes. This also implies that parameters derived from most experiments are "formal" or "conditional". At the inner interface, in the mixed-solvents zone where conditions vary sharply, particularly the dielectric constant, ion pairing evolution should be important, which does not ease verifying a theory of the liquid-liquid interface by comparison with experimental results.

As Smickler mentioned (citation above), experimental results are not really numerous; the complex situation at the interface and the lack of a good model of the interface can be part of the explanation. Much more reactants must be brought together than for conventional electrochemistry, so that designing an experiment at the liquid-liquid interface is somewhat more complicated. As an example, the study of the electron transfer requires the presence of a reversible redox sytem in the aqueous phase and of another one in the solvent, when the study of a redox couple at a solid electrode is simpler. Fully reversible redox couples are not that numerous in water and even rarer in organic solvents; moreover, for an investigation of the electron transfer at a liquid-liquid interface, the redox couple in one medium should not partition appreciably in the second phase.

In spite of these difficulties, more experiments should be designed to gather information about the structure of the interface as well as about reactions and transfers of species; these experimental observations are essential for the development of theoretical works. Meanwhile, the use of the liquid-liquid electrode as a tool for the study of some phenomena does not require that every aspect of the transfer process has been elucidated; conditional parameters are acceptable for many purposes, such as for analysis.

2   TRANSFERS OF CHARGES AT THE LIQUID-LIQUID INTERFACE

The electron transfer is here the main concern, this is however a specific case of charge transfer through the liquid-liquid interface. Thermodynamics of the transfer of charges will be discussed first, before a brief description of the kinetics of the electron transfer across the boundary between the two immiscible media.

2.1   Thermodynamics of charge transfer

The condition for the equilibrium of a species X between two phases is expressed by the equivalence of its electrochemical potentials in the two media; for the present subject, these are usually an aqueous phase (W) and a solution in an organic solvent (S):

$$\tilde{\mu}_X^W = \tilde{\mu}_X^S \tag{1}$$

The electrochemical potential includes the chemical potential and the electrical energy term which depends on the electrical charge of the species and on the inner electrical potential of the phase:

$$\tilde{\mu}_X^W = \mu_X^W + z_X F\phi^W \qquad\qquad \tilde{\mu}_X^S = \mu_X^S + z_X F\phi^S \tag{2}$$

For a molecule or an electrolyte, as $z = 0$, the electrochemical and the chemical potentials are equivalent. In most studies the transfer of charged species is concerned, so that the contribution of individual ions must be estimated; whether that makes sense or not has been a much debated issue. However, for those who watch ions at work, crossing the interface when dragged or pushed back by an applied electrical field for example, it is obvious that the application of thermodynamics to charged species is essential. As experiments are performed with electrolytes, an hypothesis is required to separate the contribution of each of the ions. Many "extrathermodynamic" hypothesis were proposed and have been reviewed, the most convenient being (i) the one stating that tetraphenylarsonium and tetraphenylborate ions will contribute equally to the transfer of their common salt and (ii) that ferrocene and its oxidized form will be equally affected by solvation. As the consequence of this latter hypothesis is that the redox potential of the ferrocenium/ ferrocene redox couple will be the same in the different solvents, this has proved to be very useful for electrochemists.

If thermodynamics can be applied to an ion (i), at equilibrium when (1) applies, the Galvani potential difference is :

$$\Delta_w^s\phi = (\phi^S - \phi^W) = \frac{1}{z_i F}(\mu_i^W - \mu_i^S) \tag{3}$$

Considering that:

$$\mu_i = \mu_i^\circ + RT\ln(a_i) \tag{4}$$

the inner potential difference can be expressed as:

$$\Delta_w^s\phi = \frac{1}{z_i F}(\mu_i^{S,\circ} - \mu_i^{W,\circ}) + \frac{RT}{z_i F}\ln(\frac{a_i^W}{a_i^S}) = \Delta_w^s\phi^\circ + \frac{RT}{z_i F}\ln(\frac{a_i^W}{a_i^S}) \tag{5}$$

$$\Delta_w^s \phi^\circ = \frac{1}{z_i F}(\mu_i^{w,\circ} - \mu_i^{s,\circ}) = \frac{1}{z_i F}\Delta_w^s G_i^\circ \qquad (6)$$

$\Delta_w^s G_i^\circ$ being the standard Gibbs energy for the transfer of i from the aqueous to the organic phase; (5) is the Nernst equation for the transfer of *i* across the liquid-liquid interface. $\Delta_w^s \phi$, the difference in the inner potentials of the two phases at equilibrium, can be calculated from (5) and (6), knowing $\Delta_w^s G_i^\circ$ or $\Delta_w^s \phi_i^\circ$ and the activities (or the concentrations) of i in the two phases. $\Delta_w^s G_i^\circ$ is not the standard Gibbs energy of transfer of *i* from pure water to the anhydrous solvent, which can be calculated from the standard solvation Gibbs energies in the pure media, because the transfer occurs between water and the solvent mutually saturated. Values of $\Delta_w^s \phi_i^\circ$ for a few ions are noted in Table 1.

The real situation is usually different from the one described above, where the presence of only one electrolyte has been considered. Usually, the aqueous and organic phases contain two different background electrolytes, chosen for their solubility and affinity for the phases; they must be added in rather high concentration to lower the ohmic drop, particularly in the organic phase. It is essential that the electrolyte has a much higher affinity for the phase where it is dissolved than for the other medium, in order that the potentials corresponding to the transfer of its ions be high, positively or negatively. That widens the windows open for exploration between two limits corresponding to the transfer of the ions of the supporting electrolytes; the liquid-liquid interface is thus polarizable over a larger domain. The best situation would be to use electrolytes exclusively soluble in one of the two phases, which is never really the case. Thermodynamics still governs the partition phenomena when different electrolytes are present in each of the two phases and the situation at equilibrium can still be calculated [25]. Tetrabutylammonium tetraphenylborate, or the salts of this anion with more lipophilic tetraalkylammonium cations, are good electrolytes for the organic media. In the aqueous phase, Li_2SO_4 is frequently used as both the cation and the anion have a much higher affinity for water than for the organic solvents. As an example, the potential domain available for voltammetry is about $0.8 - 0.9$ V when Li_2SO_4 is used as the electrolyte in the aqueous medium and tetraheptylammonium tetra(chlorophenyl)borate in dichloroethane; an estimate of the potential range open for exploration can be guessed from values like those reported in Table 1.

As the solvated electron is not stable in the liquid media, the transfer of an electron across the liquid-liquid interface occurs only between redox couples; O_1^w/R_1^w and O_2^s/R_2^s are the redox couples, respectively soluble in water and in the immiscible solvent. This is sketched in Figure 1.

Table 1. Standard ion transfer potentials $\Delta_W^S \phi_i^\circ/V$ for mutually saturated water-organic solvent system at 25 °C.

Ion	Nitrobenzene	1,2-Dichloroethane
Li^+	0.395	0.493
Na^+	0.355	0.490
H^+	0.337	
K^+	0.241	0.499
Rb^+	0.201	0.445
Cs^+	0.159	0.36
$(CH_3)_4N^+$	0.037	0.182
$(C_2H_5)_4N^+$	-0.063	0.044
$(C_3H_7)_4N^+$	-0.160	-0.091
$(C_4H_9)_4N^+$	-0.270	-0.225
$(C_6H_{13})_4N^+$	-0.472	-0.494
$(C_6H_5)_4As^+$	-0.372	-0.364
Cl^-	-0.395	-0.481
Br^-	-0.335	-0.408
NO_3^-	-0.270	
I^-	-0.195	-0.273
BF_4^-	-0.091	
ClO_4^-	-0.091	-0.178
PF_6^-	0.012	
Pic^-	0.047	-0.069
$(C_6H_5)_4B^-$	0.372	0.364

Values based on the tetraphenylarsonium tetraphenylborate extrathermodynamic assumption. Adapted from Vanýsek, P. *Electrochemistry on Liquid/Liquid Interfaces*, Springer-Verlag, Berlin, (1985) and Kakiuchi, T. in *Liquid-Liquid Interfaces, Theory and Methods*, Edit. By Volkov, A.G.; Deamer, D.W., CRC Press, Boca Raton, (1996).

Assuming that only one electron is exchanged between the redox pairs:

$$O_1^w + R_2^s \;\rightleftharpoons\; \rightarrow O_2^{+,s} + R_1^{-,w} \tag{7}$$

in such a case, the equilibrium conditions for (7) lead to the Nernst equation:

$$\Delta_w^s\phi = (\phi^s - \phi^w) = \Delta_w^s\phi_{ET}^\circ + \frac{RT}{F}\ln[\; \frac{a_{O_1}^w \, a_{R_2}^s}{a_{R_1}^w \, a_{O_2}^s} \;] \tag{8}$$

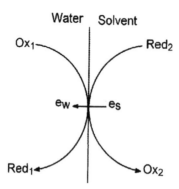

Figure 1. Electron transfer across a liquid/liquid interface.

where the driving force of the redox reaction through the interface, the standard potential of the electron transfer (ET) is:

$$\Delta_w^s \phi_{ET}^\circ = (E_{O_1/R_1}^{\circ,W} - E_{O_2/R_2}^{\circ,S}) \tag{9}$$

The evaluation of $\Delta_w^s \phi_{ET}^\circ$ requires that the two formal potentials of $(O_1/R_1)^W$ in water and $(O_2/R_2)^S$ in the solvent are known, on the same potential scale. $E_{O_1/R_1}^{\circ,W}$ is usually reported vs SHE, so that $E_{O_2/R_2}^{\circ,W}$ must be evaluated versus the same reference. When $E_{O_2/R_2}^{\circ,W}$ is known (vs SHE), the difference with $E_{O_2/R_2}^{\circ,S}$ is the result of solvation on both species of the redox couple:

$$E_{O_2/R_2}^{\circ,S} = E_{O_2/R_2}^{\circ,W} + \frac{\left(\Delta_w^s\right)G_{O_2}^\circ - \left(\Delta_w^s\right)G_{R_2}^\circ}{F} \tag{10}$$

where $\Delta_w^s G_{O_2}^\circ$ and $\Delta_w^s G_{R_2}^\circ$ are the standard Gibbs energies of transfer for O_2 and R_2, from water to the solvent.

Equilibrium (7) is the simplest case of a redox reaction at a liquid-liquid interface; when stoichiometry is different or when other species, like proton, are involved in the redox reaction, equation (8) must be adapted.

2.2 Kinetics of the electron exchange

When the potential applied to the interface is different from the value at equilibrium, electrons flow through the interface to restore concentration conditions compatible with the potential. $I=f(\Delta\phi)$ curves are obtained as in voltammetry at solid electrodes, which shapes depend on the mass transport regime. Figures 2 and 3 present examples of electron transfer voltammetry under linear diffusion conditions and steady-state mass transport conditions.

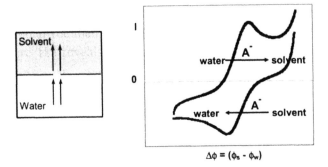

Figure 2. Voltammetry under linear diffusion conditions at a liquid/liquid interface.

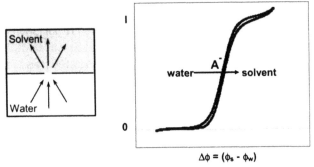

Figure 3. Voltammetry under steady-state mass transport conditions at a liquid/liquid interface.

Voltammograms like those of classical cyclic voltammetry at solid electrodes are quite common because the liquid-liquid interface is usually immobile; stationary electrodes have been used for the very first studies of the polarized interface, by chronopotentiometry. However, experimental devices with dropping aqueous electrolyte solutions have been developed for polarographic studies, the current in that case being limited by mass transport. Waves with a limited current are also observed at micro ITIES, when diffusion is radial at low potential scan rates. By analogy with electrochemistry at metal electrodes, an expression for the variation of the steady-state current with potential as been derived for the transfer of an ionic species [26]:

$$\Delta\phi = \Delta\phi^\circ + \frac{RT}{2F}\ln[A] + \frac{RT}{F}\ln[\frac{(I_{dR1}-I)(I_{dO2}-I)}{(I_{dO1}-I)(I_{dR2}-I)}];$$

$$[A]=[\frac{D_{O1}\,D_{R2}}{D_{R1}\,D_{O2}}] \tag{11}$$

From these different $I=f(\Delta\phi)$ curves, information is obtained about thermo-dynamics $(\Delta\phi_{1/2})$ or kinetics of the interfacial reaction (α, k).

The ion transfer across the LL interface obeys equations (12) and (13):

$$I/FA = k°C_i^W e^{[\frac{\alpha F(\Delta\phi-\Delta\phi°)}{RT}]} - k°C_i^S e^{[\frac{-(1-\alpha)F(\Delta\phi-\Delta\phi°)}{RT}]} \tag{12}$$

$$k_f = k°e^{[\frac{\alpha F(\Delta\phi-\Delta\phi°)}{RT}]}; \qquad k_b = k°e^{[\frac{-(1-\alpha)F(\Delta\phi-\Delta\phi°)}{RT}]} \tag{13}$$

where the parameters are similar to those of the electron transfer between a solution and a solid electrode: k_f is the forward rate constant (transfer of a univalent ion i from water to the solvent), k_b the rate constant of the reverse transfer, $k°$ the standard rate constant (when $\Delta\phi=\Delta\phi°$) and α the charge transfer coefficient [2,4,27-29].

An electron transfer at ITIES, reaction (7), is a bimolecular reaction which should obey a second-order rate law [29]:

$$I/FA = k_f C_{O1}^W C_{R2}^S - k_b C_{R1}^W C_{O2}^S \tag{14}$$

It has been verified, in many experiments, that the ion transfer obeys a Butler-Volmer's law; the situation is however not as clear for the electron transfer. Girault noted in 1991 *"Contrary to ion transfer reactions where we possess interesting experimental results ... we have interesting theories for electron transfer reactions not yet corroborated by experiments"* [2]. A few years before, the same author mentioned *"the paucity of experimental results for this type of reaction"* [3]. Since, more experimental studies have been performed, allowing much insight into electron transfer at ITIES [5,6]. The matter is however still controversial as a change in the Galvani potential difference can affect the interfacial concentration of ions (concentration polarization) as well as influence the electron transfer through a Butler-Volmer's effect. Discrepancies are observed between the reported experimental studies [6,7].

Samec was the first to consider the activation barrier to the electron transfer at the LL interface [18], following Marcus theory [30,31]. Later, others including Marcus, developed the theoretical studies [32-38]. Girault and Schiffrin [39] proposed for the electron transfer at ITIES, the precursor-successor model which as been used for other electron transfer processes [40,41]:

$$O_1 + R_2 \longleftrightarrow O_1|R_2 \longleftrightarrow \{O_1R_2\} \underset{k_{et}}{\overset{\upsilon_{eh}}{\longleftrightarrow}} \{R_1O_2\} \longleftrightarrow R_1|O_2 \longleftrightarrow R_1 + O_2 \tag{15}$$

$O_1 \mid R_2$ is the precursor complex formed after diffusion of the two reacting species; $R_1 \mid O_2$, the successor complex, is the last step of the global electron transfer, before the separation of the products and their diffusion away. The electron exchange can occur only between the reorganized precursor and successor complexes, $\{O_1R_2\}$ and $\{R_1O_2\}$; the electron can exchange at the hopping frequency υ_{eh} (at the exchange rate k_{et}), between two states of identical configurations and energies (Franck-Condon principle).

Looking at electron transfer reactions, experimentalists examine reaction (15), through k_{obs} that they measure from their experiments, but the true electron transfer is:

$$\{O_1R_2\} \;\overset{\upsilon_{eh}}{\underset{k_{et}}{\longleftrightarrow}}\; \{R_1O_2\} \tag{16}$$

These rate constants are linked:

$$k_{et} = \upsilon_{eh}\, e^{[-\Delta G^*/RT]} \tag{17}$$

$$k_{obs} = K_p k_{et} = K_p K_r \upsilon_{eh} \tag{18}$$

$$k_{obs} = k_{et}\, Z\, e^{[\frac{-W_p}{RT}]} \tag{19}$$

K_p and K_r being the equilibrium constants for the formation of the precursor and of the successor $\{O_1R_2\}$. Z is called the preexponential factor and ΔG^* is the activation energy of the electron transfer, given by the Marcus relation:

$$\Delta G^* = (\lambda + \Delta G^\circ_{et})^2 / 4\lambda \tag{20}$$

where ΔG°_{et} is the standard Gibbs energy of the electron transfer, related to ΔG°_r, the standard Gibbs energy of the overall reaction (15):

$$\Delta G^\circ_{et} = \Delta G^\circ_r + W_p + W_s \tag{21}$$

In equation (21), W_p and W_s are respectively the work for the formation of the precursor from the reactants, and for the separation of the products from the successor complex.

The reorganization energy λ, which includes the contributions of the solvent and of the vibrational reorganizations, is not easily calculated even if some solution have been proposed for the liquid-liquid interface [33,38]. Z depends on the position of the reactants at the interface, which is of course rather difficult to predict considering what has been written above about the knowledge of the interface. Lyons [20], citing Girault and Schiffrin [39], mentioned that the electron transfer rate constant could vary from 0.3 to 60 cm s^{-1} depending on the values of the radii of O_1 and R_2. Thinking of some reactants which have been used for electron transfer studies, the bisphthalocyanines of Lu(III) or of Sn(IV) for example, one should wonder what is the radius to be introduced into the calculation, knowing that in such cases the electron is exchanged with the π orbitals of the ligands of these sandwich complexes.

Following Samec suggestion, it is well accepted that the rate constant measured at the potential of zero charge of the interface (pzc) makes clearer the evaluation of k_{et} because the work terms W are equal to zero [2,18,39].

The transfer coefficient of the electron transfer represents the fraction of the Gibbs energy of the electron transfer used for activation of the ET process:

$$\Delta G^* = \frac{\lambda}{4} + \alpha \, \Delta G_{et} \tag{22}$$

Marcus theory [30] shows that:

$$\alpha = \frac{1}{2} + \frac{\Delta G_{et}}{2\lambda} \tag{23}$$

It is thus clear that the parameters of the electron transfer at the liquid-liquid interface are not straightforwardly linked to factors derived from the experiments. From the works already described, it appears that the electron transfer rate constant can depend on the variation of the Galvani potential difference [6], but there is no general agreement [42]. As mentioned by Benjamin, ion transport can mask or obscure the electron transfer when the external potential is applied to the electrodes; this could explain why the rate constant follows Butler-Volmer's relation in some experimental works and not in other reports [7]. Ion-pairing at the interface should also play a key role [36].

3   EXPERIMENTAL METHODS

Studying electron transfer through the liquid-liquid interface is not fundamentally different from examining other charge transfers, so that the methods of investigation as well as the experimental tools are identical. However, as the electron transfer occurs between molecular species (ionized or not) or coordination compounds, additional techniques are available, due to their spectroscopic properties which are different from those of the ions of most common electrolytes. The species involved in the electron exchange at the interface are redox species, so that scanning electrochemical microscopy (SECM) has been abundantly used, providing invaluable information.

3.1   Cells, electrodes and methods

Most experimental arrangements and practical points which are essential for working with liquid-liquid interfaces have been abundantly developed in the literature [4,27].

Electrochemistry at a liquid-liquid interface, whether it is for the study of ion transfer or electron exchange, requires the measurement of the potential drop across the interface and consequently the presence of two reference probes, two nonpolarizable electrodes, one on each side of the ITIES. As such electrodes cannot carry currents, two counter electrodes are also used, usually two large area platinum electrodes. That constitutes the four electrode cell for classical experiments at liquid-liquid interfaces of a few mm in diameter, or larger, which is is schematically represented on Figure 4a; the four electrode configuration is

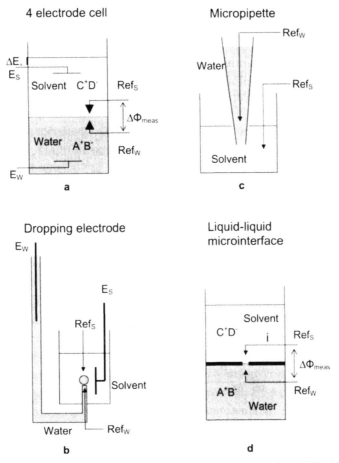

Figure 4. Experimental arrangements for electrochemistry at the liquid/liquid interface.

essential to reduce the ohmic drop. As the area of the interface is rather large, the diffusion regime at usual scan rates is linear and the I=f($\Delta\phi$) curves are similar to cyclic voltammograms at solid electrodes.

For analysis, currents limited by mass transport are very useful as shown by the success of polarography. That is why electrolyte dropping electrodes have been designed; Figure 4b is a very rough sketch of such a device, which also has four electrodes. This technique designed by Koryta [43] has been used for many investigations about ion transfer [4,27]. The electric resistance of the aqueous solution being lower, it seems more logical to introduce this phase in the longer compartment, the one at the end of which the drop forms; as the density of this phase is lower than that of the solvents most frequently used for studies at the LL interfaces, like nitrobenzene or dichloroethane, this often imposes the ascending configuration. However, hanging electrolyte (nitrobenzene) drop electrodes have also been used.

Quite recently, the dropping electrode has been combined with electrochemistry at a solid microelectrode in a technique called "microelectrochemical measurements at expanding droplets" (MEMED); the solid microelectrode placed close to the tip of a capillary, where electrolyte dropplets form, is used as a probe. This experimental method has been used for the study of electron transfer reactions through the liquid-liquid interface [44-48].

The study of oil microdroplets deposited on the surface of a solid electrode was introduced by Marken *et al.* as a way to investigate electrochemical processes in nonhomogeneous media, such as emulsions [49,50]. This concept was extended and abundantly developed by Shi and Anson for the examination of electron transfer across the liquid-liquid interface established between a thin film supported on the surface of a solid electrode immersed in an aqueous solution [51-55]. The technique has also been developed and used by other research groups [56-60]. Photoelectrochemistry can even be performed at such an interface [61]. The electron transfer at the surface of the solid electrode on which is deposited the thin organic film is controlled by the transfer of ions between the two solutions, through the LL interface, so that this is an indirect way to study ion transfer. Due to their easy manipulation, these electrodes modified by a film of an organic solvent are very convenient.

When a four electrode potentiostat is not available, a conventional three electrode configuration can be employed, provided that the working electrode potential does not drift much when current passes, or when the potential shift is tolerable; this situation is of course not frequently acceptable.

Since a long time, electrophysiologists use micropipettes which can be considered as microelectrodes; Girault developed their use for the study of liquid-liquid microinterfaces [62]. The area of the interface between the phase inside the pipette and the electrolyte outside is very low; consequently, this of course is also true for the current and the IR drop. Pipettes with an internal diameter of a few nanometers have been used. The micro current generated at such an electrode would not affect the potential of a reference electrode, and a two reference electrode configuration can be used for studies with such microinterfaces (Figure 4c). In order to prevent creeping of the aqueous solution on the external or inner wall of the pipette, which is hydrophilic, glass must be pretreated by silanization; this, not only improves the reproducibility of the measurements but also allows the use of an organic phase inside a pipette dipped in an aqueous solution [63].

Another way for the development of microinterfaces is the one represented on Figure 4d: the communication between the two immiscible electrolytes is established through a microhole drilled in a thin plate of a non conducting material (polymer, alumina, glass) [64]. These micro ITIES are conceptually close to the metallic microelectrodes and the diffusion regime is dependent on the time window of the experiment. Waves with limited currents are obtained at low scan rates, when diffusion is radial; peak-shaped curves are observed at higher rates, due to linear diffusion. A microhole electrode cannot be considered as being symmetrical, depending on the position of the interface inside the channel; this is apparent from the current evolution along the forward and backward scans of a cyclic voltammogram, at certain rates [65]. Of course, the diffusion regime at micropipettes is even less symmetrical. Their use as probes in scanning electrochemical microscopy (SECM) to examine interfaces, including the liquid-liquid interface itself, is promising; dual micropipettes have been utilised for generator-collector experiments [66,67].

Conventional electrochemical methods for electrochemistry at metal/ solution interfaces can be used for studies of ITIES too; this has been described in many publications devoted to the development of techniques suitable for electrochemistry at the liquid-liquid interface, and in reviews [4,27]. Chronopotentiometry has been used quite early for measurements of ion transfer parameters [68-70] and also for ET studies [71].

Cyclic voltammetry became popular for the study of charge transfer at ITIES after the works of Samec [18,26]. Its application to electron transfer [72,73] and to homogeneous chemical reactions coupled with ET has been discussed [74,75]. This method became as popular for ITIES studies as for other fields of electrochemistry, because the reversibility of a process is immediately visible.

The electrolyte dropping electrode has been developed in the mid-1970's, at the Heyrovsky Institute [43,76], the theory for the $I=f(\Delta\phi)$ curves for the electron transfer at this dropping electrode being developed by Samec [26]. The dropping electrode can be used to investigate ion transfer and homogeneous chemical reactions coupled with ET [77-79].

Impedance spectroscopy is a valuable technique for the measurement of the interfacial capacitance, widely used also for studies at the liquid-liquid interface [80-84]. AC-impedance analysis has been employed for the analysis of electron transfers at the liquid-liquid interface [85,86], as well as a closely related technique, modulated electroreflectance spectroscopy [87].

Scanning electrochemical microscopy (SECM) [88] is of great importance for the study of charge exchange at the liquid-liquid interface, particularly for examining the electron transfer. An ultramicroelectrode (UME), with higher resolution than conventional electrodes, can be used not only to analyse the solutions close to the interface and as a topographic tool to map concentration profiles, but also to obtain temporal information essential for the study of kinetics. The UME, a platinum disc for example, is a collector which distance from the interface can be adjusted; it can also be used as a generator to modify the local concentration of a reactant and thus to influence the reaction rate. The tip plays the role of the ring in a ring-disc electrode. The possibilities of SECM as a tool for the study of the electron transfer at ITIES has been the subject of numerous reviews [89-93]. The same technique is of course useful for studying charge transfer through bilayer lipidic membranes established at the interface between two liquid media [94,95].

In conjonction with electrochemistry, optical methods are used, in two ways, for the study of charge transfer: (i) for the photo production of reactive species which participate in an electron transfer at ITIES, (ii) as spectroscopic methods for the analysis of the interfacial region during the electron transfer.

Photoinduced charge transfer at the liquid/liquid interface, and in particular the electron transfer, has attracted investigators because of the close similarities with photosynthesis and photocatalysis processes. The different aspects of the interfacial photoreactions have been the subject of some reviews, as well as the works already published [2,96-98].

The bulk liquid phases are centrosymmetric but interfaces between two liquids are generally not, particularly when solute molecules are adsorbed. Under laser irradiation, these interfaces induce non linear optical phenomena, second harmonic generation (SHG) and sum frequency generation (SFG). These techniques are suitable for the study of the dynamics, such as diffusion of energy and matter, structure changes, rotation of molecules at the interface [99-102].

Other spectroscopic techniques have seldom been applied to the study of the interface and particularly for investigation of the electron transfer. Redox reactions very often generate species with an unpaired electron, active in EPR spectroscopy; this method has been adapted for the study of the interface during an electron transfer [103,104]. Time resolved fluorescence is ideally suited for studying photo-induced electron transfer but has rarely been utilised [105].

The combined use of electrochemistry and ellipsometry, a technique sensitive to changes of the species concentration at the interface, seems to be a very promising tool for the study of ITIES [106]; this has however not yet been applied to the study of the electron transfer.

3.2   Redox systems for the aqueous and organic phases

The electron exchange at a liquid-liquid interface is a redox reaction which involves two redox systems in different media, one soluble in the aqueous phase, the other being dissolved in the organic solvent (reaction(7), Figure 1). These redox systems must be chosen in order that they react, of course, but they must also fulfil other conditions. When the redox reaction is performed for the purpose of synthesis, the reaction does not need to be reversible; however, for studies where acquiring information about thermodynamics is the main objective, the overall process must be reversible, which requires both redox systems to be fully reversible. When the aim is the study of the electron transfer, no other reaction must be coupled with it; for example, no ion transfer must occur simultaneously, which means that both species of a redox couple must be soluble only in the medium they have been chosen for. Otherwise, the interfacial reaction is a complex process equivalent to EC or CE processes at solid electrodes. All these requirements explain why examples of electron exchange studied at the liquid-liquid interface are rather scarce.

Redox couples that can be used in the aqueous phase are not that numerous, for the reasons described above; however, a few redox systems have already been employed for SECM [88] or for electron transfer studies [6]. Redox reagents which could be used for the study of electron transfer at the LL interface could be found in the literature [107-109]. Some of these mediators already used in aqueous media are listed in Table 2. Reviews about redox agents in organic solvents, or lists of potentials, are quite rare [110,111]. The potentials for a few of them have been reported on the redox scales in Figure 5.

All the reported potentials of redox couples in the different media must be utilized with circumspection. Some of the values are standard potentials, many of them being formal values; overall, most of the time these potentials have been measured in aqueous media or nonaqueous solvents and rarely in the phases where the experiments are performed, *e.g.* a water saturated solvent. The best would be to measure the formal potentials of the redox systems in both media, as used for the study.

As explained above, the driving force of the electron exchange is:

$$\Delta_w^s \phi_{ET}^\circ = (E_{O_1/R_1}^{\circ,W} - E_{O_2/R_2}^{\circ,S}) \tag{9}$$

Table 2. Selected Oxidation/Reduction Mediators in Aqueous Solution

Mediator	E°/V (NHE)
Br_2/Br	1.0874
$Fe(phen)_3^{3+/2+}$	1.13±0.01
$Fe(bpy)_3^{3+/2+}$	1.11±0.01
$IrCl_6^{2-/3-}$	0.867
$Ru(CN)_6^{3-/4-}$	0.86±0.05
$Os(bpy)_3^{3+/2+}$	0.885
$Mo(CN)_8^{3-/4-}$	0.725
$Co(oxalate)_3^{3-/4-}$	0.57±0.02
I_3/I^-	0.536
$W(CN)_8^{3-/4-}$	0.457
$Ferrocene^{+/0}$	0.400
$Fe(CN)_6^{3-/4-}$	0.3610±0.0005
$Co(phen)_3^{3+/2+}$	0.327±0.02
$Ru(en)_3^{3+/2+}$	0.210±0.005
$Ru(NH_3)_6^{3+/2+}$	0.10±0.01
$Co(en)_3^{3+/2+}$	-0.180±0.002

From Bard, A.J.; Parsons, R.; Jordan, J. *Standard Potentials in Aqueous Solution*, IUPAC, Marcel Dekker, Inc., New York, (1985). Most of these potentials are formal, conditional, parameters.

both potentials, in the aqueous phase and in the solvent, being measured relative to the same reference electrode, SHE for example. This necessitates to evaluate the relative positions of the two redox scales in different media and requires an extrathermodynamic assumption, which validity cannot of course be verified. None of these is really satisfying, particularly when organic media must be compared with water [112-114]; however, the Strehlow's assumption based on the ferrocenium/ferrocene couple is very convenient when redox potentials measurements are involved. Moreover, IUPAC recommends to refer the potentials of redox couples in solvents to this system [115]. For these reasons, the ferrocenium/ferrocene system has been chosen for the comparison of the redox potential scales, as illustrated on Figure 5. A few redox systems in aqueous solution appear on scale 1 (Figure 5); the redox potential of Fc⁺/Fc on that scale has been measured to be 0.39 V [116] or 0.40 V vs SHE [117]. Scale 2 represents redox couples in a solvent, dichloromethane (DCE), the formal potentials being refered to that of Fc⁺/Fc. As the potential of this redox system is 0.40 V in SHE

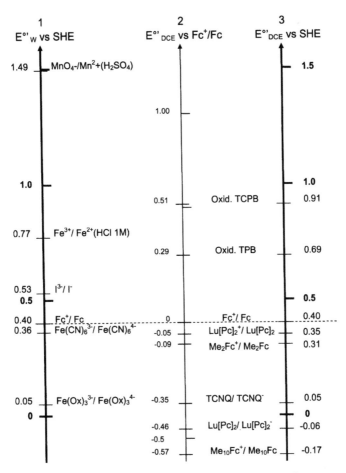

Figure 5. Potential scales for redox systems: (1) versus SHE, in water; (2) versus $E^{\circ\prime}_{Fc+/Fc}$, in DCE; (3) versus SHE, in DCE, after adopting the extrathermodynamic assumption about the invariance of $E^{\circ\prime}_{Fc+/Fc}$.

scale, the formal redox potentials of the redox couples in the solvent can be estimated versus SHE; they could have been reported on scale 1 but, for clarity, they are represented on scale 3, vs SHE in the solvent. The redox systems in the organic and the aqueous phases of an experiment can be compared from potentials on scales 1 and 3 for the evaluation of $(E^{\circ}_{O_1/R_1}{}^{W} - E^{\circ}_{O_2/R_2}{}^{S})$ in equation (9). It should constantly be kept in mind that the comparison is imprecise due to the imperfections of the extrathermodynamic assumptions; as an example, based on the assumption that solvation affects similarly the tetraphenylborate anion and the tetraphenylarsonium cation, the redox potential of Fc^+/Fc should be 0.64 V on the $E^{\circ\prime}_{DCE}$ scale vs SHE [6], but 0.40 V when the Strehlow's hypothesis is adopted.

Redox couples used in aqueous phases for the study of the interfacial ET are of different types. Some aqueous solutions have been obtained from chlorides or sulfates of redox systems like $Ce^{4+/3+}$ [48,108,118,119], $Eu^{3+/2+}$ [105], $Fe^{3+/2+}$ [120,121] or $V^{3+/2+}$ [119,122]. As these cations are acidic, the species in solution depend on pH; the formal equilibrium potential of the aqueous redox system should be measured. Complexes of these or similar metal ions should also have formal redox potentials influenced by pH and solution state: $IrCl_6^{2-/3-}$ [47,52,55,123-125], $Fe^{III/II}$(citrate) [121], $Fe(EDTA)^{-/2-}$ [119,126], $Co^{III/II}$ (sepulchrate) [119,122]. When the redox couple in the organic phase is under investigation, the electron acceptor/donor system in the aqueous phase must behave as simply as possible; that is why outer-sphere redox reagents, with an invariable coordination sphere, or supposedly so, are prefered (Table 2). The most frequently used has been $Fe(CN)_6^{3-/4-}$ [18,45,46,51,54,55,79,86,87,104,108,118,119,121-149]. However, $Ru(NH_3)_6^{3+/2+}$ [51,53,109,118,119], $Ru(CN)_6^{3-/4-}$ [46,54,55,118,119,122,126,150-153], $Mo(CN)_8^{3-/4-}$ [55,119,122,126] and $W(CN)_8^{3-/4-}$ [55,119,122,126] have also been utilized. One of the favorite redox couples for the study of electron transfer involving photoexcited states generated at the interface is $Ru(bipy)_3^{3+/2+}$, the ruthenium bispyridine complexes [154-158]. The oxalate ion [159,160] and methylviologen [161,162] were used as quencher of excited species photogenerated at the interface, in aqueous solutions. Various hydrophylic porphyrins, bearing carboxylic, sulfonated or pyridinic substituents, have also been used for the study of the photoinduced ET [148,163-170]. Even more complex natural redox centers, such as chlorophyll [171] and cytochrome c [172] have been examined at the liquid/liquid interface. A few redox couple relevant to biological redox processes, dissolved in the water phase, were studied at the interface: quinone/hydroquinone [108], NAD$^+$/NADH [173], and the reductants ascorbate or ascorbic acid [174-176], benzylnicotinamide [158], glucose [177]. Tetrachlorohydroquinone in 1,2-DCE has been oxidized by oxygen from air, the interfacial redox reaction being controlled by ion transfers [178].

Different solvents, benzene ($\varepsilon = 2.3$), toluene ($\varepsilon = 2.4$), dichloromethane ($\varepsilon = 9.1$), 1-octanol ($\varepsilon = 10.3$), 2-nonanone ($\varepsilon = 9.1$), 2-heptanone ($\varepsilon = 11.95$), 2-octanone ($\varepsilon = 10.94$), benzonitrile ($\varepsilon = 25.9$), butyronitrile ($\varepsilon = 24.8$), benzylcyanide ($\varepsilon = 17.9$) and o-nitrophenyloctylether ($\varepsilon = 24.2$) have been utilized as the organic phase, the most frequently used being dichloroethane ($\varepsilon = 10.4$) and nitrobenzene ($\varepsilon = 35.6$).

In the organic phases, various compounds have been used as electron acceptors or donors. The excited state of $[Ru(bipy)_3]^{2+}$ as been photogenerated at the interface for the reduction of viologens [161,162], as well as the electrogenerated oxidized ruthenium complexes of the bipyridine ligand to which lipophilic chains had been attached, for the oxidation of oxalate with emission of light [160]; Ru^{3+}/Ru^{2+} polypyridine complexes were examined too, for $C_2O_4^{2-}$ oxidation [159]. ET between anthracene, as a quencher in the organic phase, and Europium(III) has been investigated [105]. The redox couple quinone/hydroquinone (benzoquinone, tetrachlorobenzoquinone) has been used in the aqueous phase for a few studies [103,148,173-176,178]. Since the very first studies of electron transfer at the liquid/liquid interface, the redox couples of the bis(cyclopentadienyl) sandwich complexes of iron have been favourite species as redox agents for the organic phase: ferrocenium/ferrocene [18,26, 108,109,123,124,132,134-138,157,165-167,170,171,179,180], decamethylferrocene [51,53-55,87,109,121,124,125,127,135,136,150,172,177,180,181] and other substituted complexes [52,109,156,165-168,181]. The redox potentials of various ferrocenes in acetonitrile have been compared with aqueous redox systems, assuming that [hydroxymethylferrocene]$^{+/o}$ is unaffected by solvation, which is hardly believable [109]. The bisphthalocyanines of the lanthanide cations are neutral sandwich complexes having certain similarities with the ferrocene derivatives, as they can also be oxidized; these compounds should however offer some advantages over the ferrocenes. They are also reducible, through a reversible one-electron process, the three redox states being stable, which is not true for the ferrocenium cation. Another difference is their total insolubility in aqueous media, unless they have been substituted with highly hydrophilic groups. The bisphthalocyanines of Lu^{3+} [139,140,142,182] and of Sn^{4+} [141] have been used as electron donors in the organic phase. The fulleride anions (C_{60}^- and C_{70}^-) were oxidized by $Fe(CN)_6^{3-}$; it has been suggested that the anomalous behaviour observed could result from ET in the inverted region of Marcus theory [183]. Another compound which has been examined in a very large number of works at the liquid-liquid interface is tetracyanoquinodimethane, at least the redox system $TCNQ/TCNQ^-$ [45,46,51,79,103,104,108,119,121, 128-131,140,143-146,154,155,165,168,169,174,182,184]. Porphyrins [97], as well as the ruthenium complexes bearing pyridine as axial ligands on the metal center [140,182] and the zinc derivatives [119,122,126,140,147,149,151-153,182] have been employed as electron donors in the organic phase. The oxidation of methylanisole by aqueous Ce(IV) was studied kinetically; this reaction is an irreversible multielectron process [48]. The electron transfer between aqueous redox systems and quantum dots, "monolayer-protected clusters" (MPC) which are nanoparticles of gold stabilized by alkanethiols, in CH_2Cl_2, has been studied by SECM [118]. Some of these reactions will be detailed below.

4   DESCRIPTION OF EXPERIMENTAL ELECTRON
    TRANSFER STUDIES

Reports on electron transfer through the liquid-liquid interface have been rare before the 1970's. In reviews, an article by Guainazzi and coworkers is often cited as the first example of electron transfer through the liquid-liquid interface [185]; this describes the reduction, by a solution of $[V(CO)_6]^-$ in dichloroethane, of aqueous copper and silver salts to a film of metal at the interface. However, a biphasic oxidation by $KMnO_4$ of the methyl group of a toluene derivative to carboxylic acid has been described much earlier [10]; other biphasic redox reactions accelerated by phase-transfer catalysis have been cited [186]. To our knowledge, the first observation of a redox reaction at an interface was reported by Faraday: the production of colloidal gold from the reduction of aqueous $AuCl_4^-$ by phosphorus in CS_2 [12].

4.1   Ferrocenes

The first electrochemical study of the electron transfer at the interface between two immiscible electrolytic solutions was about the reaction:

$$Fe(CN)_6^{3-,W} + Fc^{NB} \rightleftharpoons \cdot Fe(CN)_6^{4-,W} + Fc^{+,NB} \qquad (24)$$

The electrochemical cell for the study could be represented as follows:

$$|Ag|AgCl|0.05 \text{ M TBACl}^W|0.05M \text{ TBATPB}^{NB}, 0.1M \text{ Fc}^{NB}\|$$

$$10^{-3} \text{M Fe(CN)}_6^{3-,W}, 10^{-4} \text{M Fe(CN)}_6^{4-,W}, 0.05 \text{ M LiCl}^W|Pt$$

where NB is nitrobenzene and TBATPB, tetrabutylammonium tetraphenyl-borate [132]. The electron transfer between the redox couples in the aqueous phase and in the organic solvent was evidenced for the first time, by cyclic voltammetry (Figure 6).

The process is reversible and diffusion controlled when the concentration of ferrocene in nitrobenzene is high (0.2 M); the system becomes quasi-reversible at lower concentrations of Fc^{NB} and $Fe(CN)_6^{3-,W}$ [133]. The rate constant $k^{NB \to W}$ does not change much with potential [134]. In these early reports, Samec mentioned that a simpler electron transfer reaction should be found in order to go farther on the exploration of the kinetics of ET. Obviously, some problems obscure the interfacial electron exchange, such as ion-pairing which could be neglected when nitrobenzene is the organic phase but not when dichloromethane or dichloroethane are the solvents, or any other medium with a low dielectric constant. Moreover, ferrocene as well as ferrocenium cause problem. Ferrocene, even if much less soluble in water than in solvents, does partition to water so that

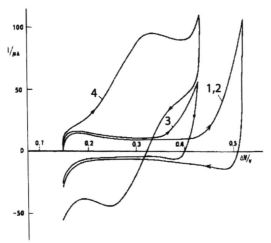

Figure 6. Cyclic voltammetry of ET between ferrocene in nitrobenzene and $Fe(CN)_6^{3-/4-}$ in water. (1): 0.05 M TBATPB in NB; (2): 0.1 M ferrocene in NB; (3): 10^{-3} M $K_3Fe(CN)_6$ + 10^{-4} M $K_4Fe(CN)_6$ in water; 4: all the reactants present. Reproduced with permissiom from Samec, Z.; Mareček, V; Weber, J. *J. Electroanal. Chem. 96*, 245 (1979). Copyright Elsevier.

the homogeneous reaction (25) could occur in the aqueous phase [187]:

$$Fe(CN)_6^{3-,W} + Fc^W \rightleftharpoons Fe(CN)_6^{4-,W} + Fc^{+,W} \tag{25}$$

More worrying is the possibility that the ion transfer

$$Fc^{+,NB} \rightleftharpoons Fc^{+,W} \tag{26}$$

accompanies the electron transfer; it has been shown that these two processes occur only 0.15 V apart [135,187]. The situation should be more favourable in the case of lipophilic ferrocene derivatives, like dimethylferrocene (Me_2Fc) and particularly for decamethylferrocene ($Me_{10}Fc$) [188]. Figure 7 illustrates the transfer of Fc^+ from water to nitrobenzene.

The transfer of the ferrocenium ion not only influences the electron transfer but can also induce following chemical reactions, for example the oxidation of tetraphenylborate ion [135]. More important, since the first studies about this redox couple, the decomposition of Fc^+ in many solvents or in the presence of coordinating ions (Cl^-, Br^-) and nucleophiles is known to produce Fe(III) and Fe(II) [189,190]. This, rarely mentioned in studies about ferrocene oxidation at LL interfaces, could explain why the formation of prussian blue-like compounds has been observed [136]. It must be noticed that, using a platinum electrode as a probe near the interface between a DCE droplet and the aqueous solution (an

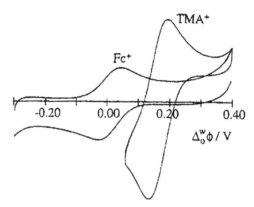

Figure 7. Cyclic voltammogram of the ferrocenium transfer across the water-DCE interface. From Fermín, D.J.; Lahtinen, R. Reproduced with permissiom from *Surfactant Science Series, 95,* 179 (2001).

experimental technique called MEMED), Unwin and coworkers have shown that $Me_{10}Fc^+$ does not cross the interface in the potential domain corresponding to the electron transfer [124]. However, a recent publication shows that $Fe(CN)_6^{3-}$ is extracted in dichloromethane, forming an ion-pair with Me_2Fc^+ [191].

The rate constant value of the bimolecular ET process between aqueous $Fe(CN)_6^{3-}$ and ferrocene in nitrobenzene is close to log $(k^{NB \to W}/ m\ s^{-1}\ M^{-1}) =$ 2.8, going down slightly with an increase in the $Fe(CN)_6^{3-}$ concentration and also in the potential; the concentration of ferrocene has no influence on it [134]. The authors conclude that the analysis of the experimental results is complicated by various factors, among which ion-pairing. From AC impedance measurements of the ET rate at the interface between water and nitrobenzene, it has been shown that the $Fe(CN)_6^{3-}$ concentration has more impact than that of ferrocene [192]. Various factors influence the electron transfer, as it has been clearly established in a study of $Me_{10}Fc$ oxidation, by SECM and MEMED [125]. The rate of the electron transfer depends on the redox couples used in both phases, as expressed by equation (9). The junction potential between the two media is also part of the driving force of the electron transfer; this can be controled by an ion common to both phases, as shown by equation (27):

$$\Delta G_{ET}^{o\prime} = - F\,[\,E_{O_1/R_1}^{o\prime,W} - E_{Fc^+/Fc}^{o\prime,DCE} + \Delta_{DCE}^W \phi\,] \tag{27}$$

When ClO_4^- is the ion determining $\Delta_{DCE}^W \phi$, the potential difference between the two phases, the bimolecular rate constant is 0.9 cm s^{-1} M^{-1} for the oxidation of ferrocene by $Fe(CN)_6^{3-}$, 1.8 cm s^{-1} M^{-1} for the reaction with $Ru(CN)_6^{3-}$, the oxidation by $Ir(Cl)_6^{2-}$ being very fast. This reflects the influence of the potential of the aqueous redox couple, $E_{O_1/R_1}^{o\prime W}$ on the driving force of the interfacial reaction. Increasing the concentration of ClO_4^- in water lowers the rate when the contrary should be observed, as that is supposed to increase the driving force; this was attributed to strong ion-pair formation with perchlorate. When tetrabutylammonium was used to fix $\Delta_{DCE}^W\phi$, a Butler-Volmer's behaviour was observed. Electrolytes in the aqueous phase have a "buffering" effect at high concentration; the rate constant is unchanged by modification of perchlorate concentration in the presence of NaCl 2 M or Li_2SO_4 1.5 M. From these results it also appears that the rate of the reaction is almost the same in DCE and NB; however, the transfer coefficient of the electron transfer is much higher in the latter solvent. The authors of the study conclude that *"There is no single theory for ET at ITIES that can explain all of the experimental observations... care should be taken when using potential-determining ions to establish the interfacial potential drop"*. They attributed these observations to the possibility of strong ion-pair formation [125]. From these works about ferrocenes, it is obvious that the influence of the addition of an electrolyte on ET at the liquid-liquid interface is not really known. The situation at the interface is extremely complicated, due to the number of electrolytes and reacting species, as well as to the lack of knowledge about the exact state of the interface; for example, the effect of ion pairing cannot be quantified.

4.2   Bisphthalocyanines

The bisphthalocyanine of lutetium(III), $Lu[Pc]_2$, a sandwich complex of a metal ion with two conjugated cyclic ligands, is very similar to ferrocene over which it has, nevertheless, some advantages. $Lu[Pc]_2$ can be oxidized to $Lu[Pc]_2^+$ but is also reducible to $Lu[Pc]_2^-$, only 0.4 V below the oxidation potential. The phthalocyanines are much larger macrocycles, so that the charge density after the one-electron oxidation is lower than on ferrocenium. The solubility of the neutral species and of its oxidized or reduced forms is undetectable in water, which means that there will be no ion transfer coupled to the electron transfer; the drawback is also a poor solubility in solvents. Grafting substituents on the peripheral benzene rings of the macrocycle is rather easy and modifies the solubility as well as the redox potential [193]. One major advantage is the high stability of these complexes. Schiffrin and coworkers studied the electron exchange between $Lu[Pc]_2$ in DCE, with a tetraphenylarsonium (TPAs$^+$) salt as the electrolyte, and $Fe(CN)_6^{3-/4-}$ in water (1.5 M Li_2SO_4). The concentration of $Fe(CN)_6^{3-}$ and $Fe(CN)_6^{4-}$ were higher than that of the bisphthalocyanine, at least 100 times, so that diffusion in the aqueous phase was not limiting [139]; another consequence was that the Fermi level of the aqueous phase remained constant along an experiment and that the aqueous phase behaved as a metallic electrode, more or less. The apparent rate constant, evaluated from cyclic voltammetry experiments, was found to be $k' = 0.9 \times 10^{-3}$ cm s^{-1}, one order of magnitude lower than the rate of Lu[Pc]$_2$ electron exchange at a metallic electrode. The transfer coefficient and rate constant for that ET, at half-wave, were also estimated from AC impedance spectroscopy: $\alpha_o = 0.65$, $k_s = 2.7 \times 10^{-3}$ cm s^{-1}, $k_o = 0.07$ cm s^{-1} M^{-1}; the potential dependence of the transfer coefficient was attributed to interfacial ion pairing between the anionic species of the aqueous phase and TPAs$^+$ from the solvent [86]. At a micro ITIES, under very similar experimental conditions, the apparent rate constant for the substituted lutetium bisphthalocyanine, Lu[(tBu)$_4$Pc]$_2$, was measured to be $k' = 2 \times 10^{-3}$ cm s^{-1} in dichloromethane, the apparent transfer coefficient for the interfacial redox reaction being $\alpha = 0.65$ [142].

From the Nernst equation (8) corresponding to the interfacial redox reaction, it can be derived that

$$RT \ln \left[\frac{a_{O_1}^{W} \, a_{R_2}^{S}}{a_{R_1}^{W} \, a_{O_2}^{S}} \right] = F \left[\Delta_w^s \phi - \Delta_w^s \phi_{ET}^{\circ} \right] \tag{28}$$

the driving force of the electron exchange being $[\Delta_W^S \phi - \Delta_W^S \phi_{ET}^{\circ}]$. The ratio a_O^W / a_R^W, can be maintained constant, using for example Fe(CN)$_6^{3-}$ and Fe(CN)$_6^{3-}$ in larger concentration than that of the redox couple in the organic phase. $\Delta_W^S \phi$, the Galvani potential difference between the two phases can be modified by electrochemical techniques or by the partition of a common ion in the two media, as expressed by (5). This has been frequently used for the study of the potential dependence of ET, by SECM, and was reported for the first time for the reaction of tin(IV) bisphthalocyanine in 1,2-dichloroethane with aqueous Fe(CN)$_6^{3-/4-}$ [141]. Figure 8 illustrates the evolution of the ratio Sn(Pc)$_2$/Sn(Pc)$_2^+$, measured by spectrophotometry in DCE, when $\Delta_W^S \phi$ is varied by modifying the concentration of tetraethylammonium or tetrapropylammonium tetraphenylborates in the solvent; the chlorides of the same cations were the electrolytes in the aqueous phase. As expected, log(Sn(Pc)$_2$/Sn(Pc)$_2^+$) is a linear function of $\Delta_W^S \phi$, the slope being close to the theoretical value of 58 mV/log .

4.3 TCNQ

Tetracyanoquinodimethane (TCNQ) was mentioned for the first time by Kihara *et al.*, as a possible oxidant in nitrobenzene, in a work where these authors reviewed different redox systems in both phases; they measured half-wave potentials from polarograms obtained at an aqueous electrolyte solution dropping electrode [108]. The half-wave potential of TCNQ / TCNQ$^{\bullet-}$ in DCE was measured by cyclic voltammetry and the rate constant by impedance spectroscopy, Fe(CN)$_6^{3-/4-}$ being the redox system in water; the authors of the

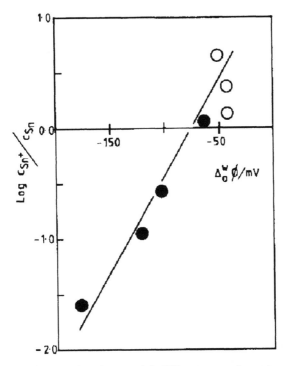

Figure 8. Influence of the Galvani potential difference on the redox reaction between Sn(Pc)$_2$ in DCE and Fe(CN)$_6^{3-/4-}$ in water. Plot of log[$C_{Sn(Pc)_2^+}$ / $C_{Sn(Pc)_2}$] vs $\Delta_S^W\phi$ established with TPA$^+$ (●) or TPE$^+$ (○) salts. Reproduced with permission from Cunnane, V.J.; Schiffrin, D.J.; Beltran, C.; Geblewicz, G.; Solomon, T. *J. Electroanal. Chem. 247*, 203 (1988). Copyright Elsevier.

study observed a potential-dependence of the transfer coefficient that they attributed to interfacial ion pairing [182]. This molecule dissolved in a lipid layer is able to shuttle the electron between two redox systems, Fe(CN)$_6^{3-/4-}$ in an aqueous phase and Ru(TPP)(py)$_2^{+/\circ}$ in a solvent, on both sides of the membrane [140]. From cyclic voltammetry at the interface between DCE and aqueous Fe(CN)$_6^{3-/4-}$, it appears that changing the ratio Fe(CN)$_6^{4-}$/ Fe(CN)$_6^{3-}$ shifts the potential at which ET occurs, but not as much as expected [121]. As shown by UV-vis spectroelectrochemistry, the reduced species produced at the interface between DCE and water has a spectrum identical to that of TCNQ$^{\cdot-}$ in acetonitrile [87,131]. The formation of the radical anion at the water/DCE interface has also been proved by EPR [103,104]. Techniques such as SECM or MEMED, at nonpolarizable interfaces where the driving force of the interfacial ET is adjusted through variations in the concentration of a common ion in the two phases, seem more adapted for the study of kinetics [143,144]. Applied to the transfer involving TCNQ / TCNQ$^{\cdot-}$, these methods clarified matters concerning

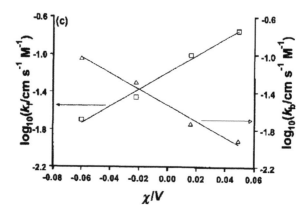

Figure 9. Dependence of the ET between TCNQ (in NB) and $Fe(CN)_6^{4-}$ (in water). Tafel plots for the forward (k_f) and backward (k_b) bimolecular rate constants. Reproduced with permission from Zhang, J.; Unwin, P.R. *Phys. Chem. Chem. Phys.* **4**, 3820 (2002).

the interfacial electron transfer. Experiments using MEMED suggest that $TCNQ^{\cdot-}$ probably transfers to water from nitrobenzene [46]. These techniques have proved that the forward and backward rates of the electron transfer between TCNQ and $Fe(CN)_6{}^{4-}$ are potential dependent, suggesting that Butler-Volmer's formalism applies at the liquid-liquid interface (Figure 9), at least for this redox system under the experimental conditions used for these studies [46,128,146]. The ET transfer coefficient, close to 0.5-0.6, is not significantly influenced by the ionic strength of the aqueous phase; the rate constants near equilibrium is k_f = k_b = 0.06 cm s^{-1} M^{-1}.

Even at the interface between DCE and a frozen aqueous phase, a Butler-Volmer's behaviour for TCNQ / $Fe(CN)_6^{4-}$ has been observed [129]. The comparison of the measured rate constants and Marcus theory applied to ITIES indicates that ET occurs through a thin interfacial boundary, on Angstrom scale [46,146].

4.4 Porphyrins

The oxidized form of zinc tetraphenyl porphyrin, $Zn(TPP)^+$, has been used by Bard's group as an acceptor of the electron exchanged at the interface with an aqueous solution of $Ru(CN)_6^{4-}$. This was one of the first studies by SECM [152], in benzene, which showed the potentiality of that technique. As the potential difference at ITIES is fixed, not by an external voltage but by a common ion in both phases, ClO_4^- in the present case, no current flows and even poorly conducting media can be studied. Tafel's plots were obtained for variation of log k_f with the interfacial potential drop; Butler-Volmer's model is valid at ITIES, in that case. As the transfer coefficient of ET is $\alpha = 0.5$, it has been concluded that the electron transfer occured at a very thin boundary, and not after penetration of the solutes in a thick mixed solvent layer. Deviations observed at higher concentration of the electrolytes were attributed to interactions between the tetrahexylammonium cation and $Ru(CN)_6^{4-}$, by precipitation at the interface rather than by interfacial adsorption [153]. Other redox systems have been used in the aqueous phase, allowing modification of the driving force of the interfacial reaction [119,122,126].

The dependence of the bimolecular rate constant on the driving force, at the interface between aqueous solutions and solvents (benzene, benzonitrile, DCE), is shown in Figure 10. Not only was the Butler-Volmer's behaviour confirmed at moderate driving force, but also, and for the first time, the existence of a Marcus inverted region for ET at the LL interface, when much more negative redox mediators like $Co(sepulchrate)^{3+/2+}$ or $V^{3+/2+}$ were used. The influence of the ionic strength on the measured kinetic parameters was demonstrated, explaining discrepancy between measurement by different groups. It was also shown that a phosphatidylcholine layer, forming at the interface between water and the organic phase, lowers the rate of ET which decreases with the length of the carbon chain of adsorbed lipids, as predicted by theory [122]; the properties of phosphatidylserine monolayers are similar [149].

About the reaction of $Zn(TPP)^{+/o}$ with an aqueous solution of $Ru(CN)_6^{3-/4-}$, it has been observed later that the reverse of the reaction studied by Tsionsky,

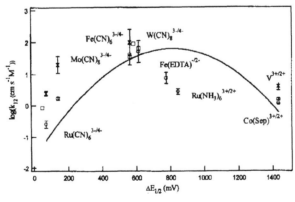

Figure 10. Dependence of the bimolecular rate constant of the electron transfer between aqueous reductants and ZnTPP+ in different solvents on the driving force of the reaction. Reproduced with permission from From Ding, Z.; Quinn, B.M.; Bard, A.J. *J. Phys. Chem. B 105*, 6367 (2001).

the oxidation of Zn(TPP)° by Ru(CN)$_6^{3-}$ in benzene and in nitrobenzene, is not potential dependent when $\Delta_w^S\phi$ is fixed by ClO$_4^-$ [149,151]. The authors explained the difference between the observations about the backward and the forward reactions from Schmikler's model for ITIES [42]: Zn(TPP)$^+$ should accumulate at the interface with the increase of $\Delta_w^S\phi$, but the concentration of Zn(TPP)$^\circ$ should be potential-independent. These authors noted that this had not been observed before, during the studies of the oxidation processes of TCNQ, Fc, Me$_{10}$Fc, Lu[Pc$_2$] or Sn[Pc$_2$]. However, when the concentration of tetrabutylammonium ion was varied in the aqueous phase, in order to modify the interfacial potential drop, a variation of the rate constant of Zn(TPP)° oxidation was observed [149]. This particular behaviour of ZnTPP at the interface was interpreted as being the result of its adsorption, influenced differently by ClO$_4^-$ which does not interfere with the interface, and TBA$^+$ which adsorbs and even forms a film at the solvent/water boundary.

4.5   Photoinduced electron transfer

Considering the concern of chemists about photosynthesis, for the production and storage of energy from solar light, the study of various heterogeneous systems like micelles, microemulsions and vesicles is not surprising; these liquid-liquid interfaces are expected to be useful for the separation of charges. In this respect, photoinduced electron transfer at liquid-liquid interfaces is of particular interest. Bearing in mind how complex is the charge transfer at ITIES, it is understandable that this, combined with a photoinduced process, will be even more difficult to analyse; this has been discussed in reviews by Kotov and Kuzmin [96,97] as well as by Fermin and Lahtinen [6]. Photocurrents detected at interfaces designed for the observation of photoinduced electron transfer could in fact be produced by the transfer of other ionic species generated by illumination. Works where irradiation of porphyrins in the organic phase induces the production of radicals from the excited states are good examples of this; these species generate then charge carriers via reaction with solutes [194,195].

The remarkable photoredox properties of the ruthenium-trisbipyridine complex explain its frequent use in such studies. A photocurrent was observed when an aqueous solution of Ru(bpy)$_3^{2+}$ was illuminated, in contact with a thin layer of a lipophilic viologen, 1,1'-diheptyl-4,4'-bipyridinium (C$_7$V^{2+}); this was attributed to the excitation followed by electron transfer [158]:

$$Ru(bpy)_{3,W}^{2+} \xrightarrow{h\nu} Ru(bpy)_{3,W}^{2+*} \tag{29}$$

$$Ru(bpy)_{3,W}^{2+*} + C_7V_{DCE}^{2+} \rightarrow Ru(bpy)_{3,W}^{3+} + C_7V_{DC}^{\bullet+} \tag{30}$$

Other similar works with methylviologen report photocurrents attributed to a light-induced interfacial electron transfer [161,162]; however, the evolution of the photocurrents casts doubts on the origin of the charge transfer which is observed [6]. By showing that ion-transfer could not occur under the experimental conditions, it has been demonstrated that the photoinduced charge separation at the liquid-liquid interface occured when TCNQ was used as the quencher for $Ru(bpy)_{3,W}^{2+*}$ [155]. The reduction of $Ru(bpy)_{3,W}^{2+*}$ to $Ru(bpy)_{3,W}^{+}$ by an interfacial layer of an amphiphilic ferrocene derivative was unambiguously established by second harmonic generation at the interface [156]. Time-resolved spectroscopy at the interface has been used to characterize the quenching of the laser-induced fluorescence of $Eu(III)_W$ by anthracene in DCE, and to measure the rate constant $(1.9 \times 10^{-4} \text{ m}^4 \text{ mol}^{-1} \text{ s}^{-1})$ [105]. Analysis of the photocurrent evolution, of its potential dependence and of the concentration influence has been used to distinguish between ET, homogeneous transfer and ion transfer. This was studied for example in the case of the quenching of the excited states of water soluble zinc porphyrins or dimers, by ferrocenes, TCNQ, benzoquinone in DCE [148,165-168,184]. The interfacial electron transfer between a lipophilic form of $Ru(bpy)_3^{3+}$ in benzonitrile and oxalate generates CO_2^{-} which, by reaction through the interface with a second $Ru(bpy)_3^{3+*}$ produces the excited state $Ru(bpy)_3^{2+*}$, which emits light:

$$Ru(bpy)_{3,BN}^{2+*} \rightarrow Ru(bpy)_{3,BN}^{2+} + h\nu \qquad (31)$$

This electrogenerated chemiluminescence process at the liquid-liquid interface suggests a Marcus inverted region behaviour [160]. Recently was examined the electron transfer between a series of ferrocene derivatives in DCE and the triplet state of a pair S of the zinc *meso-(p*-sulfonatophenyl)$_4$-porphyrin and zinc *meso-(N*-methylpyridyl)$_4$-porphyrin [166]. The heterodimer was adsorbed at the water-DCE interface and the redox potential of S*/S$^-$ was estimated to be 1.00 V vs SHE. As the the five ferrocene derivatives have redox potentials from 0.64 V to 0.07 V, $\Delta G_{ET}^{o\prime}$ the standard Gibbs energy of electron transfer (eq. 27) extends over a range of almost 1 V. The dependence of the bimolecular electron transfer rate constant, k_{ET}^{II}, on $\Delta G_{ET}^{o\prime}$ is presented in Figure 11. The reorganization energy, estimated to be $\lambda = 1.05$ eV and due primarily to solvent reorganization, corresponds to an average distance of 0.8 nm between the redox centers; the electron transfer occurs at a sharp liquid-liquid boundary. The maximum value of the rate constant is $k_{ET}^{II} = 3 \times 10^{-19}$ cm^4 s^{-1}. The mechanism of the photoreduction of quinones mediated by this porphyrin ion pair has been studied by ultrafast time-resolved spectroscopy and dynamic photoelectrochemistry [196].

Efficient decarboxylation of phenoxyacetic acid derivatives in reverse micelles (benzene), followed by reaction, was induced by flash photolysis of methylene

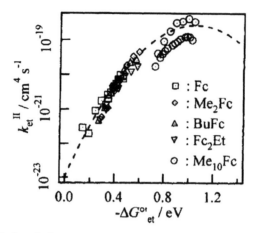

Figure 11. Photoinduced electron transfer at the water/DCE interface, between ferrocene derivatives and the heterodimer of porphyrins (ZnTPPS^{4-}/ZnTMPyP^{4+}). The dimer is adsorbed and/or aggregated at the interface and illuminated at 477 nm. Reproduced with permission from Eugster, N.; Fermín, D.J.; Girault, H.H. *J. Phys. Chem. B 106*, 3428 (2002).

blue in water; these studies suggest that the liquid-liquid interface controls the photocatalysed electron transfer [197,198].

4.6 Liquid film electrodes

A nitrobenzene film electrode was used as early as 1988 for the study of the photoassisted electron transfer between heptyl-4,4'-bipyridinium and aqueous Ru(bpy)$_3^{2+}$ [158]. Later on, the formation of a micrometer-thick layer of aqueous solution at a platinum tip electrode, in nitrobenzene, was described during a SCEM experiment [157]. These were the first reports about the use of thin organic films in contact with aqueous phases. Oxidation of tetrahexylphenylene diamine deposited as microdroplets on solid electrode surfaces has been studied by voltammetry, in an attemp to get information on the behaviour of oil droplets and emulsions [49]. Droplets of nitrobenzene or dichloroethane, containing a neutral redox active couple, have been attached to paraffin impregnated graphite electrodes, in contact with an aqueous solution of a salt. The organic medium has a low electric conductivity so that the three-phase junction at the periphery of the solvent droplet, where the two liquid phases and the electrode are in contact, is supposed to play a particular role; the redox reaction should occur there, because the charge compensation by ion transfer is faster at the junction where the solvent layer is thinner. Decamethylferrocene (or ferrocene) oxidation in nitrobenzene, or dichloroethane [56,57,59,199,200], but also iodine reduction [58,201] have been utilized to study ion transfer processes through the liquid-liquid interface between solvent droplets on platinum or carbon electrodes in contact with water. In square-wave voltammetry for example, the peak potential for the oxidation of $Me_{10}Fc$ varies linearly with $\Delta_W^S \phi_i^o$, for various anions in the aqueous phase [202]. Such devices can also be used for the observation of photocurrents [61].

Organic solvents wet a pyrolytic graphite electrode when its surface is perpendicular to the graphite planes; they spread over the electrode surface. These modified electrodes, in contact with aqueous solutions, constitute liquid-liquid thin layer cells. This has been developed by Anson and coworkers who used mostly nitrobenzene as the organic phase, in a ~30µm thick film. Ferrocene and Co(TPP), with $Fe(CN)_6^{3-}$ in water, were used to test the electrode and to demonstrate that the electron transfer can be observed at the liquid-liquid interface [54]. Various redox systems in the aqueous phase as well as in different solvents were then studied by cyclic voltammetry [109]. Microdroplets and thin films of an imidazolium tetrafluoroborate, an ionic liquid at room temperature, have been deposited on graphite electrodes and studied by voltammetry; it was shown that $Fe(CN)_6^{3-}$ accumulates from the aqueous phase in the layer, by ion exchange with the BF_4^- anion from the ionic liquid [60].

Electrodes composed on graphite particles embedded in a porous glass matrix, prepared by a sol gel process are volumic versions of the "three-phase junction electrodes". When soaked with organic solvents in which $Me_{10}Fc$ is dissolved, an anion exchange with the aqueous phase limits the electron transfer; however, the contact between the two liquid media is unknown [203,204]. When an alkanethiol gold modified electrode, covered with a very thin film of polypeptides (5-30 nm), in which are incorporated water molecules and $Fe(CN)^{3-/4-}$, is in contact with DCE, the interfacial redox process is governed by the ion exchange through the interface between the organic solvent and the film. Electron transfer occurs when the modified gold electrode is in contact with DCE containing $Me_{10}Fc$, but the factors governing ET at the interface are not clearly established [205].

Bimolecular rate constants have been measured for the electron transfer between Fc, $Me_{10}Fc$, Zn(TPP), Co(TPP) in NB and various redox reactants in water [53,55,206]. The comparison with more classical ITIES measurements is not easy because really comparable experiments are scarce. However, it can be said that the rate of ET measured under such conditions should not be sensible to ion transfer effects; this seems to be verified even in the case of proton-consuming electrode processes [207]. In experiments with Zn(TPP) at the NB/H_2O interface, it was speculated that the apparent insensitivity of the rate constant to the driving force of the reaction was a consequence of the formation of the precursor inside the diffuse layer, following Schmickler's hypothesis. In the case of

TCNQ/ $Fe(CN)^{3-/4-}$ reaction, on the contrary, the forward and reverse rate constants are influenced by the driving force of the reaction [51]. Models for voltammetry with a thin organic layer modified electrode [208] and for electron transfer across the liquid-liquid interface [209] have been studied. From these studies it appears that the experiments with thin organic films must be carefully designed and that conclusions should be drawn cautiously; after these warnings, the way to choose the proper concentrations of the coreactants has been specified [52].

This solvent layer supported on graphite is a type of a modified electrode, not fundamentally different from other electrodes with surface-films, which presents some advantages over the conventional cell for the study of the liquid-liquid interface. Provided that the comparison with measurements from more conventional techniques proves that it is an adequate tool for ET studies at ITIES, its convenience will attract people; the electrode with a thin layer of an organic solvent will at least be very useful for analytical purposes.

5   APPLICATIONS

Examples cited above illustrate that the electron transfer across the liquid-liquid interface has been the matter of many studies aimed at the knowledge of this heterogeneous exchange. However, these reactions at the interface between two different liquid media can be applied to reach goals which could not be achieved through more conventional ways; knowing how to properly take advantage of the physical barrier constituted by ITIES is a way to control processes.

Synthetic organic reactions such as biphasic redox reactions, accelerated by phase-transfer catalysis, has been cited [10,186,197,198]; electrochemistry at the liquid-liquid interface was used to study the two-phase oxidation of cyclooctene by permanganate [210]. Separation of the reactants is also useful for many other reactions and in the preparation of some materials, the interface between the two media contributes to the process of film formation. For the study of molecular magnets, $[Fe^{III}(C_5Me_5)_2]_3[Fe^{III}(CN)_6]_2$ has been prepared by ET and ion-pair extraction [191]. The deposition of metals, Ag and Cu, from the reduction of their aqueous salts by a $V(CO)_6^-$ solution in DCE is often reported as one of the first examples of ET [185]; the reduction of cations at the water/organic interface has been applied in a few studies for the preparation of films of zinc [211], nanoparticles of silver [212] and gold clusters [213]. It has already been mentioned in the present review that redox biphasic processes are utilized for the preparation of colloidal gold [11,12]. The preparation of films and particles of conducting polymers at the surface of aqueous solutions has been abundantly used [214] but, more recently, oligomers and polymer layers of pyrroles or thiophenes have been prepared by electrosynthesis [215-221].

The study of the oxidation of N,N'-tetramethyl-1,4-phenylenediamine by $Ru(NH_3)_6^{3+}$ has been examined between two immiscible liquid streams, in a microelectrochemical reactor. Stable hydrodynamic conditions have been obtained so that such a device looks very promising for the development of liquid-liquid flow injection systems for analysis [222].

As many physiological reactions such as photosynthesis and respiration involve ion or electron exchange through membranes, the study of charge transfers across the liquid-liquid interface has focused attention as a way to collect information about these biological processes. The various aspects of these fundamental problems have been examined: energetics and kinetics of multielectron processes, reverse electron transfer, coupling with proton exchange, [38,223-228]. Experiments have been done with chlorophyll adsorbed at a solvent-water interface [229-231], or porphyrins as model compounds [232], or both [233,234]. A biphasic system can mimic a respiration reaction, the oxydation of pyruvate to CO_2 and acetate by oxygen, accompanied by the transfer of Na^+ across the interface between water and nitrobenzene;   the mediators are flavin mononucleotide in water and $Me_{10}Fc$ in the solvent [235].

In the domain of energy conversion, a solar cell has been described, based on the photoinduced electron transfer between dimers of zinc porphyrins in the aqueous phase and TCNQ as a quencher in DCE. The overall solar energy conversion is low, 0.05 %, but the authors think that these very thin devices based on a liquid-liquid interface should present advantages [164]. A system based on Zn(TPP) and $Ru^{III/II}(NH_3)_5(bpy)$, presented as a reversible molecular switch, functions as a light-driven cell [236]. In fuel cells, the oxidizer and reductant are separated to prevent their direct chemical reaction and force them to exchange electrons through the external electrical circuit. The physical barrier must permit the circulation of the electrical current between the two compartments of the cell and allow the exchange of ions. In a biofuel cell based on the oxidation of glucose by an organic peroxide, it has been shown that the interface between water and dichloromethane prevents the direct electron exchange across the liquid-liquid interface but allows the current circulation by ion transfer; more attention should have been devoted to the study of this phenomenon [237].

6   CONCLUSION

During the last one or two decades, the study of the liquid-liquid interface has progressed tremendously, in the fields of the ion and of the electron transfer studies. Not only were examined interfacial reactions between various redox systems in the aqueous phase as well as in the organic solvents, but also photo-assisted charge transfers. Models of the liquid-liquid interface have been developed, with the hope to find some agreement between the experiments and the theoretical treatments. However, the most impressive development in that field of electrochemistry was certainly experimental. Many new devices have been used for electrochemical studies at ITIES, such as micropipettes, microholes, droplets and films on electrodes; the most fruitful and promising among these tools is scanning electrochemical microscopy (SECM). What is new also, in the recent improvements, is the occurrence of non electrochemical methods, such as the many spectroscopic techniques available for the observation of the liquid-liquid interface, particularly the optical developments like second harmonic generation. The situation for the experimental studies as been fundamentally improved, so that this raises hopes for a much better observation and understanding of the charge transfer reactions at the liquid-liquid interface in the future.

However, from a theoretical point of view, the agreement between experiments and models is much less satisfactory. The solvent distribution, as well as the charge repartition and the resulting potential drop between the two phases, affect the interfacial transfer reactions; these facts are not known with a very high precision. When one refers to the surface of the electrode in the case of a solid-solution interface, the separation with the solution is rather sharp and the situation does not evolve much with time, along an experiment. At the frontier between two immiscible liquid media, the situation is much more fuzzy, ambiguous. Schmickler, in 1996, noted "... *our double-layer theories for ITIES are simply not accurate enough to furnish reliable estimates*" [238]; since, the situation has evolved and the knowledge of phenomena at ITIES has been improved but not to the point where things become clear. As an example, some electron transfer reactions at the liquid-liquid interface seem to obey Butler-Volmer's law and others do not, as examplified by SECM experiments where the potential drop is established through the presence of a common ion in both phases. It seems highly probable that this can be interpreted as the fact that the interfacial potential difference does influence the electron exchange reaction, as predicted by Butler-Volmer's law, but that the effective potential drop applied to the electron transfer is not always exactly known. Indeed, due to the nature of the interfacial region leading to uncertainties about the dissociation of the electrolytes, their precipitation or adsorption, the real concentration of ions is not always known with a high accuracy so that the uncertainty about the inner potential drop at the liquid-liquid interface is large.

Experimental tools for the study of the interface between two liquid phases have recently been considerably developed and will probably hold the promise of a better and more precise information about the structure of the interfacial region. It can reasonably be expected that the knowledge about the ion and electron transfer processes through the interface between two immiscible liquid phases will be developed and become more accurate during the coming years.

REFERENCES

1. Cunnane, V. J.; Murtomäki, L., *Liquid/liquid interfaces. Theory and methods.*, A. G. Volkov and D. W. Deamer, Eds., CRC Press, Boca Raton, 1996.
2. Girault, H. H., *Modern Aspects of Electrochemistry*, J. O. M. Bockris et al., Eds., Plenum Press, New York and London, 1993.
3. Girault, H.; Schiffrin, D., *Electroanalytical Chemistry. A Series of Advances*, A. Bard, Ed., Marcel Dekker, INC., New York and Basel, 1989.
4. Koryta, J.; Vanysek, P., *Advances in Electrochemistry and Electrochemical Engineering*, *12*, 113 (1981).
5. Schmickler, W., *Annual Reports on the Progress of Chemistry, Sect. C.*, *95*, 117 (1999).
6. Fermin, D.; Lahtinen, R., *Surfactant Science Series*, *95* (*Liquid Interfaces in Chemical, Biological, and Pharmaceutical Applications*), 179 (2001).
7. Benjamin, I., *Progress in Reaction Kinetics and Mechanism - Science Reviews*, *27*, 87 (2002).
8. Gibbs, J. W., *Collected works*, Vol. 1, Langmans Green & Co, New York, 1928.
9. Nernst, W.; Riesenfeld, E. H., *Ann.Phys.*, *8*, 600 (1902).
10. Bell, R. P., *J. Phys. Chem.*, *32*, 882 (1928).
11. Brust, M.; Walker, M.; Bethell, D.; Schiffrin, D. J.; Whyman, R., *J. Chem. Soc., Chem. Commun.*, 801 (1994).
12. Faraday, M., *Philos. Trans. R. Soc. London*, *147-181*, 145 (1857).
13. Guastalla, J., *J. Chim.Phys.*, *53*, 470 (1956).
14. Dupeyrat, M.; Guastalla, J., *J. Chim. Phys.*, *53*, 469 (1956).
15. Dupeyrat, M., *J. Chim. Phys.*, *61*, 306 (1964).
16. Dupeyrat, M., *J. Chim. Phys.*, *61*, 323 (1964).
17. Gavach, C., *C. R. Acad. Sci. Paris*, *269*, 1356 (1969).
18. Samec, Z., *J. Electroanal. Chem.*, *99*, 197 (1979).
19. Koryta, J., *Anal. Chim. Acta*, *91*, 1 (1977).
20. Lyons, M. E. G., *Annual Reports on the Progress of Chemistry, Section C: Physical Chemistry (1991)*, *87*, 119 (1990).
21. Watts, A.; Van der Noot, T. J., *Liquid/liquid interfaces. Theory and methods*, A. G. Volkov and D. W. Deamer, Eds., CRC Press, Boca Raton, 1996.
22. Benjamin, I., *Annu. Rev. Phys. Chem.*, *48*, 407 (1997).
23. Benjamin, I., *Chem. Rev.*, *96*, 1449 (1996).
24. Benjamin, I., *Liquid/liquid interfaces. Theory and methods.*, A. G. Volkov and D. W. Deamer, Eds., CRC Press, Boca Raton, 1996.
25. Kakiuchi, T., *Liquid/liquid interfaces. Theory and methods*, A. G. Volkov and D. W. Deamer, Eds., CRC Press, Boca Raton, 1996.
26. Samec, Z., *J. Electroanal. Chem.*, *103*, 1 (1979).
27. Vanysek, P., *Electrochemistry on Liquid-Liquid Interfaces*, Springer-Verlag, Berlin, 1985.

28. Senda, M.; Yamamoto, Y., *Liquid/liquid interfaces. Theory and methods*, A. G. Volkov and D. W. Deamer, Eds., CRC Press, Boca Raton, 1996.
29. Samec, Z., *Liquid/liquid interfaces. Theory and methods.*, A. G. Volkov and D. W. Deamer, Eds., CRC Press, Boca Raton, 1996.
30. Marcus, R. A., *J. Chem. Phys.*, *43*, 679 (1965).
31. Albery, W., *Annu. Rev. Phys. Chem.*, *31*, 227 (1980).
32. Marcus, R. A., *J. Phys. Chem.*, *94*, 4152 (1990).
33. Marcus, R. A., *J. Phys. Chem.*, *94*, 1050 (1990).
34. Marcus, R. A., *J. Phys. Chem.*, *95*, 2010 (1991).
35. Marcus, R. A., *J. Electroanal. Chem.*, *438*, 251 (1997).
36. Marcus, R. A., *J. Phys. Chem. B*, *102*, 10071 (1998).
37. Marcus, R. A., *J. Electroanal. Chem.*, *483*, 2 (2000).
38. Kharkats, Y. I.; Volkov, A. G., *J.Electroanal.Chem.*, *184*, 435 (1985).
39. Girault, H. H.; Schiffrin, D. J., *J.Electroanal.Chem*, *244*, 15 (1988).
40. Sutin, N., *Acc. Chem. Res.*, *15*, 275 (1982).
41. Newton, M. D.; Sutin, N., *Annu. Rev. Phys. Chem.*, *35*, 437 (1984).
42. Schmickler, W., *J. Electroanal. Chem.*, *428*, 123 (1997).
43. Koryta, J.; Vanysek, P.; Brezina, M., *J. Electroanal. Chem.*, *67*, 263 (1976).
44. Slevin, C. J.; Unwin, P. R., *Langmuir*, *13*, 4799 (1997).
45. Zhang, J.; Slevin, C. J.; Murtomaki, L.; Kontturi, K. S.; Williams, D. E.; Unwin, P. R., *Langmuir*, *17*, 821 (2001).
46. Zhang, J.; Unwin, P. R., *Phys. Chem. Chem. Phys.*, *4*, 3820 (2002).
47. Slevin, C. J.; Unwin, P. R., *Langmuir*, *15*, 7361 (1999).
48. Slevin, C. J.; Zhang, J.; Unwin, P. R., *J. Phys. Chem. B*, *106*, 3019 (2002).
49. Marken, F.; Webster, R. D.; Bull, S. D.; Davies, S. G., *J. Electroanal. Chem.*, *437*, 209 (1997).
50. Fujihira, M.; Yanagisawa, M.; Kondo, T., *Bull. Chem. Soc. Jpn*, *66*, 3600 (1994).
51. Shi, C.; Anson, F. C., *J. Phys. Chem. B*, *105*, 8963 (2001).
52. Shi, C.; Anson, F. C., *J. Phys. Chem. B*, *105*, 1047 (2001).
53. Shi, C.; Anson, F. C., *J. Phys. Chem. B*, *103*, 6283 (1999).
54. Shi, C.; Anson, F. C., *Anal. Chem.*, *70*, 3114 (1998).
55. Shi, C.; Anson, F. C., *J. Phys. Chem. B*, *102*, 9850 (1998).
56. Komorsky-Lovric, S.; Lovric, M.; Scholz, F., *Collect. Czech. Chem. Commun.*, *66*, 434 (2001).
57. Komorsky-Lovric, S.; Lovric, M.; Scholz, F., *J. Electroanal. Chem.*, *508*, 129 (2001).
58. Mirceski, V.; Scholz, F., *J. Electroanal. Chem.*, *522*, 189 (2002).
59. Scholz, F.; Komorsky-Lovric, S.; Lovric, M., *Electrochem. Commun.*, *2*, 112 (2000).
60. Wadhawan, J.; Schröder, U.; Neudeck, A.; Wilkins, S.; Compton, R. G.; Marken, F.; Consorti, S.; de Souza, R.; Dupont, J., *J. Electroanal. Chem.*, *493*, 75 (2000).
61. Wadhawan, J.; Compton, R. G.; Marken, F.; Bull, S.; Davies, S., *J. Solid State Electrochem.*, *5*, 301 (2001).
62. Taylor, G.; Girault, H. H., *J. Electroanal. Chem.*, *208*, 179 (1986).
63. Shao, Y. H.; Mirkin, M. V., *Anal. Chem.*, *70*, 3155 (1998).
64. Campbell, J.; Girault, H. H., *J. Electroanal. Chem.*, *266*, 465 (1989).

65. Peulon, S.; Guillou, V.; L'Her, M., *J. Electroanal. Chem.*, *514*, 94 (2001).

66. Liu, B.; Mirkin, M. V., *Electroanalysis*, *12*, 1433 (2000).

67. Liu, B.; Mirkin, M. V., *Surfactant Science Series*, *95*, 373 (2001).

68. Homolka, D.; Hung, L. Q.; Hofmanova, A.; Khalil, M.; Koryta, J.; Marecek, V.; Samec, Z.; Sen, S.; Vanysek, P.; Weber, J.; Brezina, M.; Janda, M.; Stibor, I., *Anal. Chem.*, *52*, 1606 (1980).

69. Marecek, V.; Samec, Z., *J. Electroanal. Chem.*, *185*, 263 (1985).

70. Gavach, C.; Henry, F., *J. Electroanal. Chem.*, *54*, 361 (1974).

71. Makrlik, E., *Collect. Czech. Chem. Commun.*, *50*, 1636 (1985).

72. Stewart, A. A.; Campbell, J. A.; Girault, H. H.; Eddowes, M., *Ber. Bunsen-Ges. Phys. Chem.*, *94*, 83 (1990).

73. Makrlik, E., *Z. Phys. Chem. (Leipzig)*, *268*, 826 (1987).

74. Makrlik, E., *Z. Phys. Chem. (Leipzig)*, *268*, 59 (1987).

75. Makrlik, E., *Z. Phys. Chem. (Leipzig)*, *268*, 507 (1987).

76. Koryta, J.; Vanysek, P.; Brezina, M., *J. Electroanal. Chem.*, *75*, 211 (1977).

77. Makrlik, E., *Z. Phys. Chem. (Leipzig)*, *268*, 200 (1987).

78. Makrlik, E., *Z. Phys. Chem. (Leipzig)*, *267*, 851 (1986).

79. Maeda, K.; Kihara, S.; Suzuki, M.; Matsui, M., *J.Electroanal.Chem.*, *303*, 171 (1991).

80. Reid, J.; Vanysek, P.; Buck, R., *J. Electroanal. Chem.*, *161*, 1 (1984).

81. Reid, J.; Vanysek, P.; Buck, R., *J. Electroanal. Chem.*, *170*, 109 (1984).

82. VanderNoot, T.; Schiffrin, D. J., *J. Electroanal. Chem.*, *278*, 137 (1990).

83. Marecek, V.; Gratzl, M.; Janata, J., *J. Electroanal. Chem.*, *296*, 537 (1990).

84. Vanysek, P., *Chimica Oggi*, *8*, 47 (1990).

85. Chen, Q. Z.; Iwamoto, K.; Seno, M., *Electrochim. Acta*, *36*, 291 (1991).

86. Cheng, Y.; Schiffrin, D. J., *J. Chem. Soc., Faraday Trans.*, *89*, 199 (1993).

87. Ding, J.; Fermin, D.; Brevet, P.-F.; Girault, H. H., *J. Electroanal. Chem.*, *458*, 139 (1998).

88. Bard, A. J.; Fan, F. R. F.; Mirkin, M. V., *Electroanalytical Chemistry*, A. J. Bard, Ed., Marcel Dekker, Inc., New York, 1994.

89. Mirkin, M. V.; Tsionsky, M., *Scanning Electrochemical Spectroscopy*, A. J. Bard and M. V. Mirkin, Eds., Marcel Dekker, Inc., New York, 2001.

90. Mirkin, M. V.; Horrocks, B., *Anal. Chim. Acta*, *406*, 119 (2000).

91. Martin, R.; Unwin, P. R., *J. Electroanal. Chem.*, *439*, 123 (1997).

92. Mirkin, M. V., *Mikrochim. Acta*, *130*, 127 (1999).

93. Barker, A.; Slevin, C.; Unwin, P. R.; Zhang, J., *Surfactant Science Series*, *95*, 283 (2001).

94. Tsionsky, M.; Zhou, J.; Amemiya, S.; Fan, F. R. F.; Bard, A. J.; Dryfe, R., *Anal. Chem.*, *71*, 4300 (1999).

95. Amemiya, S.; Ding, Z.; Zhou, J.; Bard, A. J., *J. Electroanal. Chem.*, *483*, 7 (2000).

96. Kuzmin, M. G.; Soboleva, I. V.; Kotov, N. A., *Anal. Sci.*, *15*, 3 (1999).

97. Kotov, N. A.; Kuzmin, M. G., *Liquid/liquid interfaces. Theory and methods*, A. Volkov and D. Deamer, Eds., CRC Press, Boca Raton, 1996.

98. De Armond, M. K.; De Armond, A. H., *Liquid/liquid interfaces. Theory and methods.*, A. Volkov and D. Deamer, Eds., CRC Press, Boca Raton, 1996.

99. Eisenthal, K. B., *Chem. Rev.*, *96*, 1343 (1996).

100. Eisenthal, K. B., *J.Phys.Chem*, *100*, 12997 (1996).

101. Brevet, P. F.; Girault, H. H., *Liquid/liquid interfaces. Theory and methods.*, A. Volkov and D. Deamer, Eds., CRC Press, Boca Raton, 1996.

102. Brevet, P. F.; Girault, H. H., *Prog. Colloid Polym. Sci.*, *103*, 1 (1997).

103. Webster, R. D.; Dryfe, R. A. W.; Coles, B. A.; Compton, R. G., *Anal. Chem.*, *70*, 792 (1998).

104. Dryfe, R. A. W.; Webster, R. D.; Coles, B. A.; Compton, R. G., *Chem. Commun.*, 779 (1997).

105. Dryfe, R. A. W.; Ding, Z.; Wellington, R. G.; Brevet, P. F.; Kuznetzov, A. M.; Girault, H. H., *J. Phys. Chem. A*, *101*, 2519 (1997).

106. Webster, R., *Phys. Chem. Chem. Phys.*, *2*, 5660 (2000).

107. Bard, A. J.; Parsons, R.; Jordan, J., Marcel Dekker, Inc., New York, 1985.

108. Kihara, S.; Suzuki, M.; Maeda, K.; Ogura, K.; Matsui, M.; Yoshida, Z., *J. Electroanal. Chem.*, *271*, 107 (1989).

109. Shafer, H.; Derback, T.; Koval, C., *J. Phys. Chem. B*, *104*, 1025 (2000).

110. Bard, A. J.; Faulkner, L., *Electrochemical Methods. Fundamentals and Applications*, John Wiley & Sons, New York, 1980.

111. Connelly, N. G.; Geiger, W. E., *Chem. Rev.*, *96*, 877 (1996).

112. Cox, B. G.; Parker, A. J., *J. Am. Chem. Soc.*, *95*, 402 (1973).

113. Alfenaar, M., *J. Phys. Chem.*, *79*, 2200 (1975).

114. Kolthoff, I. M.; Chantooni, M. K., *J. Am. Chem. Soc.*, *93*, 7104 (1971).

115. Gritzner, G.; Kuta, J., *Pure Appl. Chem.*, *56*, 461 (1984).

116. De Ligny, C. L.; Alfenaar, M.; Van Der Veen, N. G., *Recl. Trav. Chim. Pays-Bas*, *87*, 585 (1968).

117. Bond, A. M.; McLennan, E. A.; Stojanovic, R. S.; Thomas, F. G., *Anal. Chem.*, *59*, 2853 (1987).

118. Quinn, B. M.; Liljeroth, P.; Kontturi, K., *J. Am. Chem. Soc.*, *124*, 12915 (2002).

119. Ding, Z.; Quinn, B. M.; Bard, A. J., *J. Phys. Chem. B*, *105*, 6367 (2001).

120. Ulmeanu, S.; Lee, H.; Fermin, D.; Girault, H. H.; Shao, Y., *Electrochem. Commun.*, *3*, 219 (2001).

121. Quinn, B.; Kontturi, K. S., *J. Electroanal. Chem.*, *483*, 124 (2000).

122. Tsionsky, M.; Bard, A. J.; Mirkin, M. V., *J. Am. Chem. Soc.*, *119*, 10785 (1997).

123. Selzer, Y.; Mandler, D., *J. Electroanal. Chem.*, *409*, 15 (1996).

124. Zhang, J.; Slevin, C. J.; Unwin, P. R., *Chem. Commun.*, 1501 (1999).

125. Zhang, J.; Barker, A.; Unwin, P. R., *J. Electroanal. Chem.*, *483*, 95 (2000).

126. Barker, A.; Unwin, P. R.; Amemiya, S.; Zhou, J.; Bard, A. J., *J. Phys. Chem. B*, *103*, 7260 (1999).

127. Georganopoulou, D.; Strutwolf, J.; Pereira, C.; Silva, F.; Unwin, P. R.; Williams, D., *Langmuir*, *17*, 8348 (2001).

128. Zhang, J.; Unwin, P. R., *J. Phys. Chem. B*, *104*, 2341 (2000).

129. Zhang, Z. Q.; Ye, J. Y.; Sun, P.; Yuan, Y.; Tong, Y. H.; Hu, J. M.; Shao, Y. H., *Anal. Chem.*, *74*, 1530 (2002).

130. Zhang, Z. Q.; Tong, Y. H.; Sun, P.; Shao, Y. H., *Chemical journal of chinese Universities Chineses*, *22*, 206 (2001).

131. Ding, Z.; Brevet, P.-F., *Chem. Commun.*, 2059 (1997).
132. Samec, Z.; Marecek, V.; Weber, J., *J. Electroanal. Chem.*, *96*, 245 (1979).
133. Samec, Z.; Marecek, V.; Weber, J., *J. Electroanal. Chem.*, *103*, 11 (1979).
134. Samec, Z.; Marecek, V.; Weber, J.; Homolka, D., *J. Electroanal. Chem.*, *126*, 105 (1981).
135. Cunnane, V.; Geblewicz, G.; Schiffrin, D. J., *Electrochim. Acta*, *40*, 3005 (1995).
136. Quinn, B.; Lahtinen, R.; Murtomaki, L.; Kontturi, K. S., *Electrochim. Acta*, *44*, 47 (1998).
137. Hotta, H.; Akagi, N.; Sugihara, T.; Ichikawa, S.; Osakai, T., *Electrochem. Commun.*, *4*, 472 (2002).
138. Yuan, Y.; Gao, Z.; Zhang, M. Q.; Zhang, Z. Q.; Shao, Y. H., *Sci. China, Ser. B*, *45*, 494 (2002).
139. Geblewicz, G.; Schiffrin, D. J., *J. Electroanal. Chem.*, *244*, 27 (1988).
140. Cheng, Y.; Schiffrin, D. J., *J. Chem. Soc., Faraday Trans.*, *90*, 2517 (1994).
141. Cunnane, V.; Schiffrin, D. J.; Beltran, C.; Geblewicz, G.; Solomon, T., *J.Electroanal.Chem*, *247*, 203 (1988).
142. L'Her, M.; Rousseau, R.; L'Hostis, E.; Roue, L.; Laouenan, A., *C. R. Acad. Sci. Paris, t.322, serie II b*, 55 (1996).
143. Solomon, T.; Bard, A. J., *Anal. Chem.*, *67*, 2787 (1995).
144. Solomon, T.; Bard, A. J., *J.Phys.Chem*, *99*, 17487 (1995).
145. Kihara, S.; Maeda, K.; Suzuki, M.; Sohrin, Y.; Shirai, O.; Matsui, M., *Anal. Sci.*, *7*, 1415 (1991).
146. Barker, A.; Unwin, P. R.; Zhang, J., *Electrochem. Commun.*, *3*, 372 (2001).
147. Delville, M. H.; Tsionsky, M.; Bard, A. J., *Langmuir*, *14*, 2774 (1998).
148. Lahtinen, R.; Fermin, D.; Konturri, K. S.; Girault, H. H., *J. Electroanal. Chem.*, *483*, 81 (2000).
149. Liu, B.; Mirkin, M. V., *J. Phys. Chem. B*, *106*, 3933 (2002).
150. Zhang, J.; Unwin, P. R., *J. Electroanal. Chem.*, *494*, 47 (2000).
151. Liu, B.; Mirkin, M. V., *J. Am. Chem. Soc.*, *121*, 8352 (1999).
152. Tsionsky, M.; Bard, A. J.; Mirkin, M. V., *High-Resolution Study of Charge Transfer Processes at the Liquid/liquid Interface*, 1996, p 16.
153. Tsionsky, M.; Bard, A. J.; Mirkin, M. V., *J. Phys. Chem.*, *100*, 17881 (1996).
154. Ding, Z.; Wellington, R. G.; Brevet, P.-F.; Girault, H. H., *J. Phys. Chem.*, *100*, 10658 (1996).
155. Brown, A. R.; Yellowlees, L. J.; Girault, H. H., *J. Chem. Soc., Faraday Trans.*, *89*, 207 (1993).
156. Kott, K. L.; Higgins, D. A.; McMahon, R. J.; Corn, R. M., *J. Am. Chem. Soc.*, *115*, 5342 (1993).
157. Wei, C.; Bard, A. J.; Mirkin, M. V., *J. Phys. Chem.*, *99*, 16033 (1995).
158. Thomson, F. L.; Yellowlees, L. J.; Girault, H. H., *J. Chem. Soc., Chem. Commun.*, 1547 (1988).
159. Kanoufi, F.; Cannes, C.; Zu, Y.; Bard, A. J., *J. Phys. Chem. B*, *105*, 8951 (2001).
160. Zu, Y. B.; Fan, F. R. F.; Bard, A. J., *J. Phys. Chem. B*, *103*, 6272 (1999).
161. Dvorak, O.; De Armond, A. H.; De Armond, M. K., *Langmuir*, *8*, 508 (1992).
162. Marecek, V.; De Armond, A. H.; De Armond, M. K., *J.Am.Chem.Soc.*, *111*, 2561 (1989).

163. Osakai, T.; Muto, K., *J. Electroanal. Chem.*, *496*, 95 (2001).

164. Fermin, D. J.; Duong, H. D.; Ding, Z.; Brevet, P. F.; Girault, H. H., *Electrochem. Commun.*, *1*, 29 (1999).

165. Fermin, D. J.; Duong, H. D.; Ding, Z.; Brevet, P.-F.; Girault, H. H., *J.Am.Chem.Soc.*, *121*, 10203 (1999).

166. Eugster, N.; Fermin, D. J.; Girault, H. H., *J. Phys. Chem. B*, *106*, 3428 (2002).

167. Fermin, D. J.; Ding, Z.; Duong, H. D.; Brevet, P.-F.; Girault, H. H., *J. Phys. Chem. B*, *102*, 10334 (1998).

168. Fermin, D. J.; Ding, Z.; Duong, H. D.; Brevet, P.-F.; Girault, H. H., *Chem. Commun.*, 1125 (1998).

169. Duong, H. D.; Brevet, P.-F.; Girault, H. H., *J. Photochem. Photobiol. A*, *117*, 27 (1998).

170. Jensen, H.; Khakkassery, J.; Nagatani, H.; Fermin, D. J.; Girault, H. H., *J. Am. Chem. Soc.*, *122*, 10943 (2000).

171. Jensen, H.; Fermin, D. J.; Girault, H. H., *Phys. Chem. Chem. Phys.*, *3*, 2503 (2001).

172. Lillie, G. C.; Holmes, S. M.; Dryfe, R. A. W., *J. Phys. Chem. B*, *106*, 12101 (2002).

173. Ohde, H.; Maeda, K.; Yoshida, Y.; Kihara, S., *Electrochim. Acta*, *44*, 23 (1998).

174. Osakai, T.; Jensen, H.; Nagatani, H.; Fermin, D. J.; Girault, H. H., *J. Electroanal. Chem.*, *510*, 43 (2001).

175. Osakai, T.; Akagi, N.; Hotta, H.; Ding, J.; Sawada, S., *J. Electroanal. Chem.*, *490*, 85 (2000).

176. Suzuki, M.; Umetani, S.; Matsui, M.; Kihara, S., *J. Electroanal. Chem.*, *420*, 119 (1997).

177. Georganopoulou, D. G.; Caruana, D. J.; Strutwolf, J.; Williams, D. E., *Faraday Discuss.*, 109 (2000).

178. Ohde, H.; Maeda, K.; Yoshida, Y.; Kihara, S., *J. Electroanal. Chem.*, *483*, 108 (2000).

179. Zhang, Z. Q.; Yuan, Y.; Sun, P.; Su, B.; Guo, J. D.; Shao, Y. H.; Girault, H. H., *J. Phys. Chem. B*, *106*, 6713 (2002).

180. Donten, M.; Stojek, Z.; Scholz, F., *Electrochem. Commun.*, *4*, 324 (2002).

181. Noviandri, I.; Brown, K. N.; Fleming, D. S.; Gulyas, P. T.; Lay, P. A.; Masters, A. F.; Phillips, L., *J. Phys. Chem. B*, *103*, 6713 (1999).

182. Cheng, Y.; Schiffrin, D. J., *J.Chem.Soc., Faraday Trans.*, *89*, 199 (1993).

183. Zhang, J.; Unwin, P. R., *J.Chem.Soc., Perkin Trans. 2*, 1608 (2001).

184. Fermin, D. J.; Duong, H. D.; Ding, Z. F.; Brevet, P.-F.; Girault, H. H., *Phys. Chem. Chem. Phys.*, *1*, 1461 (1999).

185. Guainazzi, M.; Silvestri, G.; Serravalle, G., *J.Chem.Soc.,Chem.Commun.*, 200 (1975).

186. Dehmlow, E. V., *Angew. Chem., Int. Ed. Engl.*, *13*, 170 (1974).

187. Hanzlik, J.; Samec, Z.; Hovorka, J., *J. Electroanal. Chem.*, *216*, 303 (1987).

188. Barker, A.; Unwin, P. R., *J. Phys. Chem. B*, *105*, 12019 (2001).

189. Prins, R.; Korswagen, A. R.; Kortbeek, A. G. T. G., *J. Organometal. Chem.*, *39*, 335 (1972).

190. Page, J. A.; Wilkinson, G., *J. Am. Chem. Soc.*, *74*, 6149 (1952).

191. Vaucher, S.; Charmant, J.; Sorace, L.; Gatteschi, D.; Mann, S., *Polyhedron*, *20*, 2467 (2001).

192. Chen, Q.-Z.; Iwamoto, K.; Seno, M., *Electrochim. Acta*, *36*, 291 (1991).

193. L'Her, M.; Pondaven, A., *The Porphyrin Handbook; vol. 16; Phthalocyanines: spectroscopic and electrochemical characterization.*, Academic Press-Elsevier, Amsterdam, 2003.

194. Kotov, N. A.; Kuzmin, M. G., *J. Electroanal. Chem.*, *338*, 99 (1992).

195. Kotov, N. A.; Kuzmin, M. G., *J.Electroanal.Chem*, *341*, 47 (1992).

196. Eugster, N.; Fermin, D. J.; Girault, H. H., *J. Am. Chem. Soc.*, *125*, 4862 (2003).

197. Das, S.; Thanulingama, T. L.; Rajesha, C. S.; George, M. V., *Tetrahedron Lett.*, *36*, 1337 (1995).

198. Rajesh, C.; Thanulingam, T.; Das, S., *Tetrahedron*, *53*, 16817 (1997).

199. Aoki, K.; Tasakorn, P.; Chen, J., *J. Electroanal. Chem*, *542*, 51 (2003).

200. Tasakorn, P.; Chen, J.; Aoki, K., *J. Electroanal. Chem.*, *533*, 119 (2002).

201. Mirceski, V.; Gulaboski, R.; Scholz, F., *Electrochem. Commun.*, *4*, 814 (2002).

202. Komorsky-Lovric, S.; Lovric, M.; Scholz, F., *Collect. Czech. Chem. Commun.*, *66*, 434 (2001).

203. Saczek-Maj, M.; Opallo, M., *Electroanalysis*, *14*, 1060 (2002).

204. Opallo, M.; Saczek-Maj, M., *Electrochem. Commun.*, *3*, 206 (2001).

205. Cheng, Y.; Murtomäki, L.; Corn, R., *J. Electroanal. Chem.*, *483*, 88 (2000).

206. Steiger, B.; Anson, F. C., *Inorg. Chem.*, *39*, 4579 (2000).

207. Chung, T.; Anson, F. C., *Anal. Chem.*, *73*, 337 (2001).

208. Myland, J. C.; Oldham, K. B., *J. Electroanal. Chem.*, *530*, 1 (2002).

209. Barker, A.; Unwin, P. R., *J. Phys. Chem.*, *104*, 2330 (2000).

210. Forssten, C.; Strutwolf, J.; Williams, D. E., *Electrochem. Commun.*, *3*, 619 (2001).

211. Nakabayashi, S.; Aogaki, R.; Karantonis, A.; Iguchi, U.; Ushida, K.; Nawa, M., *J. Electroanal. Chem.*, *473*, 54 (1999).

212. Sathaye, S.; Patil, K.; Paranjape, D.; Padalkar, S., *Mater. Res. Bull.*, *36*, 1149 (2001).

213. Cheng, Y.; Schiffrin, D. J., *J. Chem. Soc., Faraday Trans.*, *92*, 3865 (1996).

214. Kassim, A.; Abdullah, A. H.; Idris, Z., *Ultra Scientist of Physical Sciences*, *13*, 306 (2001).

215. Cunnane, V. J.; Evans, U., *Chem. Commun.*, 2163 (1998).

216. Gorgy, K.; Fusalba, F.; Evans, U.; Kontturi, K. S.; Cunnane, V. J., *Synth. Met.*, *125*, 365 (2002).

217. Johans, C.; Clohessy, J.; Fantini, S.; Kontturi, K. S.; Cunnane, V. J., *Electrochem. Commun.*, *4*, 227 (2002).

218. Johans, C.; Kontturi, K. S.; Schiffrin, D. J., *J. Electroanal. Chem.*, *526*, 29 (2002).

219. Johans, C.; Lahtinen, R.; Konturri, K. S.; Schiffrin, D. J., *J. Electroanal. Chem.*, *488*, 99 (2000).

220. Johans, C.; Liljeroth, P.; Kontturi, K. S., *Phis. Chem. Chem. Physics*, *4*, 1067 (2002).

221. Maeda, K.; Jänchenova, H.; Lhotsky, A.; Stibor, I.; Budka, J.; Marecek, V., *J. Electroanal. Chem.*, *516*, 103 (2001).

222. Yunus, K.; Marks, C. B.; Fisher, A. C.; Allsopp, D. W. E.; Ryan, T. J.; Dryfe, R. A.

W.; Hill, S. S.; Roberts, E. P. L.; Brennan, C. M., *Electrochem. Commun.*, *4*, 579 (2002).
223. Solomon, T., *Bull. Chem. Soc. Ethiop.*, *11*, 79 (1997).
224. Volkov, A. G.; Kharkats, Y., *Kinetika i Kataliz*, *26*, 1322 (1985).
225. Volkov, A. G.; Gugeshashvili, M.; Deamera, D., *Electrochim. Acta*, *40*, 2849 (1995).
226. Kharkats, Y.; Volkov, A. G., *Biochimica et Biophysica Acta*, *891*, 56 (1987).
227. Volkov, A. G., *J. Electroanal. Chem.*, *483*, 150 (2000).
228. Volkov, A. G., *Electrochimica Acta*, *44*, 139 (1998).
229. Boguslavskii, L.; Volkov, A. G.; Kandelaki, M., *FEBS Lett.*, *65*, 155 (1976).
230. Boguslavskii, L.; Volkov, A. G.; Kandelaki, M., *Bioelectrochem. Bioenerg.*, *4*, 68 (1977).
231. Kharkats, Y.; Volkov, A. G.; Boguslavskii, L., *J.Theor.Biol.*, *65*, 379 (1977).
232. Volkov, A. G.; Mironov, A.; Boguslavskii, L., *Elektrokhimiya*, *12*, 1326 (1976).
233. Volkov, A. G.; Deamer, D. W., *Prog. Colloid Polym. Sci.*, *103*, 21 (1997).
234. Boguslavskii, L.; Volkov, A. G.; Kandelaki, M.; Nizhnikovskii, E.; Bibikova, M., *Biofizika*, *22*, 223 (1977).
235. Maeda, K.; Nishihara, M.; Ohde, H.; Kihara, S., *Anal. Sci.*, *14*, 85 (1998).
236. Sortino, S.; Petralia, S.; Di Bella, S., *J. Am. Chem. Soc.*, *125*, 5610 (2003).
237. Katz, E.; Filanovsky, B.; Willner, I., *New J. Chem.*, *23*, 481 (1999).
238. Schmickler, W., *Interfacial electrochemistry*, Oxford University Press, Oxford, 1996.

Index

Milton Keynes UK
Ingram Content Group UK Ltd.
UKHW051850071024
449327UK00025B/1910